08-12

PEARSON

Foundations for College Mathematics 12

Chad Coene

Bonnie Edwards

Glenn Gadsby

Catherine Heideman

Duncan LeBlanc

Robert Sherk

Karen Timson

Frank Torti

Gordon Cooke

Ed Haines

Carolyn Kieran

Simona Matei

Margaret Sinclair

PEARSON

Publisher
Mike Czukar

Research and Communications Manager
Barbara Vogt

Editorial Team
Claire Burnett
Nirmala Nutakki
Sarah Mawson
Lynda Cowan
Stephanie Kleven
Ioana Gagea
Margaret McClintock
Ellen Davidson
Jane Schell
Karen Alley
Cheri Westra
Judy Wilson

Editorial Contributor
Bob Berglind

Photo Research
Carrie Gleason

Design
Word & Image Design Studio Inc.

Composition
Lapiz Digital Services, India

Copyright ©2009 Pearson Education Canada, a division of Pearson Canada Inc.

ISBN-13: 978-0-321-49367-5
ISBN-10: 0-321-49367-2

Printed and bound in Canada

1 2 3 4 5 – TCP – 12 11 10 09 08

All rights reserved. This publication is protected by copyright, and permission should be obtained from the publisher prior to any prohibited reproduction, storage in a retrieval system, or transmission in any form or by any means, electronic, mechanical, photocopying, recording, or likewise. For information regarding permission, write to the Permissions Department.

The information and activities presented in this book have been carefully edited and reviewed. However, the publisher shall not be held liable for any damages resulting, in whole or in part, from the reader's use of this material.

Consultants and Advisers

Technology Consultants

Carolyn Kieran
Université de Québec à Montréal

Duncan LeBlanc
Toronto District School Board

Advisers

Charlotte Cutajar
Toronto District School Board

Peggy Leroux
Waterloo Catholic District School Board

Sandra G. McCarthy
Ottawa-Carleton District School Board

Gizele Price
Dufferin-Peel Catholic District School Board

Dwight Stead
Dufferin-Peel Catholic District School Board

Reviewers

Tina Bawa
London District Catholic School Board

Gerry Bossy
Lambton Kent District School Board

Mandy Clark
Limestone District School Board

Gordon Doctorow
Independent Math Consultant, Toronto

Heather L. Edwards
Upper Grand District School Board

Veronica Gervais
Ottawa-Carleton Catholic School Board

Tina Grandy
Peel District School Board

Gary Greer
Limestone District School Board

Ed Haines
Formerly District School Board of Niagara

Judi Hanta
Hamilton-Wentworth Catholic District School Board

Paul Higgins
Greater Essex County District School Board

Victoria Kudrenski
Hamilton-Wentworth District School Board

Susan Lavigne
Formerly Thames Valley District School Board

Thuy Leu
Waterloo Region District School Board

Frieda Leung
Peel District School Board

Susanne Ling
York Region District School Board

William Lozinski
Formerly Windsor-Essex Catholic District School Board

Linda Lu
Hamilton-Wentworth District School Board

Simona Matei
Peel District School Board

Linda McLaren
Catholic District School Board of Eastern Ontario

Jason Murray
Halton District School Board

Chris Schleihauf
Lambton Kent District School Board

Linda Vardy
Formerly Kawartha Pine Ridge District School Board

Justin Veiga
Toronto Catholic District School Board

Jim Vincent
Peel District School Board

Ann Winacott
Durham District School Board

Kevin Wong
Halton Catholic District School Board

Michael Zahra
Algonquin and Lakeshore Catholic District School Board

Table of Contents

Chapter 1: Trigonometry

Activate Prior Knowledge	2
1.1 Trigonometric Ratios in Right Triangles	4
1.2 **Technology:** Investigating the Sine, Cosine, and Tangent of Obtuse Angles	13
1.3 Sine, Cosine, and Tangent of Obtuse Angles	20
Game: Trigonometric Search	25
Mid-Chapter Review	26
1.4 The Sine Law	27
Transitions: Collecting Important Ideas	34
1.5 The Cosine Law	35
1.6 Problem Solving with Oblique Triangles	42
1.7 **Research:** Occupations Using Trigonometry	50
Study Guide	53
Chapter Review	54
Practice Test	57
Chapter Problem: Designing a Stage	58

Chapter 2: Geometry

Activate Prior Knowledge	60
Transitions: Managing Your Time	66
2.1 Area Applications	67
2.2 Working with Composite Objects	76
Mid-Chapter Review	86
2.3 Optimizing Areas and Perimeters	87
2.4 **Technology:** Optimizing Area and Perimeter Using a Spreadsheet	97
2.5 **Technology:** Dynamic Investigations of Optimal Measurements	103
2.6 Optimizing Volume and Surface Area	105
Puzzle: Cube Creations	114
2.7 **Technology:** Optimizing Surface Area Using a Spreadsheet	115
Study Guide	119
Chapter Review	120
Practice Test	123
Chapter Problem: A Winning Design	124

Chapter 3: Two-Variable Data

Activate Prior Knowledge	126
Transitions: Getting Extra Practice	129
3.1 One- and Two-Variable Data	130
3.2 Using Scatter Plots to Identify Relationships	138
3.3 Line of Best Fit	148
Mid-Chapter Review	158
3.4 **Technology:** Analysing Data Using a Graphing Calculator	159
Game: Give It to Me Straight	165
3.5 **Technology:** Analysing Data Using a Spreadsheet	166
3.6 **Technology:** Analysing Data Using *Fathom*	172
3.7 **Exploration:** Conducting an Experiment to Collect Two-Variable Data	180
Study Guide	185
Chapter Review	186
Practice Test	189
Chapter Problem: Temperatures around the Globe	190

Chapter 4: Statistical Literacy

Activate Prior Knowledge	192
Transitions: Ethical Conduct	195
4.1 Interpreting Statistics	196
4.2 **Research:** Statistics in the Media	206
4.3 Surveys and Questionnaires	208
4.4 **Exploration:** Conducting a Survey to Collect Two-Variable Data	218
Mid-Chapter Review	222
4.5 The Use and Misuse of Statistics	223
4.6 Understanding Indices	233
4.7 **Research:** Indices and E-STAT	241
4.8 **Research:** Statistical Literacy and Occupations	246
Game: Concept Clues	249
Study Guide	250
Chapter Review	251
Practice Test	254
Chapter Problem: To Ban or Not to Ban?	256
Project A: For the Birds!	258
Project B: Making Headlines with Statistics	260
Cumulative Review Chapters 1–4	262

v

Chapter 5: Graphical Models

Activate Prior Knowledge	**266**
Transitions: Teaching Yourself	**268**
5.1 Trends in Graphs	269
5.2 Rate of Change	278
5.3 Linear Models	288
5.4 Quadratic Models	298
Mid-Chapter Review	**308**
Game: Curves of Concentration	**309**
5.5 Exponential Models	310
5.6 Selecting a Regression Model for Data	319
5.7 **Technology:** Applying Trends in Data	328
Study Guide	**331**
Chapter Review	**332**
Practice Test	**335**
Chapter Problem: Modelling the Price of Diamonds	**336**

Chapter 6: Algebraic Models

Activate Prior Knowledge	**338**
Transitions: Learning with Others	**341**
6.1 Using Formulas to Solve Problems	342
6.2 Rearranging Formulas	350
6.3 Laws of Exponents	358
6.4 **Technology:** Patterns in Exponents	366
Mid-Chapter Review	**371**
6.5 Rational Exponents	372
Game: Power Dominoes	**379**
6.6 Exponential Equations	380
6.7 Applications of Exponential Equations	387
6.8 **Research:** Occupations Using Mathematical Modelling	396
Study Guide	**399**
Chapter Review	**400**
Practice Test	**403**
Chapter Problem: A Butterfly Conservatory	**404**

Chapter 7: Annuities and Mortgages

Activate Prior Knowledge	**406**
Transitions: Evaluations	**408**
7.1 The Amount of an Annuity	409
7.2 The Present Value of an Annuity	419
7.3 The Regular Payment of an Annuity	426
7.4 **Technology:** Using a Spreadsheet to Investigate Annuities	433
7.5 Saving for Education and Retirement	439
Mid-Chapter Review	**442**
7.6 **Research:** What Is a Mortgage?	443
7.7 **Technology:** Amortizing a Mortgage	446
Game: Mortgage Tic-Tac-Toe	**455**
7.8 **Technology:** Using Technology to Generate an Amortization Table	456
7.9 **Technology:** Reducing the Interest Costs of a Mortgage	461
Study Guide	**467**
Chapter Review	**468**
Practice Test	**471**
Chapter Problem: Planning Ahead	**472**

Chapter 8: Budgets

Activate Prior Knowledge	**474**
Transitions: Your Financial Future	**476**
8.1 **Research:** Choosing a Home	477
8.2 The Costs of Owning or Renting a Home	479
8.3 **Research:** Estimating Living Costs	488
Mid-Chapter Review	**492**
8.4 Designing Monthly Budgets	493
8.5 **Technology:** Creating a Budget Using a Spreadsheet	505
Game: Budget Shuffle Challenge	**510**
8.6 **Research:** Making Decisions about Buying or Renting	511
8.7 **Research:** Occupations Involving Finance	516
Study Guide	**519**
Chapter Review	**520**
Practice Test	**523**
Chapter Problem: Preparing with Financing	**524**

Project C: Will Women Eventually Run as Fast as Men?	**526**
Project D: A Home of Their Own	**528**
Cumulative Review Chapters 1–8	**530**
Glossary	**535**
Answers	**543**
Technology Index	**587**
Index	**589**
Acknowledgments	**593**
Course Study Guide (See inside back cover.)	

About
Pearson Foundations for College Mathematics 12

Whether you are planning for an apprenticeship program, for college admissions, or for work, *Foundations for College Mathematics 12* will help you achieve your goals.

This book organizes the course into four major mathematical topics. By studying each topic separately, you can make the best use of your time and efforts, focus on new concepts for an extended period, and develop greater understanding, and better recall.

- **Chapter 1, Trigonometry**
- **Chapter 2, Geometry**

This mathematical strand is important in such careers as carpentry, plumbing, landscape design, theatre design, or commercial arts.

- **Chapter 3, Two-Variable Data**
- **Chapter 4, Statistical Literacy**

Are you interested in running your own business? Researching and analysing data would be part of preparing a sound business plan.

- **Chapter 5, Graphical Models**
- **Chapter 6, Algebraic Models**

Knowing how to model situations using graphs, tables, or formulas, is a key skill for technologists, financial planners, and business managers.

- **Chapter 7, Annuities and Mortgages**
- **Chapter 8, Budgets**

Understanding the principles of investments, loans, and budgets has applications in your personal life, and in careers in banking, insurance, and business.

Your Book Highlights

In addition to the Chapters, this book supports your learning with these additional tools.

- A **Glossary** of key math terms
- **Answers** so that you can check your work
- A **Technology Index** to reference specific calculator or software instructions
- An **Index**

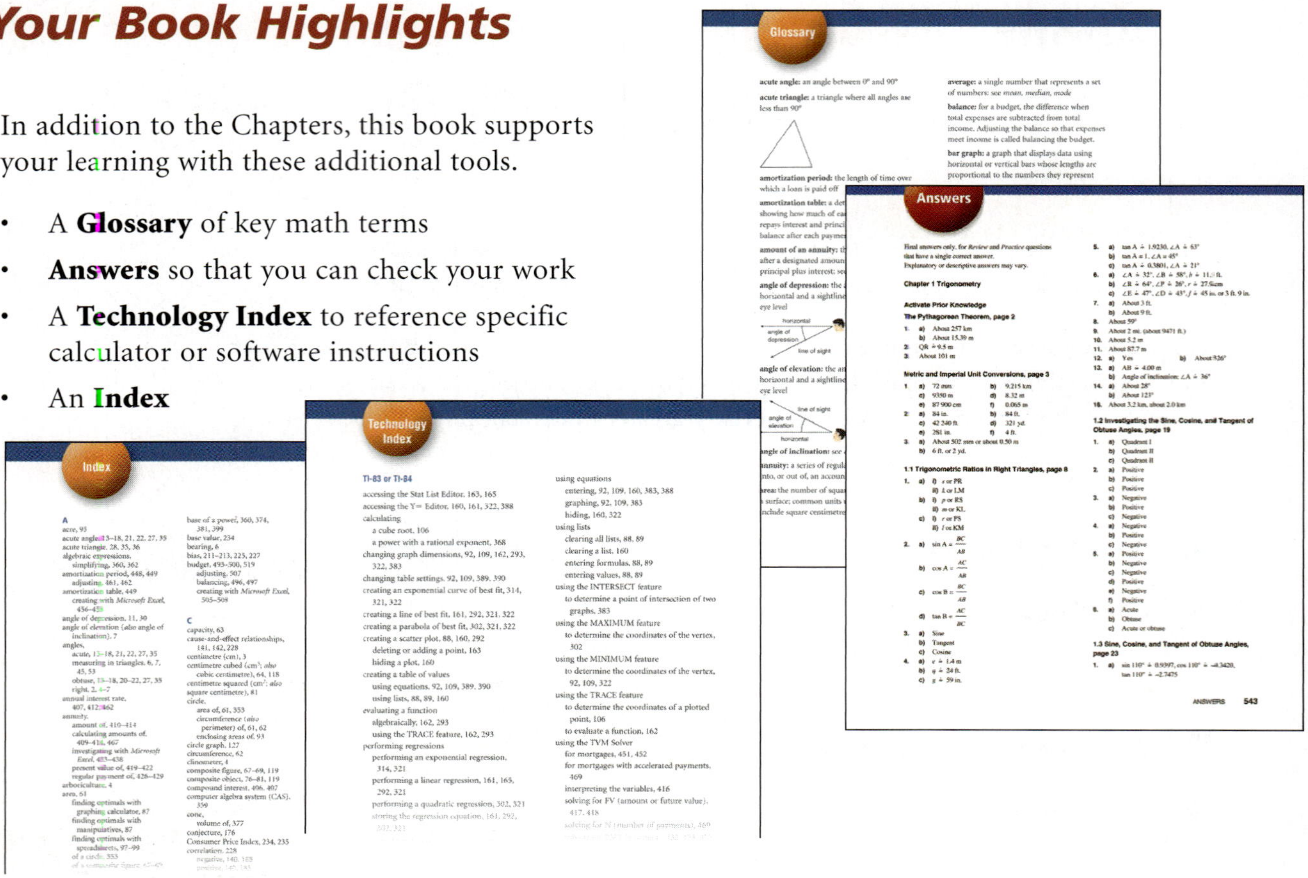

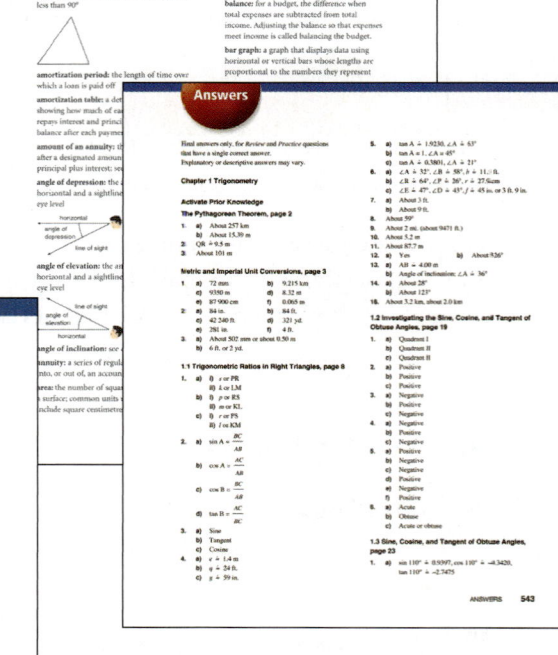

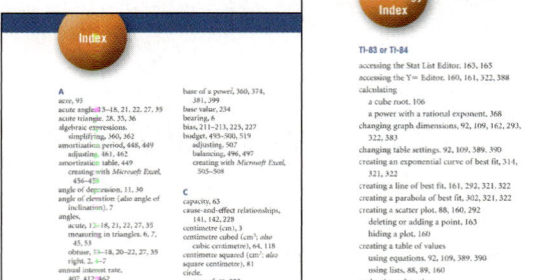

- A **Course Study Guide**, printed on the inside back cover, as a quick reference for key formulas, theorems, or properties.

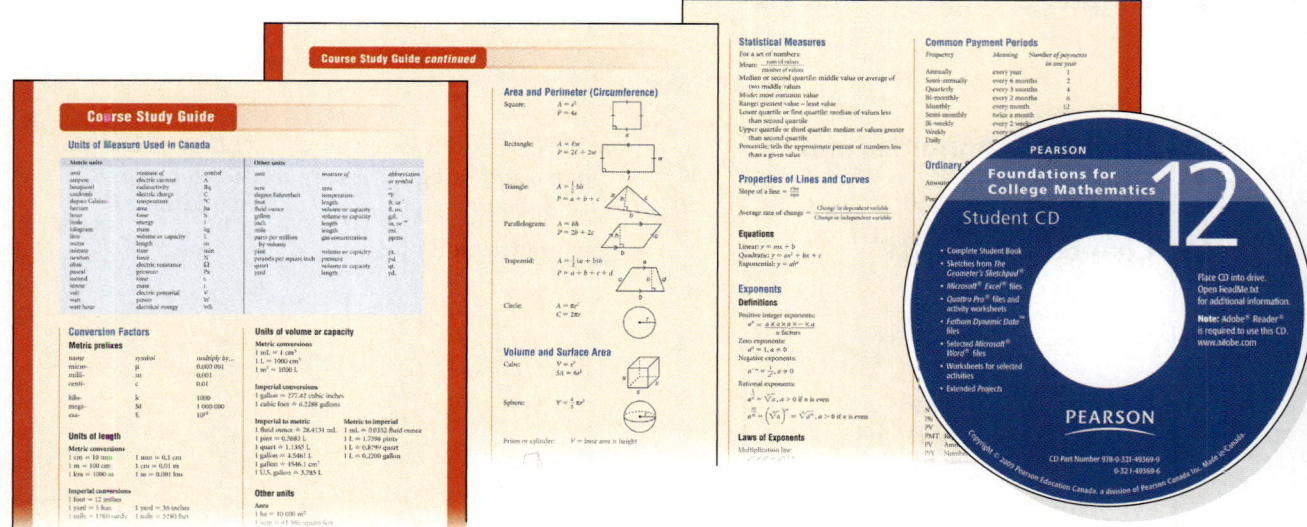

- A **Student CD-ROM** with software files for selected activities in this book.

viii

Chapter Structure

Each chapter starts with a page like this...

> See **What You'll Learn** and **Why**.
>
> Check the list of **Key Words**.

...followed by a review feature.

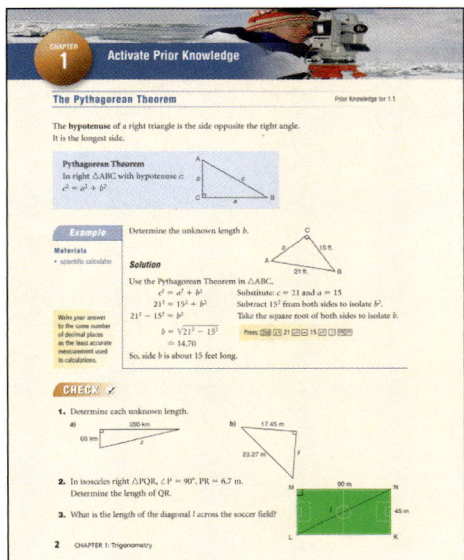

Activate Prior Knowledge lets you review knowledge from previous grades, or from earlier chapters.

Each chapter also has a **Transitions** feature, and a **Game** or **Puzzle**.

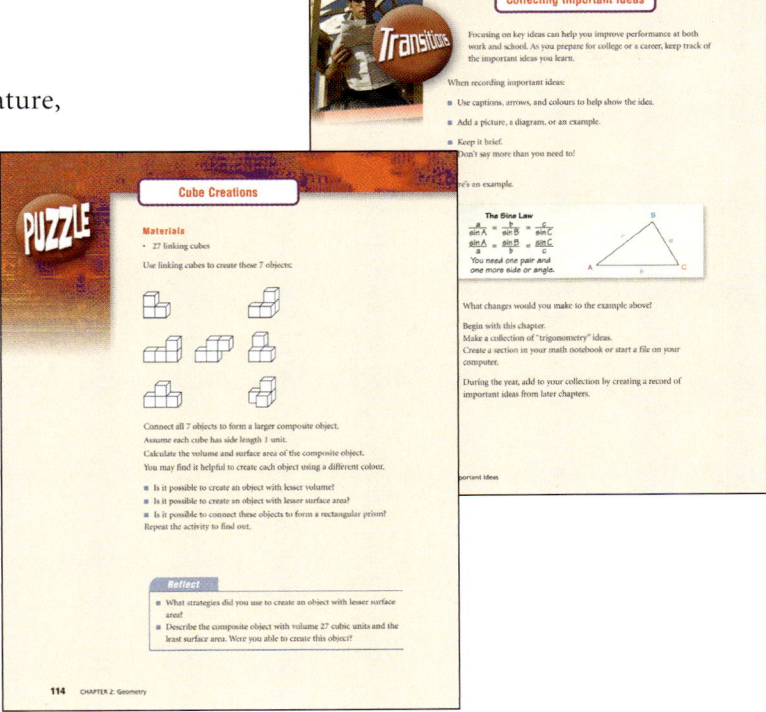

> Each **Transitions** page highlights a key principle related to independent study, or your choice of pathway beyond grade 12.

> Each **Game** or **Puzzle** is an opportunity to reinforce the work of the chapter.

ix

Here is how a typical section in a chapter works.

Connect the Ideas to consolidate your learning.

Investigate a problem to develop your understanding of a new concept.

Reflect on your investigation results with other students.

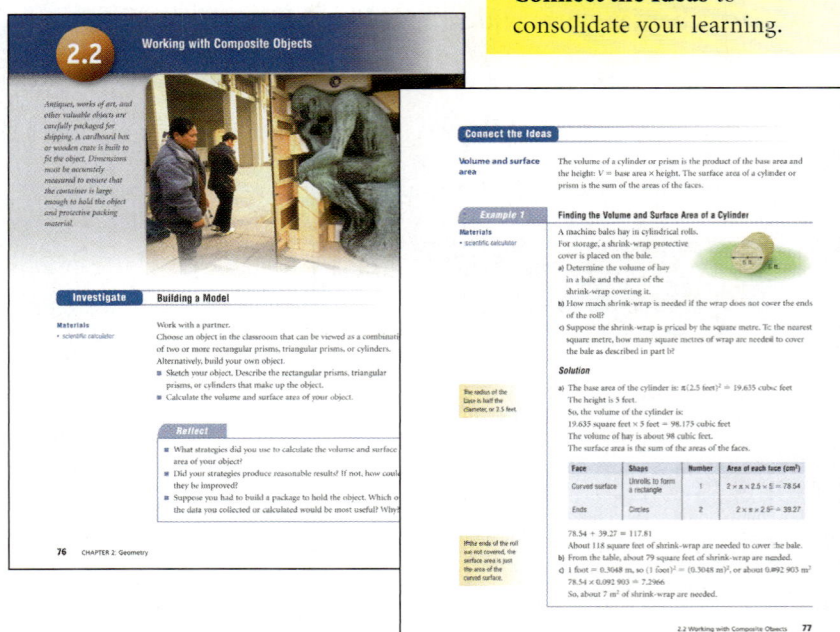

Worked **Examples** show how the concept can be applied, in general cases and in specific applications.

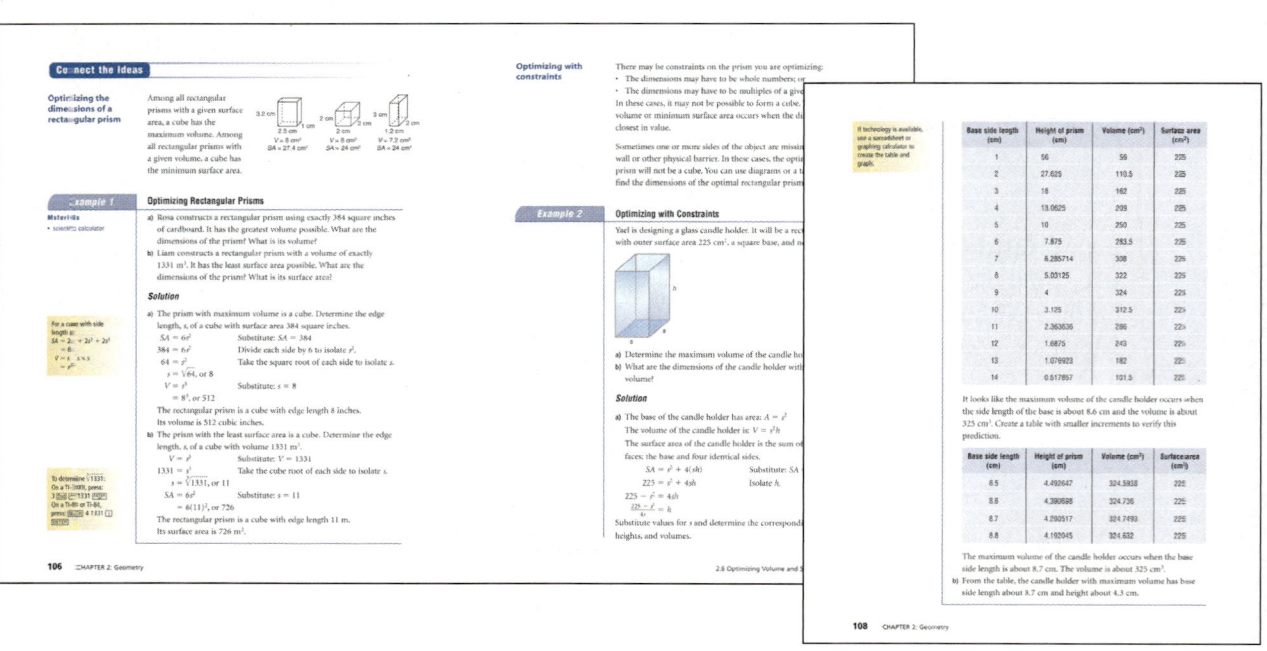

x

Practice questions reinforce the math.

Questions progress in level of difficulty.
Try the **A-level** questions to check your basic understanding.

Margin notes direct you to **Examples** that model solutions for selected exercises.

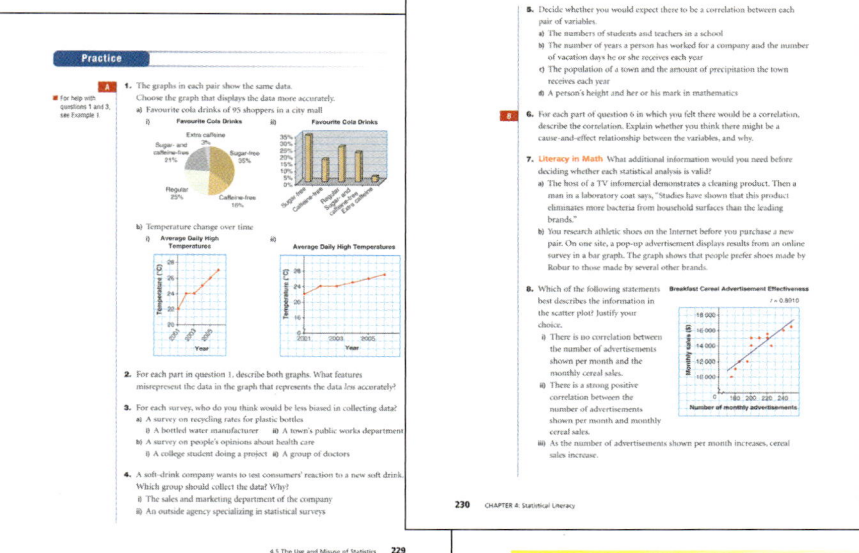

Solve the **B-level** questions to consolidate understanding, and to see how you can apply the concepts in a variety of situations.

Try the **C-level** questions to extend your thinking.

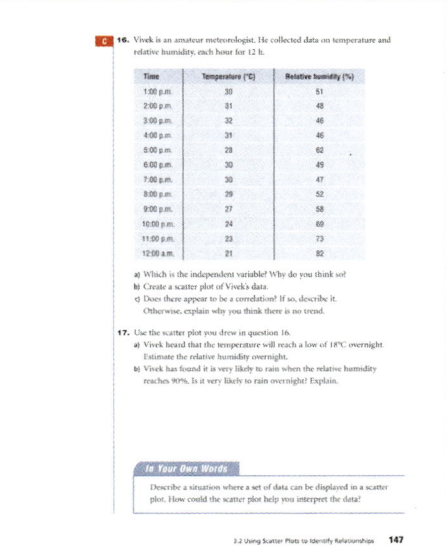

Describe your learning **In Your Own Words**.

xi

Some sections take you deeper into the content.

Inquire about a problem.

You may be
- using technology to investigate relationships
- conducting an experiment
- performing research

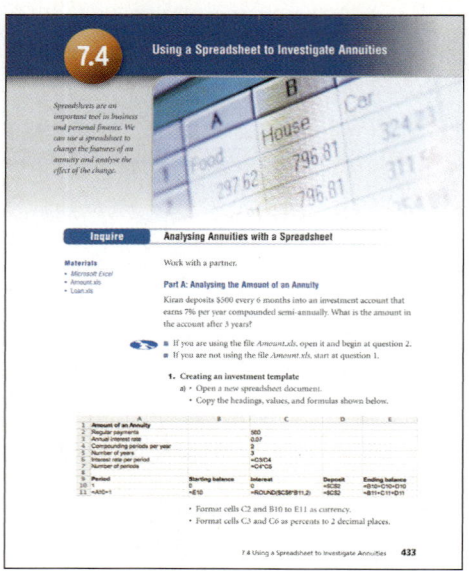

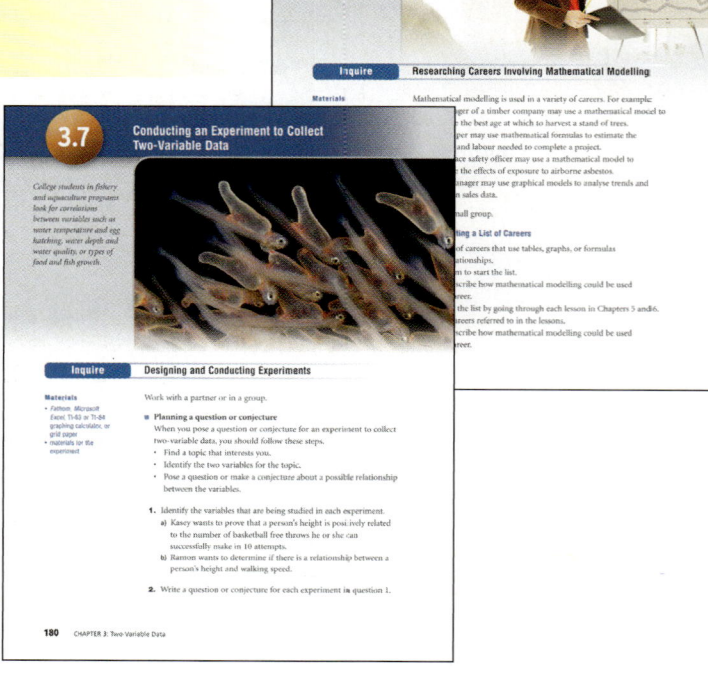

These lessons often provide **Practice**.
Try the exercises to see if your observations and generalizations carry through.

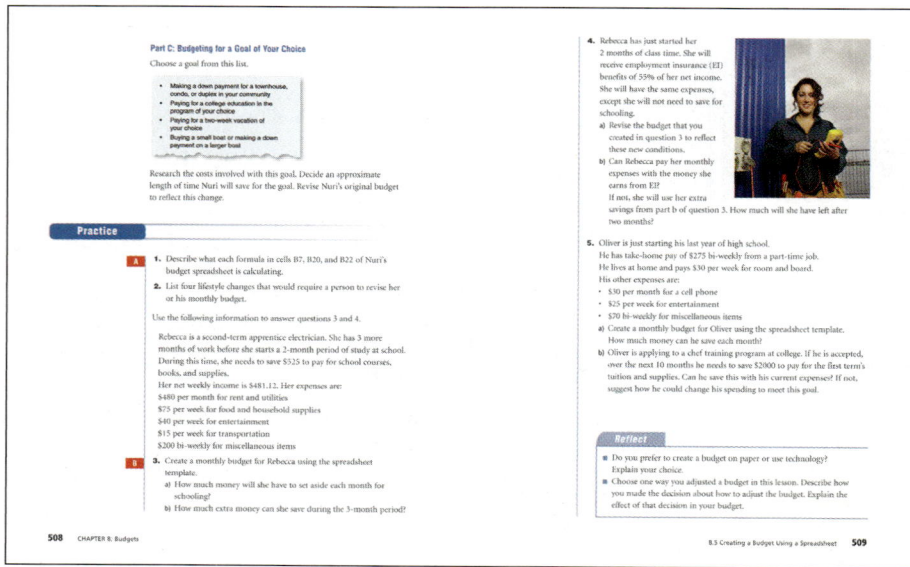

Reflect on your results with other students.

xii

There are regular opportunities to review in each chapter.

Use the **Mid-Chapter Review** to consolidate key concepts in the first part of the chapter. You'll build on these concepts as you continue through the chapter.

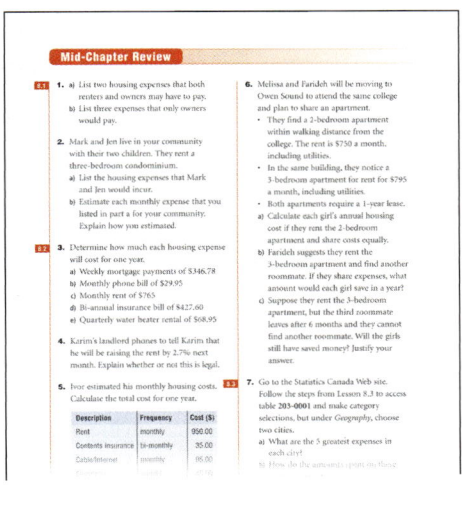

Also see the inside back cover for a comprehensive **Course Study Guide**.

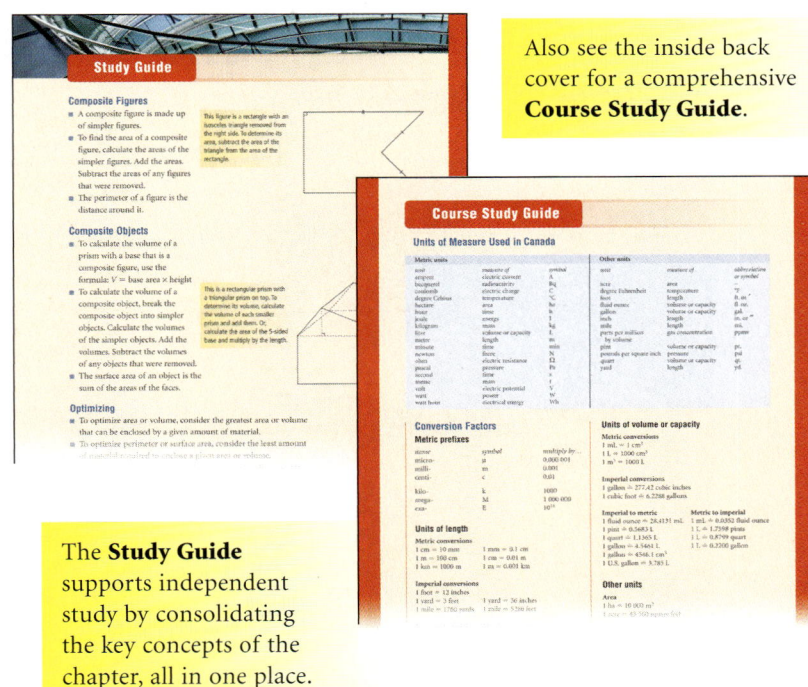

Use the **Chapter Review** to check your learning. If you can't do one of the review exercises, go to the lesson that's referenced in the margin to look for similar worked **Examples**.

The **Study Guide** supports independent study by consolidating the key concepts of the chapter, all in one place.

The **Practice Test** models the kinds of questions you might see on a class test.

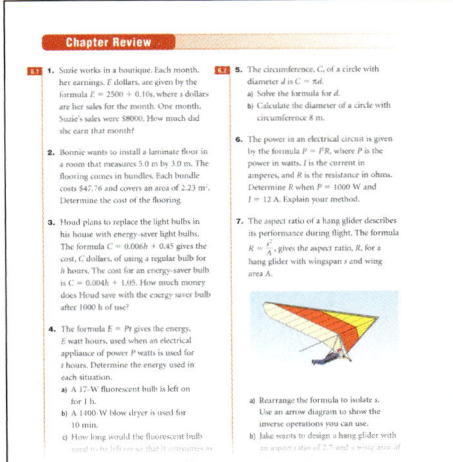

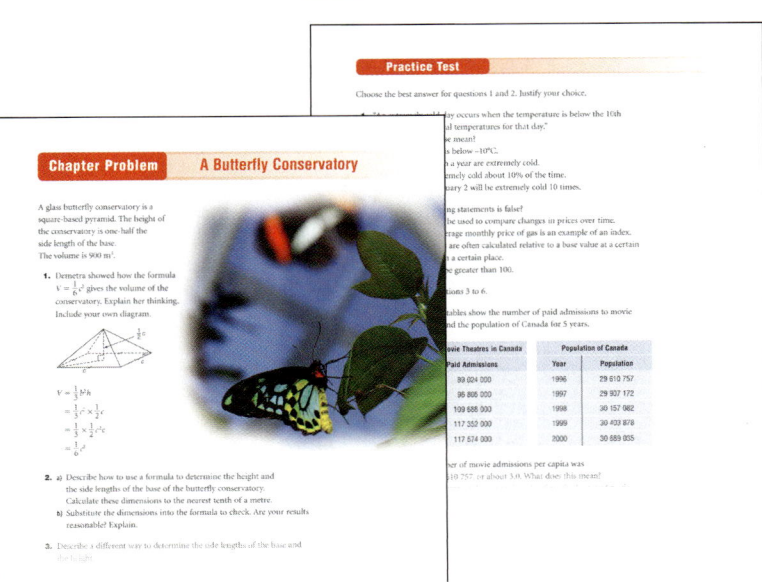

The **Chapter Problem** allows you to apply the concepts of the chapter.

xiii

Your book also has four **Projects**, two after Chapter 4, and two after Chapter 8.

Each **Project** is an opportunity to apply mathematical concepts as you solve a relevant, open-ended, problem.

Your teacher may provide masters for an **Extended Project**. Or, find more ideas for the projects on the **Student CD-ROM**.

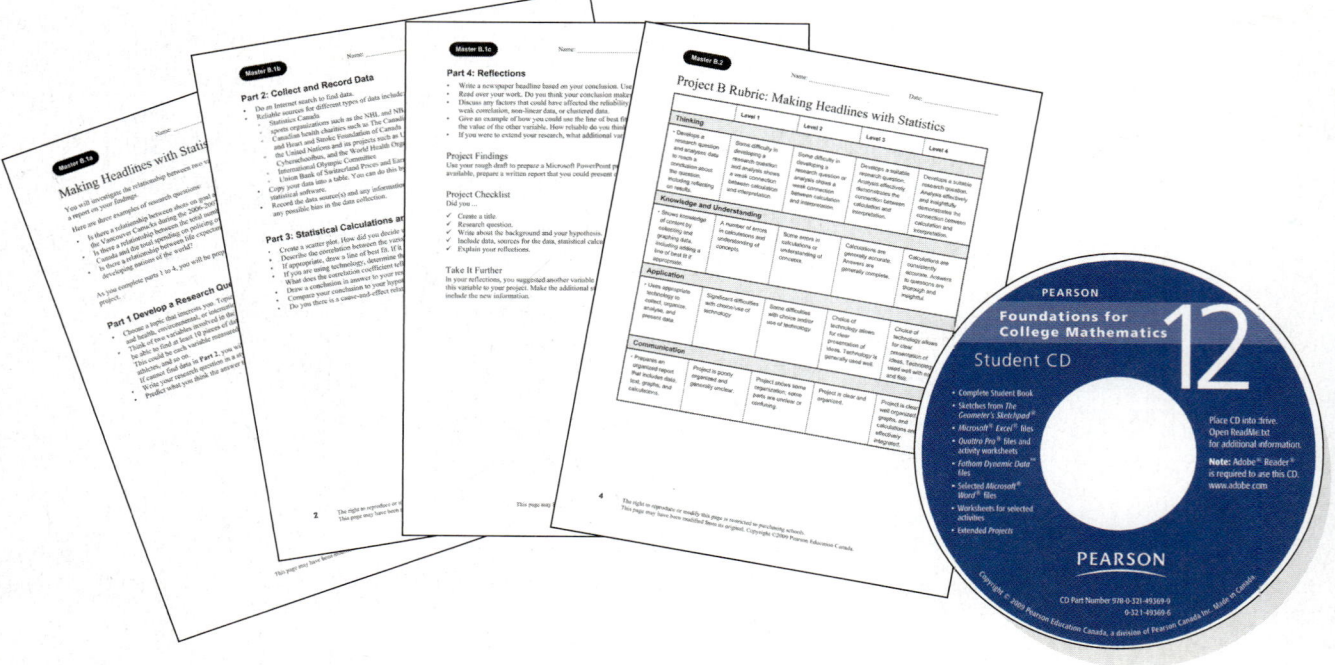

1 Trigonometry

What You'll Learn
To determine the measures of sides and angles in right, acute, and obtuse triangles and to solve related problems

And Why
Applications of trigonometry arise in land surveying, navigation, cartography, computer graphics, machining, medical imaging, and meteorology, where problems call for calculations involving angles, lengths, and distances using indirect measurements.

Key Words
- primary trigonometric ratios
- sine
- cosine
- tangent
- angle of inclination
- angle of depression
- acute triangle
- obtuse triangle
- oblique triangle
- Sine Law
- Cosine Law

CHAPTER 1

Activate Prior Knowledge

The Pythagorean Theorem

Prior Knowledge for 1.1

The **hypotenuse** of a right triangle is the side opposite the right angle. It is the longest side.

Pythagorean Theorem
In right $\triangle ABC$ with hypotenuse c:
$c^2 = a^2 + b^2$

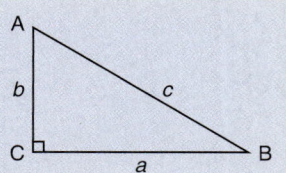

Example

Determine the unknown length b.

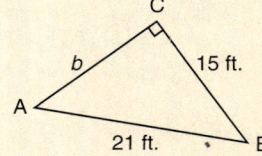

Materials
- scientific calculator

Solution

Use the Pythagorean Theorem in $\triangle ABC$.

$c^2 = a^2 + b^2$ Substitute: $c = 21$ and $a = 15$
$21^2 = 15^2 + b^2$ Subtract 15^2 from both sides to isolate b^2.
$21^2 - 15^2 = b^2$ Take the square root of both sides to isolate b.
$b = \sqrt{21^2 - 15^2}$ Press: [2nd] [x²] 21 [x²] [−] 15 [x²] [)] [ENTER]
$\doteq 14.70$

So, side b is about 15 feet long.

Write your answer to the same number of decimal places as the least accurate measurement used in calculations.

CHECK ✓

1. Determine each unknown length.

a) 250 km, 60 km, z

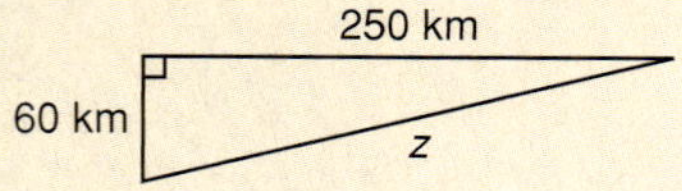

b) 17.45 m, 23.27 m, y

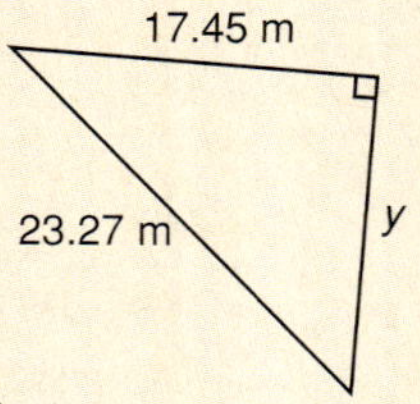

2. In isosceles right $\triangle PQR$, $\angle P = 90°$, $PR = 6.7$ m. Determine the length of QR.

3. What is the length of the diagonal l across the soccer field?

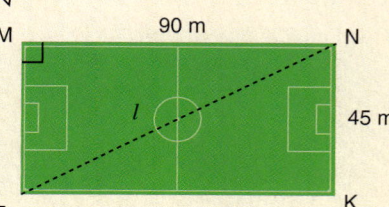

2   CHAPTER 1: Trigonometry

Metric and Imperial Unit Conversions

Prior Knowledge for 1.1

The metric system is based on powers of 10.

Metric conversions
1 cm = 10 mm   1 m = 100 cm   1 km = 1000 m
1 mm = 0.1 cm   1 cm = 0.01 m   1 m = 0.001 km

The most common imperial units of length are the inch, foot, yard, and mile.

Imperial conversions
1 foot = 12 inches   1 yard = 3 feet   1 mile = 5280 feet
                       1 yard = 36 inches   1 mile = 1760 yards

Example

Write each pair of measures using the same unit.
a) 54 cm, 3.8 m
b) 22 inches, 12 feet 4 inches

Solution

a) 1 m = 100 cm; so, 3.8 m = 3.8 × 100 = 380 cm
   Alternatively, 1 cm = 0.01 m, so 54 cm = 54 × 0.01 = 0.54 m

b) 1 foot = 12 inches, so 12 feet = 144 inches;
   12 feet 4 inches = 144 inches + 4 inches = 148 inches
   Alternatively, 22 inches = 1 foot 10 inches

CHECK ✓

1. Convert each metric measure to the unit indicated.
   a) 7.2 cm to millimetres
   b) 9215 m to kilometres
   c) 9.35 km to metres
   d) 832 cm to metres
   e) 879 m to centimetres
   f) 65 mm to metres

2. Convert each imperial measure to the unit indicated.
   a) 7 feet to inches
   b) 28 yards to feet
   c) 8 miles to feet
   d) 963 feet to yards
   e) 23 feet 5 inches to inches
   f) 48 inches to feet

3. Determine q. If you need to convert measurements to a different unit, explain why.

   a)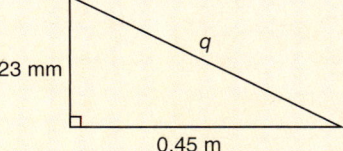
   b)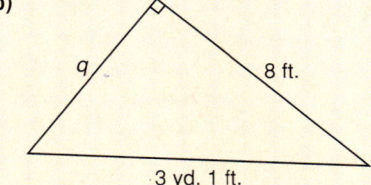

Activate Prior Knowledge   3

1.1 Trigonometric Ratios in Right Triangles

Specialists in forestry and arboriculture apply trigonometry to determine heights of trees. They may use a clinometer, an instrument for measuring angles of elevation.

Investigate — Choosing Trigonometric Ratios

Materials
- scientific calculator

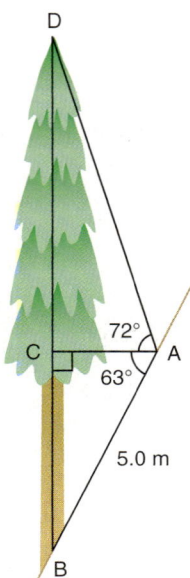

Work with a partner.

An arborist uses a clinometer to determine the height of a tree during a hazard evaluation. This diagram shows the arborist's measurements.

- Use △ABC.
  Determine the lengths of BC and AC.
- Use △ACD.
  Determine the length of CD.
- What is the height of the tree?

> For accuracy, keep more decimal places in your calculations than you need in the final answer.

Reflect

- Describe the strategies you used to determine the height of the tree. What angles and trigonometric ratios did you use?
- Compare your results and strategies with another pair. How are they similar? How are they different?

Connect the Ideas

Primary trigonometric ratios

The word "trigonometry" means "measurement of a triangle."
- The **primary trigonometric ratios** are sine, cosine, and tangent.

Each vertex is labelled with a capital letter. Each side is labelled with the lowercase letter of the opposite vertex.

The primary trigonometric ratios
For acute $\angle A$ in right $\triangle ABC$:

$$\sin A = \frac{\text{length of side opposite } \angle A}{\text{length of hypotenuse}} = \frac{a}{c}$$

$$\cos A = \frac{\text{length of side adjacent to } \angle A}{\text{length of hypotenuse}} = \frac{b}{c}$$

$$\tan A = \frac{\text{length of side opposite } \angle A}{\text{length of side adjacent to } \angle A} = \frac{a}{b}$$

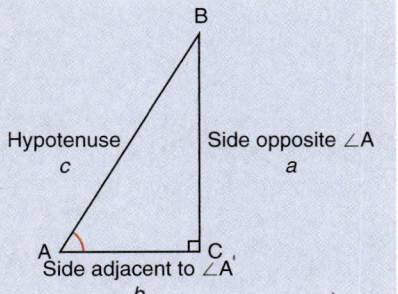

- We can use the primary trigonometric ratios or combinations of these ratios to determine unknown measures.

Example 1

Determining Side Lengths

Materials
- scientific calculator

Determine the length of p in $\triangle MNP$.

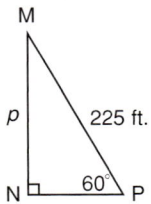

Set your calculator to degree mode before using sine, cosine, or tangent.

Solution

In $\triangle MNP$:
- The length of the hypotenuse is given.
- The measure of acute $\angle P$ is given.
- p is opposite $\angle P$.

So, use the sine ratio.

$$\sin P = \frac{MN}{MP} \quad \text{Substitute: } MN = p, \angle P = 60°, \text{ and } MP = 225$$

$$\sin 60° = \frac{p}{225} \quad \text{Multiply both sides by 225 to isolate } p.$$

$$\sin 60° \times 225 = p \quad \text{Press: } \boxed{\text{SIN}}\ 60\ \boxed{)}\ \boxed{\times}\ 225\ \boxed{\text{ENTER}}$$

$$p \doteq 194.86$$

The length of p is about 195 feet.

Write the length of p to the nearest foot because the length of n is to the nearest foot.

1.1 Trigonometric Ratios in Right Triangles **5**

Inverse ratios

It the key strokes shown here do not work on your calculator, refer to the user manual.

- You can use the **inverse ratios** $\sin^{-1}$, $\cos^{-1}$, and $\tan^{-1}$ to determine the measure of an angle when its trigonometric ratio is known. Press 2nd SIN, 2nd COS, or 2nd TAN to access the inverse ratios on a scientific calculator.

> **Inverse ratios**
> For acute $\angle A$ in right $\triangle ABC$:
> $\sin^{-1}(\sin A) = A$
> $\cos^{-1}(\cos A) = A$
> $\tan^{-1}(\tan A) = A$

- In navigation and land surveying, direction is described using a **bearing**. The bearing is given as a three-digit angle between 000° and 360° measured clockwise from the north line.

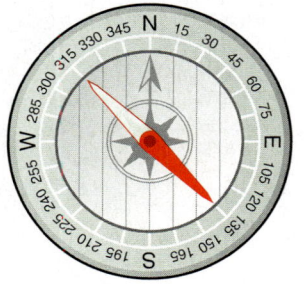

Example 2 — Determining Angle Measures

Materials
- scientific calculator

Michelle is drawing a map of a triangular plot of land.
a) Determine the angles in the triangle.
b) Determine the bearing of the third side of the plot, AB.

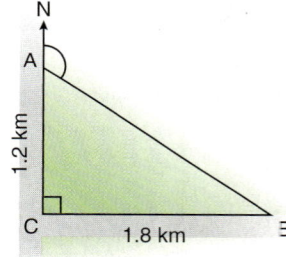

Solution

a) $\tan A = \dfrac{BC}{AC}$ Substitute: $BC = 1.8$ and $AC = 1.2$
   $= \dfrac{1.8}{1.2}$
   $\angle A = \tan^{-1}\left(\dfrac{1.8}{1.2}\right)$ Press: 2nd TAN 1.8 ÷ 1.2) ENTER
   $\doteq 56.31°$

The measure of $\angle A$ is about 56°.
$\angle B = 180° - \angle C - \angle A$
$= 180° - 90° - 56°$
$= 34°$

The measure of $\angle B$ is about 34°.

The sum of the angles in a triangle is 180°.

b) $\angle BAC$ and $\angle TAB$ form a straight line. So:
$\angle TAB = 180° - 56°$
$= 124°$

The bearing of the third side of the plot is about 124°.

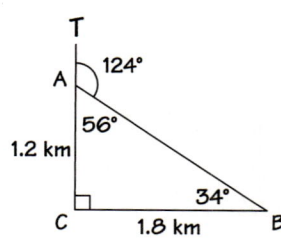

6   CHAPTER 1: Trigonometry

Example 3

Materials
- scientific calculator

An angle of elevation is also called an angle of inclination.

Solving Problems

Jenny and Nathan want to determine the height of the Pickering wind turbine. Jenny stands 60.0 feet from the base of the turbine. She measures the angle of elevation to the top of the turbine to be 81°. On the other side of the turbine, Nathan measures an angle of elevation of 76°. Jenny and Nathan each hold their clinometers about 4.8 feet above the ground when measuring the angle of elevation.

a) What is the height of the wind turbine?
b) How far away from the base is Nathan?

Solution

Sketch and label a diagram. Drawing a diagram may help you visualize the information in the problem. Your diagram does not need to be drawn to scale. For example, this diagram is not drawn to scale.

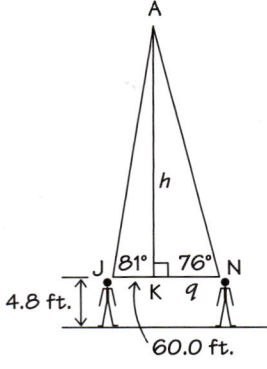

a) Use △AKJ to find the length of AK.

$\tan J = \frac{AK}{JK}$ Substitute: $AK = h$, $\angle J = 81°$, and $JK = 60.0$

$\tan 81° = \frac{h}{60.0}$ Multiply both sides by 60.0 to isolate h.

$h = \tan 81° \times 60.0$ Press: TAN 81) × 60.0 ENTER

$h \doteq 378.83$

The turbine is about 378.8 feet above eye level.

378.8 feet + 4.8 feet = 383.6 feet

The height of the wind turbine is about 384 feet.

b) In right △AKN, the measure of ∠N and the length of the opposite side AK are known.

$\tan N = \frac{AK}{KN}$ Substitute: $KN = q$, $\angle N = 76°$, and $AK \doteq 378.8$

$\tan 76° = \frac{378.8}{q}$ Rearrange the equation to isolate q.

$q = \frac{378.8}{\tan 76°}$ Press: 378.8 ÷ TAN 76) ENTER

$q \doteq 94.4$

Nathan is about 94 feet away from the base of the turbine.

This solution assumes that Jenny and Nathan are standing on level ground. Can you explain why?

1.1 Trigonometric Ratios in Right Triangles   **7**

Practice

A

1. For each triangle, name each side in two different ways.
   a) Hypotenuse
   b) Side opposite the marked angle
   c) Side adjacent to the marked angle

   i)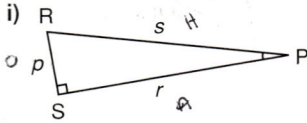

   ii)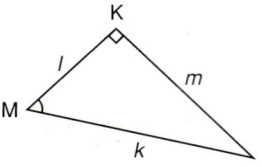

2. Write each trigonometric ratio as a ratio of sides.
   a) sin A   b) cos A   c) cos B   d) tan B

   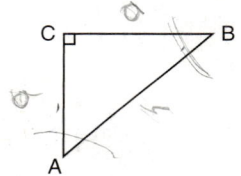

■ For help with questions 3 and 4, see Example 1.

3. Which primary trigonometric ratio can you use to calculate the length of each indicated side?

   a) 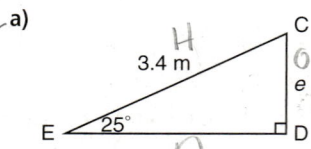   b)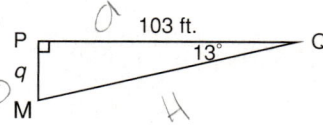

   c)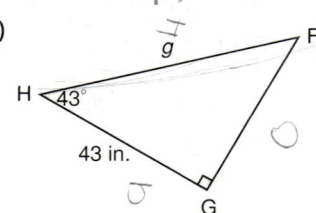

Give answers to the same number of decimal places as the least accurate measurement used in calculations.

4. Use the ratios you found in question 3 to calculate the length of each indicated side.

8   CHAPTER 1: Trigonometry

■ For help with question 5, see Example 2.

5. For each triangle, determine tan A. Then, determine the measure of ∠A.

a)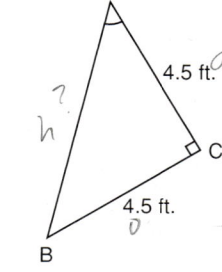

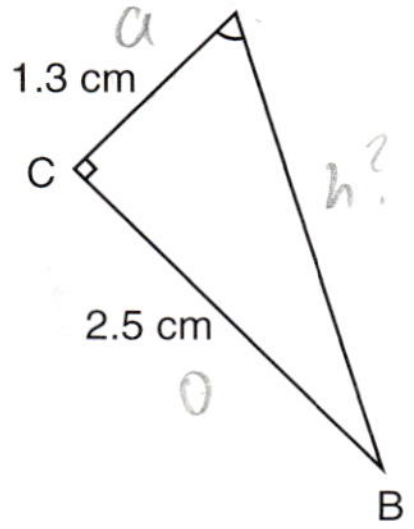

b)

c)

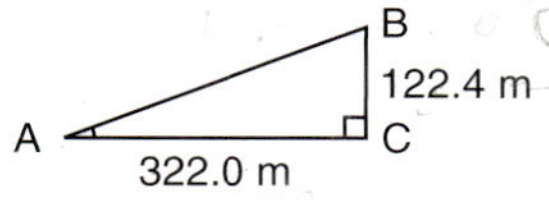

Give each angle measure to the nearest degree.

To *solve a triangle* means to determine the measures of all its sides and angles.

6. Use trigonometric ratios and the Pythagorean Theorem to solve each triangle.

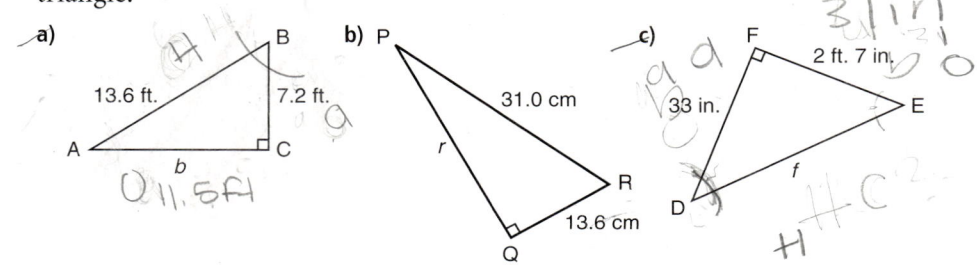

Use the *Course Study Guide* at the end of the book to recall any measurement conversions.

B

7. A ladder 10 feet long is leaning against a wall at a 71° angle.
  a) How far from the wall is the foot of the ladder?
  b) How high up the wall does the ladder reach?

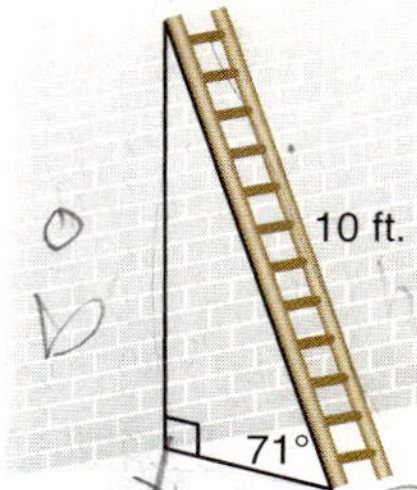

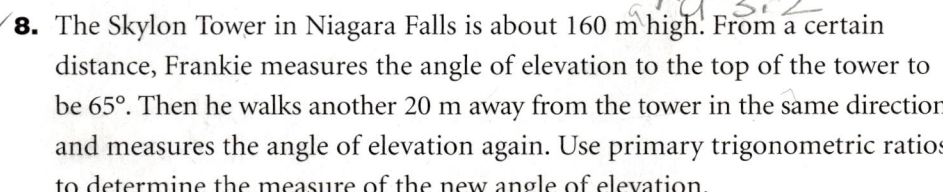

■ For help with questions 8 and 9, see Example 3.

8. The Skylon Tower in Niagara Falls is about 160 m high. From a certain distance, Frankie measures the angle of elevation to the top of the tower to be 65°. Then he walks another 20 m away from the tower in the same direction and measures the angle of elevation again. Use primary trigonometric ratios to determine the measure of the new angle of elevation.

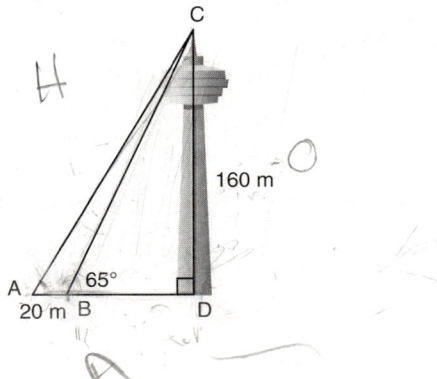

1.1 Trigonometric Ratios in Right Triangles   **9**

■ For help with question 9, see Example 3.

9. A rescue helicopter is flying horizontally at an altitude of 1500 feet over Georgian Bay toward Beausoleil Island. The angle of depression to the island is 9°. How much farther must the helicopter fly before it is above the island? Give your answer to the nearest mile.

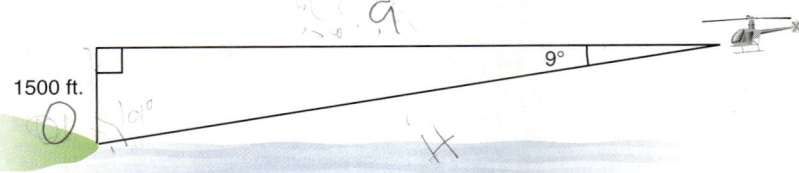

10. A theatre lighting technician adjusts the light to fall on the stage 3.5 m away from a point directly below the lighting fixture. The technician measures the angle of elevation from the lighted point on the stage to the fixture to be 56°. What is the height of the lighting fixture?

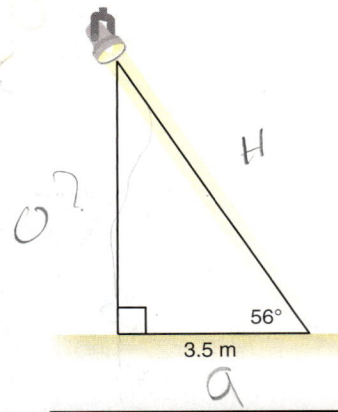

10 CHAPTER 1: Trigonometry

11. Kenya's class is having a contest to find the tallest building in Ottawa. Kenya chose the Place de Ville tower. Standing 28.5 m from the base of the tower, she measured an angle of elevation of 72° to its top. Use Kenya's measurements to determine the height of the tower.

> Assume Kenya measures the angle of elevation from the ground to the top of the tower.

12. A ship's chief navigator is plotting the course for a tour of three islands. The first island is 12 miles due west of the second island. The third island is 18 miles due south of the second island.
    a) Do you have enough information to determine the bearing required to sail directly back from the third island to the first island? Explain.
    b) If your answer to part a is yes, describe how the navigator would determine the bearing.

> An *angle of inclination* measures an angle above the horizontal. An *angle of depression* measures an angle below the horizontal.

13. **Assessment Focus** A carpenter is building a bookshelf against the sloped ceiling of an attic.
    a) Determine the length of the sloped ceiling, AB, used to build the bookshelf.
    b) Determine the measure of ∠A. Is ∠A an angle of inclination or an angle of depression? Why?
    c) Describe another method to solve part b. Which method do you prefer? Why?

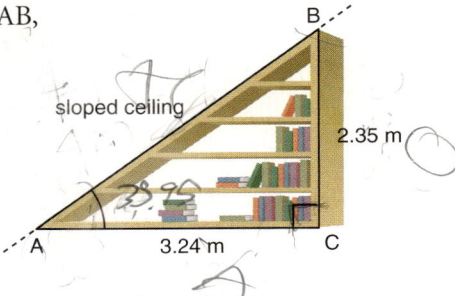

1.1 Trigonometric Ratios in Right Triangles   **11**

14. A roof has the shape of an isosceles triangle.

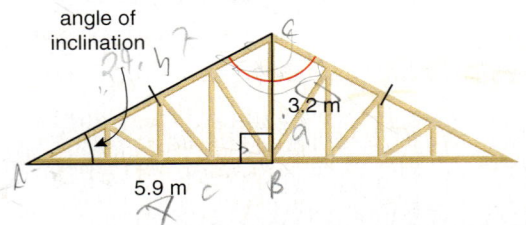

a) What is the measure of the angle of inclination of the roof?
b) What is the measure of the angle marked in red?
c) Write your own problem about the roof.
   Make sure you can use the primary trigonometric ratios to solve it.
   Solve your problem.

15. **Literacy in Math** In right △ABC with ∠C = 90°, sin A = cos B. Explain why.

16. Two boats, F and G, sail to the harbour, H. Boat F sails 3.2 km on a bearing of 176°. Boat G sails 2.5 km on a bearing of 145°. Determine the distance from each boat straight to the shore.

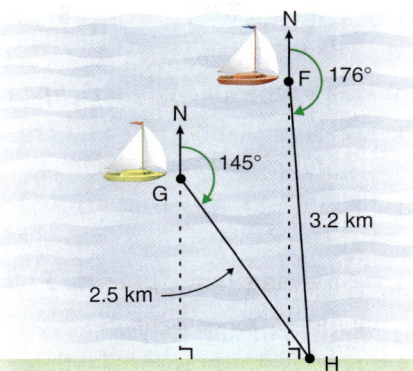

17. Use paper and a ruler. Draw a right △ABC where:
a) sin A = cos B = 0.5
b) tan A = tan B = 1.0
Describe your method.

> **In Your Own Words**
>
> Explain why someone might need to use primary trigonometric ratios in daily life or a future career.

12   CHAPTER 1: Trigonometry

1.2 Investigating the Sine, Cosine, and Tangent of Obtuse Angles

To create a proper joint between two pieces of wood, a carpenter needs to measure the angle between them. When corners meet at a right angle, the process is relatively simple. When pieces of wood are joined at an acute or an obtuse angle, the task of creating a proper joint is more difficult.

Inquire

Exploring Trigonometric Ratios

Materials
- The Geometer's Sketchpad or grid paper and protractor
- TechSinCosTan.gsp
- scientific calculator

The intersection of the *x*-axis and *y*-axis creates four regions, called **quadrants**, numbered counterclockwise starting from the upper right.

	y	
Quadrant II		Quadrant I
		x
Quadrant III		Quadrant IV

The angle made by line segment *r* with the positive *x*-axis is labelled ∠A.

Choose *Using The Geometer's Sketchpad* or *Using Pencil and Paper.*

Using *The Geometer's Sketchpad*

Work with a partner.

Part A: Investigating Trigonometric Ratios Using Point P(*x*, *y*)

- Open the file *TechSinCosTan.gsp*.
  Make sure your screen looks like this.

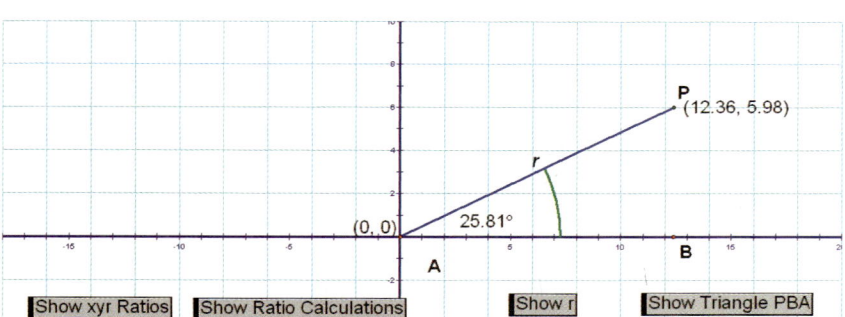

1. Use the **Selection Arrow** tool.

   Click on **Show Triangle PBA**, then deselect the triangle.

   > To deselect, use the **Selection Arrow** and click anywhere on the screen.

2. Move point P(x, y) around in Quadrant I.

   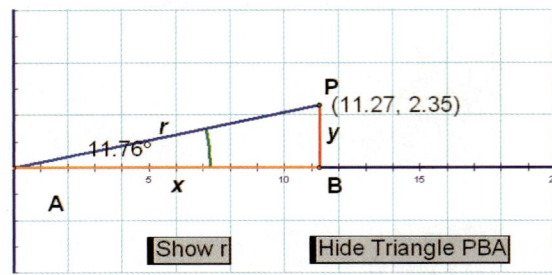

   In right △PBA, how can you use the Pythagorean Theorem and the values of x and y to determine the length of side r?
   Click on **Show r**. Compare your method with the formula on the screen.

3. Choose a position for point P in Quadrant I.
   In right △PBA:
   a) What is the measure of ∠A?
   b) Which side is opposite ∠A? Adjacent to ∠A? Which side is the hypotenuse?
   c) Use x, y, and r. Write each ratio.
      i) sin A   ii) cos A   iii) tan A
   Click on **Show xyr Ratios**.
   Compare your answers with the ratios on the screen.

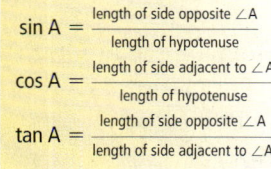

$$\sin A = \frac{\text{length of side opposite } \angle A}{\text{length of hypotenuse}}$$

$$\cos A = \frac{\text{length of side adjacent to } \angle A}{\text{length of hypotenuse}}$$

$$\tan A = \frac{\text{length of side opposite } \angle A}{\text{length of side adjacent to } \angle A}$$

4. Choose a position for point P in Quadrant II.
   a) What is the measure of ∠A?
   b) In right △PBA, what is the measure of ∠PAB?
   c) Is the *x*-coordinate of P positive or negative?
   d) Is the *y*-coordinate of P positive or negative?

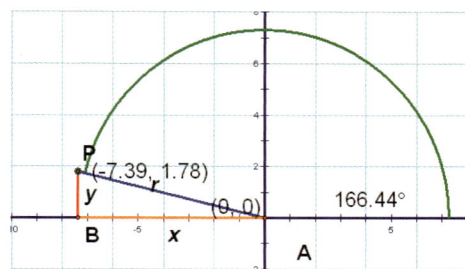

5. Copy the table. Use a scientific calculator for parts a and b.

	Angle measure	sin A	cos A	tan A
Acute ∠A				
Obtuse ∠A				

> An acute angle is less than 90°. An obtuse angle is between 90° and 180°.

   a) Use the measure of ∠A from question 3. Complete the row for acute ∠A.
   b) Use the measure of ∠A from question 4. Complete the row for obtuse ∠A.
   c) Click on **Show Ratio Calculations**. Compare with the results in the table.

Part B: Determining Signs of Trigonometric Ratios

■ Use the **Selection Arrow** tool.
  Move point P around Quadrants I and II.

6. a) Which type of angle is ∠A if point P is in Quadrant I?
   b) Which type of angle is ∠A if point P is in Quadrant II?

7. a) Can $r = \sqrt{x^2 + y^2}$ be negative? Why or why not?
   b) When ∠A is acute, is *x* positive or negative?
   c) When ∠A is obtuse, is *x* positive or negative?
   d) When ∠A is acute, is *y* positive or negative?
   e) When ∠A is obtuse, is *y* positive or negative?

1.2 Investigating the Sine, Cosine, and Tangent of Obtuse Angles

8. Suppose ∠A is acute. Use your answers from question 7.
   a) Think about sin A = $\frac{y}{r}$.
      Is sin A positive or negative? Explain why.
   b) Think about cos A = $\frac{x}{r}$.
      Is cos A positive or negative? Explain why.
   c) Think about tan A = $\frac{y}{x}$.
      Is tan A positive or negative? Explain why.
   d) Move point P around in Quadrant I. What do the sine, cosine, and tangent calculations show about your work for parts a, b, and c?

9. Suppose ∠A is obtuse. Use your answers from question 7.
   a) Think about sin A = $\frac{y}{r}$.
      Is sin A positive or negative? Explain why.
   b) Think about cos A = $\frac{x}{r}$.
      Is cos A positive or negative? Explain why.
   c) Think about tan A = $\frac{y}{x}$.
      Is tan A positive or negative? Explain why.
   d) Move point P around in Quadrant II. What do the sine, cosine, and tangent calculations on screen show about your work for parts a, b, and c?

Using Pencil and Paper

Work with a partner.

Part A: Investigating Trigonometric Ratios Using Point P(x, y)

1. On grid paper, draw a point P(x, y) in Quadrant I of a coordinate grid. Label the sides and vertices as shown.

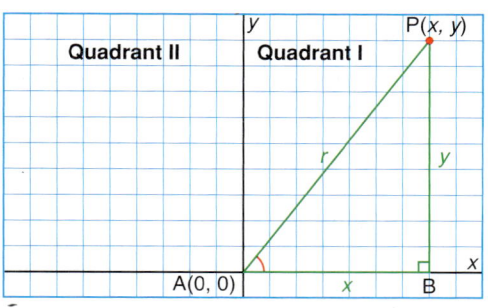

2. In right △PBA, how can you use the Pythagorean Theorem and the values of x and y to determine the number of units for side r?

$$\sin A = \frac{\text{length of side opposite } \angle A}{\text{length of hypotenuse}}$$

$$\cos A = \frac{\text{length of side adjacent to } \angle A}{\text{length of hypotenuse}}$$

$$\tan A = \frac{\text{length of side opposite } \angle A}{\text{length of side adjacent to } \angle A}$$

3. In right △PBA:
   a) What is the measure of ∠A?
   b) Which side is opposite ∠A? Adjacent to ∠A? Which side is the hypotenuse?
   c) Use x, y, and r. Write each ratio.
      i) sin A   ii) cos A   iii) tan A

4. On the same grid, choose a second position for point P in Quadrant II. Label as shown.

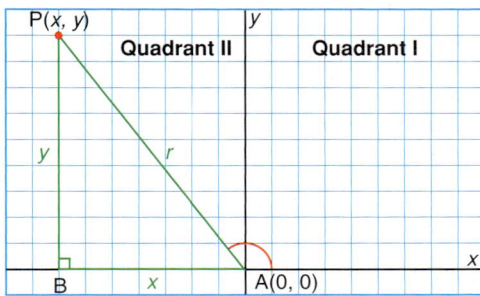

   a) What is the measure of ∠A?
   b) In right △PBA, what is the measure of ∠PAB?
   c) Is the x-coordinate of P positive or negative?
   d) Is the y-coordinate of P positive or negative?

1.2 Investigating the Sine, Cosine, and Tangent of Obtuse Angles

5. Copy the table.

	Angle measure	sin A	cos A	tan A
Acute ∠A				
Obtuse ∠A				

Use a scientific calculator.
a) Use the measure of ∠A and the number of units for each side from question 3. Complete the row for acute ∠A.
b) Use the measure of ∠A and the number of units for each side from question 4. Complete the row for obtuse ∠A.

Part B: Determining Signs of the Trigonometric Ratios

> An acute angle is less than 90°. An obtuse angle is between 90° and 180°.

6. a) Which type of angle is ∠A if point P is in Quadrant I?
b) Which type of angle is ∠A if point P is in Quadrant II?

7. a) Can $r = \sqrt{x^2 + y^2}$ be negative? Why or why not?
b) When ∠A is acute, is x positive or negative?
c) When ∠A is obtuse, is x positive or negative?
d) When ∠A is acute, is y positive or negative?
e) When ∠A is obtuse, is y positive or negative?

8. Suppose ∠A is acute. Use your answers from question 7.
a) Think about $\sin A = \frac{y}{r}$.
Is sin A positive or negative? Explain why.
b) Think about $\cos A = \frac{x}{r}$.
Is cos A positive or negative? Explain why.
c) Think about $\tan A = \frac{y}{x}$.
Is tan A positive or negative? Explain why.

9. Suppose ∠A is obtuse. Use your answers from question 7.
a) Think about $\sin A = \frac{y}{r}$.
Is sin A positive or negative? Explain why.
b) Think about $\cos A = \frac{x}{r}$.
Is cos A positive or negative? Explain why.
c) Think about $\tan A = \frac{y}{x}$.
Is tan A positive or negative? Explain why.

Practice

A

1. Use your work from *Inquire*. For each angle measure, is point P in Quadrant I or Quadrant II?
   a) 35°
   b) 127°
   c) 95°

2. Is the sine of each angle positive or negative?
   a) 45°
   b) 67°
   c) 153°

3. Is the cosine of each angle positive or negative?
   a) 168°
   b) 32°
   c) 114°

4. Is the tangent of each angle positive or negative?
   a) 123°
   b) 22°
   c) 102°

B

5. Is each ratio positive or negative? Justify your answers.
   a) tan 40°   b) cos 120°
   c) tan 150°   d) sin 101°
   e) cos 98°   f) sin 13°

6. Is ∠A acute, obtuse, or either? Justify your answers.
   a) cos A = 0.35
   b) tan A = −0.72
   c) sin A = 0.99

Reflect

- What did your results show about the measure of ∠A and its trigonometric ratios when point P is in Quadrant I?
- What did your results show about the measure of ∠A and its trigonometric ratios when point P is in Quadrant II?

1.3 Sine, Cosine, and Tangent of Obtuse Angles

Surveyors and navigators often work with angles greater than 90°. They need to know how to interpret their calculations.

Investigate — Exploring Supplementary Angles

Materials
- scientific calculator

The sum of the measures of two supplementary angles is 180°.

Work with a partner.

■ Copy and complete the table for $\angle A = 25°$.

	$\angle A = 25°$	Supplementary angle
sin A		
cos A		
tan A		

■ Repeat for $\angle B = 105°$ and for $\angle C = 150°$.
■ Examine your tables for $\angle A$, $\angle B$, and $\angle C$. What relationships do you see between
  • The sines of supplementary angles?
  • The cosines of supplementary angles?
  • The tangents of supplementary angles?

20 CHAPTER 1: Trigonometry

Reflect

- If you know the trigonometric ratios of an acute angle, how can you determine the ratios of its supplementary angle?
- If you know the trigonometric ratios of an obtuse angle, how can you determine the ratios of its supplementary angle?

Connect the Ideas

xyr definition

The trigonometric ratios can be defined using a point P(x, y) on a coordinate grid.

Trigonometric ratios
In right $\triangle$PBA:
$\sin A = \frac{y}{r}$ $(r \neq 0)$
$\cos A = \frac{x}{r}$ $(r \neq 0)$
$\tan A = \frac{y}{x}$ $(x \neq 0)$

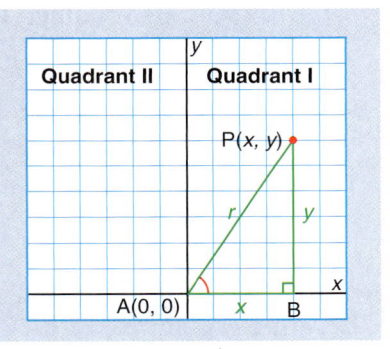

When $\angle A$ is obtuse, x is negative.

Signs of the primary trigonometric ratios

When point P is in Quadrant I, $\angle A$ is acute.
- sin A is positive.
- cos A is positive.
- tan A is positive.

When point P is in Quadrant II, $\angle A$ is obtuse.
- sin A is positive.
- cos A is negative.
- tan A is negative.

Example 1

Determining Trigonometric Ratios of an Obtuse Angle

Materials
- scientific calculator

Suppose $\angle C = 123°$.
Determine each trigonometric ratio for $\angle C$, to 4 decimal places.
a) sin C   b) cos C   c) tan C

Solution

Use a calculator.

a) $\sin C = \sin 123°$
   $\doteq 0.8387$

b) $\cos C = \cos 123°$
   $\doteq -0.5446$

c) $\tan C = \tan 123°$
   $\doteq -1.5399$

Supplementary angles

In Lesson 1.2, we investigated relationships between trigonometric ratios of an acute angle and its supplement. We can use these relationships to determine the measure of an obtuse angle.

> The sum of the measures of two supplementary angles is 180°.

Properties of supplementary angles

Given an acute angle, A, and its supplementary obtuse angle $(180° − A)$:
- $\sin A = \sin(180° − A)$
- $\cos A = −\cos(180° − A)$
- $\tan A = −\tan(180° − A)$

Example 2 — Determining the Measure of an Obtuse Angle

Materials
- scientific calculator

Write the measure of each supplementary obtuse angle when:
a) The sine of acute $\angle P$ is 0.65.
b) The cosine of acute $\angle R$ is 0.22.
c) The tangent of acute $\angle S$ is 0.44.

Solution

Method 1

First determine the measure of the acute angle.

a) $\sin P = 0.65$
$\angle P = \sin^{-1}(0.65)$
$\angle P \doteq 40.5°$
$180° − \angle P = 180° − 40.5°$
$= 139.5°$

b) $\cos R = 0.22$
$\angle R = \cos^{-1}(0.22)$
$\angle R \doteq 77.3°$
$180° − \angle R = 180° − 77.3°$
$= 102.7°$

c) $\tan S = 0.44$
$\angle S = \tan^{-1}(0.44)$
$\angle S \doteq 23.7°$
$180° − \angle R = 180° − 23.7°$
$= 156.3°$

Method 2

First determine the trigonometric ratio of the obtuse angle.

a) $\sin(180° − P) = \sin P$
$= 0.65$
Using a calculator:
$\sin^{-1}(0.65) \doteq 40.5°$
$180° − 40.5° = 139.5°$

b) $\cos(180° − R) = −\cos R$
$= −0.22$
Using a calculator:
$\cos^{-1}(−0.22) \doteq 102.7°$

c) $\tan(180° − S) = −\tan S$
$= −0.44$
Using a calculator:
$\tan^{-1}(−0.44) \doteq −23.7°$
$180° + (−23.7°) = 156.3°$

> Since different angles have the same trigonometric ratios, your calculator may not return the measure of the obtuse angle when you use an inverse trigonometric ratio.
>
> To determine the measure of an obtuse angle, A, using a calculator:
> - $\angle A = 180° − \sin^{-1}(\sin A)$
> - $\angle A = \cos^{-1}(\cos A)$
> - $\angle A = 180° + \tan^{-1}(\tan A)$
>
> See how these properties are applied in *Method 2*.

Practice

A

■ For help with question 1, see Example 1.

1. Determine the sine, cosine, and tangent ratios for each angle. Give each answer to 4 decimal places.
   a) 110°   b) 154°   c) 102°

2. Is each trigonometric ratio positive or negative? Use a calculator to check your answer.
   a) sin 35°   b) tan 154°   c) cos 134°

3. Each point on this coordinate grid makes a right triangle with the origin, A, and the x-axis. Determine the indicated trigonometric ratio in each triangle.

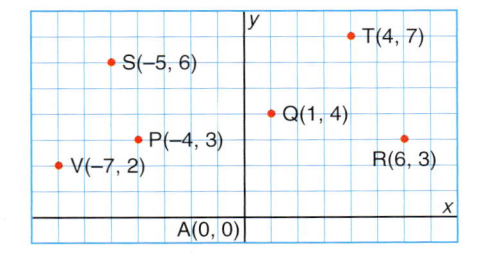

Points shown: S(−5, 6), T(4, 7), P(−4, 3), Q(1, 4), V(−7, 2), R(6, 3), A(0, 0)

   a) sin A and point P
   b) cos A and point Q
   c) tan A and point R
   d) cos A and point S
   e) sin A and point T
   f) tan A and point V

■ For help with questions 4, 5, and 6, see Example 2.

4. Suppose ∠P is an obtuse angle. Determine the measure of ∠P for each sine ratio. Give each angle measure to the nearest degree.
   a) 0.23   b) 0.98   c) 0.57   d) 0.09

5. Determine the measure of ∠R for each cosine ratio. Give each angle measure to the nearest degree.
   a) −0.67   b) 0.56   c) −0.23   d) −0.25

6. Determine the measure of ∠Q for each tangent ratio. Give each angle measure to the nearest degree.
   a) 0.46   b) −1.60   c) −0.70   d) −1.53

B

7. ∠M is between 0° and 180°. Is ∠M acute or obtuse? How do you know?
   a) cos M = 0.6   b) cos M = −0.6   c) sin M = 0.6

8. For each trigonometric ratio, identify whether ∠Y could be between 0° and 180°. Justify your answer.
   a) cos Y = −0.83
   b) sin Y = −0.11
   c) tan Y = 0.57
   d) tan Y = 0.97

1.3 Sine, Cosine, and Tangent of Obtuse Angles

9. **Literacy in Math** Create a table to organize information from this lesson about the sine, cosine, and tangent ratios of supplementary angles. Use *Connect the Ideas* to help you.

	Acute angle	Obtuse angle
Sine	positive	positive
Cosine		

10. $\angle G$ is an angle in a triangle. Determine all measures of $\angle G$.
    a) $\sin G = 0.62$
    b) $\cos G = -0.85$
    c) $\tan G = 0.21$
    d) $\tan G = -0.32$
    e) $\cos G = -0.71$
    f) $\sin G = 0.77$

11. The measure of $\angle Y$ is between $0°$ and $180°$. Which equations result in two different values for $\angle Y$? How do you know?
    a) $\sin Y = 0.32$
    b) $\sin Y = 0.23$
    c) $\cos Y = -0.45$
    d) $\cos Y = 0.38$
    e) $\tan Y = -0.70$
    f) $\tan Y = 0.77$

12. **Assessment Focus** Determine all measures of $\angle A$ in a triangle, given each ratio. Explain your thinking.
    a) $\sin A = 0.45$
    b) $\cos A = -0.45$
    c) $\tan A = 0.45$

13. The cosine of an obtuse angle is -0.45. Calculate the sine of this angle to 4 decimal places.

14. The sine of an obtuse angle is $\frac{12}{13}$. Calculate the cosine of this angle to 4 decimal places.

C

15. a) Determine the values of $\sin 90°$ and $\cos 90°$.
    b) Explain why $\tan 90°$ is undefined.

> **In Your Own Words**
>
> Is knowing a trigonometric ratio of an angle enough to determine the measure of the angle? If so, explain why. If not, what else do you need to know?

Trigonometric Search

Play with a partner.

- Each player:
  - Cuts out two 20 by 12 grids. Draws and labels each grid as shown.
  - Draws △ABC on the first grid. Writes the *x*- and *y*-coordinates of each vertex.
  - Joins vertices A, B, and C with the origin, O. Measures the angle made by each vertex with the positive *x*-axis.
  - Calculates a trigonometric ratio of their choice for the angle measured.
  - In turn, states the *x*- or *y*-coordinate and the calculated trigonometric ratio for the angle measured.

- The other player:
  - Uses the ratio to calculate the angle made with the positive *x*-axis.
  - Uses the *x*- or *y*-coordinate to plot the point.
  - Uses primary trigonometric ratios to find the second coordinate of the vertex.
  - Marks the findings on the second grid.

Materials
- grid paper
- scissors
- protractor
- scientific calculator

Rules for drawing △ABC
- Draw all vertices on intersecting grid lines.
- Do not draw any vertex on an axis.

Player A:
$\sin A = 0.16$
and $x = 6$ units

What are the coordinates of vertex A?

Player B:
- $\angle A = \sin^{-1}(0.16) \doteq 9°$
- $\tan 9° = \dfrac{y}{6}$
- $y = \tan 9° \times 6 \doteq 0.95$

So: A(6, 1)

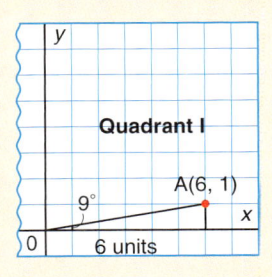

The player who first solves the other player's triangle wins the round. Repeat the game with different triangles.

Reflect

- What strategy did you use to determine the other coordinate of each vertex?
- What are the most common mistakes one can make during the game? How can you avoid them?

Mid-Chapter Review

1.1

1. Solve each triangle.

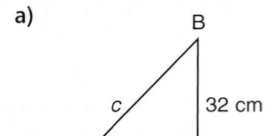

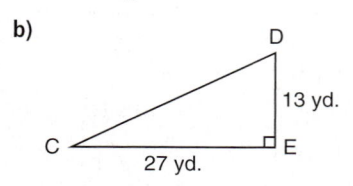

a) Triangle with B, C, A; BC = 32 cm, angle A = 47°, side c, side b, right angle at C.

b) Triangle with C, D, E; DE = 13 yd, CE = 27 yd, right angle at E.

2. Solve each right $\triangle XYZ$. Sketch a diagram.
   a) $\angle X = 53°$, $\angle Z = 90°$, $x = 3.6$ cm
   b) $z = 3$ feet 5 inches, $x = 25$ inches, $\angle Z = 90°$

3. A flight of stairs has steps that are 14 inches deep and 12 inches high. A handrail runs along the wall in line with the steps. What is the angle of elevation of the handrail?

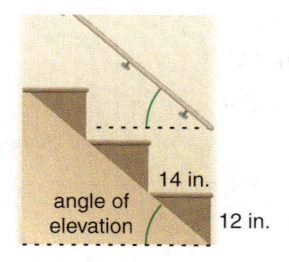

4. Cables stretch from each end of a bridge to a 4.5 m column on the bridge. The angles of elevation from each end of the bridge to the top of the column are 10° and 14°. What is the length of the bridge?

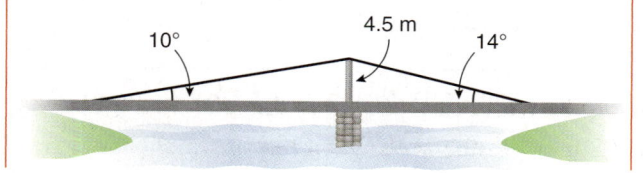

5. John hikes 2.5 miles due west from Temagami fire tower, then 6 miles due north.

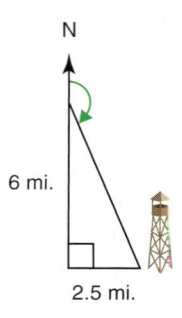

   a) How far is he from the fire tower at the end of the hike?
   b) What bearing should he use to return to the fire tower?

Answer questions 6 and 7 without using a calculator.

1.2

6. Determine whether the sine, cosine, and tangent of the angle is positive or negative. Explain how you know.
   a) 27°   b) 95°   c) 138°

7. Is $\angle P$ acute or obtuse? Explain.
   a) $\cos P = 0.46$ b) $\tan P = -1.43$
   c) $\sin P = 0.5$ d) $\cos P = -0.5877$

1.3

8. Determine each measure of obtuse $\angle P$.
   a) $\sin P = 0.22$ b) $\cos P = -0.98$
   c) $\tan P = -1.57$ d) $\sin P = 0.37$

9. $\angle G$ is an angle in a triangle. Determine all possible values for $\angle G$.
   a) $\sin G = 0.53$ b) $\cos G = -0.42$
   c) $\tan G = 0.14$ d) $\sin G = 0.05$

26 CHAPTER 1: Trigonometry

1.4 The Sine Law

The team of computer drafters working on the Michael Lee-Chin Crystal, at the Royal Ontario Museum, needed to know the precise measures of angles and sides in triangles that were not right triangles.

Investigate

Relating Sine Ratios in Triangles

Materials
- protractor
- scientific calculator

An acute triangle has three acute angles.
An obtuse triangle has one obtuse angle.

Work with a partner.

■ Draw two large triangles: one acute and one obtuse. Label the vertices of each triangle A, B, and C.

■ Copy and complete the table for each triangle.

Angle	Angle measure	Sine of angle	Length of opposite side	Ratios	
∠A			$a =$	$\dfrac{a}{\sin A} =$	$\dfrac{\sin A}{a} =$
∠B			$b =$	$\dfrac{b}{\sin B} =$	$\dfrac{\sin B}{b} =$
∠C			$c =$	$\dfrac{c}{\sin C} =$	$\dfrac{\sin C}{c} =$

■ Describe any relationships you notice in the tables.

Reflect

Compare your results with other pairs. Are the relationships true for all triangles? How does the *Investigate* support this?

Connect the Ideas

An **oblique triangle** is any triangle that is not a right triangle.

The Sine Law

The **Sine Law** relates the sides and the angles in any oblique triangle.

An acute triangle is an oblique triangle. An obtuse triangle is an oblique triangle.

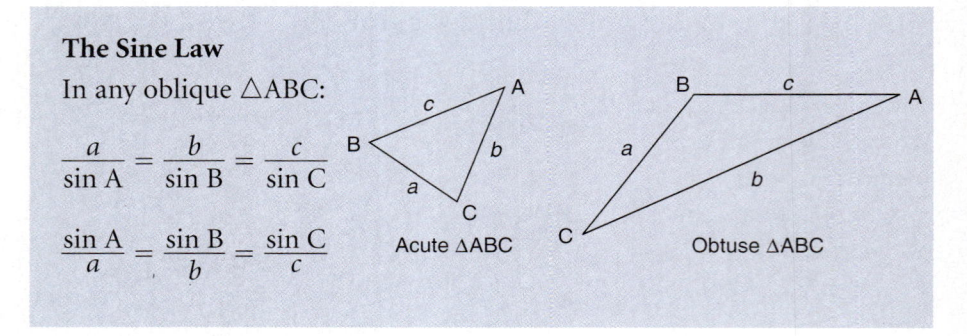

The Sine Law
In any oblique $\triangle ABC$:

$$\frac{a}{\sin A} = \frac{b}{\sin B} = \frac{c}{\sin C}$$

$$\frac{\sin A}{a} = \frac{\sin B}{b} = \frac{\sin C}{c}$$

Acute $\triangle ABC$ Obtuse $\triangle ABC$

Angle-Angle-Side (AAS)

When we know the measures of two angles in a triangle and the length of a side opposite one of the angles, we can use the Sine Law to determine the length of the side opposite the other angle.

Example 1

Determining the Length of a Side Using the Sine Law

Materials
- scientific calculator

What is the length of side d in $\triangle DEF$?

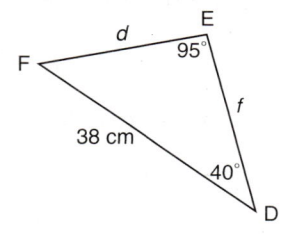

Solution

Write the Sine Law for $\triangle DEF$:
$$\frac{d}{\sin D} = \frac{e}{\sin E} = \frac{f}{\sin F}$$

Use the two ratios that include the known measures.

$\frac{d}{\sin D} = \frac{e}{\sin E}$ Substitute: $\angle D = 40°$, $\angle E = 95°$, and $e = 38$

$\frac{d}{\sin 40°} = \frac{38}{\sin 95°}$ Multiply each side by $\sin 40°$ to isolate d.

$d = \frac{38}{\sin 95°} \times \sin 40°$

$\doteq 24.52$

Press: 38 ÷ SIN 95) × SIN 40) ENTER

So, side d is about 25 cm long.

CHAPTER 1: Trigonometry

Angle-Side-Angle (ASA)

When we know the measure of two angles in a triangle and the length of the side between them, we can determine the measure of the unknown angle using the sum of the angles in a triangle, then we can use the Sine Law to solve the triangle.

Example 2 — Determining the Measure of an Angle Using the Sine Law

Materials
- scientific calculator

a) Calculate the measure of ∠Z in △XYZ.
b) Determine the unknown side lengths x and y.

[Diagram: Triangle XYZ with ∠X = 83°, ∠Y = 35°, side XY = 2 ft. 2 in., side opposite Y labeled y, side opposite X labeled x.]

Solution

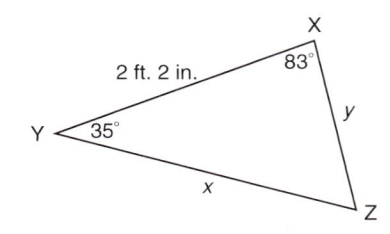

The sum of the angles in a triangle is 180°.

a) ∠Z = 180° − 83° − 35°
   = 62°

So, ∠Z is 62°.

1 foot = 12 inches

b) Write the length of side z in inches: 2 feet 2 inches = 26 inches

Write the Sine Law for △XYZ: $\dfrac{x}{\sin X} = \dfrac{y}{\sin Y} = \dfrac{z}{\sin Z}$

To find x, use the first and the third ratios.

$\dfrac{x}{\sin X} = \dfrac{z}{\sin Z}$ Substitute: ∠X = 83°, ∠Z = 62°, and $z = 26$

$\dfrac{x}{\sin 83°} = \dfrac{26}{\sin 62°}$ Multiply each side by sin 83° to isolate x.

$x = \dfrac{26}{\sin 62°} \times \sin 83°$

$\doteq 29.23$

So, side x is about 30 inches, or about 2 feet 6 inches, long.

To find y, use the second and third ratios in the Sine Law for △XYZ.

$\dfrac{y}{\sin Y} = \dfrac{z}{\sin Z}$ Substitute: ∠Y = 35°, ∠Z = 62°, and $z = 26$

$\dfrac{y}{\sin 35°} = \dfrac{26}{\sin 62°}$ Multiply each side by sin 35° to isolate y.

$y = \dfrac{26}{\sin 62°} \times \sin 35°$

$\doteq 16.89$

So, side y is about 17 inches, or about 1 foot 5 inches, long.

Example 3 Applying the Sine Law

Materials
- scientific calculator

A plane is approaching a 7500 m runway.
The angles of depression to the ends of the runway are 9° and 16°.
How far is the plane from each end of the runway?

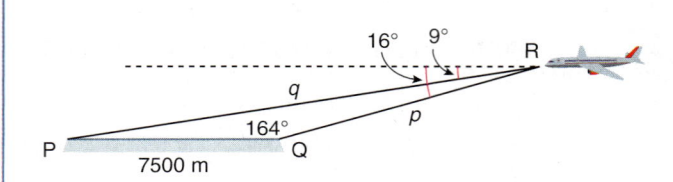

Solution

To determine $\angle R$ in $\triangle PQR$, use the angles of depression.
$\angle PRQ = 16° - 9° = 7°$
Write the Sine Law for $\triangle PQR$:
$$\frac{p}{\sin P} = \frac{q}{\sin Q} = \frac{r}{\sin R}$$
To determine q, use the second and third ratios.

$\frac{q}{\sin Q} = \frac{r}{\sin R}$ Substitute: $\angle R = 7°$, $\angle Q = 164°$, and $r = 7500$

$\frac{q}{\sin 164°} = \frac{7500}{\sin 7°}$ Multiply each side by sin 164° to isolate q.

$q = \frac{7500}{\sin 7°} \times \sin 164°$

$\doteq 16\,963.09$

$\angle P = 180° - 164° - 7°$
$= 9°$

To determine p, use the first and third ratios in the Sine Law for $\triangle PQR$.

$\frac{p}{\sin P} = \frac{r}{\sin R}$ Substitute: $\angle R = 7°$, $\angle P = 9°$, and $r = 7500$

$\frac{p}{\sin 9°} = \frac{7500}{\sin 7°}$ Multiply each side by sin 9° to isolate p.

$p = \frac{7500}{\sin 7°} \times \sin 9°$

$\doteq 9627.18$

So, the plane is about 16 963 m from one end and about 9627 m from the other end of the runway.

Practice

A

For help with questions 1 and 2, see Example 1.

1. Write the Sine Law for each triangle. Circle the ratios you would use to calculate each indicated length.

   a) [Triangle XYZ with side 2.5 cm opposite Z, angle X = 52°, angle Z = 55°, side x opposite X]

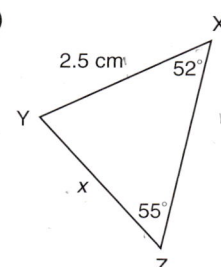

   b) [Triangle URT with angle U = 105°, angle T = 23°, side t, side UR = 3.0 ft]

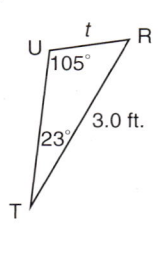

   c) [Triangle MNP with angle N = 19°, angle M = 33°, side NP = 23 m, side n opposite N]

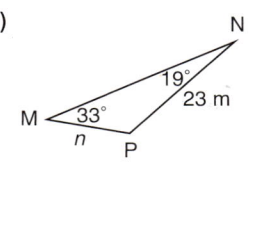

2. For each triangle in question 1, calculate each indicated length.

3. Use the Sine Law to determine the length of each indicated length.

   a) [Triangle GEF with angle G = 27°, angle F = 67°, side EF = 8.2 m, side f]

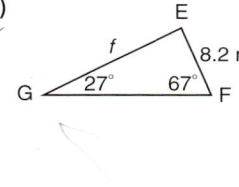

   b) [Triangle KJH with angle K = 23°, angle J = 121°, side JH = 230 in., side j]

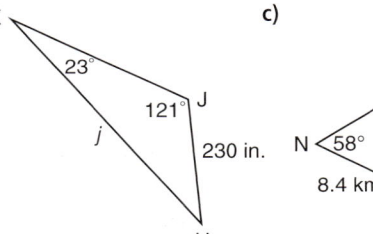

   c) [Triangle MNP with angle M = 42°, angle N = 58°, side NP = 8.4 km, side n]

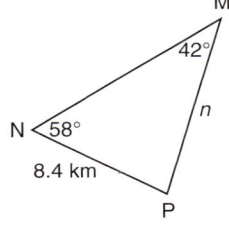

For help with questions 4 and 5, see Example 2.

4. a) Write the Sine Law for each triangle.
   b) Circle the ratios you would use to find each unknown side.
   c) Solve each triangle.

   i) [Triangle XYZ with angle X = 21°, angle Y = 21°, side XY = 5 in.]

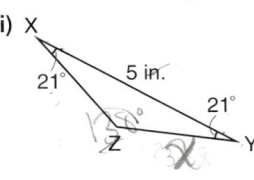

   ii) [Triangle GEF with angle E = 120°, angle F = 11°, side EF = 2.4 km]

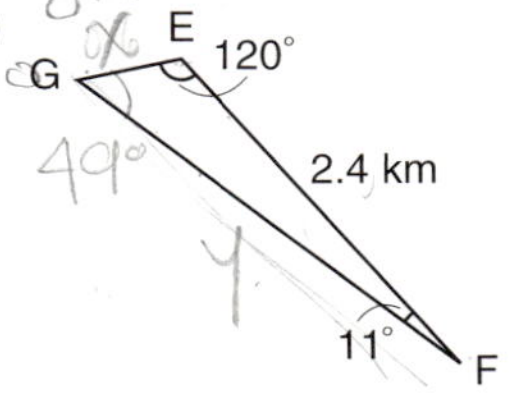

   iii) [Triangle KML with angle K = 69°, angle M = 76°, side KM = 2.2 m]

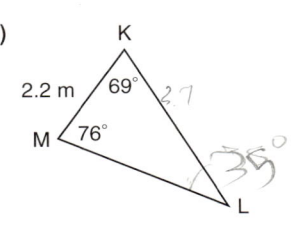

5. Solve each triangle.

   a) [Triangle XYZ with angle Z = 163°, angle Y = 10°, side XZ = 5.5 m]

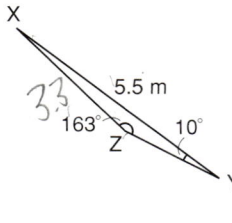

   b) [Triangle XYZ with angle X = 125°, angle Y = 25°, side XY = 14.2 mi.]

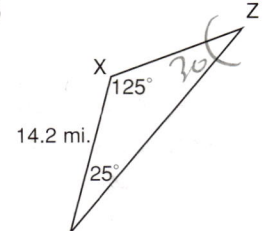

   c) [Triangle XYZ with angle Y = 39°, angle X = 121°, side YX = 13 cm]

1.4 The Sine Law **31**

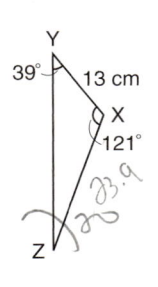

6. Solve each triangle.
   a) [Triangle XYZ with X = 128°, XY = 4 cm, angle at Y = 11°]
   b) [Triangle XYZ with Y = 30°, YX = 1 ft. 2 in., X = 120°]
   c) [Triangle XYZ with Y = 58°, XY = 55 mm, X = 102°]

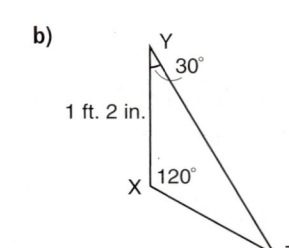

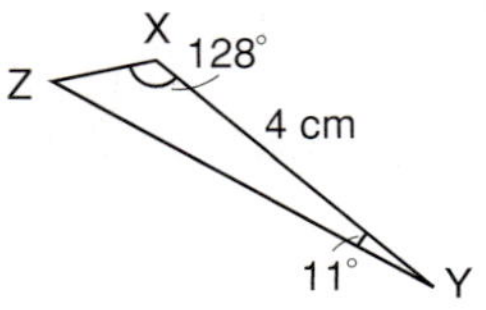

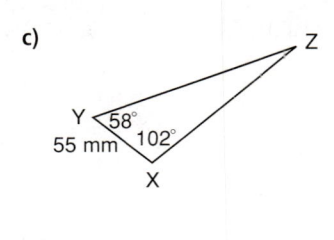

B 7. Choose one part from question 6. Write to explain how you solved the triangle.

8. Could you use the Sine Law to determine the length of side a in each triangle? If not, explain why not.
   a) [Triangle ABC with AC = 5.4 cm, A = 32°, B = 20°, side a opposite A]
   b) [Triangle ABC with A = 66°, AB = 9 ft., AC = 8 ft., side a]
   c) [Triangle ABC with right angle at C, A = 60°, AB = 23 m, side a]

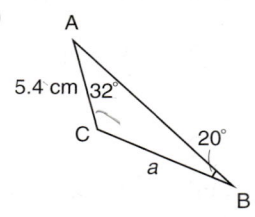

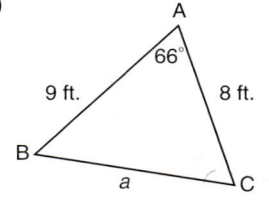

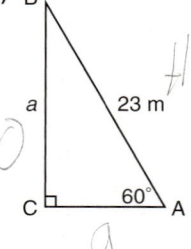

9. In question 8, determine a where possible using the Sine Law.

10. a) What is the measure of $\angle T$ in $\triangle TUV$?
    b) Determine the length of side a.

    [Triangle TUV with UV = 15 yd., U = 79°, V = 88°, side a = UT]

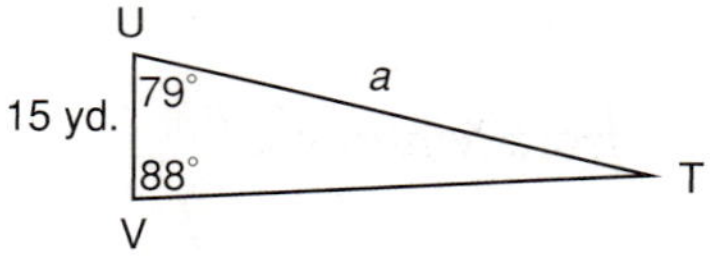

11. a) Use the Sine Law to determine the length of side b.
    b) Explain why you would use the Sine Law to solve this problem.
    c) Which pair of ratios did you use to solve for b? Explain your choice.

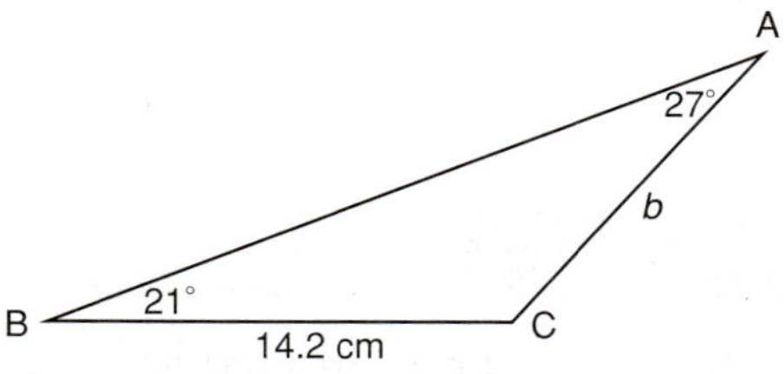

    [Triangle ABC with A = 27°, B = 21°, BC = 14.2 cm, side b = AC]

32 CHAPTER 1: Trigonometry

12. In △XYZ:
   a) Determine the lengths of sides x and z.
   b) Explain your strategy.

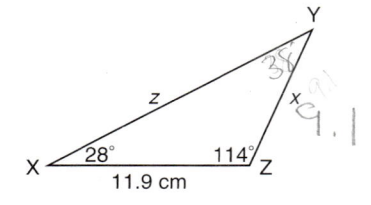

13. In △DEF, ∠D = 29°, ∠E = 113°, and f = 4.2 inches.
   a) Determine the length of side e.
   b) Explain how you solved the problem.

14. Literacy in Math Describe the steps for solving question 13. Draw a diagram to show your work.

For help with question 15, see Example 3.

15. A welder needs to cut this triangular shape from a piece of metal.

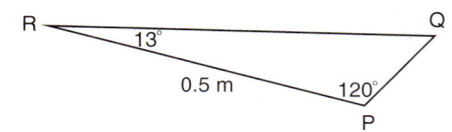

Determine the measure of ∠Q and the side lengths PQ and QR.

16. Assessment Focus A surveyor is mapping a triangular plot of land. Determine the unknown side lengths and angle measure in the triangle. Describe your strategy.

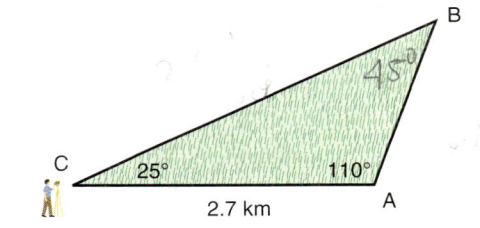

17. How far are ships R and S from lighthouse L?

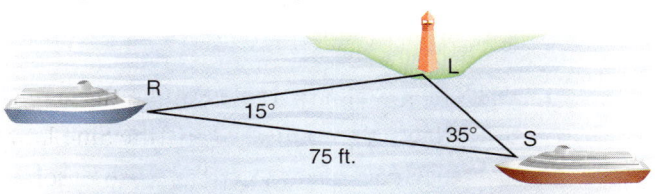

In Your Own Words

What information do you need in a triangle to be able to use the Sine Law? How do you decide which pair of ratios to use?

1.4 The Sine Law

Collecting Important Ideas

Focusing on key ideas can help you improve performance at both work and school. As you prepare for college or a career, keep track of the important ideas you learn.

When recording important ideas:

- Use captions, arrows, and colours to help show the idea.

- Add a picture, a diagram, or an example.

- Keep it brief.
  Don't say more than you need to!

Here's an example.

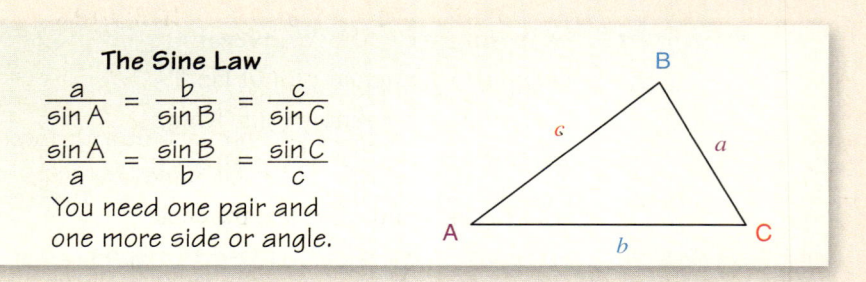

The Sine Law

$$\frac{a}{\sin A} = \frac{b}{\sin B} = \frac{c}{\sin C}$$

$$\frac{\sin A}{a} = \frac{\sin B}{b} = \frac{\sin C}{c}$$

You need one pair and one more side or angle.

1. What changes would you make to the example above?

2. Begin with this chapter.
   Make a collection of "trigonometry" ideas.
   Create a section in your math notebook or start a file on your computer.

3. During the year, add to your collection by creating a record of important ideas from later chapters.

34 Transitions: Collecting Important Ideas

1.5 The Cosine Law

A furniture designer's work begins with a concept, developed on paper or on a computer, then built to test its practicality and functionality.
The designs may then be mass-produced. Precise angle measurements are required at all times.

Investigate

Using Cosine Ratios in Triangles

Materials
- protractor
- scientific calculator

Work with a partner.
- Construct a △ABC for each description.
  - ∠C = 90°
  - All angles are acute
  - ∠C is obtuse
- Copy and complete the table.

△ABC	a	a^2	b	b^2	c	c^2	cos C	2ab cos C	$a^2 + b^2 - 2ab$ cos C
Right triangle									
Acute triangle									
Obtuse triangle									

- What patterns do you notice in the table?
- Compare your results with other pairs. Are your results true for all triangles?

Reflect

The Cosine Law is: $c^2 = a^2 + b^2 - 2ab \cos C$.
How are the Cosine Law for triangles and the Pythagorean Theorem the same? How are they different? How does the information in your table show this?

Connect the Ideas

The Cosine Law

The Sine Law, although helpful, has limited applications. We cannot use the Sine Law to solve an oblique triangle unless we know the measures of at least two angles. We need another method when:
- We know the lengths of two sides and the measure of the angle between them (SAS)
- We know all three side lengths in a triangle (SSS)

The **Cosine Law** can be used in both of these situations.

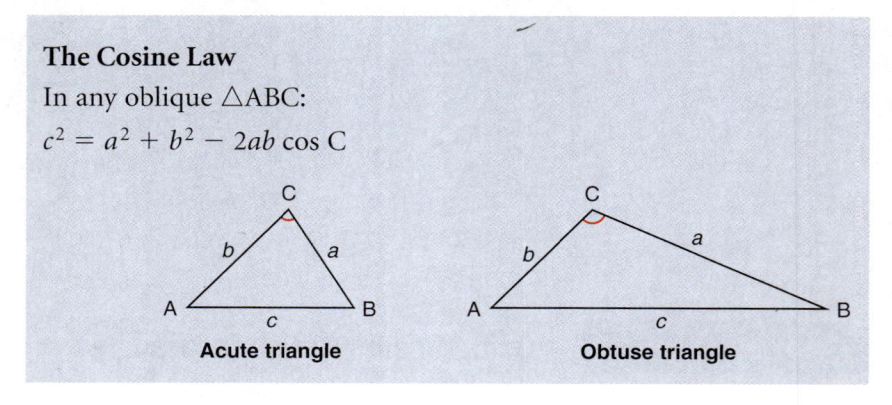

> **The Cosine Law**
> In any oblique $\triangle ABC$:
> $c^2 = a^2 + b^2 - 2ab \cos C$
>
> Acute triangle   Obtuse triangle

Side-Angle-Side (SAS)

In any triangle, given the lengths of two sides and the measure of the angle between them, we can use the Cosine Law to determine the length of the third side.

Example 1

Determining a Side Length Using the Cosine Law

Materials
- scientific calculator

Determine the length of n in $\triangle MNP$.

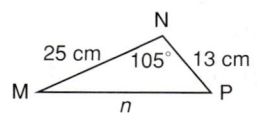

(Triangle with M, N, P; MN = 25 cm, NP = 13 cm, angle N = 105°, MP = n)

The Sine Law cannot be used here, because no pair of ratios includes the three given measures, m, p, and ∠N.

Solution

Write the Cosine Law for $\triangle MNP$.
$n^2 = m^2 + p^2 - 2mp \cos N$ Substitute: $m = 13$, $p = 25$, and $\angle N = 105°$
$ = 13^2 + 25^2 - 2 \times 13 \times 25 \times \cos 105°$
$ \doteq 962.2323$
$n = \sqrt{962.2323}$
$ \doteq 31.02$

n is about 31 cm long.

Apply the order of operations.

Press: 13 [x²] [+] 25 [x²] [−]
2 [×] 13 [×] 25 [×]
[COS] 105 [ )] [ENTER]

36 CHAPTER 1: Trigonometry

Side-Side-Side (SSS)

We can use the Cosine Law to calculate the measure of an angle in a triangle when the lengths of all three sides are known.

Example 2 | **Determine an Angle Measure Using the Cosine Law**

Materials
- scientific calculator

Determine the measure of $\angle C$ in $\triangle BCD$.

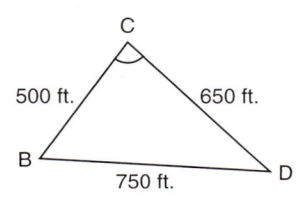

(Triangle BCD with BC = 500 ft., CD = 650 ft., BD = 750 ft.)

Solution

Write the Cosine Law for $\triangle BCD$ using $\angle C$.

$c^2 = b^2 + d^2 - 2bd \cos C$ Substitute: $c = 750, b = 650, d = 500$

$750^2 = 650^2 + 500^2 - 2 \times 650 \times 500 \times \cos C$

$562\,500 = 422\,500 + 250\,000 - 650\,000 \times \cos C$

$-110\,000 = -650\,000 \times \cos C$ Isolate $\cos C$.

$\cos C = \dfrac{-110\,000}{-650\,000}$ Isolate C.

$\angle C = \cos^{-1}\left(\dfrac{-110\,000}{-650\,000}\right)$

$\angle C \doteq 80.2569$

$\angle C$ is about $80°$.

Example 3 | **Navigating Using the Cosine Law**

Materials
- scientific calculator

An air-traffic controller at T is tracking two planes, U and V, flying at the same altitude. How far apart are the planes?

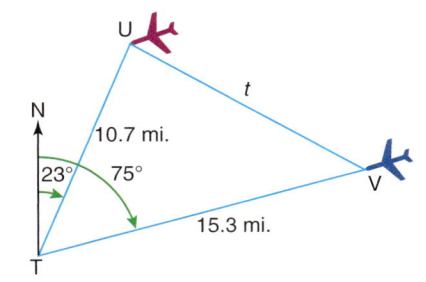

(Diagram: planes U and V with T at vertex; TU = 10.7 mi., TV = 15.3 mi., bearings 23° and 75°, side t between U and V.)

Solution

$\angle UTV = 75° - 23° = 52°$

Write the Cosine Law for $\triangle TUV$ using $\angle T$.

$t^2 = u^2 + v^2 - 2uv \cos T$ Substitute: $u = 15.3, v = 10.7, \angle T = 52°$

$= 15.3^2 + 10.7^2 - 2 \times 15.3 \times 10.7 \times \cos 52°$

$\doteq 147.0001$

$t = \sqrt{147.0001}$

$\doteq 12.12$

The planes are about 12.1 miles apart.

> The number of decimal places in your answer should match the given measures.

1.5 The Cosine Law **37**

Practice

A

■ For help with questions 1 and 2, see Example 1.

1. Write the Cosine Law you would use to determine each indicated side length.

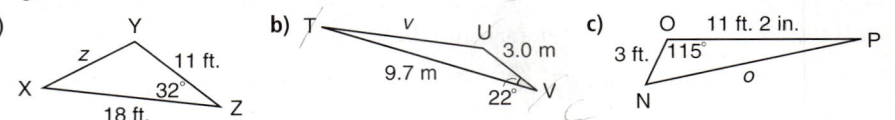

a) 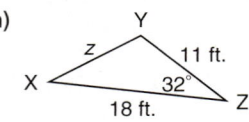 Triangle XYZ: XY = z, YZ = 11 ft, XZ = 18 ft, angle at Y = 32°

b) Triangle TUV: TU = v, TV = 9.7 m, UV = 3.0 m, angle at V = 22°

c) Triangle NOP: NO = 3 ft, OP = 11 ft. 2 in., angle at O = 115°, side o

2. Determine each unknown side length in question 1.

3. Determine each unknown side length.

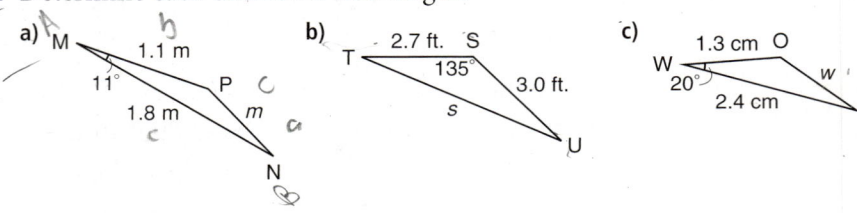

a) Triangle MNP: MP = 1.1 m, MN = 1.8 m, angle at M = 11°, side m... (n)

b) Triangle STU: TS = 2.7 ft, SU = 3.0 ft, angle at S = 135°, side s

c) 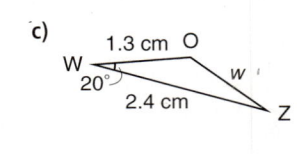 Triangle WOZ: WO = 1.3 cm, OZ = 2.4 cm, angle at W = 20°, side w

4. Determine the length of *a*.

a) 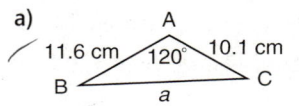 Triangle ABC: AB = 11.6 cm, AC = 10.1 cm, angle A = 120°, side a

b) 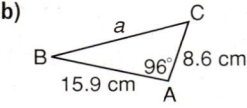 Triangle ABC: BC = a, AB = 15.9 cm, AC = 8.6 cm, angle C = 96°

c) Triangle ABC: AB = 3.7 in., AC = 5.7 in., angle A = 72°, side a

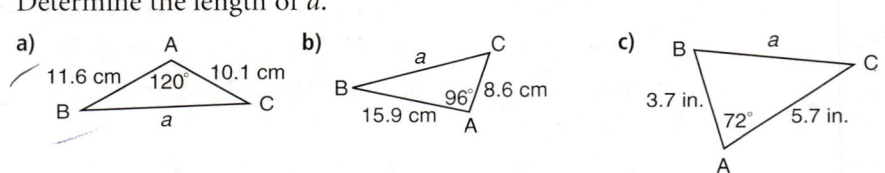

■ For help with questions 5 and 6, see Example 2.

5. Write the Cosine Law you would use to determine the measure of the marked angle in each triangle.

a) 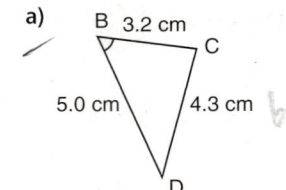 Triangle BCD: BC = 3.2 cm, CD = 4.3 cm, BD = 5.0 cm, marked angle at B

b) Triangle PQR: PQ = 145 yd., PR = 111 yd., RQ = 35 yd., marked angle at R

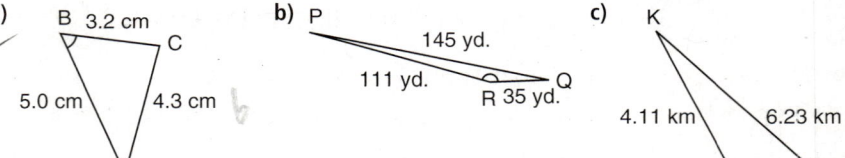

c) 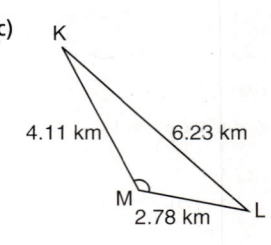 Triangle KLM: KM = 4.11 km, KL = 6.23 km, ML = 2.78 km, right angle at M

6. Determine the measure of each marked angle in question 5.

7. Determine the measure of each unknown angle.

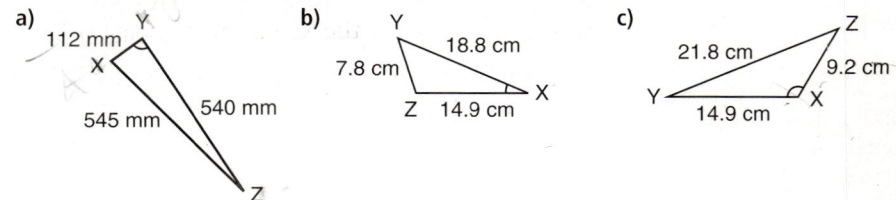

a) Triangle XYZ: XY = 112 mm, XZ = 540 mm, YZ = 545 mm

b) 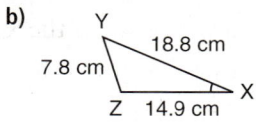 Triangle XYZ: XY = 18.8 cm, YZ = 7.8 cm, ZX = 14.9 cm

c) 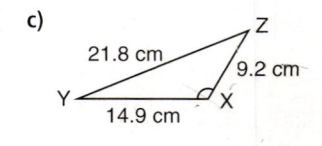 Triangle XYZ: YZ = 21.8 cm, ZX = 9.2 cm, YX = 14.9 cm

38 CHAPTER 1: Trigonometry

8. Determine the measure of each unknown angle.
   Start by sketching the triangle.
   a) In △XYZ, $z = 18.8$ cm, $x = 24.8$ cm, and $y = 9.8$ cm
      Determine ∠X.
   b) In △GHJ, $g = 23$ feet, $h = 25$ feet, and $j = 31$ feet
      Determine ∠J.
   c) In △KLM, $k = 12.9$ yards, $l = 17.8$ yards, $m = 14.8$ yards
      Determine ∠M.

B 9. Would you use the Cosine Law or the Sine Law to determine the labelled length in each triangle? Explain your reasoning.
   a) [Triangle ABC with AC = 14.6 m, CB = 15.9 m, ∠C = 47°, side c labelled]
   b) [Triangle ABC with AC = 16 in., ∠A = 65°, ∠C = 39°, side a labelled]
   c) [Triangle ABC with CA = 39.75 km, ∠A = 22°, ∠B = 105°, side a labelled]

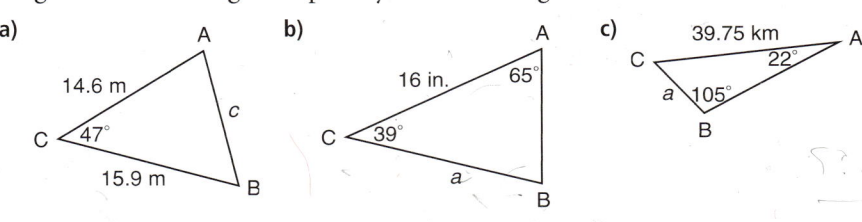

■ For help with questions 10 and 11, see Example 3.

10. A harbour master uses a radar to monitor two ships, B and C, as they approach the harbour, H. One ship is 5.3 miles from the harbour on a bearing of 032°. The other ship is 7.4 miles away from the harbour on a bearing of 295°.
   a) How are the bearings shown in the diagram?
   b) How far apart are the two ships?

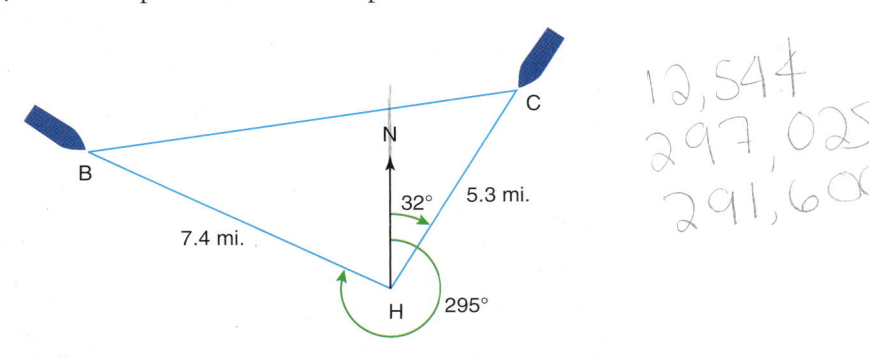

[Diagram showing ships B and C with harbour H, BH = 7.4 mi, CH = 5.3 mi, angle at H = 32°, bearing 295°, N arrow]

12,544
297,025
291,600

11. A theatre set builder's plans show a triangular set with two sides that measure 3 feet 6 inches and 4 feet 9 inches. The angle between these sides is 45°. Determine the length of the third side.

1.5 The Cosine Law   39

12. A telescoping ladder has a pair of aluminum struts, called ladder stabilizers, and a base. What is the angle between the base and the ladder?

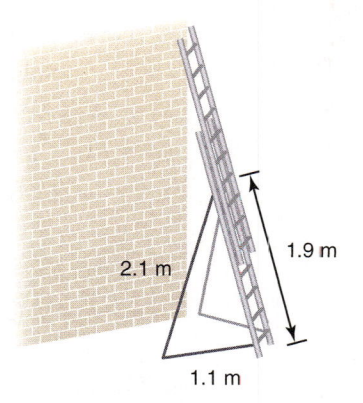

13. A hydro pole needs two guy wires for support. What angle does each wire make with the ground?

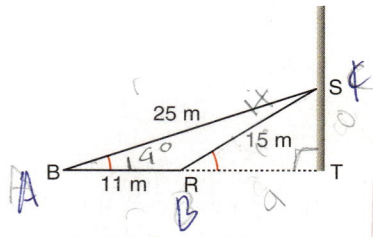

14. A land survey shows that a triangular plot of land has side lengths 2.5 miles, 3.5 miles, and 1.5 miles. Determine the angles in the triangle. Explain how this problem could be done in more than one way.

15. Assessment Focus An aircraft navigator knows that town A is 71 km due north of the airport, town B is 201 km from the airport, and towns A and B are 241 km apart.
   a) On what bearing should she plan the course from the airport to town B? Include a diagram.
   b) Explain how you solved the problem.

16. **Literacy in Math** Write a problem that can be represented by this diagram. Solve your problem.

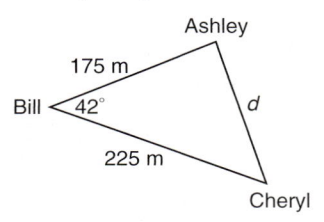

17. Use this diagram of a roof truss.

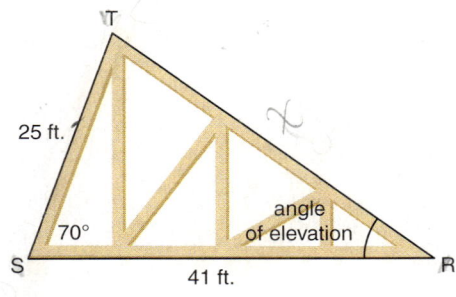

   a) Determine the length of TR.
   b) What is the angle of elevation? Record your answer to the nearest degree.

18. Marie wants to determine the height of an Internet transmission tower. Due to several obstructions, she has to use indirect measurements to determine the tower height. She walks 50 m from the base of the tower, turns 110°, and then walks another 75 m. Then she measures the angle of elevation to the top of the tower to be 25°.

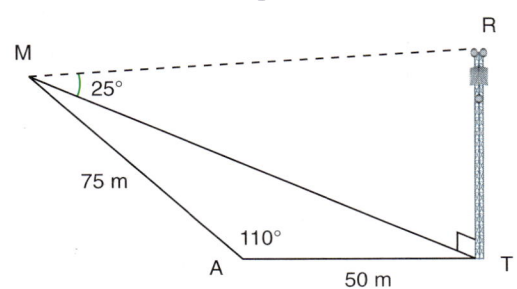

   a) What is the height of the tower?
   b) What assumptions did you make? Do you think it is reasonable to make these assumptions? Justify your answer.

Indirect measurements are used to determine inaccessible distances and angles, which cannot be measured directly.

In Your Own Words

How can the Pythagorean Theorem help you remember the Cosine Law? Include a diagram with your explanation.

1.6 Problem Solving with Oblique Triangles

Campfires, a part of many camping experiences, can be dangerous if proper precautions are not taken. Forest rangers advise campers to light their fires in designated locations, away from trees, tents, or other fire hazards.

Investigate: Choosing Sine Law or Cosine Law

Materials
- protractor
- scientific calculator

Work with a partner.
Two forest rangers sight a campfire, F, from their observation towers, G and H. How far is the fire from each observation deck?

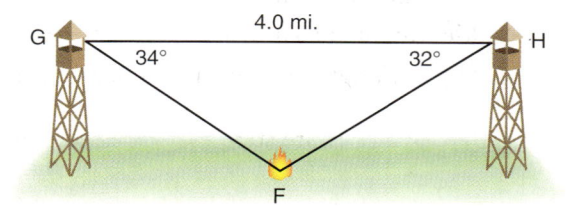

Decide whether to use the Sine Law or the Cosine Law to determine each distance.

G — 4.0 mi. — H
∠G = 34°, ∠H = 32°

Reflect

- Compare strategies with another pair. How are they different? How are they the same?
- How did you decide which law or laws to use? Justify your choices.
- Could you have used another law? Explain why or why not.

42   CHAPTER 1: Trigonometry

Connect the Ideas

Sine Law

In any triangle, we can use the Sine Law to solve the triangle when:
- We know the measures of two angles and the length of any side (AAS or ASA)

Cosine Law

In any triangle, we can use the Cosine Law to solve the triangle when:
- We know the lengths of two sides and the measure of the angle between them (SAS)
- We know the lengths of three sides (SSS)

Solving triangle problems

When solving triangle problems:
- Read the problem carefully.
- Sketch a diagram if one is not given. Record known measurements on your diagram.
- Identify the unknown and known measures.
- Use a triangle relationship to determine the unknown measures.

Often, two or more steps may be needed to solve a triangle problem.

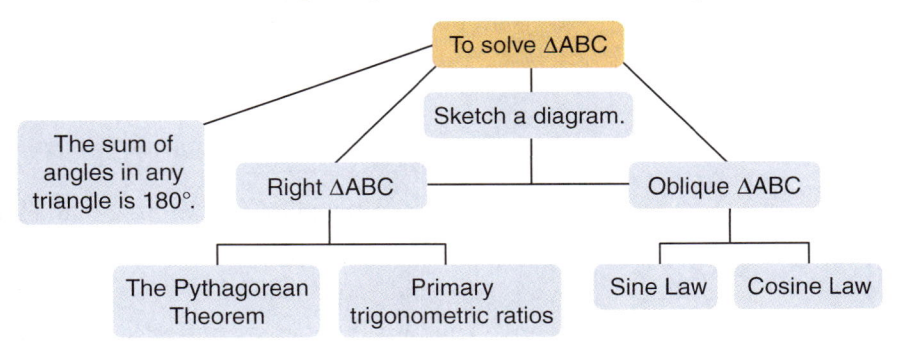

```
                    To solve △ABC
                         |
                  Sketch a diagram.
                         |
    ┌────────────────────┼────────────────────┐
The sum of           Right △ABC           Oblique △ABC
angles in any           │                      │
triangle is 180°.       │                      │
                  ┌─────┴─────┐          ┌─────┴─────┐
              The Pythagorean  Primary   Sine Law  Cosine Law
                 Theorem    trigonometric
                              ratios
```

1.6 Problem Solving with Oblique Triangles   **43**

Example 1 — Determining Side Lengths

Materials
- scientific calculator

Decide whether to use the Sine Law or Cosine Law to determine m and p in $\triangle MNP$. Then, determine each side length.

[Diagram: Triangle MNP with angle M = 15°, side from M to P = 33 yd, angle P = 23°, side p opposite P between M and N, side m opposite M between N and P]

Solution

> Since $\triangle MNP$ is an oblique triangle, we need to use the Sine Law or the Cosine Law to solve it.

In oblique $\triangle MNP$, we know the measures of two angles and the length of the side between them (ASA). So, use the Sine Law.

Write the Sine Law for $\triangle MNP$:

$$\frac{m}{\sin M} = \frac{n}{\sin N} = \frac{p}{\sin P}$$

$\angle N = 180° - 15° - 23° = 142°$

To find m, use the first two ratios.

$\dfrac{m}{\sin M} = \dfrac{n}{\sin N}$ Substitute: $n = 33$, $\angle M = 15°$, and $\angle N = 142°$

$\dfrac{m}{\sin 15°} = \dfrac{33}{\sin 142°}$ Multiply both sides by $\sin 15°$ to isolate m.

$m = \dfrac{33}{\sin 142°} \times \sin 15°$

$m \doteq 13.87$

m is about 14 yards.

To find p, use the second and third ratios.

$\dfrac{n}{\sin N} = \dfrac{p}{\sin P}$ Substitute: $n = 33$, $\angle N = 142°$, and $\angle P = 23°$

$\dfrac{33}{\sin 142°} = \dfrac{p}{\sin 23°}$ Multiply both sides by $\sin 23°$ to isolate p.

$p = \dfrac{33}{\sin 142°} \times \sin 23°$

$p \doteq 20.94$

p is about 21 yards.

Example 2

Materials
- scientific calculator

Determining Angle Measures

Write the equation you would use to determine the cosine of each angle in $\triangle ABC$. Determine each angle measure.

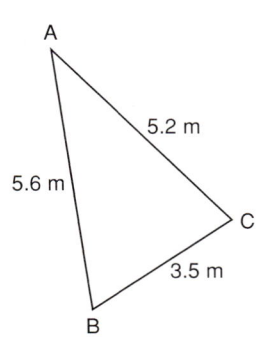

[Triangle ABC with AB = 5.6 m, AC = 5.2 m, BC = 3.5 m]

Solution

In oblique $\triangle ABC$, we are given the lengths of all three sides (SSS). So, use the Cosine Law.

Write the Cosine Law using $\angle C$:

$$c^2 = a^2 + b^2 - 2ab \cos C \qquad \text{Isolate } \cos C.$$

$$\cos C = \frac{a^2 + b^2 - c^2}{2ab} \qquad \text{Isolate } C.$$

$$C = \cos^{-1}\left(\frac{a^2 + b^2 - c^2}{2ab}\right) \qquad \text{Substitute: } a = 3.5, b = 5.2, c = 5.6$$

$$= \cos^{-1}\left(\frac{3.5^2 + 5.2^2 - 5.6^2}{2 \times 3.5 \times 5.2}\right)$$

$$\doteq 77.42°$$

So, the measure of $\angle C$ is about 77°.

Write the Cosine Law using $\angle B$:

$$b^2 = a^2 + c^2 - 2ac \cos B \qquad \text{Isolate } \cos B.$$

$$\cos B = \frac{a^2 + c^2 - b^2}{2ac} \qquad \text{Isolate } B.$$

$$B = \cos^{-1}\left(\frac{a^2 + c^2 - b^2}{2ac}\right) \qquad \text{Substitute: } a = 3.5, b = 5.2, c = 5.6$$

$$= \cos^{-1}\left(\frac{3.5^2 + 5.6^2 - 5.2^2}{2 \times 3.5 \times 5.6}\right)$$

$$\doteq 64.99°$$

So, the measure of $\angle B$ is about 65°.

$\angle A = 180° - 77° - 65° = 38°$

So, the measure of $\angle A$ is about 38°.

Example 3 — Solving Two Oblique Triangles

Materials
- scientific calculator

Determine lengths a and b.

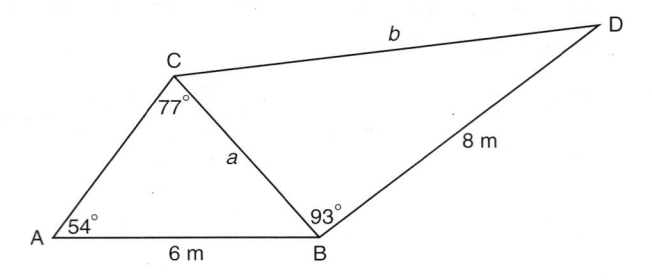

Solution

In oblique △ABC, we know the measures of two angles and the length of the side opposite one of them (AAS). So, use the Sine Law.

Write the Sine Law for △ABC:
$$\frac{a}{\sin A} = \frac{b}{\sin B} = \frac{c}{\sin C}$$

To find a, use the first and the third ratios.

$\dfrac{a}{\sin A} = \dfrac{c}{\sin C}$ Substitute: $\angle C = 77°$, $\angle A = 54°$, and $c = 6$

$\dfrac{a}{\sin 54°} = \dfrac{6}{\sin 77°}$ Multiply both sides by sin 54° to isolate a.

$a = \dfrac{6}{\sin 77°} \times \sin 54°$

$\doteq 4.98$

So, a is about 5 m long.

In oblique △BCD, we now know the lengths of two sides and the measure of the angle between them (SAS). So, use the Cosine Law.

Write the Cosine Law for △BCD using $\angle B$.

$b^2 = a^2 + c^2 - 2ac \cos B$ Substitute: $a = 4.98$, $c = 8$, $\angle B = 93°$

$ = 4.98^2 + 8^2 - 2 \times 4.98 \times 8 \times \cos 93°$

$ \doteq 92.97$

$b = \sqrt{92.97}$

$ \doteq 9.64$

So, b is about 10 m long.

> To ensure accurate results when calculating b, use the value for a before rounding.

Practice

A

For help with questions 1 and 2, see Example 1.

1. Decide whether you would use the Sine Law or the Cosine Law to determine each indicated length.

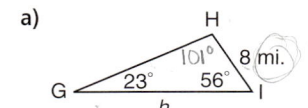

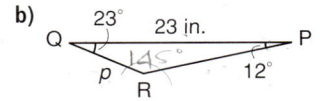

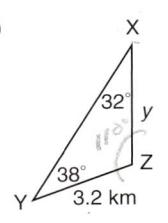

   a) Triangle GHI: ∠G = 23°, ∠H = 101°, ∠I = 56°, HI = 8 mi, GI = h

   b) Triangle QRP: ∠Q = 23°, ∠R = 145°, ∠P = 12°, QP = 23 in., QR = p

   c) Triangle XYZ: ∠X = 32°, ∠Y = 38°, YZ = 3.2 km, XZ = y

2. Determine each indicated length in question 1.

3. Use the Sine Law or the Cosine Law to determine each indicated length.

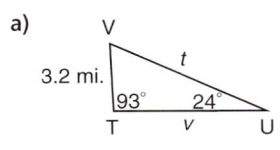

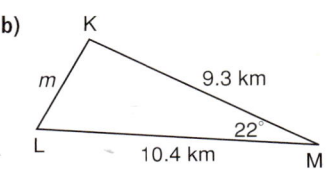

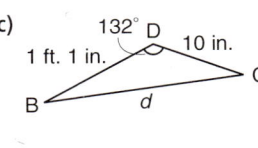

   a) Triangle TVU: VT = 3.2 mi, ∠T = 93°, ∠U = 24°, TU = t, UV = v

   b) Triangle KLM: LM = 10.4 km, KM = 9.3 km, ∠M = 22°, KL = m

   c) Triangle BDC: ∠D = 132°, BD = 1 ft. 1 in., DC = 10 in., BC = d

For help with questions 4 and 5, see Example 2.

4. Decide whether you would use the Sine Law or the Cosine Law to determine each angle measure.

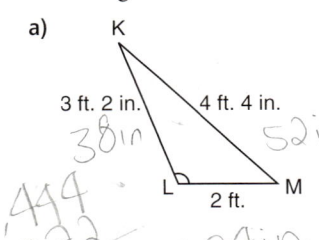

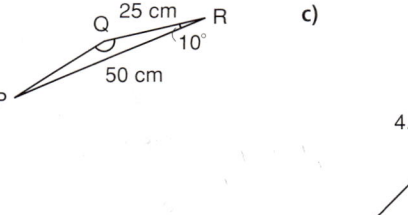

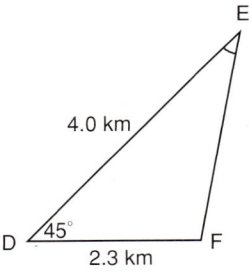

   a) Triangle KLM: KL = 3 ft. 2 in., KM = 4 ft. 4 in., LM = 2 ft., ∠L = 90°

   b) Triangle QPR: QR = 25 cm, QP = 50 cm, ∠Q = 10°

   c) Triangle DEF: DE = 4.0 km, DF = 2.3 km, ∠D = 45°

5. Determine the measure of each angle in question 4.

6. Use the Sine Law or the Cosine Law to determine each measure for ∠B.

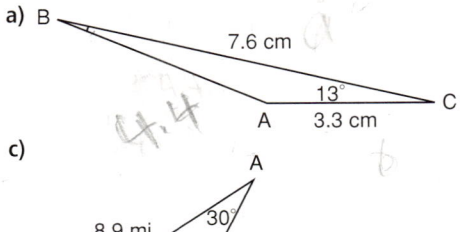

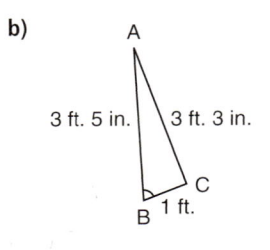

   a) Triangle BAC: BA = 7.6 cm, AC = 3.3 cm, ∠A = 13°

   b) Triangle ABC: AB = 3 ft. 5 in., AC = 3 ft. 3 in., BC = 1 ft.

   c) Triangle CAB: CA = 8.9 mi, AB = 6.0 mi, ∠A = 30°

1.6 Problem Solving with Oblique Triangles   **47**

■ For help with questions 7 and 8, see Example 3.

7. Phoebe and Holden are on opposite sides of a tall tree, 125 m apart. The angles of elevation from each to the top of the tree are 47° and 36°. What is the height of the tree?

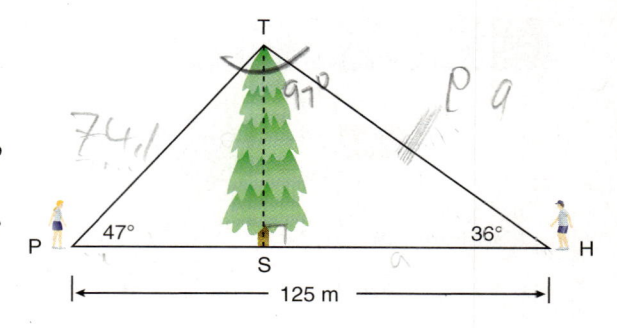

B

8. Carrie says she can use the Cosine Law to solve △XYZ. Do you agree? Justify your answer.

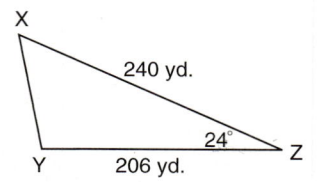

9. Use what you know from question 8. Write your own question that can be solved using the Sine Law or the Cosine Law. Show your solution.

10. A hobby craft designer is designing this two-dimensional kite.
   a) What is the angle measure between the longer sides?
   b) What is the angle measure between the shorter sides? Explain your strategies.

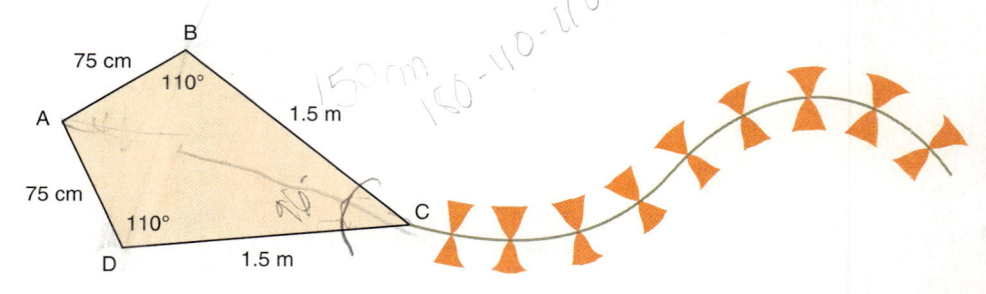

The sum of the angles in a quadrilateral is 360°.

11. Use this diagram of the rafters in a greenhouse.
   a) What angle do the rafters form at the peak of the greenhouse?
   b) What angle do they form with the sides of the greenhouse? Solve this problem two ways: using the Cosine Law and using primary trigonometric ratios.

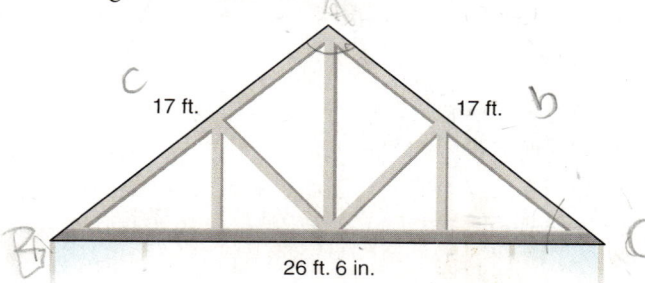

48 CHAPTER 1: Trigonometry

12. A boat sails from Meaford to Christian Island, to Collingwood, then to Wasaga Beach.
    a) What is the total distance the boat sailed?
    b) What is the shortest distance from Wasaga Beach to Christian Island?

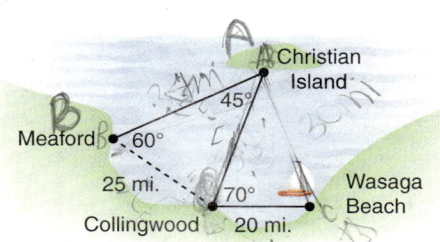

13. A triangle has side lengths measuring 5 inches, 10 inches, and 7 inches. Determine the angles in the triangle.

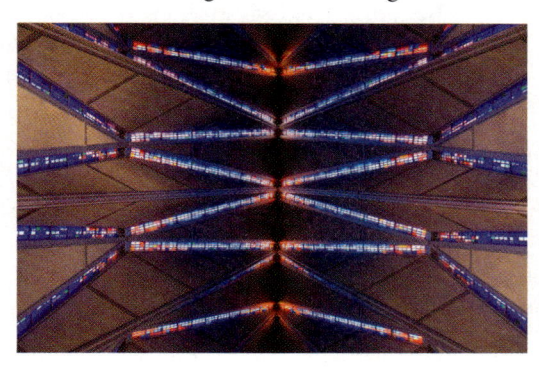

14. **Assessment Focus** Roof rafters and truss form oblique △PQR and △SQT.
    a) Describe two different methods that could be used to determine ∠SQT.
    b) Determine ∠SQT, using one of the methods described in part a.
    c) Explain why you chose the method you used in part b.

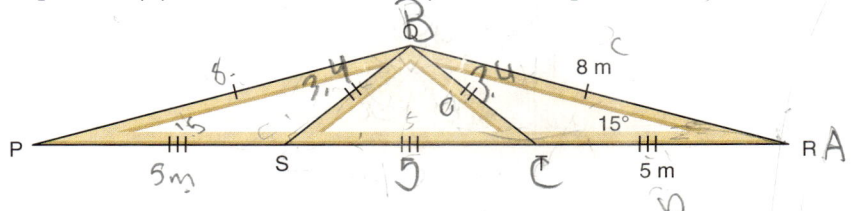

15. **Literacy in Math** Explain the flow chart in *Connect the Ideas* in your own words. Add any other important information.

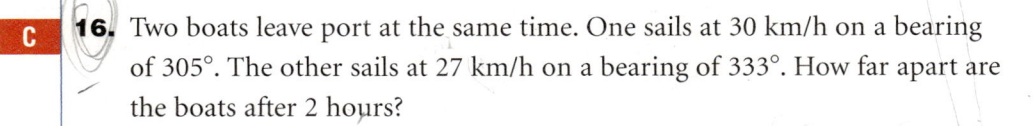

16. Two boats leave port at the same time. One sails at 30 km/h on a bearing of 305°. The other sails at 27 km/h on a bearing of 333°. How far apart are the boats after 2 hours?

In Your Own Words

What mistakes can someone make while solving problems with the Sine Law and the Cosine Law? How can they be avoided?

1.6 Problem Solving with Oblique Triangles **49**

1.7 Occupations Using Trigonometry

Professional tools, such as tilt indicators on boom trucks, lasers, and global positioning systems (GPS) perform trigonometric computations automatically.

Inquire: Researching Applications of Trigonometry

Materials
- computers with Internet access

It may be helpful to invite an expert from a field that uses trigonometry or an advisor from a college or apprenticeship program.

Work in small groups.

Part A: Planning the Research

- Brainstorm a list of occupations that involve the use of trigonometry. Include occupations you read about during this chapter.

- Briefly describe how each occupation you listed involves the use of trigonometry. Use these questions to guide you.
  - What measures and calculations might each occupation require?
  - What tools and technologies might each occupation use for indirect measurements? How do these tools and technologies work?

- Find sources about career guidance. Write some information that might be of interest to you or to someone you know.

Part B: Gathering Information

- Choose one occupation from your list. Investigate as many different applications of trigonometry as you can in the occupation you chose. Include answers to questions such as:
  - What measurement system is used: metric, imperial, or both?
  - How important is the accuracy of measurements and calculations?
  - What types of communication are used: written, graphical, or both?
  - What other mathematics does this occupation involve?
  - What is a typical wage for an entry-level position in this career?
  - What is a typical wage for someone with experience in this occupation?
  - Are employees paid on an hourly or a salary basis?

Think of any key words or combinations of key words that might help you with your research on the Internet.

- Read about the occupation you chose on the Internet or in printed materials.
- If possible, interview people who work in the occupation you chose. Prepare a list of questions you would ask them about trigonometry.

Search words
- ☐ occupational information
- ☐ working conditions
- ☐ other qualifications
- ☐ earnings, salaries
- ☐ related occupations

1.7 Occupations Using Trigonometry

Search words
- ☐ training
- ☐ education information
- ☐ apprenticeship programs

Part C: Researching Educational Requirements

■ Research to find out the educational requirements for the occupation you chose. On the Internet, use search words related to education. You might decide to use the same words you used before.
  - Find out about any pre-apprenticeship training programs or apprenticeship programs available locally.
  - Go to Web sites of community colleges, or other post-secondary institutions.
  - Use course calendars of post-secondary institutions or other information available through your school's guidance department.

■ How can you get financial help for the required studies? How could you find out more?

Part D: Presenting Your Findings

■ Prepare a presentation you might give your class, another group, or someone you know. Think about how you can organize and clearly present the information and data you researched. Use diagrams or graphic organizers if they help.

Reflect

■ Write about a problem you might encounter in the occupation you researched. What would you do to solve it? How does your solution involve trigonometry?

■ Why do you think it is important to have a good understanding of trigonometry in the occupation you researched?

Study Guide

Primary Trigonometric Ratios

When ∠A is an acute angle in a right △ABC:

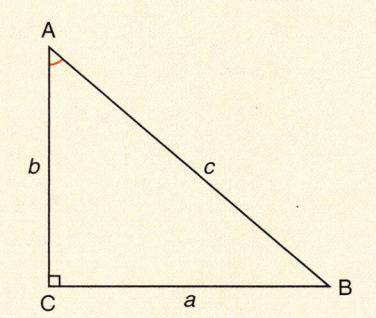

$$\sin A = \frac{\text{length of side opposite } \angle A}{\text{length of hypotenuse}} = \frac{a}{c}$$

$$\cos A = \frac{\text{length of side adjacent to } \angle A}{\text{length of hypotenuse}} = \frac{b}{c}$$

$$\tan A = \frac{\text{length of side opposite } \angle A}{\text{length of side adjacent to } \angle A} = \frac{a}{b}$$

The inverse ratios are $\sin^{-1}$, $\cos^{-1}$, and $\tan^{-1}$.

- $\sin^{-1}(\sin A) = A$
- $\cos^{-1}(\cos A) = A$
- $\tan^{-1}(\tan A) = A$

Trigonometric Ratios of Supplementary Angles

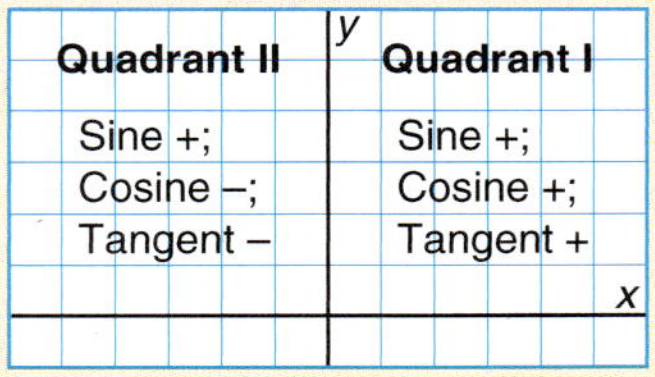

Two angles are supplementary if their sum is 180°.
For an acute angle, A, and its supplementary obtuse angle (180° − A):

- $\sin A = \sin(180° - A)$
- $\cos A = -\cos(180° - A)$
- $\tan A = -\tan(180° - A)$

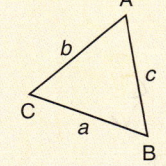

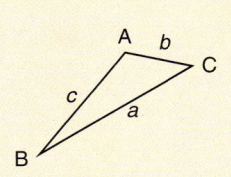

Quadrant II	Quadrant I
Sine +;	Sine +;
Cosine −;	Cosine +;
Tangent −	Tangent +

The Sine Law

In any △ABC: $\frac{a}{\sin A} = \frac{b}{\sin B} = \frac{c}{\sin C}$

To use the Sine Law, you must know at least one side-angle pair (for example, a and ∠A).
To determine a side length, you must know an additional angle measure:

- angle−angle−side (AAS)
- angle−side−angle (ASA)

The Cosine Law

In any △ABC: $c^2 = a^2 + b^2 - 2ab \cos C$

To determine a side length using the Cosine Law, you must know:

- side−angle−side (SAS)

To determine an angle measure using the Cosine Law, you must know:

- side−side−side (SSS)

Chapter Review

1.1

1. Determine each indicated measure.

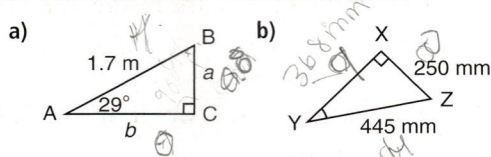

a) Triangle with A = 29°, AB = 1.7 m, right angle at C, side b = AC

b) Triangle XYZ with right angle at X, XZ = 250 mm, YZ = 445 mm

2. Danny is building a ski jump with an angle of elevation of 15° and a ramp length of 4.5 m. How high will the ski jump be?

3. Determine the measure of ∠CBD.

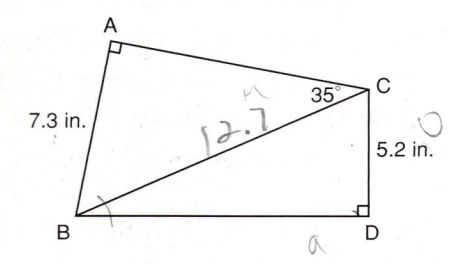

Quadrilateral with right angle at A, AB = 7.3 in., ∠ACB region with 35° at C, CD = 5.2 in., right angle at D.

4. A triangular lot is located at the intersection of two perpendicular streets. The lot extends 350 feet along one street and 450 feet along the other street.
  a) What angle does the third side of the lot make with each road?
  b) What is the perimeter of the lot? Explain your strategy.

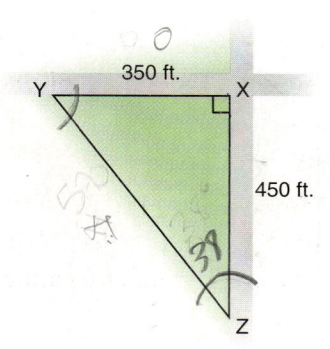

Right triangle with right angle at X, YX = 350 ft., XZ = 450 ft.

5. A ship navigator knows that an island harbour is 20 km north and 35 km west of the ship's current position. On what bearing could the ship sail directly to the harbour?

Answer questions 6 to 8 without using a calculator.

1.2

6. Is each trigonometric ratio positive or negative? Explain how you know.
  a) tan 53°   b) cos 96°   c) sin 132°

7. Is ∠B acute or obtuse? Explain.
  a) tan B = 1.6   b) cos B = −0.9945
  c) cos B = 0.35   d) sin B = 0.7

8. a) cos A = −0.94
     Determine cos (180° − A).
  b) sin A = 0.52
     Determine sin (180° − A).
  c) tan A = 0.37
     Determine tan (180° − A).

1.3

9. Determine the measure of each obtuse angle.
  a) sin R = 0.93   b) cos D = −0.56

10. 0° < ∠G < 180°. Is ∠G acute or obtuse? How do you know?
  a) tan G = −0.2125   b) sin G = 0.087

11. Determine the measure of ∠A. If you get more than one answer for the measure of ∠A, explain why.
  a) tan A = −0.1746   b) sin A = 0.3584

12. ∠Z is an angle in a triangle. Determine all possible values for ∠Z.
  a) cos Z = 0.93   b) sin Z = 0.73

54 CHAPTER 1: Trigonometry

1.4 **13.** Use the Sine Law to determine *x* and *y*.

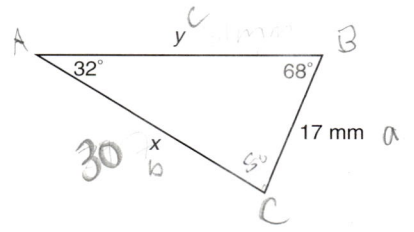

(Triangle ABC: angle A = 32°, angle B = 68°, side y = AB, side x = AC, BC = 17 mm)

14. a) Determine the lengths of sides *p* and *q*.
  b) Explain your strategy.

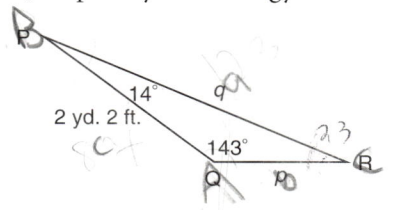

(Triangle with angle 14° at B, side 2 yd. 2 ft., angle 143°)

15. Solve each triangle. Sketch a diagram first.
  a) △LMN, ∠M = 48°, ∠N = 105°, and *l* = 17 m
  b) △HIJ, ∠H = 21°, ∠J = 57°, and *h* = 9 feet 4 inches

16. Lani received these specifications for two different triangular sections of a sailboat sail.
  a) *a* = 5.5 m, *b* = 1.0 m, and ∠C = 134°. Determine ∠A.
  b) *d* = 7.75 m, *e* = 9.25 m, and ∠F = 45°. Determine ∠E.

(Diagram of sailboat sail with triangles labeled with sides a, b, c, d, e, f and vertices A, B, C, D, E, F)

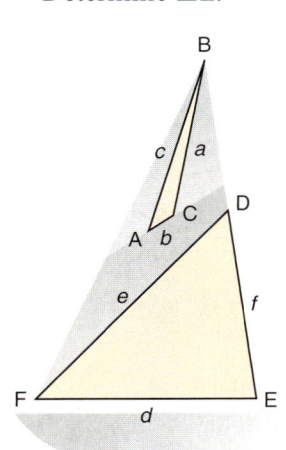

17. One side of a triangular lot is 2.6 m long. The angles in the triangle at each end of the 2.6-m side are 38° and 94°. Determine the lengths of the other two sides of the lot.

1.5 **18. a)** Explain why you would use the Cosine Law to determine *q*.

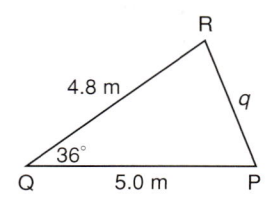

(Triangle PQR: QR = 4.8 m, QP = 5.0 m, angle Q = 36°, side q = RP)

  b) Which version of the Cosine Law would you use to solve this problem? Explain how you know the version you chose is correct.
  c) Determine the length of side *q*.

19. Sketch and label △BCD with *b* = 7.5 km, *d* = 4.3 km, and ∠C = 131°. Solve △BCD.

20. Determine the measure of ∠Q.

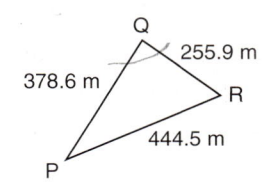

(Triangle PQR: PQ = 378.6 m, QR = 255.9 m, PR = 444.5 m)

21. a) Sketch each triangle.
  b) Determine the measure of the specified angle.
    i) In △MNO, *m* = 3.6 m, *n* = 10.7 m, and *o* = 730 cm. Determine ∠N.
    ii) In △CDE, *c* = 66 feet, *d* = 52 feet, and *e* = 59 feet. Determine ∠D.

Chapter Review **55**

22. Jane is drawing an orienteering map that shows the location of three campsites. Determine the two missing angle measures.

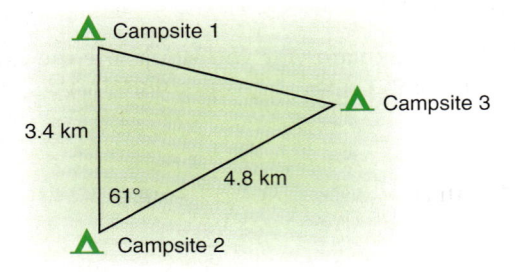

23. A machinist is cutting out a large triangular piece of metal to make a part for a crane. The sides of the piece measure 4 feet 10 inches, 3 feet 10 inches, and 5 feet 2 inches. What are the angles between the sides?

24. Renée and Andi volunteered to help scientists measure the heights of trees in old-growth forests in Algonquin Park. The two volunteers are 20 m apart on opposite sides of an aspen. The angle of elevation from one volunteer to the top of the tree is 65°, and from the other, 75°. What is the height of the tree? What assumptions did you make?

25. Determine the length, x, of the lean-to roof attached to the side of the cabin.

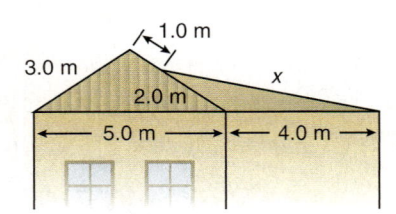

26. Two ferries leave dock B at the same time. One travels 2900 m on a bearing of 098°. The other travels 2450 m on a bearing of 132°.
  a) How are the bearings shown on the diagram?
  b) How far apart are the ferries?

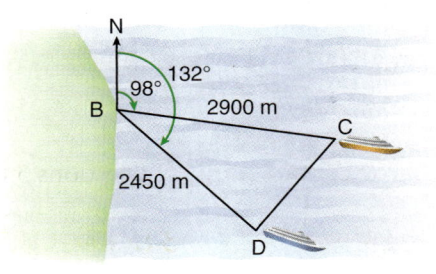

27. Determine the lengths of sides s and t.

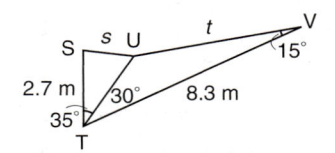

28. a) Write a word problem that can be solved using the Sine Law. Explain your strategy.
  b) Write a word problem that can be solved with the Cosine Law. Explain your strategy.

29. Describe one occupation that uses trigonometry. Give a specific example of a calculation a person in this occupation might perform as part of their work.

30. Tell about an apprenticeship program or post-secondary institution that offers the training required by a person in the career you identified in question 29.

Practice Test

Multiple Choice: Choose the correct answer for questions 1 and 2. Justify each choice:

1. Which could you use to determine the measure of an angle in an oblique triangle if you only know the lengths of all three sides?
   A. Sine Law   B. Cosine Law   C. Tangent ratio   D. Sine ratio

2. If ∠A is obtuse, which is positive?
   A. sin A   B. cos A   C. tan A   D. none

Show all your work for questions 3 to 6.

3. **Knowledge and Understanding** Solve △ABC.

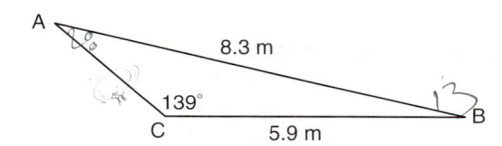

   (Triangle with A at top left, C at bottom left with 139°, B at right. AB = 8.3 m, CB = 5.9 m)

4. **Communication** How do you decide when to use the Sine Law or the Cosine Law to solve a triangle problem? Give examples to illustrate your explanation.

5. **Application** A car windshield wiper is 22 inches long. Through which angle did the blade in this diagram rotate?

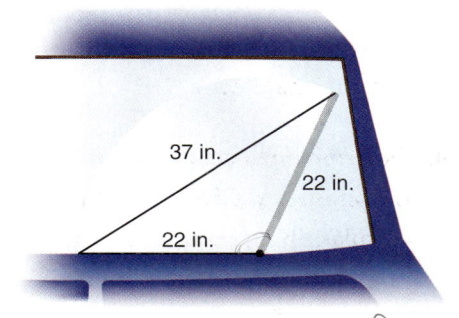

   (Diagram showing triangle with sides 37 in., 22 in., and 22 in.)

6. **Thinking** A sailboat leaves Port Hope and sails 23 km due east, then 34 km due south.
   a) On what bearing will the boat travel on its way back to the starting point?
   b) How far is the boat from the starting point?
   c) What assumptions did you make to answer parts a and b?

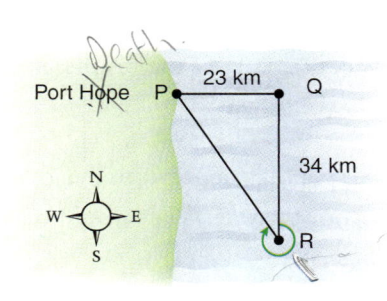

   (Map diagram: Port Hope at P, 23 km east to Q, 34 km south to R)

Chapter Problem: Designing a Stage

You have been asked to oversee the design and construction of a concert and theatre stage in your local community park.

1. Design a stage using at least two right triangles and at least two oblique triangles. Make decisions about the stage.
   - What will be the shape of the stage?
   - Will it have a roof?
   - Where will the stairs be? What will they look like? Will you include a ramp?
   - Include any further details you consider important for your stage.

2. Estimate reasonable angle measures and side lengths. Mark them on your drawing.

   What tools did you use?

3. a) Describe two calculations for designing your stage that include a right triangle. Show your calculations.
   b) Describe two calculations for designing your stage that include an oblique triangle. Show your calculations.
   c) How could you use the results of your calculations from parts a and b? Justify your answer.

4. Tell what you like about your stage. Why is your stage appropriate for a concert or theatre? What would make someone choose your design for a stage?

58 CHAPTER 1: Trigonometry

2 Geometry

What You'll Learn

To apply measurement and geometry concepts to composite figures and objects and to investigate optimization problems in the real world

And Why

Area, volume, and surface area calculations are important when estimating the supplies needed for painting, pouring concrete, and other construction jobs. Knowledge of surface area and volume can also help manufacturers reduce the amount of material used to package items.

Key Words

- perimeter
- area
- capacity
- volume
- surface area
- composite figure
- composite object
- optimization

CHAPTER 2

Activate Prior Knowledge

Metric and Imperial Units of Length

Prior Knowledge for 2.1

Use conversion factors to change imperial units to metric units, or vice versa.

Imperial to Metric	Metric to Imperial
1 inch = 2.54 cm	1 cm ≐ 0.3937 inch
1 foot = 30.48 cm	1 m ≐ 39.37 inches
1 foot = 0.3048 m	1 m ≐ 3.2808 feet
1 mile ≐ 1.609 km	1 km ≐ 0.6214 mile

The symbol ' represents feet and the symbol " represents inches.

Example

Materials
- scientific calculator

Round each answer to the same degree of accuracy as the given measurement.

Convert each length.
a) $5\frac{1}{4}$ inches to centimetres
b) 7.3 km to miles

Solution

a) First, express $5\frac{1}{4}$ as a decimal: 5.25
   Each inch is 2.54 cm.
   $5.25 \times 2.54 = 13.335$
   So, $5\frac{1}{4}$ inches is about 13.34 cm.

b) Each kilometre is about 0.6214 mile.
   $7.3 \times 0.6214 = 4.53622$
   So, 7.3 km is about 4.5 miles.

CHECK ✓

1. Convert each length.
   a) A van is 17 feet long. Convert to metres.
   b) A regular soccer pitch must be between 90 m and 120 m long. Convert to feet.
   c) A car travels 120 km. Convert to miles.
   d) A piece of paper is 21.6 cm wide. Convert to inches.

2. Room dimensions are often written using feet and inches instead of decimals.
   a) A room measures 3.5 m by 4.2 m. Convert to feet.
   b) Express the decimal parts of your answer to part a in inches, rounded to the nearest inch. Explain your method.

1 foot = 12 inches

60 CHAPTER 2: Geometry

Perimeter and Area

Prior Knowledge for 2.1

The **perimeter** of a figure is the distance around it.
The **area** of a figure is the number of square units needed to cover it.

Figure	Perimeter	Area
Rectangle 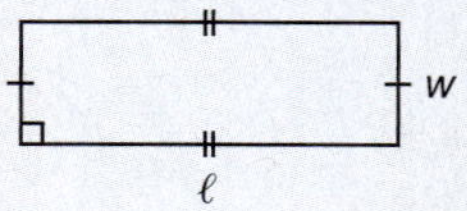	$P = 2\ell + 2w$ or $P = 2(\ell + w)$	$A = \ell w$
Triangle 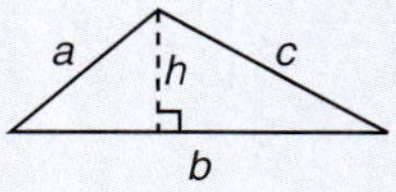	$P = a + b + c$	$A = \frac{1}{2}bh$
Parallelogram 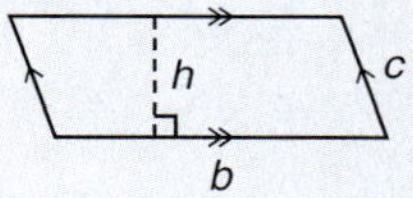	$P = 2b + 2c$	$A = bh$
Trapezoid 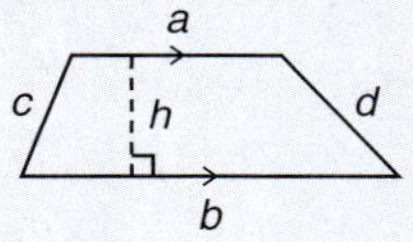	$P = a + b + c + d$	$A = \frac{1}{2}(a + b)h$
Circle 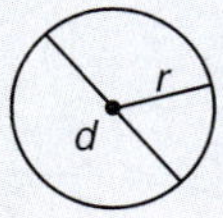	$C = \pi d$ or $C = 2\pi r$	$A = \pi r^2$

Example

Materials
- scientific calculator

Determine the perimeter and area of each figure.

a) 3.6 cm, 4.2 cm (square)

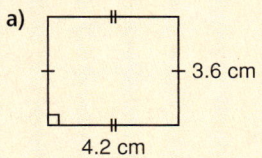

b) 2.3 m, 1.6 m, 2.5 m, 3.6 m (triangle)

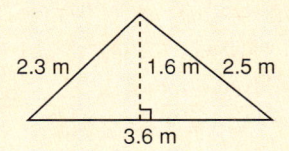

c) 3 in. (circle radius)

Activate Prior Knowledge

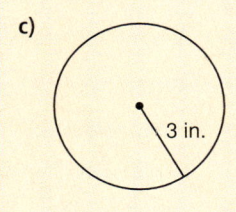

Solution

Round the answer to the same degree of accuracy as the least accurate measurement used in the calculation.

a) $P = 2\ell + 2w$
   $= 2(4.2) + 2(3.6)$
   $= 15.6$
   The perimeter is 15.6 cm.

   $A = \ell w$
   $= (4.2)(3.6)$
   $= 15.12$
   The area is about 15.1 cm^2.

b) $P = a + b + c$
   $= 2.3 + 2.5 + 3.6$
   $= 8.4$
   The perimeter is 8.4 m.

   $A = \frac{1}{2}bh$
   $= \frac{1}{2}(3.6)(1.6)$
   $= 2.88$
   The area is about 2.9 m^2.

c) $C = 2\pi r$
   $= 2\pi(3)$
   $\doteq 18.850$
   The circumference is about 19 inches.

   $A = \pi r^2$
   $= \pi(3)^2$
   $\doteq 28.274$
   The area is about 28 square inches.

CHECK ✓

1. Determine the perimeter and area of each figure.

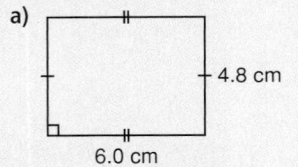

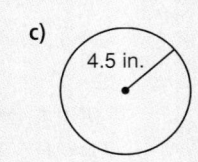

a) 4.8 cm, 6.0 cm (rectangle)
b) 8.3 m, 6.2 m, 9.3 m, 12.4 m (triangle)
c) 4.5 in. (circle)

2. Determine the perimeter and area of a football field that is 120 yards by 52 yards.

3. The circle at centre ice on a hockey rink has a diameter of 15 feet. Determine its circumference and area.

4. The roof of a house is made from trusses. The frame of each truss is an isosceles triangle, as shown.

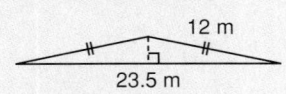

12 m, 23.5 m

a) Use the Pythagorean Theorem to determine the height of the triangle.
b) Determine the perimeter of the truss and the area it encloses.
c) Which measure from part b would be helpful when estimating how much wood will be needed for the trusses? Justify your answer.

CHAPTER 2: Geometry

Metric and Imperial Units of Capacity

Prior Knowledge for 2.2

Capacity is a measure of how much liquid a container can hold. Use conversion factors to change imperial units to metric units, or vice versa.

Imperial to Metric	Metric to Imperial
1 fluid ounce ≐ 28.413 mL	1 mL ≐ 0.0352 fluid ounce
1 pint ≐ 0.568 L	1 L ≐ 1.7598 pints
1 quart ≐ 1.1365 L	1 L ≐ 0.8799 quart
1 gallon ≐ 4.546 L	1 L ≐ 0.22 gallon

The capacity units used in the U.S. differ from those used in Canada. Unless stated otherwise, use Canadian units.

Example

Materials
- scientific calculator

Round each answer to the same degree of accuracy as the given measurement.

Convert each measure.

a) 8.0 gallons of gas to litres

b) 170 mL of water to fluid ounces

Solution

a) Each gallon is about 4.546 L.
   8.0 × 4.546 = 36.368
   8.0 gallons are about 36.4 L.

b) Each millilitre is about 0.0352 fluid ounce.
   170 × 0.0352 = 5.984
   170 mL are about 6 fluid ounces.

CHECK ✓

1. Convert each measure.
   a) A container holds 5 pints. Convert to litres.
   b) A gas tank can hold 80 L. Convert to gallons.
   c) A bottle contains 4.2 fluid ounces of perfume. Convert to millilitres.
   d) A large metal garbage can holds 40 gallons. Convert to litres.
   e) A can of pasta in tomato sauce contains 398 mL. Covert to fluid ounces.

2. In 2005, Canadians consumed on average 94.7 L of milk per person; Americans consumed on average 21.2 U.S. gallons per person.
   Each U.S. gallon is equivalent to 3.785 L.
   Which country had the greater milk consumption per person?
   Justify your answer.

Activate Prior Knowledge

Volumes of Prisms and Cylinders

Prior Knowledge for 2.2

Volume is the amount of space occupied by an object, measured in cubic units. The volume of a cylinder or prism is the product of the base area and the height.

$V = $ base area $\times$ height

An object's height is measured perpendicular to its base.

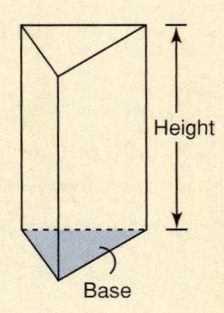

Example

Materials
- scientific calculator

Identify the base of each object and calculate its area.
Use the base area and height to calculate the volume of the object.

a) 1.0 ft, 3.1 ft, 2.4 ft, 5.1 ft, 4.0 ft

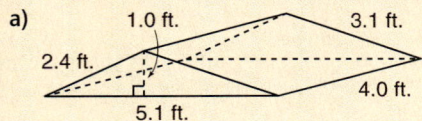

b) 2.5 cm, 9.6 cm

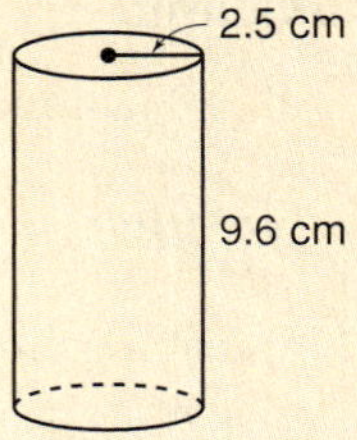

Solution

Round the final answer to the same degree of accuracy as the least accurate measurement used in the calculations.

a) The base is a triangle with base 5.1 feet and height 1 foot.
Its area is: $\frac{1}{2}bh = \frac{1}{2}(5.1)(1)$
$= 2.55$ square feet.
The height of the prism is 4 feet.
$V = $ base area $\times$ height
$= 2.55 \times 4$
$= 10.2$
The volume is 10.2 cubic feet.

b) The base is a circle with radius 2.5 cm.
Its area is: $\pi r^2 = \pi (2.5)^2$,
or about 19.635 cm^2.
The height of the cylinder is 9.6 cm.
$V = $ base area $\times$ height
$= 19.635 \times 9.6$
$= 188.496$
The volume is about 188.5 cm^3.

CHECK ✓

1. Identify the base of each object and calculate its area.
Use the base area and height to calculate the volume of the object.

a) 1.5 in., 11.5 in., 4.5 in.

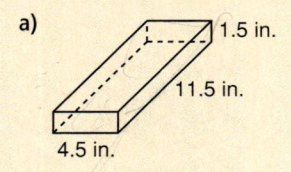

b) 6 cm, 5 cm, 3 cm, 4 cm

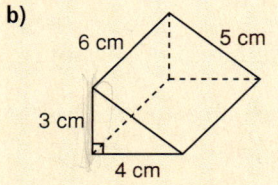

c) 4.5 m, 8.4 m

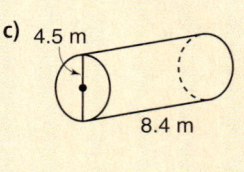

2. How do you know which face is the base of a triangular prism?

64 CHAPTER 2: Geometry

Surface Areas of Prisms and Cylinders

Prior Knowledge for 2.2

Surface area is the total area of the surface of an object.

Example

Materials
- scientific calculator

Describe the faces of each object. Then determine the surface area.

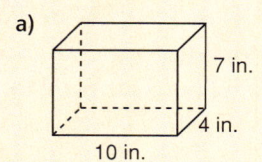

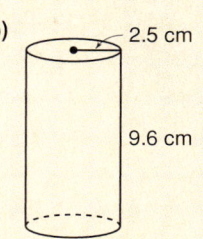

a) [rectangular prism: 10 in. by 4 in. by 7 in.]

b) [cylinder: radius 2.5 cm, height 9.6 cm]

Solution

a)

Face	Number	Area of each	Total area (square inches)
10-in. by 4-in. rectangle	2	10 × 4 = 40	2 × 40 = 80
4-in. by 7-in. rectangle	2	4 × 7 = 28	2 × 28 = 56
10-in. by 7-in. rectangle	2	10 × 7 = 70	2 × 70 = 140

SA = 80 + 56 + 140 = 276; the surface area is 276 square inches.

b)

Face	Number	Area of each	Total area (cm²)
Circle with radius 2.5 cm	2	$\pi(2.5)^2 \doteq 19.635$	2 × 19.635 = 39.270
One curved surface that can be unrolled to form a rectangle with length $2\pi(2.5)$, or 5π cm, and width 9.6 cm	1	$5\pi \times 9.6 \doteq 150.796$	150.796

SA = 39.270 + 150.796; the surface area is about 190.1 cm².

CHECK ✓

1. Use the art in question 1 on page 64. For each object, describe the faces of the object and determine its surface area.

2. How many faces does a rectangular prism have? A triangular prism?

Activate Prior Knowledge

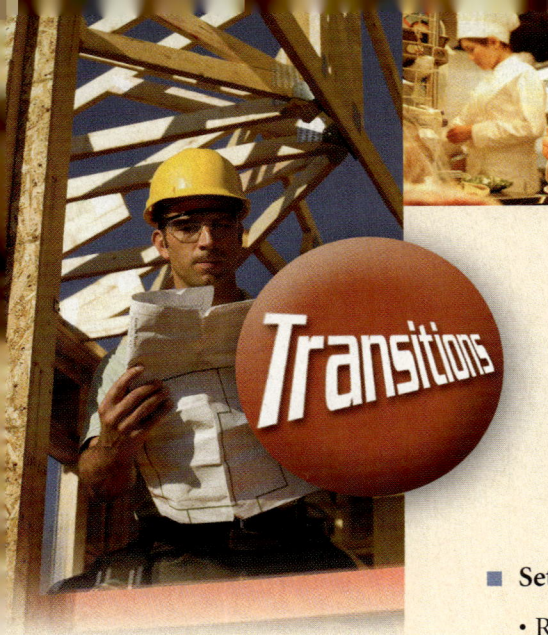

Managing Your Time

In college, or in a job or apprenticeship, you will be expected to manage your workload in the time designated for the task and to meet deadlines. You will need to set priorities, create a realistic plan for completing them, and follow your plan, making adjustments when necessary.

- **Set priorities**
  - Rank the importance of the tasks on your to-do list.
  - Decide which tasks must be done as soon as possible. Plan when the others must or should be done.

- **Create and follow a schedule**
  - Prioritize your tasks to help you decide what to do in a day, week, month, or year.
  - Be realistic. Allow time for socializing, travelling to and from work or school, eating, sleeping, and completing other responsibilities and activities.
  - Get an early start on longer tasks and work regularly on them so you have enough time to adapt your schedule to any unexpected delays.

- Complete questions 2 to 4 when it is appropriate as you work through this chapter.

1. What is another strategy you would suggest for managing your time?
2. Apply strategies for managing your time as you prepare for the chapter test or work on the chapter problem.
3. Compare the schedule you have now with a schedule you may have in college or an apprenticeship. Use the information below, or research information about courses or apprenticeships that interest you.
   - Some college students spend about 15 h per week in class and, for each hour in class, about 2 h on independent work and study.
   - Required workplace and classroom hours for apprenticeship programs vary. Assume you will spend 30 h on the job and 5 h in the classroom each week.
4. Describe a situation in a future apprenticeship or job in which you might use time-management strategies.

2.1 Area Applications

Shantel has a summer job working at a miniature golf course. Before the course can open in the spring, the old carpeting at each hole is inspected. If the carpeting is too worn, Shantel calculates how much carpeting is needed and replaces it.

Investigate

Determining the Area of a Composite Figure

Materials
- scientific calculator

The design for one hole of a miniature golf course is shown. The curve is a semicircle.

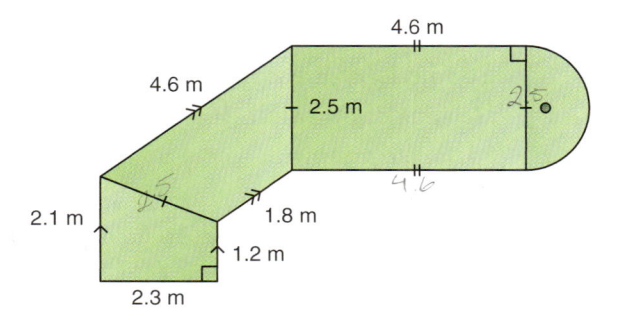

- Determine the area of this miniature golf hole.
- Trade solutions with a classmate. Check your classmate's solution. Did you solve the problem the same way?
- Compare areas. If you have different answers, find out why.

> **Reflect**
>
> - Suppose the length of the rectangular part of the hole were only 3.6 m. How would this affect the area? Justify your answer.
> - Suppose this hole needs new carpeting. How would knowing the area help Shantel decide how much carpet to buy? What other information does she need to know about how the carpeting is sold?

Connect the Ideas

Composite figures

A figure that is made up from other simpler figures is called a **composite figure**.

Example 1

Describing a Composite Figure

Describe the figures that make up this composite figure.

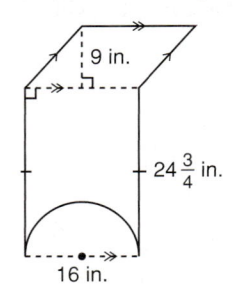

[Figure: composite figure with 9 in., $24\frac{3}{4}$ in., and 16 in. labeled]

Solution

The composite figure is made up of these figures:
- A rectangle with width 16 inches and height $24\frac{3}{4}$ inches.
- A parallelogram with height 9 inches and base 16 inches, on top of the rectangle.
- A semicircle with diameter 16 inches, removed from the base of the rectangle.

Area of a composite figure

To determine the area of a composite figure:
- Break it into simpler figures for which you know how to calculate the area.
- Calculate the area of each part.
- Add the areas.
- Subtract the areas of any parts removed from the figure.

68 CHAPTER 2: Geometry

Example 2 — Calculating the Area of a Composite Figure

Materials
- scientific calculator

Determine the area of the composite figure in *Example 1*.

Solution

Determine the area of each simple figure.
- The rectangle has dimensions 16 inches by $24\frac{3}{4}$ inches.

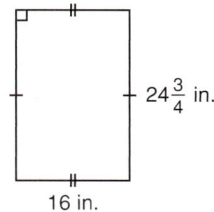

Express $24\frac{3}{4}$ as a decimal: 24.75

$A_{rectangle} = \ell w$ Substitute: $\ell = 16$ and $w = 24.75$
$\phantom{A_{rectangle}} = 16 \times 24.75$, or 396

The area of the rectangle is 396 square inches.

- The parallelogram has base length 16 inches and height 9 inches.

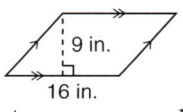

$A_{parallelogram} = bh$ Substitute: $b = 16$ and $h = 9$
$\phantom{A_{parallelogram}} = 16 \times 9$, or 144

The area of the parallelogram is 144 square inches.

- The radius of the semicircle is half the diameter:

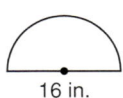

16 inches $\div$ 2 = 8 inches

The area of the semicircle is half the area of a circle with the same radius.

$A_{semicircle} = \frac{1}{2}\pi r^2$ Substitute: $r = 8$
$\phantom{A_{semicircle}} = \frac{1}{2}\pi(8)^2$
$\phantom{A_{semicircle}} \doteq 100.531$

Press: 1 [÷] 2 [×] [π] [×] 8 [x²] [ENTER]

The area of the semicircle is about 100.531 square inches.

$A_{figure} = A_{rectangle} + A_{parallelogram} - A_{semicircle}$
$\phantom{A_{figure}} = 396 \text{ square inches} + 144 \text{ square inches} - 100.531 \text{ square inches}$
$\phantom{A_{figure}} = 439.469 \text{ square inches}$

The area of the composite figure is about 439 square inches.

Use subscripts to organize calculations of similar properties. For example, $A_{rectangle}$ refers to the area of the rectangle and A_{figure} refers to the area of the composite figure.

Round the final answer to the same degree of accuracy as the least accurate measurement used in the calculations.

Finding missing lengths

Sometimes you need to use trigonometric ratios to determine missing lengths before you can calculate the area of a composite figure.

2.1 Area Applications   **69**

Example 3   Using Trigonometry to Determine an Unknown Length

Materials
- scientific calculator

Carpenters have constructed the frame for a house and will nail pressboard over the frame. Determine the area of pressboard they need for the back wall of the house.

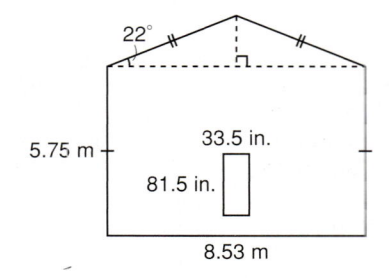

Solution

This composite figure is made up of a large rectangle, with an isosceles triangle on top and a rectangular door cut out.
Determine the area of each simple figure.

- $A_{rectangle} = \ell w$ Substitute: $\ell = 8.53$ and $w = 5.75$
  $= 8.53 \times 5.75$, or 49.0475

  The area of the rectangle is about 49.05 m.

- The isosceles triangle is made up of two congruent right triangles, each with one 22° angle and the same height as the isosceles triangle. The base of each right triangle is half the width of the wall:
  8.53 m ÷ 2 = 4.265 m

  $\tan 22° = \dfrac{h}{4.265}$

  $h = 4.265 \times \tan 22°$, or about 1.72

  The height of the triangle is about 1.72 m.

  $A_{triangle} = \dfrac{1}{2} bh$ Substitute: $b = 8.53$ and $h = 1.72$
  $= 0.5 \times 8.53 \times 1.72$, or 7.3358

  The area of the triangle is about 7.34 m.

- The dimensions of the door are in inches.
  $A_{door} = \ell w$ Substitute: $\ell = 33.5$ and $w = 81.5$
  $= 33.5 \times 81.5$, or 2730.25

  Convert the dimensions to square metres.
  1 inch = 2.54 cm, or 0.0254 m
  So, (1 inch)² = (0.0254 m)², or about 0.000 645 2 m².
  $2730.25 \times 0.000\ 645\ 2 \doteq 1.7616$

  The area of the rectangular door is about 1.76 m.

$A_{total} = A_{rectangle} + A_{triangle} - A_{door}$
$= 49.0475\ m + 7.3358\ m - 1.7616\ m$, or $54.6217\ m$

They need about 54.6 m² of pressboard for this wall of the house.

$\tan A = \dfrac{\text{length of side opposite } \angle A}{\text{length of side adjacent to } \angle A}$

Practice

A

■ For help with questions 1 to 3, see Example 1.

1. Describe the figures that make up each composite figure. The curve is a semicircle.

a)   b) 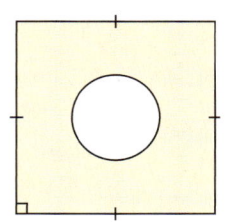   c)

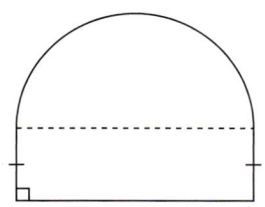

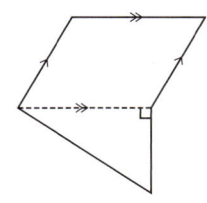

2. Describe how you would determine the area of each composite figure in question 1.

3. Sketch the figures that make up each composite figure. Include measurements in your sketches. All curves are circles or semicircles.

a) 40 cm, 30 cm, 60 cm, 25 cm

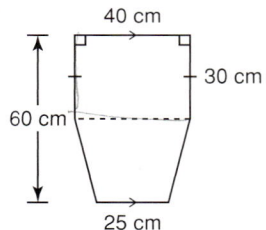

b) 3.5 m, 7.6 m, 8.7 m

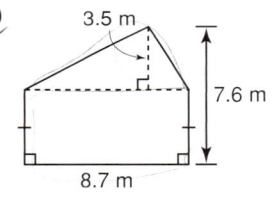

1 foot = 12 inches

c) $6\frac{3}{4}$ in., $1\frac{1}{2}$ in.

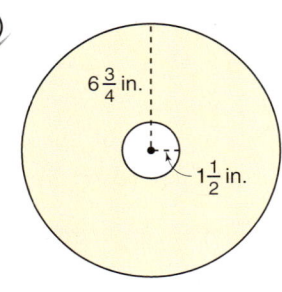

d) 3 ft., 15 in.

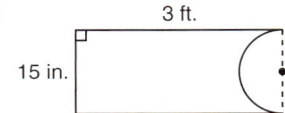

B

■ For help with question 4, see Example 2.

4. Determine the area of each composite figure in question 3.

2.1 Area Applications   **71**

5. Two students are calculating the area of this figure.

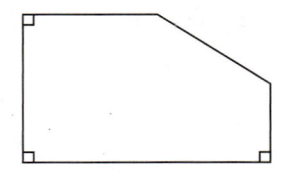

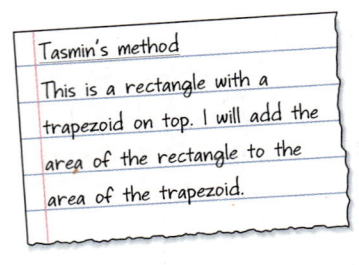

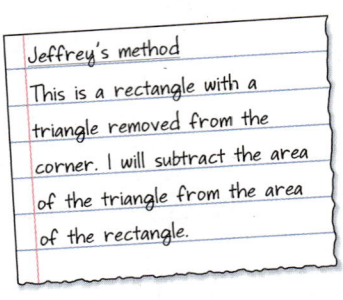

   Tasmin's method
   This is a rectangle with a trapezoid on top. I will add the area of the rectangle to the area of the trapezoid.

   Jeffrey's method
   This is a rectangle with a triangle removed from the corner. I will subtract the area of the triangle from the area of the rectangle.

   Who is correct? Justify your answer.
   Include diagrams in your explanation.

6. Describe the figures that make up each composite figure. Determine the area of each composite figure. All curves are semicircles.

   a) 2 ft., 8 ft.
   b) 20 in., 15 in., 3 ft.
   c) 18 cm, 10 cm, 12 cm

   1 foot = 12 inches

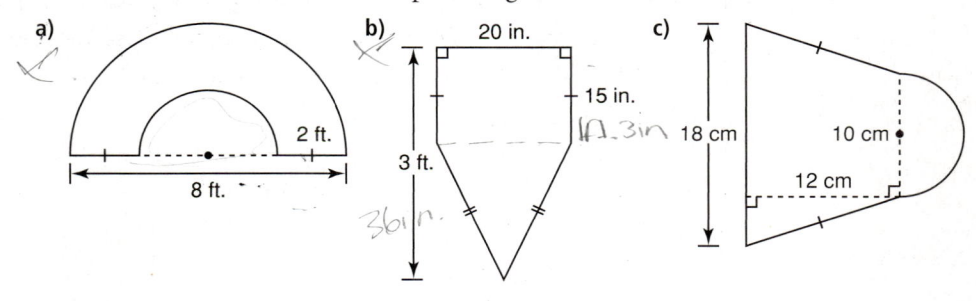

7. A grocery store display is built using cans of juice, stacked in layers. The display has 10 layers. The top three layers and an overhead view of each of these layers are shown.

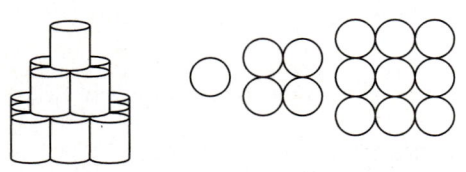

   a) Describe the pattern in the number of cans in each layer.
   b) What shape is each layer of the display?
   c) How many cans are in the tenth layer?
   d) Each can has diameter 12.5 cm. Determine the amount of floor space needed for the display. What assumption are you making?

72   CHAPTER 2: Geometry

8. The display in question 7 is changed to have a triangular base. The top three layers are shown. The cans are still stacked 10 layers high. 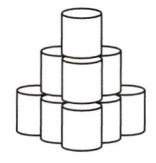
   a) Draw and describe the pattern in the number of cans in each layer.
   b) How many cans are in the bottom layer?
   c) Determine the amount of floor space needed for the display. What assumption are you making?
   d) Will the display hold more or fewer cans than the display in question 7? Justify your answer.

9. A decorator is painting this wall of an attic room. The window measures 0.6 m by 0.5 m.

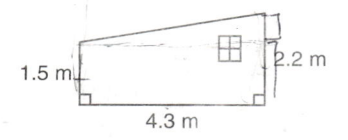

   1.5 m, 2.2 m, 4.3 m

   1 m ≐ 3.2808 feet

   a) What is the area of the wall in square metres and square feet?
   b) The paint is sold in 1-pint containers. Each container should cover between 50 square feet and 60 square feet. How many cans of paint should the decorator buy?

10. **Assessment Focus** The running track in this diagram consists of two parallel sections with semicircular sections at each end. Determine the area of the track.

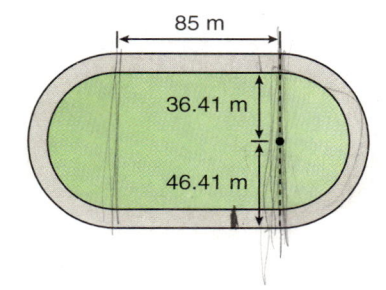

    85 m, 36.41 m, 46.41 m

11. **Literacy in Math** Write a step-by-step description of how to determine the area of a composite figure. Include an example.

2.1 Area Applications

■ For help with question 12, see Example 3.

12. An outdoor garage is being built on a farm to house vehicles and equipment. The front has two congruent garage door entrances and a round window at the top.

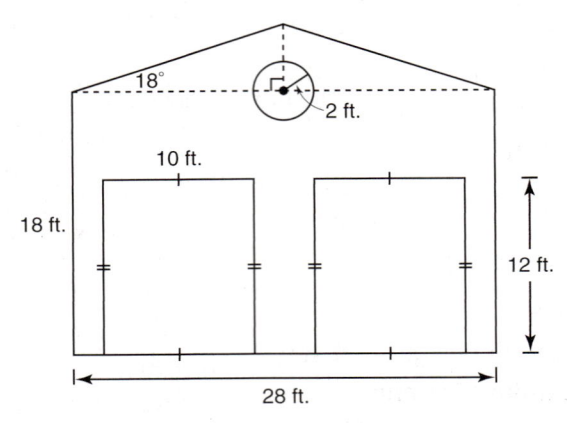

a) The front wall will be covered in sheet metal. Determine the amount of sheet metal needed, to the nearest square foot.

b) Suppose the sheet metal is priced by the square metre. How many square metres will be needed for this project?

1 foot ≐ 0.3048 m

The short lines on each side mean that the sides have equal length.

13. a) Describe how to use composite figures to determine the area of a regular octagon. Use your technique to determine the area of a regular octagon with side length 20 cm.

b) Compare your work in part a with another student. If your answers are different, try to determine why. If you used different methods, which do you think is easier?

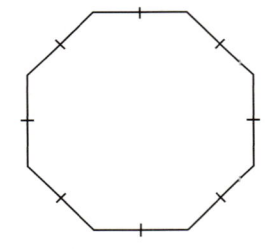

14. The design for a backyard deck is shown. It will be built using plastic lumber made from recycled materials.

a) Determine the area of the deck.

b) A circular hot tub with diameter 2 m is to be installed in the octagonal portion of the deck. How much wood needs to be cut out to make room for the hot tub?

c) The backyard is a rectangle measuring 65 feet by 45 feet. What is the area of the backyard not covered by the deck?

1 m ≐ 3.2808 feet

74   CHAPTER 2: Geometry

15. Each year, car manufacturers create concept cars to try out new technology and design ideas. For an auto show, a manufacturer wishes to display three electric concept cars on a raised platform.
- a four-door, four-passenger car with length 4856 mm and width 1915 mm
- a four-passenger sport wagon with length 4390 mm and width 1743 mm
- a two-seater sport utility vehicle with length 3885 mm and width 1598 mm

a) Choose a figure for the platform. Sketch how the cars should be arranged on the platform. Explain the decisions you made about the figure and the arrangement of the cars.

b) Determine reasonable dimensions for the platform in part a. Include an explanation of how you decided how much space to leave between the cars.

c) How much carpeting should be ordered to cover the top of the display platform you have designed?

C **16.** Monique wants to make a regular equilateral pyramid with a vertical height of 10 cm.

a) Draw a net for this object.
b) Determine the area of the net.

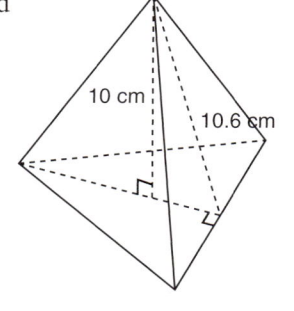

> A net is a drawing that can be cut out and folded to create a three-dimensional object.

In Your Own Words

Suppose you are hired to paint the walls of your classroom. Develop a plan describing how you would measure and calculate the area to be painted. What other information would you need to include in your plan?

2.1 Area Applications

2.2 Working with Composite Objects

Antiques, works of art, and other valuable objects are carefully packaged for shipping. A cardboard box or wooden crate is built to fit the object. Dimensions must be accurately measured to ensure that the container is large enough to hold the object and protective packing material.

Investigate   Building a Model

Materials
- scientific calculator

Work with a partner.
Choose an object in the classroom that can be viewed as a combination of two or more rectangular prisms, triangular prisms, or cylinders. Alternatively, build your own object.
- Sketch your object. Describe the rectangular prisms, triangular prisms, or cylinders that make up the object.
- Calculate the volume and surface area of your object.

Reflect

- What strategies did you use to calculate the volume and surface area of your object?
- Did your strategies produce reasonable results? If not, how could they be improved?
- Suppose you had to build a package to hold the object. Which of the data you collected or calculated would be most useful? Why?

Connect the Ideas

Volume and surface area

The volume of a cylinder or prism is the product of the base area and the height: $V = $ base area $\times$ height. The surface area of a cylinder or prism is the sum of the areas of the faces.

Example 1

Finding the Volume and Surface Area of a Cylinder

Materials
- scientific calculator

A machine bales hay in cylindrical rolls. For storage, a shrink-wrap protective cover is placed on the bale.

a) Determine the volume of hay in a bale and the area of the shrink-wrap covering it.

b) How much shrink-wrap is needed if the wrap does not cover the ends of the roll?

c) Suppose the shrink-wrap is priced by the square metre. To the nearest square metre, how many square metres of wrap are needed to cover the bale as described in part b?

Solution

The radius of the base is half the diameter, or 2.5 feet.

a) The base area of the cylinder is: $\pi(2.5 \text{ feet})^2 \doteq 19.635$ cubic feet
The height is 5 feet.
So, the volume of the cylinder is:
19.635 square feet $\times$ 5 feet $=$ 98.175 cubic feet
The volume of hay is about 98 cubic feet.
The surface area is the sum of the areas of the faces.

Face	Shape	Number	Area of each face (cm²)
Curved surface	Unrolls to form a rectangle	1	$2 \times \pi \times 2.5 \times 5 \doteq 78.54$
Ends	Circles	2	$2 \times \pi \times 2.5^2 \doteq 39.27$

$78.54 + 39.27 = 117.81$
About 118 square feet of shrink-wrap are needed to cover the bale.

If the ends of the roll are not covered, the surface area is just the area of the curved surface.

b) From the table, about 79 square feet of shrink-wrap are needed.

c) 1 foot $= 0.3048$ m, so $(1 \text{ foot})^2 = (0.3048 \text{ m})^2$, or about $0.092\,903$ m²
$78.54 \times 0.092\,903 \doteq 7.2966$
So, about 7 m² of shrink-wrap are needed.

2.2 Working with Composite Objects

Composite objects

When a structure or object is made up from several simple objects, it is called a **composite object**.

The house in the photograph can be thought of as a rectangular prism with a triangular prism on top and another rectangular prism for the chimney.

Other objects, such as this sewer pipe, can be thought of as simple three-dimensional objects with a piece removed.

Determining the volume

To determine the volume of a composite object:
- Calculate the volume of each part of the object.
- Add the volumes.
- Subtract the volume of any parts that were removed.

Sometimes, the composite object can be viewed as a prism whose base is a composite object. In these cases, you can use the formula:
$V =$ base area $\times$ height

Example 2

Finding the Volume of a Composite Object

Materials
- scientific calculator

Determine the volume of this shed in cubic metres.

79.0 cm
202.0 cm
289.5 cm
310.0 cm

78 CHAPTER 2: Geometry

Solution

Method 1

Think of the shed as a triangular prism on top of a rectangular prism.

Calculate the volume of each part:
- The base area of the rectangular prism is:

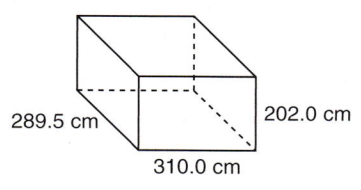

  310.0 cm × 289.5 cm
  = 89 745 cm²

  The height is 202.0 cm.
  So, $V_{rectangular\ prism}$
  = 89 745 cm² × 202.0 cm
  = 18 128 490.0 cm³

- The base area of the triangular prism is:

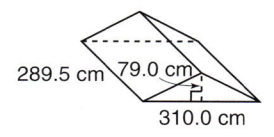

  $\frac{1}{2}$ × 310.0 cm × 79.0 cm
  = 12 245 cm²

  The height is 289.5 cm.
  So, $V_{triangular\ prism}$
  = 12 245 cm² × 289.5 cm
  = 3 544 927.5 cm³

$V_{shed} = V_{rectangular\ prism}$
$\qquad\quad + V_{triangular\ prism}$
$\qquad = 18\ 128\ 490.0\ cm^3$
$\qquad\quad + 3\ 544\ 927.5\ cm^3$
$\qquad = 21\ 673\ 417.5\ cm^3$

The volume is 21 673 417.5 cm³.

21 673 417.5 ÷ 1 000 000 ≐ 21.67
The volume of the shed is approximately 21.67 m³.

Method 2

Think of the shed as a prism with this 5-sided base:

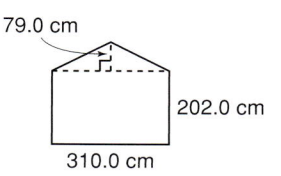

Calculate the area of each part of the base:
- $A_{rectangle}$
  = 310.0 cm × 202.0 cm
  = 62 620 cm²
- $A_{triangle}$
  = $\frac{1}{2}$ × 310.0 cm × 79.0 cm
  = 12 245 cm²

So, the base area of the prism is:
62 620 cm² + 12 245 cm²
= 74 865 cm²

The height of the prism is 289.5 cm.
The volume of the prism is:
base area × height
  = 74 865 cm² × 289.5 cm
  = 21 673 417.5 cm³

The volume is 21 673 417.5 cm³.

---

1 m = 100 cm
So, (1 m)³ = (100 cm)³,
or 1 000 000 cm³

Determining the surface area

When you determine the surface area of a composite object, include only those faces that are faces of the composite object. That is, the faces that are part of the surface of the object.

Example 3

Materials
- scientific calculator

Finding the Surface Area of a Composite Object

Determine the surface area of the shed in *Example 2* in square metres. Assume the shed has a floor that you should include in your calculations.

79.0 cm
202.0 cm
289.5 cm
310.0 cm

Solution

Before you can determine the area of the roof panels, you have to determine their width.
Draw a sketch showing a roof panel and half of the triangular panel from the front of the shed.
Use the Pythagorean Theorem.

$155.0^2 + 79.0^2 = w^2$
$30\,266 = w^2$
$w \doteq 173.97$

w, 289.5 cm, 155.0 cm, 79.0 cm

The width is about 173.97 cm.
Use a table to keep track of the faces included in the surface area.

Face	Shape	Number	Area of each face (cm²)
Roof panels	Rectangle	2	$289.5 \times 173.97 \doteq 50\,364.3$
Front and back triangular panels	Triangle	2	$\frac{1}{2}(310.0 \times 79.0) = 12\,245$
Front and back	Rectangle	2	$310.0 \times 202.0 = 62\,620$
Sides	Rectangle	2	$289.5 \times 202.0 = 58\,479$
Floor	Rectangle	1	$310.0 \times 289.5 = 89\,745$

CHAPTER 2: Geometry

$$SA_{total} = 2(50\ 364.3 + 12\ 245 + 62\ 620 + 58\ 479) + 89\ 745$$
$$= 367\ 416.6 + 89\ 745$$
$$= 457\ 161.6$$

1 m = 100 cm
So, $(1\ m)^2 = (100\ cm)^2$, or 10 000 cm²
Divide by 10 000 to convert the area to square metres.
The surface area of the shed is about 45.72 m².

Practice

A

For help with questions 1 or 2, see Example 1.

1. Describe the simple three-dimensional objects that make up each cake.
a) 
b)

2. Describe the simple three-dimensional objects that make up each object.
a) 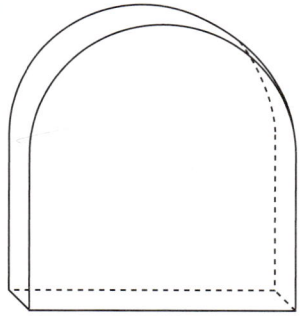
b)

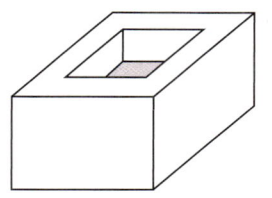

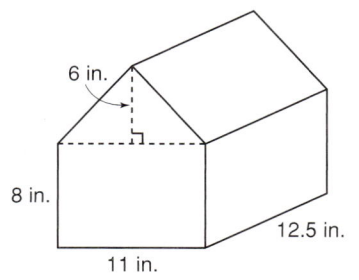

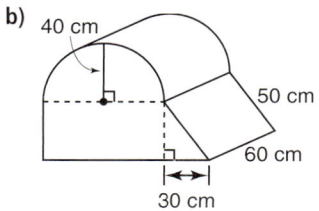

3. Describe the simple three-dimensional objects that make up each object.
a) 6 in., 8 in., 11 in., 12.5 in.
b) 40 cm, 50 cm, 60 cm, 30 cm

2.2 Working with Composite Objects

4. A fruit drink is sold in a box that contains 10 drink pouches. The dimensions of the box are shown. 
   a) Determine the surface area of the cardboard used for the box.
   b) Each drink pouch uses about 340 cm² of material. How much is used for the 10 pouches?
   c) Each drink pouch contains 200 mL. What is the total amount of drink in the package?

   14.5 cm
   9.0 cm
   28.5 cm

5. A section of water trough for a poultry farm is shown on the right. The triangular face is a right triangle with a base of 3 inches and height of 5 inches. The trough runs the length of the barn, which is 120 feet long.
   a) Determine the amount of sheet metal required to build the trough.
   b) Determine the volume of the trough in cubic inches.
   c) A cubic inch is about 0.004 33 gallons. About how many litres of water can the trough hold?

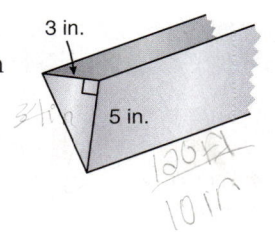

   3 in.
   5 in.

1 foot = 12 inches

1 gallon ≐ 4.546 L

1 cubic foot ≐ 6.23 gallons

6. In many places, regulations state that all the milk a dairy farmer has in a holding tank must be picked up in one trip. A milk truck has a cylindrical tank with radius 9 feet and length 14 feet. Is the tank large enough to pick up all the milk from a full 6000-gallon holding tank? Justify your answer.

7. Joshua is calculating the volume and surface area of this quarter cylinder. 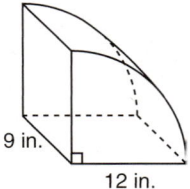
   a) How would its volume compare to the volume of a cylinder with the same radius and height?
   b) Calculate the object's volume.
   c) Complete the calculations in the table Joshua has created. Then add areas to determine the total surface area.

   9 in.
   12 in.

Face	Shape	Number	Area of each face (square inches)
Front and back	Quarter circle	2	$\frac{1}{4}(\pi \times 12^2) =$
Bottom and side	Rectangle	2	$9 \times 12 =$
Curved surface	Unrolls to form rectangle	1	$\frac{1}{4}(2\pi \times 12 \times 9) =$

   d) Explain how Joshua developed the expressions for the area of each face.

82 CHAPTER 2: Geometry

B

8. The bottom layer of the cake in part a of question 1 has diameter 39 cm and height 10 cm. The top layer has diameter 26 cm and height 10 cm. Assume that the entire top of the bottom layer is iced, but only the portion that can be seen is decorated.
   a) Determine the volume of each layer and the total volume of the two layers.
   b) Determine the area that is iced.
   c) Determine the area that is decorated.

9. The bottom layer of the cake in part b of question 1 has length and width 14 inches and height 4 inches. The top layer has length and width 10 inches and height 4 inches. Assume that the entire top of the bottom layer is iced, but only the portion that can be seen is decorated.
   a) Determine the volume of each layer and the total volume of the two layers.
   b) Determine the surface area that is iced.
   c) Determine the area that is decorated.

1 inch = 2.54 cm

10. Suppose you could buy either of the cakes shown in question 1 for the same price. Use your answers to questions 8 and 9 to decide which cake is the better deal. Justify your answer.

■ For help with question 11, see Example 2.

11. The front and back faces of the roof of this barn are isosceles triangles.
    a) Determine the volume of the barn.
    b) Would the barn be large enough to store 5000 cubic feet of hay? Justify your answer.

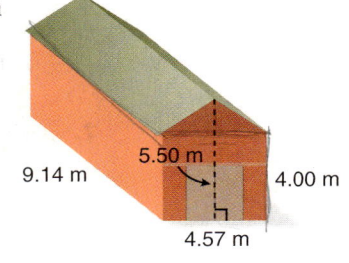

1 m ≐ 3.2808 feet

5.50 m
9.14 m
4.00 m
4.57 m

1 inch = 2.54 cm

12. Two different mailboxes are shown, one wooden, one made out of sheet metal. Which mailbox has the greater volume? Justify your answer.

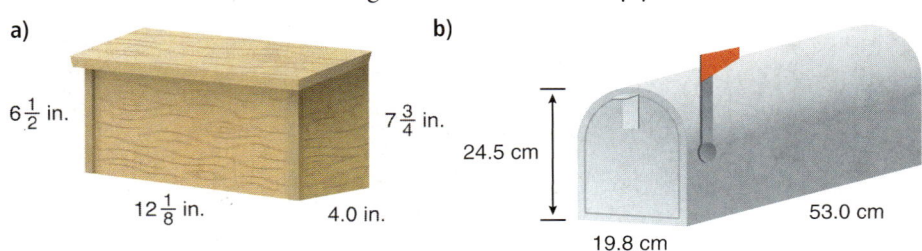

a) $6\frac{1}{2}$ in.
$12\frac{1}{8}$ in.
$7\frac{3}{4}$ in.
4.0 in.

b) 24.5 cm
19.8 cm
53.0 cm

2.2 Working with Composite Objects **83**

13. Literacy in Math Create a flow chart that outlines the steps to follow when calculating the volume of a composite object.

14. Olivia owns a furniture store. She has sold a half-cylinder console table like the one shown here. Olivia needs to build a crate to ship the table to the customer.
   a) What dimensions would you recommend for the shipping crate? Justify your answer.
   b) What will be the volume of the shipping crate?
   c) How much empty space will there be around the table for protective packing material?

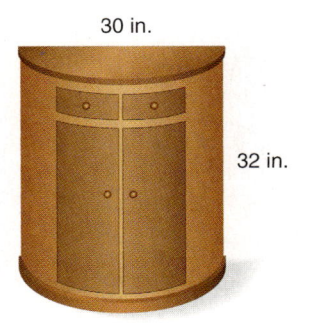

30 in.

32 in.

15. A can of peas has a diameter of 11.0 cm and a height of 7.4 cm. The cans are packed for shipping in a box. They are arranged in 2 layers of 3 rows by 4. The box is constructed to fit the cans snugly. Determine the amount of empty space in the box.

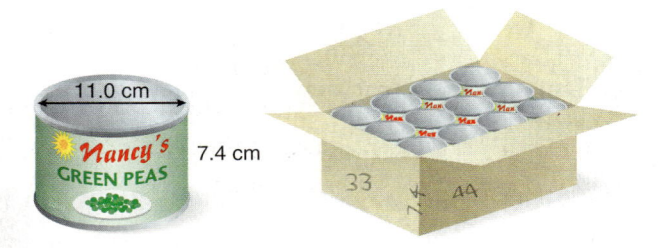

11.0 cm

7.4 cm

■ For help with question 16, see Example 3.

Amutha has built a birdhouse, and decides to paint it. Determine the surface area that requires painting.

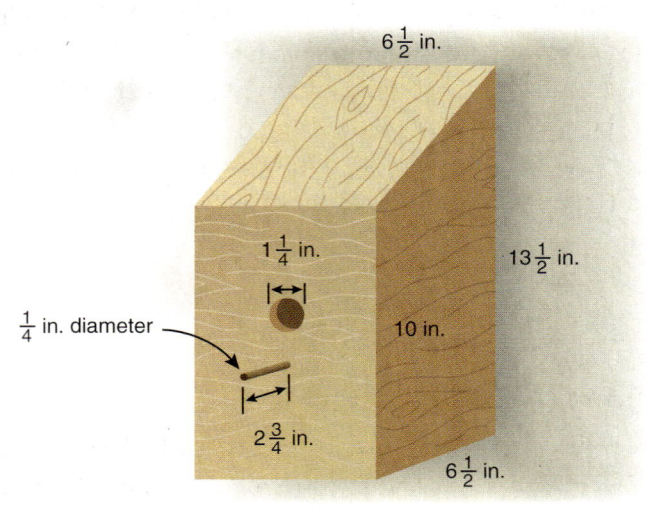

$6\frac{1}{2}$ in.

$1\frac{1}{4}$ in.

$13\frac{1}{2}$ in.

$\frac{1}{4}$ in. diameter

10 in.

$2\frac{3}{4}$ in.

$6\frac{1}{2}$ in.

84 CHAPTER 2: Geometry

17. A tire manufacturer sells tires in packages of four. The tires are shrink-wrapped together, without covering the hole in the centre. Determine the amount of shrink-wrap required for each package, in square centimetres.

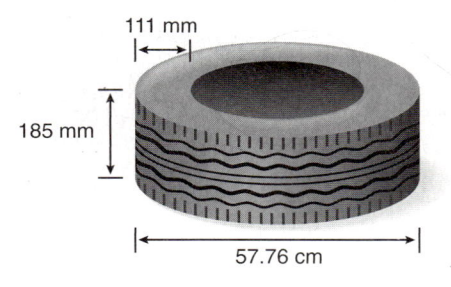

111 mm
185 mm
57.76 cm

18. Two different mailboxes are shown in question 12. Which mailbox has the lesser surface area? What is the difference in surface areas?

1 inch = 2.54 cm

19. Assessment Focus
   a) Suppose the objects in question 3 are to be moulded using concrete. Determine the volume of concrete required to make each object.
   b) Which object has the greater volume?
   c) Suppose the objects in question 3 are to be made out of sheet metal. Determine the area of metal required to build each object.
   d) Which object has the greater surface area?

C

20. A manufacturing company uses sheet metal and a press to cut out washers. The sheet metal is $\frac{1}{16}$-inch thick. The washers have outer diameter 1 inch and inner diameter $\frac{1}{4}$ inch. The sheet has length 8 yards and width 3 yards.
   a) Determine how many washers can be cut from one sheet.
   b) Calculate the volume of material not used. Include the material cut from the centre of each washer.
   c) The washers are to be sprayed with a protective coating. Determine the surface area of the washers from one sheet.

1 yard = 3 feet
1 foot = 12 inches

In Your Own Words

Suppose you are given an object and asked to design a shipping carton. Describe the steps you would follow. Include an example with diagrams in your explanation.

2.2 Working with Composite Objects   **85**

Mid-Chapter Review

2.1

1. Describe the figures that make up each composite figure. Then determine the area of each composite figure.

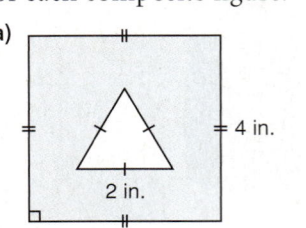

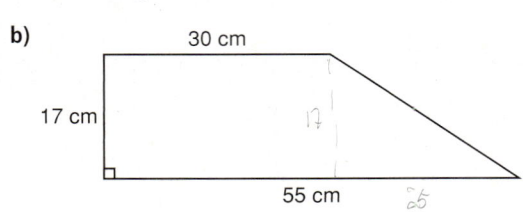

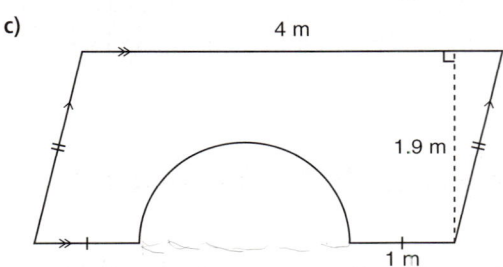

a) Square with side 4 in., containing a triangle with base 2 in.

b) Trapezoid with parallel sides 30 cm (top) and 55 cm (bottom), height 17 cm.

c) Parallelogram with top 4 m, height 1.9 m, bottom portion 1 m, with a semicircle cut out of the bottom.

2. A winter cover will be made for this swimming pool. The cover must extend 1 foot beyond the edges along the perimeter.

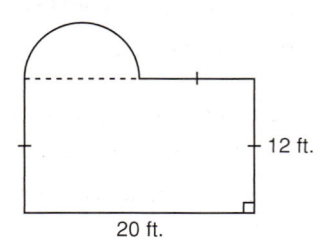

Figure: rectangle 20 ft by 12 ft with a semicircle on top.

a) What will be the area of the cover in square feet?

1 m ≐ 39.37 inches

b) The material for the cover sells for $2.50 per square metre. How many square metres are needed and how much will they cost?

2.2

3. Quarters are packaged in rolls of 40. Each quarter has a diameter of 2.4 cm, and is 1.5 mm thick. Determine the volume and surface area of a roll of quarters.

4. Describe the simple objects that make up this object. Then determine its volume and surface area.

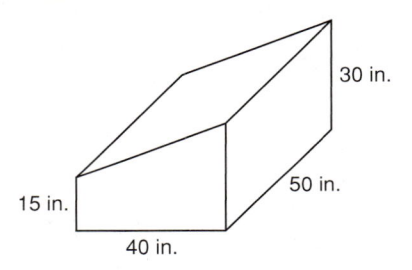

Figure: a 3D object with dimensions 15 in., 40 in., 50 in., and 30 in.

5. Max needs to make 3 copies of the object shown in question 4 from cement. He has 2 m³ of cement. Will this be enough? Justify your answer.

1 m ≐ 39.37 inches

6. Determine the volume and surface area of this sunglasses case.

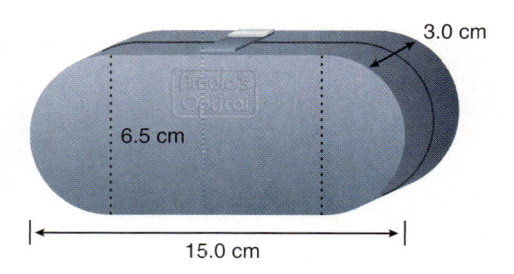

Figure: sunglasses case with dimensions 6.5 cm, 15.0 cm, and 3.0 cm.

86   CHAPTER 2: Geometry

2.3 Optimizing Areas and Perimeters

A gardener wants to determine the greatest rectangular area that can be enclosed by a given length of edging.

A dog breeder wants to find the rectangle with a given area that has the least perimeter. Both are **optimization** problems.

Investigate

Finding Maximum Areas and Minimum Perimeters

Materials
- grid paper
- 24 toothpicks
- 36 square tiles

or
- TI-83 or TI-84 graphing calculator

Choose *Using Manipulatives* or *Using a Graphing Calculator*.
Work with a partner.

Using Manipulatives

Part A: Investigating Optimal Areas

Suppose you have twenty-four 1-m sections of edging to enclose a rectangular garden. What is the maximum area that you can enclose?

■ Each toothpick represents a section of edging. Construct as many rectangles as you can using all 24 toothpicks. For each rectangle, record the dimensions, area, and perimeter in a table like this.

Length (m)	Width (m)	Area (m²)	Perimeter (m)

Use only whole toothpicks for the sides.

2.3 Optimizing Areas and Perimeters **87**

- Create a graph of area versus length. Draw a smooth curve through the points.
- What are the dimensions of the garden with the maximum area? Did you use the table or the graph?

Part B: Investigating Optimal Perimeters

Suppose you want to build a rectangular patio with thirty-six 1-m² patio tiles. What is the minimum perimeter the patio can have?

- Each square tile represents a patio tile. Construct as many rectangles as you can using all 36 square tiles. For each rectangle, record the dimensions, area, and perimeter in a table.
- Create a graph of perimeter versus length. Draw a smooth curve through the points.
- What are the dimensions of the patio with the minimum perimeter?

Using a Graphing Calculator

Part A: Investigating Optimal Areas

Suppose you have twenty-four 1-m sections of edging to enclose a rectangular garden.

What is the maximum area that you can enclose?

Steps	Display	Notes
Press [STAT] 1. • In L1, list possible lengths for the rectangle. Enter whole numbers from 1 to 11. • Press [▶] [▲] to move onto the list name L2. Press: 12 [−] [2nd] 1 [ENTER] • Move onto the list name L3. Press: [2nd] 1 [×] [2nd] 2 [ENTER]	L1: 1,2,3,4,5,6,7 L2: 11,10,9,8,7,6,5 L3: 11,20,27,32,35,36,35 L3(1)=11	If necessary, press [2nd] [+] 4 [ENTER] to clear the lists in the list editor. The formula **12 − L1** calculates the corresponding width for each length in L1. The formula **L1 × L2** calculates the area of each rectangle.
Create a graph of length and area. Press [2nd] [Y=] 1 and change the settings as shown.	Plot1 Plot2 Plot3 On Off Type: ... Xlist: L1 Ylist: L3 Mark: □ + .	Length will be plotted on the horizontal axis and area on the vertical axis.

88 CHAPTER 2: Geometry

Press ZOOM 9.
Press TRACE and use the arrow keys to see the coordinates of each point.

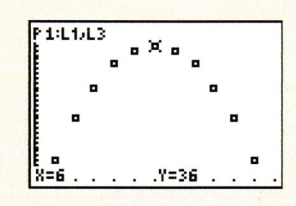

- What are the dimensions of the garden with the maximum area?

Part B: Investigating Optimal Perimeters
Suppose you want to build a rectangular patio with thirty-six 1-m² patio tiles. What is the minimum perimeter the patio can have?

Steps	Display	Notes
Clear the list editor. Press: 2nd [+] 4 ENTER Press: STAT 1 • In L1, list possible lengths for the rectangle. Enter the whole number factors of 36: 1, 2, 3, 4, 6, 9, 12, 18, 36. • Move onto the list name L2. Press: 36 ÷ 2nd 1 ENTER • Move onto the list name L3. Press: 2 × (2nd 1 + 2nd 2) ENTER		The formula **36 ÷ L1** calculates the corresponding width for each length in L1. The formula **2 × (L1 + L2)** calculates the perimeter of each rectangle.

- Create a graph of length and perimeter.
- What are the dimensions of the patio with the minimum perimeter?

> Restrictions on the possible dimensions or shape of an object are called *constraints*.

Reflect

- What constraints limited the number of rectangles you could construct in each situation?
- Did you use a table of values or a graph to determine the maximum area or the minimum perimeter? Explain your choices.

Connect the Ideas

Optimizing the dimensions of a rectangle

Among all rectangles with a given perimeter, a square has the maximum area. Among all rectangles with a given area, a square has the minimum perimeter.

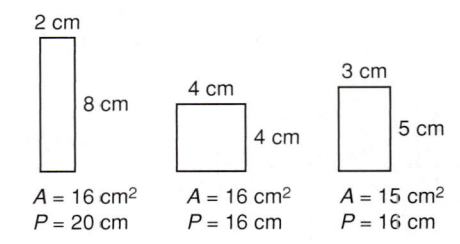

2 cm / 8 cm
$A = 16$ cm²
$P = 20$ cm

4 cm / 4 cm
$A = 16$ cm²
$P = 16$ cm

3 cm / 5 cm
$A = 15$ cm²
$P = 16$ cm

Example 1 — Finding Dimensions of Optimal Rectangles

Materials
- scientific calculator

a) What are the dimensions of a rectangle with perimeter 20 m and the maximum area? What is the maximum area?

b) What are the dimensions of a rectangle with area 45 m² and the minimum perimeter? What is the minimum perimeter?

Solution

a) The maximum area occurs when the rectangle is a square. Determine the side length, s, of a square with perimeter 20 m.
$s = P \div 4$
$= 20 \div 4$, or 5
$A = 5 \times 5$
$= 25$
The rectangle is a square with side length 5 m.
Its area is 25 m².

b) The minimum perimeter occurs when the rectangle is a square. Determine the side length, s, of a square with area 45 m².
$s = \sqrt{A}$
$= \sqrt{45}$, or about 6.71
$P = 4 \times 6.71$
$= 26.84$
The rectangle is a square with side length about 6.7 feet.
Its perimeter is about 26.8 feet.

Optimizing with restrictions

There may be restrictions on the rectangle you are optimizing:
- The length and width may have to be whole numbers; or
- The length and width may have to be multiples of a given number.

In these cases, it may not be possible to form a square. The maximum area or minimum perimeter occurs when the length and width are closest in value.

90 CHAPTER 2: Geometry

Sometimes one or more sides of the area to be enclosed are bordered by a wall or other physical barrier. In these cases, the optimal rectangle will not be a square. You can use diagrams or a table and graph to find the dimensions of the optimal rectangle.

Example 2

Materials
- TI-83 or TI-84 graphing calculator

Optimizing with Constraints

A rectangular garden is to be fenced using the wall of a house as one of side of the garden. The garden should have an area of 40 m². Determine the minimum perimeter and dimensions of the garden in each case:

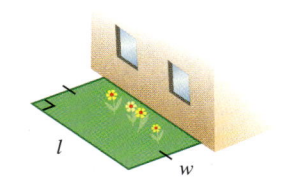

a) The dimensions must be whole numbers of metres.

b) The dimensions can be decimals.

Solution

One length of the garden is along the wall.
Only three sides of the garden need to be fenced. So, $P = \ell + 2w$
The area is 40 m².

$A = \ell w$ Substitute $A = 40$
$40 = \ell w$ Isolate w.
$w = 40 \div \ell$

Use a table or a graph to determine the minimum perimeter and dimensions of the garden.

a) Substitute values into the equations above. Record possible dimensions for the garden in a table. The minimum perimeter for whole number dimensions occurs twice: when (length, width) is (10 m, 4 m) or (8 m, 5 m). The perimeter in both cases is 18 m.

> Since the dimensions are whole numbers, ℓ is a factor of 40.

Area 40 m²

Length (m)	Width (m)	Perimeter (m)
40	1	42
20	2	24
10	4	18
8	5	18
5	8	21
4	10	24
2	20	42
1	40	81

2.3 Optimizing Areas and Perimeters

> We use a different procedure from that in *Investigate* since the dimensions can be decimal lengths.

> You could also use a scientific calculator to guess and check as in part a.

> Press [2nd] [GRAPH] to view the data in a table. To change the starting value of the table, or the increment, press [2nd] [WINDOW] and adjust TblStart and ΔTbl.

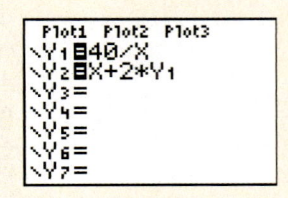

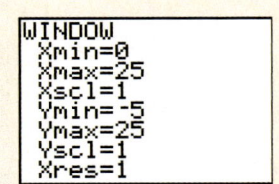

b) Use a graphing calculator to create a graph of length and perimeter.

Press [Y=].
- Use X to represent the length.
- Use Y1 to represent the width. Press: 40 [÷] [X,T,Θ,n] [ENTER]
- Use Y2 to represent the perimeter. Press: [X,T,Θ,n] [+] 2 [×] [VARS] [▶] 1 1 [ENTER]

Press [WINDOW].
Change the window settings as shown.

Press [GRAPH]. Press [2nd] [TRACE] 3 to determine the minimum of the graph.
Use the arrow keys to place the cursor on the graph of Y2. Then move the cursor to the left of the minimum and press [ENTER], move to the right of the minimum and press [ENTER], and close to the minimum and press [ENTER].
The Y-value is the minimum perimeter.

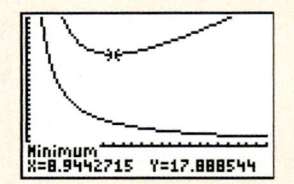

Press [▼].
This Y-value is the width.

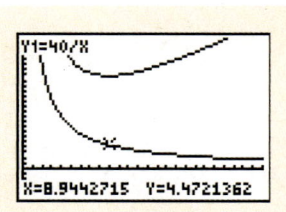

The minimum perimeter is about 17.9 m and occurs when the length along the house wall is about 8.9 m and the width is about 4.5 m.

Example 3

Materials
- scientific calculator

Enclosing Non-Rectangular Areas

A hobby farmer is creating a fenced exercise yard for her horses. She has 900 m of flexible fencing and wishes to maximize the area. She is going to fence a rectangular or a circular area. Determine which figure encloses the greater area.

Solution

Rectangular Area

The rectangle with perimeter 900 m and greatest area is a square.
Substitute $P = 900$ in the formula for the perimeter of a square.
$P = 4s$
$900 = 4s$ Solve for s.
$s = \frac{900}{4}$
$ = 225$
The side length of the square is 225 m.
Substitute $s = 225$ in the formula for the area of a square.
$A = s^2$
$ = 225^2$, or 50 625
The greatest rectangular area that can be enclosed is 50 625 m².

Circular Area

Substitute $P = 900$ in the formula for the circumference of a circle.
$C = 2\pi r$
$900 = 2\pi r$ Solve for r.
$r = 900 \div 2\pi$
$ \doteq 143.24$
The radius of the circle is about 143.2 m.
Substitute $r = 143.24$ in the formula for the area of a circle.
$A = \pi r^2$
$ \doteq \pi \times 143.24^2$
$ \doteq 64\ 458.25$
The greatest circular area that can be enclosed is about 64 458 m².

The circular pen encloses the greater area.

2.3 Optimizing Areas and Perimeters   **93**

Practice

A

For help with questions 1 to 3, see Example 1.

1. For each perimeter, what are the dimensions of the rectangle with the maximum area? What is the area?
   a) 40 cm
   b) 110 feet
   c) 25 m
   d) 87 inches

2. For each area, what are the dimensions of the rectangle with the minimum perimeter? What is the perimeter?
   a) 25 square feet
   b) 81 m²
   c) 144 cm²
   d) 169 square inches

3. For each area, what are the dimensions of the rectangle with the minimum perimeter? What is the perimeter? Round your answers to one decimal place.
   a) 30 square feet
   b) 65 m²
   c) 124 cm²
   d) 250 square inches

4. A gardener uses 24 m of fencing to enclose a rectangular vegetable garden. Some possible rectangles are shown. Determine the missing dimension for each diagram.
   a) 10 m
   b) 4 m
   c) 7.5 m

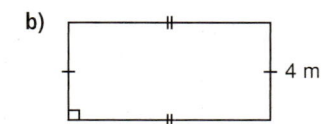

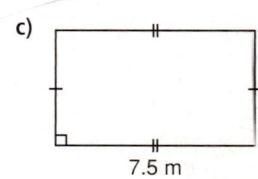

5. Calculate the area of each garden shown in question 4.

6. A farmer has 400 feet of fencing. Determine the greatest rectangular area that he can enclose with the fencing.

B

7. At an outdoor festival, 2-m sections of fencing are used to enclose an area for food sales. There are 100 sections of fencing available.
   a) How many metres of fencing are available altogether?
   b) Determine the maximum rectangular area that could be enclosed. How does the fact that the fencing is in sections affect your answer?

8. Lindsay has 20 pipe cleaners, each measuring 8 inches. She attaches them end to end to build a frame. Determine the greatest area that can be enclosed by the frame. Describe any assumptions you make.

9. A rectangular patio is to be constructed from 100 congruent square tiles.
   a) What arrangement of tiles would give the minimum perimeter?
   b) Suppose each tile has side length 50 cm. What would be the minimum perimeter? What would be the area of the patio?

94 CHAPTER 2: Geometry

10. A rectangular patio is to be constructed from 80 congruent square tiles.
   a) What arrangement of tiles would give the minimum perimeter? Justify your answer.
   b) Suppose each tile has side length 50 cm. What would be the minimum perimeter? What would be the area of the patio?

11. A farmer has 650 feet of fencing. Does she have enough fencing to enclose a rectangular area of half an acre? Justify your answer.

1 acre = 43 560 square feet

■ For help with question 12, see Example 2.

12. A rectangular section of a field is to be fenced. Because one side of the field is bordered by a creek, only 3 sides need to be fenced. The fenced section should have an area of 60 m². Determine the minimum perimeter and dimensions of the fenced area in each case:
   a) The dimensions must be whole numbers of metres.
   b) The dimensions can be decimal lengths.

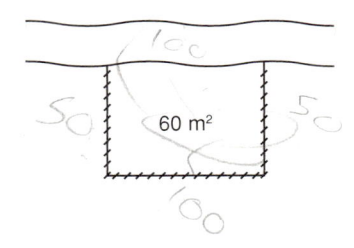

13. A lifeguard is roping off a rectangular swimming area using the beach as one side. She has 200 m of rope.
   a) Determine the greatest area she can rope off and its dimensions.
   b) Is the area in part a greater or less than 50 000 square feet? Justify your answer.

1 m ≐ 3.2808 feet

2.3 Optimizing Areas and Perimeters **95**

For help with question 14, see Example 3.

14. John buys 20 m of fencing to create a dog pen. How much more area will the dog have if John builds a circular pen rather than a square one?

15. Sasha is planning to create a garden with area 30 m². She could use a rectangular, triangular, or a circular design. Sasha decides to use the design that requires the least edging material. Which design should she use? How much edging will it require?

16. Assessment Focus The Tengs are adding a sunroom to their house. The perimeter of the sunroom will be 45 feet, not including the wall that is part of the house.
  a) One design is for a rectangular sunroom. Determine the maximum possible area of the room and the dimensions that give this area.
  b) Another design is in the shape of a semicircle, where the straight edge is attached to the house. Determine the diameter and area of the room.
  c) Which design has the greater area? How much greater is it?

17. Most of the heat loss for outdoor swimming pools is due to surface evaporation. So, the greater the area of the surface of the pool, the greater the heat loss. For a given perimeter, which surface shape would be more efficient at retaining heat: a circle or a rectangle? Justify your answer.

C

18. A farmer has 1800 m of fencing. He needs to create two congruent rectangular fields, as shown. Determine the maximum possible area of each field.

19. Twelve 2-m sections of metal fencing will be used to enclose an area. The area can have any shape, including triangle, hexagon, rectangle, and so on. The pieces do not bend, so they cannot form a circle. Determine the shape that maximizes the area. Justify your answer.

In Your Own Words

A friend has missed math class. Use an example to demonstrate that a square is the rectangle with the maximum area for a given perimeter and the minimum perimeter for a given area.

2.4 Optimizing Area and Perimeter Using a Spreadsheet

A contractor is designing a rectangular deck. One side of the deck will be against a house wall, but the other three sides will require a railing. The homeowners want the deck to have an area of 200 square feet. They also want to minimize the length of the railing.

Inquire Area and Perimeter Problems

Materials
- Microsoft Excel
- areaopt.xls
- peropt.xls

Part A: Maximum Area for a Given Perimeter

A park worker has 32 m of fencing to build a rectangular pen for rabbits. What is the maximum area that she can provide for the rabbits?

- Open the file *areaopt.xls*, or start a new spreadsheet file and enter the data and formulas shown here.

	A	B	C	D
1	Maximum Area of Pen with Perimeter 32 m			
2				
3	Length (m)	Width (m)	Area (m²)	Perimeter (m)
4	0.5	=16-A4	=A4*B4	=2*(A4+B4)
5	=A4+0.5			

1. a) What is the initial length of one side of the pen?

b) By what increment does the side length increase?

- Select cells A5 to A34. **Fill Down** to show lengths up to 15.5 m. Select cells B4 to D34. **Fill Down** to calculate the corresponding widths, areas, and perimeters.

2. Suppose the fencing comes in 0.5-m lengths. What are the dimensions of the rectangular pen with the maximum area? Justify your answer.

- Use the **Chart** feature to graph length and area.
   - From the **Insert** menu, select **Chart**.
   - In the **Chart type:** box, click on **XY(Scatter)**.
   - Select chart sub-type *Scatter with data points connected by smoothed lines.*

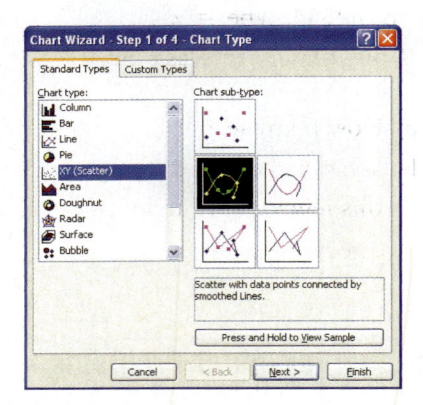

   - Click **Next >**, and select the **Series** tab.
     Click **Add**.
     Place the cursor in the **Name:** box, then click cell A1.
     Place the cursor in the **X Values** box, then select cells A4 to A34.
     Highlight ={1} in the **Y Values** box, then select cells C4 to C34.

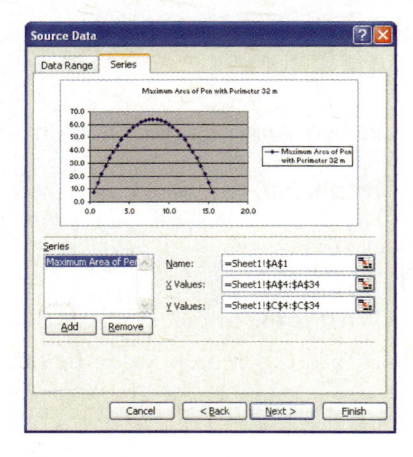

   - Click **Next >**, and enter appropriate titles for the axes.
     Click **Finish**.

CHAPTER 2: Geometry

3. a) Describe the shape of the graph.
   b) Suppose the fencing can be cut to decimal lengths. How can you determine the dimensions of the rectangular pen with the greatest area using the graph?
   c) How could you change the initial length and increment to zoom in on the region around the maximum?

■ Use the **Convert** function to express the dimensions in the spreadsheet in feet and square feet.
   - In cell E4, type: = CONVERT(A4,"m","ft")
   - In cell F4, type: = CONVERT (B4,"m","ft")
   - In cell G4, type: = E4*F4
   - In cell H4, type: = 2*(E4+F4)
   - Copy these formulas down to row 34.

4. a) How many feet of fencing does the park worker have?
   b) What is the area of the greatest rectangle she can enclose with this fencing? What are its dimensions?

Part B: Minimum Perimeter for a Given Area

A park worker is to build a rectangular pen for rabbits with an area of 24 m². What is the minimum length of fencing he needs for this project?

- Open the file *peropt.xls*, or start a new spreadsheet file and enter the data and formulas shown here.

	A	B	C	D
1	Minimum Perimeter of Pen with Area 24 m²			
2				
3	Length (m)	Width (m)	Area (m²)	Perimeter (m)
4	0.5	=24/A4	=A4*B4	=2*(A4+B4)
5	=A4+0.5			

5. a) What is the initial length of one side of the pen?
   b) By what increment does the side length increase?

- Select cells A5 to A99. **Fill Down** to show lengths up to 48.0 m. Select cells B4 to D99. **Fill Down** to calculate the corresponding widths, areas, and perimeters.

6. Suppose the fencing comes in 0.5-m lengths. What are the dimensions of the rectangular pen with the minimum perimeter? Justify your answer.

7. Use the **Chart** feature to graph length and perimeter.
   a) Describe the shape of the graph.
   b) Suppose the fencing can be cut to decimal lengths. How can you determine the dimensions of the rectangular pen with the least perimeter using the graph?
   c) How could you change the initial length and increment to zoom in on the region around the minimum?

8. Express the dimensions in the spreadsheet in feet and square feet.
   a) How many square feet must the park worker enclose?
   b) What is the minimum length of fencing he will need to enclose the rectangular area? What are the dimensions of the rectangle?

Practice

A 1. For each change to the situation in Part A of *Inquire*, describe the changes you would make to the spreadsheet.
   a) The park worker has 20 m of fencing.
   b) The fencing comes in 1-m lengths.
   c) The park worker has 38 m of fencing.
   d) The fencing comes in 0.25-m lengths.

B 2. Cy plans to create a flower garden using 0.25-m edging. He has 28 pieces of edging, and he wants to use all of it without any overlap.
   a) How many metres of edging does Cy have in total?
   b) By what increment should the side length increase? Justify your answer.
   c) Use *areaopt.xls* to determine the maximum area of the garden. What changes do you have to make to the spreadsheet?

3. You have 40 m of fencing to build a rectangular dog pen. Use *areaopt.xls* to determine the maximum area of the dog pen.

4. Asma is planning a vegetable garden with area 144 square feet. She will fence the garden to keep out rabbits, and she wants to minimize the length of fencing she must buy. Use *peropt.xls* to determine the dimensions of the rectangle with the minimum perimeter. What changes do you have to make to the spreadsheet?

5. A rectangular patio is to be constructed from 48 congruent square tiles.
   a) Use *peropt.xls* to determine the minimum perimeter of a patio built using all of the tiles.
   b) Explain why the patio with minimum area is not a square.

2.4 Optimizing Area and Perimeter Using a Spreadsheet **101**

6. Refer to the deck design problem in the lesson opener on page 97. Change *peropt.xls* to investigate this situation. If you were the contractor, what dimensions would you use for the deck? What changes do you have to make to the spreadsheet?

7. Jace has 12 m of rope to enclose a rectangular display. Use *areaopt.xls* to determine the maximum area in each situation. Justify the changes you make to the spreadsheet in each case.
   a) The rope is used for all four sides of the display.
   b) A wall is on one side, so the rope is only used for three sides.
   c) The display is in a corner of the room, so the rope is only used for two sides.

8. Tyresse uses rope to enclose a 400-m² rectangular swimming area on a lake. Use *peropt.xls* to determine the minimum length of rope needed in each situation. Justify the changes you make to the spreadsheet in each case.
   a) The rope is used for all four sides of the swimming area.
   b) One side is along the beach, so the rope is only used for three sides.
   c) One side is along the beach and an adjacent side is along a dock, so the rope is only used for two sides.

> **Reflect**
>
> - Why is a spreadsheet a good tool for solving perimeter and area optimization problems?
> - How did you decide what value to use for the length in cell A4 and how far down to copy the formulas in each spreadsheet?

102   CHAPTER 2: Geometry

2.5 Dynamic Investigations of Optimal Measurements

Dynamic training software can be used to help train pilots and drivers. It simulates what may happen in various situations. Dynamic geometry software can help you visualize what happens to objects as their dimensions change.

Inquire

Exploring Optimal Measures

Materials
- The Geometer's Sketchpad
- DynamicVolume Investigations.gsp

Work with a partner.

Open the file: *DynamicVolumeInvestigations.gsp*

Move to any page by clicking on the tabs in the lower left corner or move to the next page by clicking the **Link** button.

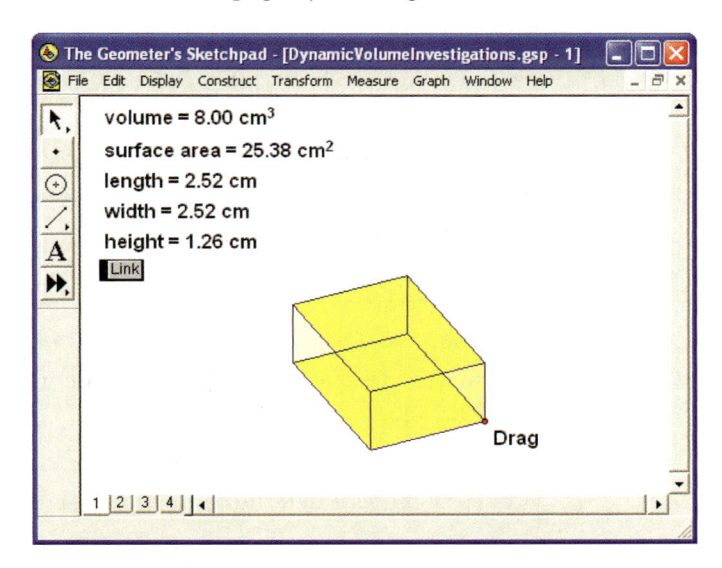

volume = 8.00 cm³
surface area = 25.38 cm²
length = 2.52 cm
width = 2.52 cm
height = 1.26 cm

2.5 Dynamic Investigations of Optimal Measurements   **103**

1. Move to page 1. Drag the labelled point and observe the changes in the object and the measured values. Answer these questions.
   a) What three-dimensional object is shown on the page? What measurements are given?
   b) As you drag the point, what measurements change? Which measurement remains the same?
   c) Make some predictions about the object with the least possible surface area or the greatest possible volume.
   d) Drag the point to create the object with the least possible surface area or the greatest possible volume. Record the optimal value of the surface area or volume and the dimensions of the object.

2. Move to page 2. Drag the labelled point and observe the changes in the object and the measured values. Repeat question 1.

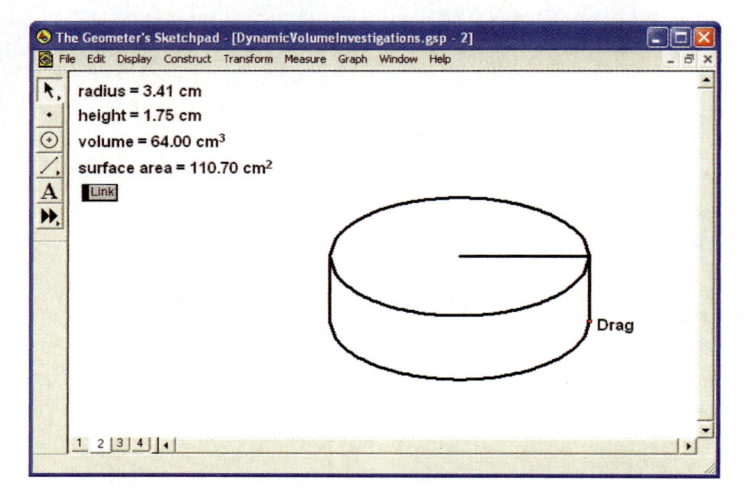

3. Move to page 3. Drag the labelled point and observe the changes in the object and the measured values. Repeat question 1.

4. Move to page 4. Drag the labelled point and observe the changes in the object and the measured values. Repeat question 1.

Reflect

- Describe a rectangular prism that is optimized for volume or surface area. How does this compare to your prediction?
- Describe a cylinder that is optimized for volume or surface area. How does this compare to your prediction?
- How did you determine the optimal measurements?

2.6 Optimizing Volume and Surface Area

A container's shape affects both the volume it can hold and the amount of material used to make it. To reduce waste and costs, a packaging designer may create a container that holds the desired volume using the least material. Other factors will also affect the design.

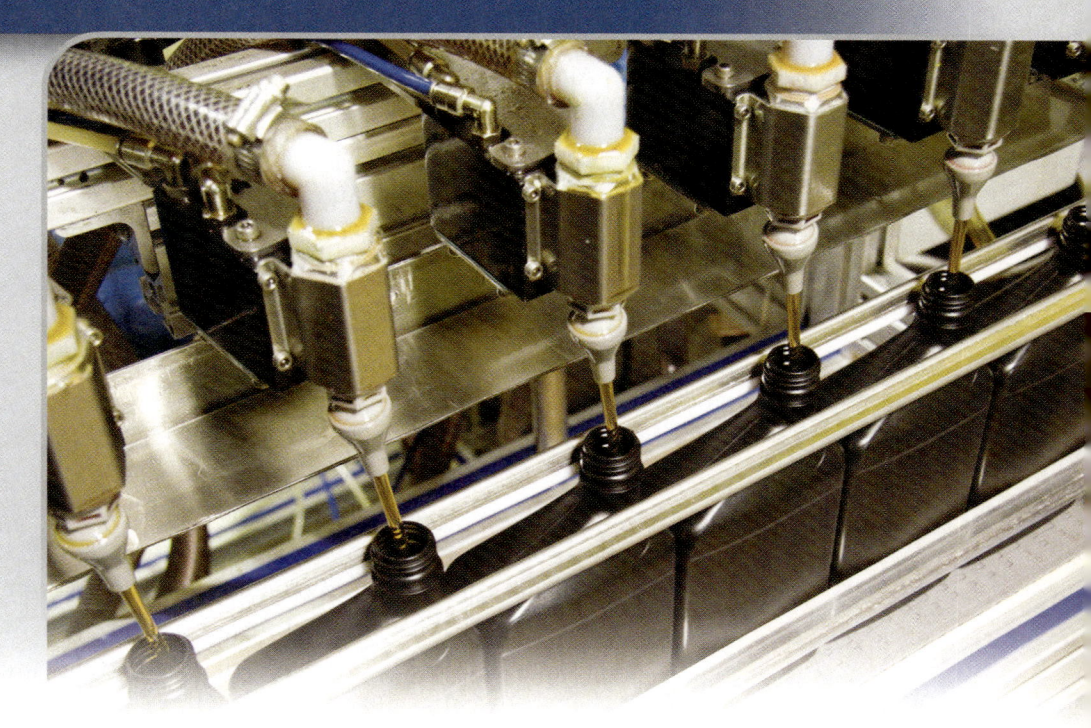

Investigate

Minimizing the Surface Area of a Rectangular Prism

Materials
- 1-cm grid paper or light cardboard
- scissors
- tape

Work with a partner.

Design and build three boxes with volume 1000 cm^3.
All the boxes must be rectangular prisms.
- Calculate the surface area of each box.
- Which box has the least surface area?
  What are its dimensions?

Reflect

- How did you determine the dimensions of each box?
- Compare boxes with classmates. What are the dimensions of the box with the least surface area? Describe the box.
- How does this compare with the two-dimensional problem of minimizing the perimeter of a rectangle with a given area?

Connect the Ideas

Optimizing the dimensions of a rectangular prism

Among all rectangular prisms with a given surface area, a cube has the maximum volume. Among all rectangular prisms with a given volume, a cube has the minimum surface area.

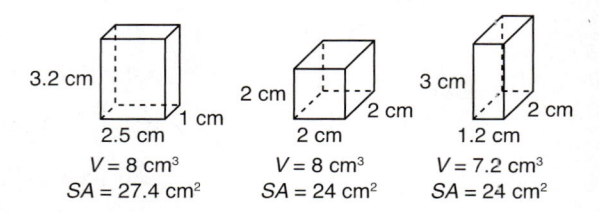

3.2 cm, 2.5 cm, 1 cm
$V = 8$ cm³
$SA = 27.4$ cm²

2 cm, 2 cm, 2 cm
$V = 8$ cm³
$SA = 24$ cm²

3 cm, 2 cm, 1.2 cm
$V = 7.2$ cm³
$SA = 24$ cm²

Example 1 — Optimizing Rectangular Prisms

Materials
- scientific calculator

a) Rosa constructs a rectangular prism using exactly 384 square inches of cardboard. It has the greatest volume possible. What are the dimensions of the prism? What is its volume?

b) Liam constructs a rectangular prism with a volume of exactly 1331 m³. It has the least surface area possible. What are the dimensions of the prism? What is its surface area?

Solution

For a cube with side length s:
$SA = 2s^2 + 2s^2 + 2s^2$
$= 6s^2$
$V = s \times s \times s$
$= s^3$

a) The prism with maximum volume is a cube. Determine the edge length, s, of a cube with surface area 384 square inches.

$SA = 6s^2$ Substitute: $SA = 384$
$384 = 6s^2$ Divide each side by 6 to isolate s^2.
$64 = s^2$ Take the square root of each side to isolate s.
$s = \sqrt{64}$, or 8
$V = s^3$ Substitute: $s = 8$
$= 8^3$, or 512

The rectangular prism is a cube with edge length 8 inches. Its volume is 512 cubic inches.

b) The prism with the least surface area is a cube. Determine the edge length, s, of a cube with volume 1331 m³.

$V = s^3$ Substitute: $V = 1331$
$1331 = s^3$ Take the cube root of each side to isolate s.
$s = \sqrt[3]{1331}$, or 11
$SA = 6s^2$ Substitute: $s = 11$
$= 6(11)^2$, or 726

To determine $\sqrt[3]{1331}$:
On a TI-30XII, press:
3 [2nd] [^] 1331 [ENTER]
On a TI-83 or TI-84, press: [MATH] 4 1331 [)] [ENTER]

The rectangular prism is a cube with edge length 11 m. Its surface area is 726 m².

Optimizing with constraints

There may be constraints on the prism you are optimizing:
- The dimensions may have to be whole numbers; or
- The dimensions may have to be multiples of a given number.

In these cases, it may not be possible to form a cube. The maximum volume or minimum surface area occurs when the dimensions are closest in value.

Sometimes one or more sides of the object are missing or bordered by a wall or other physical barrier. In these cases, the optimal rectangular prism will not be a cube. You can use diagrams or a table and graph to find the dimensions of the optimal rectangular prism.

Example 2

Optimizing with Constraints

Yael is designing a glass candle holder. It will be a rectangular prism with outer surface area 225 cm², a square base, and no top.

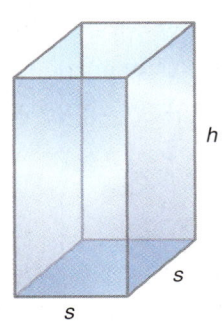

a) Determine the maximum volume of the candle holder.
b) What are the dimensions of the candle holder with the maximum volume?

Solution

a) The base of the candle holder has area: $A = s^2$

The volume of the candle holder is: $V = s^2 h$

The surface area of the candle holder is the sum of the areas of the faces: the base and four identical sides.

$SA = s^2 + 4(sh)$ Substitute: $SA = 225$
$225 = s^2 + 4sh$ Isolate h.
$225 - s^2 = 4sh$
$\frac{225 - s^2}{4s} = h$

Substitute values for s and determine the corresponding lengths, heights, and volumes.

2.6 Optimizing Volume and Surface Area   **107**

> If technology is available, use a spreadsheet or graphing calculator to create the table and graph.

Base side length (cm)	Height of prism (cm)	Volume (cm³)	Surface area (cm²)
1	56	56	225
2	27.625	110.5	225
3	18	162	225
4	13.0625	209	225
5	10	250	225
6	7.875	283.5	225
7	6.285714	308	225
8	5.03125	322	225
9	4	324	225
10	3.125	312.5	225
11	2.363636	286	225
12	1.6875	243	225
13	1.076923	182	225
14	0.517857	101.5	225

It looks like the maximum volume of the candle holder occurs when the side length of the base is about 8.6 cm and the volume is about 325 cm³. Create a table with smaller increments to verify this prediction.

Base side length (cm)	Height of prism (cm)	Volume (cm³)	Surface area (cm²)
8.5	4.492647	324.5938	225
8.6	4.390698	324.736	225
8.7	4.290517	324.7493	225
8.8	4.192045	324.632	225

The maximum volume of the candle holder occurs when the base side length is about 8.7 cm. The volume is about 325 cm³.

b) From the table, the candle holder with maximum volume has base side length about 8.7 cm and height about 4.3 cm.

108 CHAPTER 2: Geometry

Example 3

Materials
- TI-83 or TI-84 graphing calculator

1 mL = 1 cm³

Optimizing Other Objects

Naveed is designing a can with volume 350 mL. What is the minimum surface area of the can? Determine the dimensions of a can with the minimum surface area.

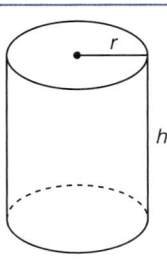

Solution

Since 1 mL = 1 cm³, the volume will be 350 cm³.

The surface area of the can is: $SA = 2\pi r^2 + 2\pi rh$

The volume of the can is:

$V = \pi r^2 h$ Divide both sides by πr^2 to isolate h.

$\dfrac{V}{\pi r^2} = h$ Substitute: $V = 350$

$\dfrac{350}{\pi r^2} = h$

Use a graphing calculator to create a graph of radius and surface area.

Press [Y=].
- Use X to represent the radius.
- Use Y1 to represent the height.
  Press: 350 [÷] [(] [2nd] [^] [×] [X,T,Θ,n] [^] 2 [)] [ENTER]
- Use Y2 to represent the surface area. Press: 2 [2nd] [^] [×] [X,T,Θ,n] [^] 2 [+] 2 [2nd] [^] [×] [X,T,Θ,n] [×] [VARS] [▶] 1 1 [ENTER]

```
Plot1 Plot2 Plot3
\Y1■350/(π*X^2)
\Y2■2π*X^2+2π*X*
Y1
\Y3=
\Y4=
\Y5=
\Y6=
```

Press [WINDOW].
Change the window settings as shown.

```
WINDOW
 Xmin=0
 Xmax=6
 Xscl=1
 Ymin=-50
 Ymax=350
 Yscl=50
 Xres=1
```

2.6 Optimizing Volume and Surface Area   **109**

Press [2nd] [GRAPH] to view the data in a table. To change the starting value of the table, or the increment, press [2nd] [WINDOW] and adjust TblStart and ΔTbl.

Press [GRAPH].
Press [2nd] [TRACE] 3, move the cursor onto the graph of Y2, and answer the prompts to determine the minimum surface area of the can.

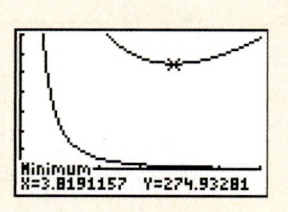

Press [▼]. Read the coordinates to determine the dimensions of the can with the minimum surface area.

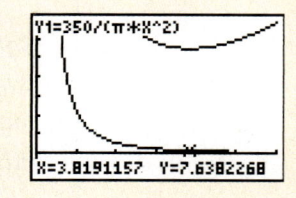

The minimum surface area of the can is about 275 cm². It occurs when the radius is about 3.8 cm and the height is about 7.6 cm.

Practice

A

For help with questions 1 or 2, see Example 1.

1. Yasmin is constructing a rectangular prism using exactly 96 cm² of cardboard. The prism will have the greatest possible volume.
   a) Describe the prism. What will be its dimensions?
   b) What will be its volume?

2. Mathew is constructing a rectangular prism with volume exactly 729 cubic inches. It will have the least possible surface area.
   a) Describe the prism. What will be its dimensions?
   b) What will be its surface area?

3. The dimensions of two rectangular prisms with volume 240 cm³ are given. Sketch each prism and predict which will have less surface area. Check your prediction.
   a) 10 cm by 6 cm by 4 cm
   b) 12 cm by 10 cm by 2 cm

4. Krikor has to design and build a box with the greatest volume possible. The box is a rectangular prism. For each surface area, what will be the dimensions of the box?
   a) 600 square inches
   b) 1350 cm²
   c) 2400 square inches

110 CHAPTER 2: Geometry

5. Tanya is designing a storage box. It will be a rectangular prism with the least possible surface area. For each volume, what will be the dimensions of the box?
   a) 1 m³
   b) 125 000 cm³
   c) 8 cubic feet

6. Jude is designing a plush activity toy for a baby. The toy will be a rectangular prism with surface area 864 cm².
   a) Determine the maximum volume of the toy.
   b) What are the dimensions of the toy with maximum volume?

7. Camping supply stores sell collapsible plastic containers for storing water. The containers are often rectangular prisms with rounded corners. Reducing the amount of plastic helps the container fold as small as possible.
   a) Convert the capacity of a 5-gallon container to cubic inches.
   b) Determine the minimum surface area of a container holding 5 gallons of water. What would be the dimensions of the container?

1 gallon ≐ 277.42 cubic inches

8. An electrical transformer box is a rectangular prism constructed from sheet metal. It must have volume at least 274 625 cm³ to hold all the necessary equipment.
   a) What dimensions for the box require the least area of sheet metal?
   b) What area of sheet metal is needed to build the box?
   c) Tony has 20 square feet of sheet metal. Will this be enough to construct the box? Justify your answer.

1 foot = 30.48 cm

For help with question 9, see Example 2.

9. Tori is designing a hanging shelf. It has volume 400 cubic inches and depth 4 inches. She will paint a design that will cover the four outside faces.
   a) Determine the minimum area she will paint.
   b) What are the dimensions of the shelf with the minimum area to paint?

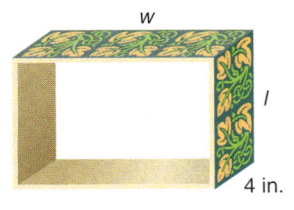

10. A company packages sugar cubes in cardboard boxes containing 144 cubes. The cubes are arranged in 2 layers, with 12 rows of 6 cubes in each layer. The company wishes to design a box that uses less packaging, but holds the same number of cubes. How could you arrange the cubes so the least amount of cardboard is used?

2.6 Optimizing Volume and Surface Area   **111**

11. **Literacy in Math** A company considers packaging cereal in cubic boxes.
    a) What advantages does a cube have over the tall, narrow boxes currently in use?
    b) What problems might this new shape create for each of these people?
        i) A consumer
        ii) A grocery store owner
        iii) A designer planning the package labelling

■ For help with question 12, see Example 3.

12. Filip is designing a can for a new vegetable product. The can should hold 750 mL of vegetables. To reduce waste, he wants the surface area of the can to be as small as possible.
    a) What dimensions should Filip use?
    b) What will the surface area be?

13. A cylindrical storage tank holds 1800 cubic feet of gasoline. Determine the minimum amount of material needed to build this tank.

14. Look at the dimensions of the optimal cylinders in *Example 3* and questions 12 and 13. How do the diameter and height appear to be related?

1 mL = 1 cm³

15. **Assessment Focus** A beverage company is investigating containers that can hold 512 mL of juice. They are debating whether to use a rectangular prism or a cylinder. Which object would require less material? Justify your answer.

16. Courtney is designing a gift box. It will be a triangular prism with surface area 220 cm². She decides the box should have a right isosceles triangular base to make it easy to package.
    a) Determine the length of the hypotenuse of the base, s.
    b) Determine the maximum possible volume of the box.
    c) What are the dimensions of the box with the maximum volume?

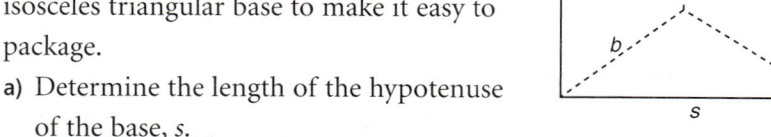

17. Kimmia and Jan thought triangular prisms might be a good shape for juice boxes. They learned from research that to minimize surface area, the base should be an equilateral triangle. They created this spreadsheet to explore the dimensions of a prism that could hold 250 mL of juice.

	A	B	C	D	E	F
1	Minimum Surface Area of a Triangular Prism with Volume 250 cm³					
2	Base: Equilateral triangle					
3	Side length of base (cm)	Base and top area (cm²)	Area of sides (cm²)	Height of prism (cm)	Volume (cm³)	Surface area (cm²)
4	9.4	76.5	184.3	6.5	250.0	260.8
5	9.6	79.8	180.4	6.3	250.0	260.2
6	9.8	83.2	176.7	6.0	250.0	259.9
7	10.0	86.6	173.2	5.8	250.0	259.8
8	10.2	90.1	169.8	5.5	250.0	259.9
9	10.4	93.7	166.5	5.3	250.0	260.2

   a) The students used this formula to determine the base area of the prism.
   Base area = $\frac{\sqrt{3}s^2}{4}$
   Use trigonometry to verify that the formula calculates the area of an equilateral triangle with side length s.

   b) What are the dimensions of the triangular prism with the least surface area? How do you know?

   c) What is the surface area of the prism in part b?

   d) How do the side length and height of the prism seem to be related?

18. A pop manufacturer creates a can with volume 355 cm³. Twelve cans are then packaged in cardboard boxes for distribution.
    a) Determine the dimensions of a can with minimum surface area.
    b) Determine the minimum amount of cardboard that can be used for each case of pop.

19. A manufacturer is designing a new shipping container for powdered chemicals. The container could be a rectangular prism, triangular prism, or cylinder. The surface area for each design is to be 1 square yard.
    a) Determine the dimensions of each container with the maximum volume.
    b) Which container would be able to hold the most material?
    c) Which container would you recommend? Justify your answer. Remember to think about other factors such as ease of use and ease of manufacturing.

1 yard = 36 inches

The triangular prism with the greatest volume will have an equilateral triangle as the base.

In Your Own Words

Describe a situation when it would be important to design an object with a minimal surface area for its volume. Explain whether you would recommend a cylinder or rectangular prism in this situation.

PUZZLE

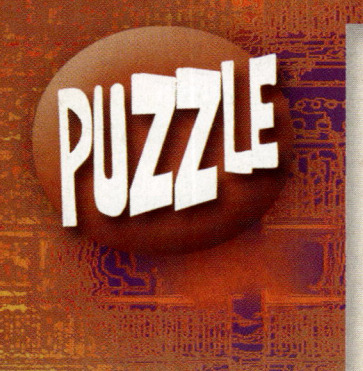

Cube Creations

Materials
- 27 linking cubes

Use linking cubes to create these 7 objects:

Connect all 7 objects to form a larger composite object.

Assume each cube has side length 1 unit.

Calculate the volume and surface area of the composite object.

You may find it helpful to create each object using a different colour.

- Is it possible to create an object with lesser volume?
- Is it possible to create an object with lesser surface area?
- Is it possible to connect these objects to form a rectangular prism? Repeat the activity to find out.

Reflect

- What strategies did you use to create an object with lesser surface area?
- Describe the composite object with volume 27 cubic units and the least surface area. Were you able to create this object?

CHAPTER 2: Geometry

2.7 Optimizing Surface Area Using a Spreadsheet

Two objects are constructed from the same material and have the same volume. The object with the greater surface area will lose heat more quickly. Engineers often try to minimize surface area when designing objects where heat loss is a problem.

Inquire

Comparing Prisms and Cylinders

Materials
- *Microsoft Excel*
- *saopt.xls*

Work with a partner.

A container is being designed with a volume of 500 cm^3.
What are the dimensions of the container with the minimum surface area?
The container could be a rectangular prism, cylinder, or triangular prism.

- Open the file *saopt.xls*. There are three sheets in the file. Move to any sheet by clicking on the tabs in the lower left corner.

1. Move to sheet **Rect. Prism**.
   a) What is the initial side length of the base of the prism?
   b) By what increment does the side length increase?
   c) Verify the formulas for calculating the height, volume, and surface area of the prism in cells B4, C4, and D4.

	A	B	C	D
1	Minimum Surface Area of a Rectangular Prism with Volume 500 cm^3			
2	Base: Square			
3	Side length of base (cm)	Height of prism (cm)	Volume (cm^3)	Surface area (cm^2)
4	1	=500/(A4^2)	=A4^2*B4	=2*A4^2+4*B4*A4
5	=A4+1			

- Select cells A5 to A23.
  Fill Down to show side lengths up to 20 cm.
  Select cells B4 to D23.
  Fill Down to calculate the corresponding heights, volumes, and surface areas.

2. **a)** What are the approximate dimensions of the rectangular prism with the minimum surface area?
   b) What is the approximate surface area?

- For more precise dimensions, adjust the initial value and change the formula in cell A5 to $= A4 + 0.1$. Copy this formula down through row 23.

> You will probably want to use an initial value slightly less than the side length you calculated in question 2, part a.

3. **a)** What are the approximate dimensions of the rectangular prism with the minimum surface area?
   b) What is the approximate surface area?

- Use the **Chart** feature to create a scatter plot of length and surface area. Connect the points with a smooth curve.

> Refer to page 98 for more instructions on using the **Chart** feature.

4. **a)** Describe the shape of the graph.
   b) How does the graph show which edge length will produce a rectangular prism with the least surface area?

5. Move to sheet **Cylinder**.
   a) What is the initial radius of the cylinder?
   b) By what increment does the radius increase?
   c) Verify the formulas for calculating the height, volume, and surface area of the cylinder in cells B4, C4, and D4.

	A	B	C	D
1	\multicolumn{4}{c}{Minimum Surface Area of a Cylinder with Volume 500 cm³}
2	**Base: Circle**			
3	**Radius (cm)**	**Height of cylinder (cm)**	**Volume (cm³)**	**Surface area (cm²)**
4	1	=500/(PI()*A4^2)	=PI()*A4^2*B4	=2*PI()*A4^2+(PI()*2*A4*B4)
5	=A4+1			

116 CHAPTER 2: Geometry

- Select cells A5 to A23.

    Fill Down to show radii up to 20 cm.

    Select cells B4 to D23.

    Fill Down to calculate the corresponding heights, volumes, and surface areas.

6. Repeat questions 2, 3, and 4 for the cylinder.

7. Move to sheet **Tri. Prism**.
    a) What is the initial side length of the base of the prism?
    b) By what increment does the side length increase?
    c) Verify the formulas for calculating the base area, height, volume, and surface area of the prism in cells B4, C4, D4, and E4.

	A	B	C	D	E
1	Minimum Surface Area of a Triangular Prism with Volume 500 cm³				
2	Base: Equilateral triangle				
3	Side length of base (cm)	Area of base (cm)	Height of prism (cm)	Volume (cm³)	Surface area (cm²)
4	3	=A4^2*SQRT(3)/4	=500/B4	=B4*C4	=2*B4+3*A4*C4
5	=A4+1				

- Select cells A5 to A21.

    Fill Down to show side lengths up to 20 cm.

    Select cells B4 to D21.

    Fill Down to calculate the corresponding base areas, heights, volumes, and surface areas.

8. Repeat questions 2, 3, and 4 for the triangular prism.

9. Compare your answers to questions 3, 6, and 8.

    Which of these objects can enclose 500 cm³ using the least surface area?

    Which requires the most surface area?

10. A container is being designed with volume 750 cm³. The container could be a rectangular prism, cylinder, or triangular prism. Change the spreadsheet to determine the dimensions for a container of each shape with the minimum surface area.

11. Refer to the dimensions of the optimal containers you calculated in questions 3, 6, and 8.
   a) How do the length, width, and height of a rectangular prism with the minimum surface area appear to be related?
   b) How do the diameter and height of a cylinder with the minimum surface area appear to be related?
   c) How do the side length of the base and the height of a triangular prism with the minimum surface area appear to be related?

12. Minimizing heat loss is an important factor in the design of water heaters.
   a) Modify the spreadsheet to determine the least surface area possible for a 50-gallon water heater tank if it is a rectangular prism, a cylinder, or a triangular prism.
   b) Which object would you recommend for a water heater tank?
   c) A typical water heater tank is a cylinder with height almost three times the diameter. How does this compare to the object you recommended?
   Why might the taller, narrower tank be preferred?

1 gallon ≐ 4546 cm^3

Reflect

- What are the advantages of using a spreadsheet to optimize surface areas?
- Suppose you are designing a container to hold a given volume. The container will be made from sheets of an expensive metal. Would you design a rectangular prism, cylinder, or triangular prism? Justify your answer.

Study Guide

Composite Figures
- A composite figure is made up of simpler figures.
- To find the area of a composite figure, calculate the areas of the simpler figures. Add the areas. Subtract the areas of any figures that were removed.
- The perimeter of a figure is the distance around it.

This figure is a rectangle with an isosceles triangle removed from the right side. To determine its area, subtract the area of the triangle from the area of the rectangle.

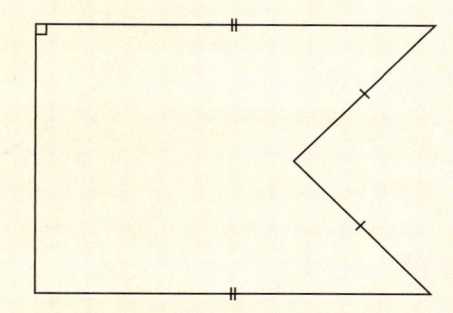

Composite Objects
- To calculate the volume of a prism with a base that is a composite figure, use the formula: $V = $ base area $\times$ height
- To calculate the volume of a composite object, break the composite object into simpler objects. Calculate the volumes of the simpler objects. Add the volumes. Subtract the volumes of any objects that were removed.
- The surface area of an object is the sum of the areas of the faces.

This is a rectangular prism with a triangular prism on top. To determine its volume, calculate the volume of each smaller prism and add them. Or, calculate the area of the 5-sided base and multiply by the length.

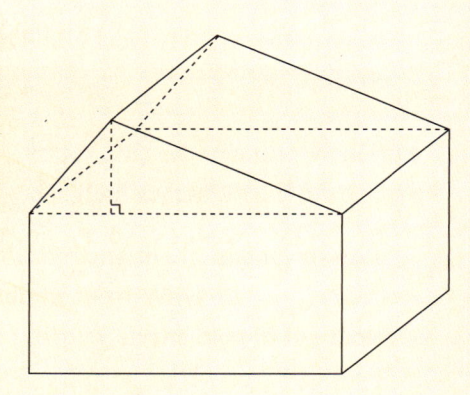

Optimizing
- To optimize area or volume, consider the greatest area or volume that can be enclosed by a given amount of material.
- To optimize perimeter or surface area, consider the least amount of material required to enclose a given area or volume.
- When there are no constraints, the optimal rectangle is a square and the optimal rectangular prism is a cube.

Square
$A = s^2$
$P = 4s$

Cube
$V = s^3$
$SA = 6s^2$

Study Guide

Chapter Review

2.1

1. Determine the area of each composite figure. The curve is a semicircle.

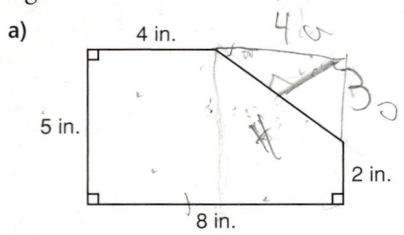

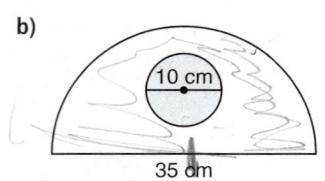

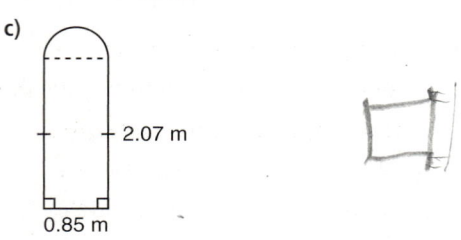

   a) [figure: trapezoid with dimensions 4 in. (top), 5 in. (left), 2 in. (right), 8 in. (bottom)]

   b) [figure: semicircle with 35 dm diameter containing a circle of 10 cm]

   c) [figure: shape with semicircle top, 2.07 m height, 0.85 m width]

2. The window in this door is 25.5 inches wide and 35.0 inches high. The hole for the lock is 2.0 inches in diameter.

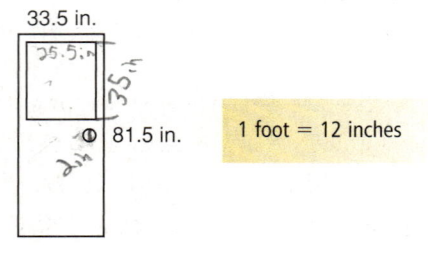

   [figure: door 33.5 in. wide, 81.5 in. tall, with window and lock hole]

   1 foot = 12 inches

   a) Determine the area of the wooden part of the door, in square inches.
   b) Salim is painting the front and back of four of these doors. He will apply two coats of paint. Each can of paint will cover 125 square feet. How many cans should he buy? Justify your answer.

3. The owner of a small art gallery is framing a painting for a client. The painting measures $15\frac{1}{2}$ inches by $15\frac{1}{2}$ inches. She leaves a border around the painting that is $2\frac{1}{2}$ inches wide on each side and $3\frac{1}{2}$ inches wide on the top and bottom.

   a) Sketch the situation. What will be the length and width of the painting and border together?
   b) Determine the area of glass that will be needed to cover the painting and border.

2.2

4. A cylindrical tank has radius 4.5 m and height 6.1 m.

   a) Determine the volume of liquid it can hold, in litres. **1 m³ = 1000 L**
   b) Jamal is painting the tank with rust paint. Determine the surface area he must paint, in square metres. Explain any assumptions you make.
   c) Suppose each can of rust paint will cover 650 square feet. How many cans should Jamal buy? Justify your answer. **1 m ≐ 3.2808 feet**

5. Shipping containers are used to transport goods across the sea, over the rail system, or by transport truck. The outer dimensions of a container are 40 feet by 8 feet by 8 feet 6 inches. The walls are 2 feet thick on all sides.

   a) The outside of a shipping container requires painting. Determine the surface area that needs to be painted.
   b) Determine the maximum volume of storage space.

CHAPTER 2: Geometry

6. Concrete parking curbs are often used in parking lots.
   a) Determine the volume of this curb. 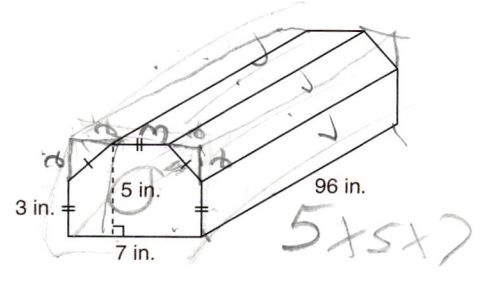

   b) Suppose Leo has to make 100 curbs like this. How many cubic yards of concrete will he need?

   > 1 yard = 36 inches

   > 1 foot = 12 inches

   c) Curbs are sometimes painted to make them more visible to drivers. Determine the surface area to be painted on one curb. Explain which face you would omit in your calculation.
   d) A can of paint covers 600 square feet of concrete. How many cans would Leo need to paint 100 curbs? Justify your answer.

7. Nasmin will construct this object from sheet metal, then coat it with enamel.
   a) Determine the volume of the object.
   b) Determine the surface area to be covered with enamel. All exposed faces are to be covered.

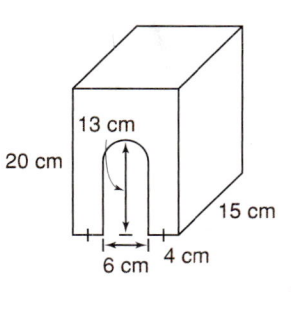

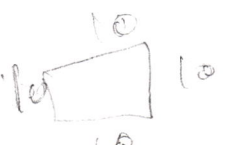

8. Amrit builds this storage unit under the stairs. The unit has depth 3 feet. He will paint the front of the unit.

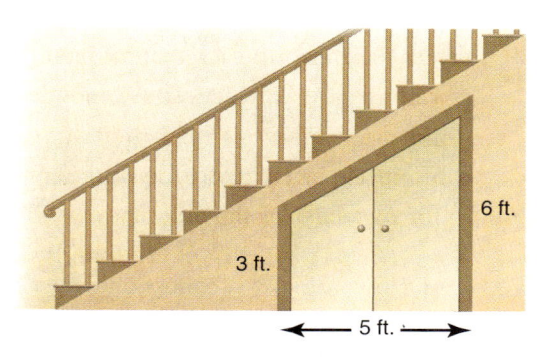

   a) What is the volume of air inside the storage unit?
   b) What area will Amrit paint?

9. For each perimeter, what are the dimensions of the rectangle with the maximum area? What is the area?
   a) 40 cm
   b) 110 feet
   c) 25 m
   d) 87 inches

10. For each area, what are the dimensions of the rectangle with the minimum perimeter? What is the perimeter?
    a) 25 square feet
    b) 81 m^2
    c) 144 cm^2
    d) 169 square inches

11. A car dealership fences in a rectangular area behind their building to secure unsold vehicles. One length will be the back wall of the dealership. What is the maximum parking area that can be created if they have 2 km of fencing to use?

Chapter Review   **121**

12. Jim is setting up a rectangular dog run in his backyard. He buys six 3-foot sections of fencing and a 3-foot wide gate. What are the dimensions of the dog run with the greatest area in each situation?
   a) Jim uses the yard fence for one side of the dog run.
   b) Jim uses the corner of the yard fence for two sides of the dog run.

13. A marine biologist is collecting data. She has 100 m of rope with buoys to outline a rectangular or circular research area on the surface of the water.
   a) Which figure will enclose a greater area?
   b) How much extra area will be enclosed by using the more efficient shape?

14. A cube is the rectangular prism with the least surface area for a given volume. What do you think a rectangular prism with the greatest surface area for a given volume would look like? Use the file *DynamicVolumeInvestigations.gsp* to explore your predictions.

15. Yasmin is constructing a rectangular prism with volume exactly 216 m³. It will have the least possible surface area.
   a) Describe the prism. What will be its dimensions?
   b) What will be its surface area?

16. Linda is a potter. She has a slab of clay with area 150 cm². She will make a ceramic box from this slab by cutting out and joining six rectangular faces. Linda wants the volume of the box to be as great as possible.
   a) Describe the prism. What will be its dimensions?
   b) What will its volume be?

17. Jake designs personalized self-adhesive notes. Each note is rectangular. The notes come in blocks that are rectangular prisms.
   a) Determine the minimum surface area of a block with volume 8 cubic inches.
   b) Determine the dimensions of each note.
   c) Each note is 0.13 mm thick. Determine the number of notes in a block with minimum surface area.

18. Giulia is designing glass storage jars with surface area 675 cm². The jars are cylinders. She wants to maximize the volume of each jar to save on the materials cost.
   a) What dimensions should Giulia use?
   b) What will the volume be?

19. Armin is designing a bottle for a new shampoo. The bottle should have a volume of 450 cm³. Armin wants to minimize the amount of plastic used in the bottle. Determine the dimensions of the bottle with the minimum surface area in each case. Explain your method.
   a) The bottle is a cylinder.
   b) The bottle is a triangular prism with a base that is an equilateral triangle.

122 CHAPTER 2: Geometry

Practice Test

Multiple Choice: Choose the correct answer for questions 1 and 2. Justify each choice.

1. The approximate area of this composite figure is: 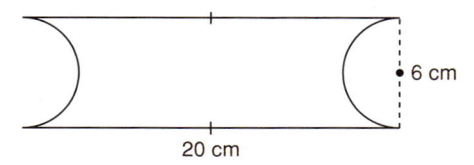
   - A. 63 cm²
   - B. 92 cm²
   - C. 120 cm²
   - D. 148 cm²

2. These objects have the same volume. Which one has the least surface area?
   - A. 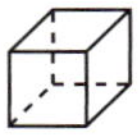
   - B. 
   - C. 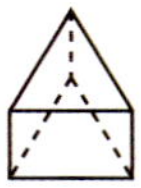
   - D.

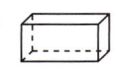

Show your work for questions 3 to 6.

3. **Knowledge and Understanding** Determine the volume and surface area of each three-dimensional object.
   a) 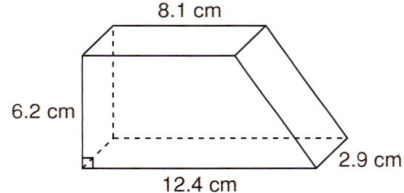 8.1 cm, 6.2 cm, 2.9 cm, 12.4 cm
   b) 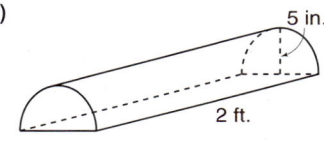 5 in., 2 ft.

4. **Communication** A rectangular patio is being built against the side of a house using 60 congruent square tiles. Determine the arrangement of tiles that requires the least amount of edging. Justify your answer, including a description of any constraints that affected your solution strategy.

5. **Application** A flowerbed is a rectangle with a semicircle at each end. The total length of the flowerbed is 50 feet and the width is 14 feet.
   a) Sketch the flowerbed and determine its area.
   b) A gardener can plant 10 plants per square yard. How many plants should he buy for this flowerbed?
   c) The fertilizer he will apply should be spread at a rate of 2 kg per 30 m². About how much fertilizer will he need for this garden?

   1 m ≐ 3.2808 feet

6. **Thinking** A dairy sells ice cream in 11.4-L cardboard containers.
   a) If the dairy uses a rectangular container optimized for surface area, what would the dimensions and surface area be?
   b) If the dairy uses a cylindrical container optimized for surface area, what would the dimensions and surface area be?
   c) How much cardboard is saved by using the more efficient container?

   1 L = 1000 cm³

Chapter Problem: A Winning Design

A company is developing a new power drink.

Imagine you are trying to win the design contract for the product. You must create a name for the product and design a container, label, and shipping carton. Here are the guidelines you must follow:

- The name of the drink should appeal to people in your age group and suggest that the drink boosts energy and tastes good.

- The container of an individual drink can be made from aluminum, plastic, or glass, or it can be a Tetra Pak™. To minimize material costs, the container must use no more than 80 square inches of material. Describe the constraints and explain how you chose the shape and dimensions of the container and the type of packaging material.

- The label must display the product name and capacity, and have room for the ingredients, UPC code, and a 2-cm by 5-cm rectangular area for the nutrition facts table.

- The shipping carton must be a cardboard case that holds between 18 and 30 individual containers. Calculate the amount of cardboard used per case and the total amount required to package 720 containers. Justify your choices.

The winner will be chosen from those designs that provide optimal packaging solutions in interesting and appealing ways.

3 Two-Variable Data

What You'll Learn

To distinguish types of data, and to analyse and represent two-variable data from primary and secondary sources

And Why

Analysing data to look for relationships is part of many college courses and professions. Fishery and forestry managers, sports trainers, medical researchers, and lab technicians all work with two-variable data.

Key Words

- variable
- one-variable data
- two-variable data
- scatter plot
- dependent variable
- independent variable
- correlation
- cause-and-effect relationship
- line of best fit
- outlier
- interpolation
- extrapolation
- non-linear data
- linear correlation
- correlation coefficient
- conjecture

CHAPTER 3

Activate Prior Knowledge

Interpreting Data Graphs

Prior Knowledge for 3.1

Data can be presented in a variety of ways. Graphical representations can include bar graphs, histograms, scatter plots, and circle graphs.

Example

a) What do the red bars in this graph represent? The blue bars?
b) For which program did the tuition change the most?
c) Estimate the tuition for photography in 2007–2008; in 2008–2009.

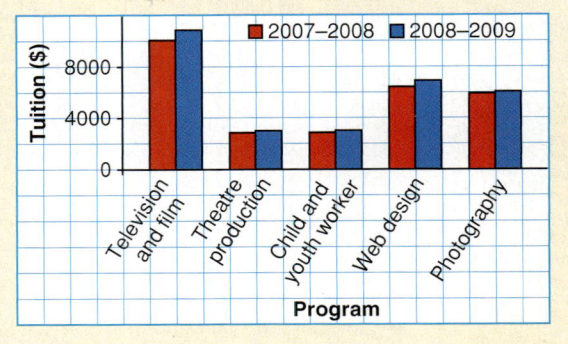

Solution

a) The red bars represent the tuition for programs at an Ontario community college in 2007–2008. The blue bars represent the tuition for the same programs in 2008–2009.
b) Look at the graph for the pair of bars with the greatest difference in height: television and film is the program with the greatest increase in tuition.
c) The tuition for photography was about $5900 in 2007–2008; about $6100 in 2008–2009.

CHECK ✓

1. The histogram shows the training heart rates, in beats per minute (bpm), of 16 runners.
   a) How many runners had heart rates between 140 and 149 bpm?
   b) What interval was the most common?
   c) What were the minimum and maximum heart rates any of these runners could have had?

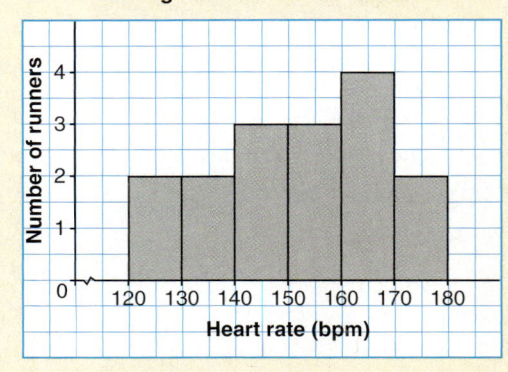

126   CHAPTER 3: Two-Variable Data

2. Miguel surveyed 250 high school students about how they usually get to school. He displayed the data in this circle graph.
   a) What does the light blue area represent? Is it more or less than half the graph?
   b) What conclusions might you make from this graph?

Usual Transportation for High School Students

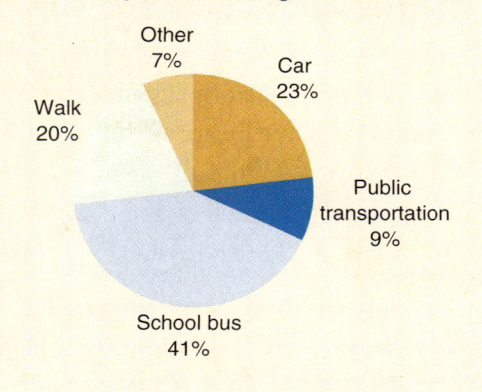

- Other 7%
- Car 23%
- Public transportation 9%
- School bus 41%
- Walk 20%

3. This scatter plot compares the mass and price of a selection of laptops.

   Price and Mass of Laptop Computers

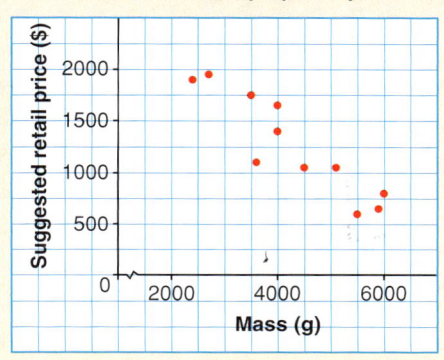

   a) What does each point show?
   b) How many of the laptops have mass less than 4000 g? How many of these cost less than $1250?
   c) How many of the laptops have mass greater than 4000 g? How many of these cost less than $1250?

4. One hundred students were asked to identify their leisure activities.

Activity	Number of students
Reading	12
Playing sports	32
Watching TV	83
Visiting friends or family	55
Hobbies	41

   a) What is the sum of the numbers in the second column of the table?
   b) Why would a circle graph not be a good tool for displaying these data?
   c) What type of graph would be good for these data? Why do you think so?

Activate Prior Knowledge   **127**

Working with Slope and Line Graphs

Prior Knowledge for 3.3

Slope is a measure of the steepness of a line.
- A positive slope means the line goes up to the right.
- A negative slope means the line goes down to the right.

To determine the slope of a line, we need the coordinates of two points on the line.

A line with slope m and y-intercept b can be represented by the equation $y = mx + b$.

Example

a) Determine the slope of the line shown.
b) Write the equation of the line in the form $y = mx + b$.

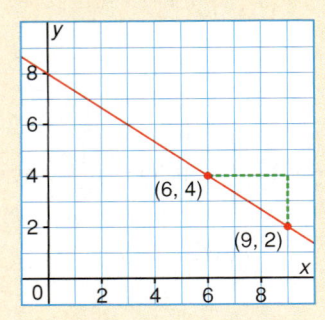

Solution

a) Choose two points on the line: (6, 4) and (9, 2)
Slope: $\frac{\text{rise}}{\text{run}} = \frac{2-4}{9-6}$, or $-\frac{2}{3}$

b) From the graph, the y-intercept is 8. The slope is $-\frac{2}{3}$.
So, the equation of the line is $y = -\frac{2}{3}x + 8$.

CHECK ✓

1. Determine the slope of each line. Write the equation of each line in the form $y = mx + b$.

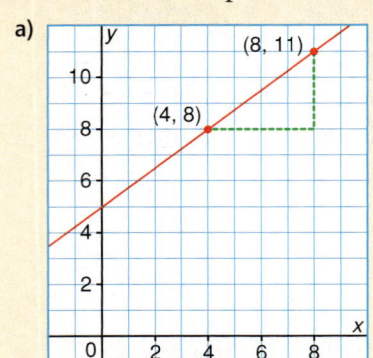

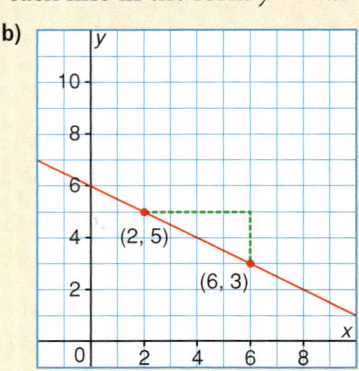

a) [graph showing points (4, 8) and (8, 11)]

b) [graph showing points (2, 5) and (6, 3)]

128 CHAPTER 3: Two-Variable Data

Transitions

Getting Extra Practice

People in diverse fields have shown that regular practice is a key part of their success. Consider the practice associated with:

- sports
- music
- acting
- firefighting
- computer applications
- aviation

Practice helps you learn to carry out procedures efficiently and accurately.

Strategies for Success

- Practise often, every day if possible.
- Push yourself beyond your current level of competence.
- Focus as you practise.
- Reflect on what you are doing well, and where you want to improve.

To practise the concepts and skills presented in your math class:

- Try a variety of questions to deepen your understanding.
- Once you feel confident doing questions at one level, try some at the next level.
- Create new problems, then solve them or ask classmates to solve them.
- Spend some time practising with other people. Compare answers to discover new approaches.

To get ready for a test or for the next chapter:

- Read the *Study Guide* at the end of the chapter.
- Try the *Practice Test*.
- Choose questions from each lesson, mixing up the order.
- Use the Internet to find some interactive mathematics questions. Try search words related to the math and technology in the chapter.
- Find data and graphs in newspapers, magazines, and Internet news sites. Create and solve your own questions about them. Trade questions and compare solutions.

Search words
- ☐ Virtual manipulative
- ☐ Java applet
- ☐ Fathom tutorial

Transitions: Getting Extra Practice   **129**

3.1 One- and Two-Variable Data

One application may have a wide variety of data that are helpful to gather, record, and analyse. What sorts of data do you think about when following a sport like women's hockey?

Investigate: Interpreting and Comparing Data

Work with a partner.

The graphs and table provide information about Canadian women's university hockey points leaders in the 2006–2007 season.

- Create a question that can be answered:
  - from the bar graph
  - from the scatter plot
  - from the table

  Record each question and your answer.

Penalty Minutes for Top Scorers in Women's University Hockey

Players (in order): Davis, Provost, Schriver, Davidson, Barry, Ernst, Allgood, Barber, McAlpine, Ferguson, Boisclair, Podloski

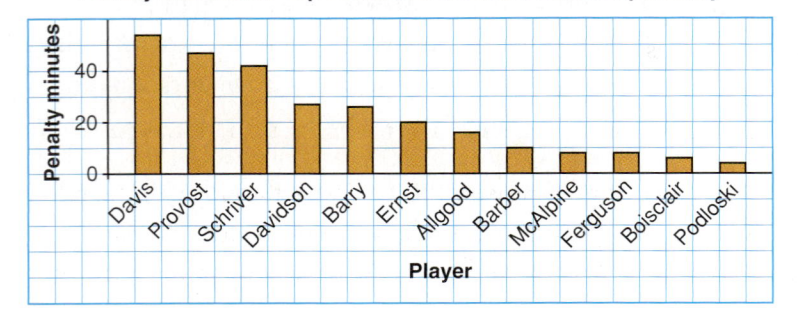

Penalty Minutes and Goals Scored (scatter plot: Penalty minutes vs Goals scored)

130 CHAPTER 3: Two-Variable Data

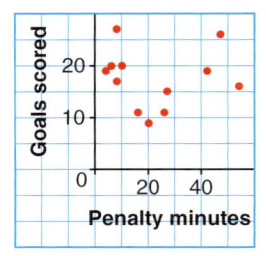

Player	Games played	Goals	Assists	Points	Penalty minutes
Lindsay McAlpine	24	27	30	57	8
Tarin Podloski	22	19	31	50	4
Mariève Provost	21	26	21	47	47
Valerie Boisclair	21	20	21	41	6
Jenna Barber	24	20	20	40	10
Courtney Schriver	21	19	16	35	42
Christina Davis	20	16	17	33	54
Candice Ernst	18	9	24	33	20
Kate Allgood	24	11	20	31	16
Brayden Ferguson	19	17	14	31	27
Vanessa Davidson	17	15	16	31	27
Taryn Barry	24	11	19	30	26

Reflect

- Exchange questions with another pair of students. Answer each others' questions.
- Only one graph displays two-variable data. Which one do you think it is? Why?

Connect the Ideas

One-variable and two-variable data

In statistics, a **variable** is an attribute that can be measured.

One-variable data sets give measures of one attribute. You can recognize one-variable situations when you see:
- Tally charts
- Frequency tables
- Bar graphs
- Histograms
- Pictographs
- Circle graphs

One-variable data can be analysed using mean, median, or mode.

Two-variable data sets give measures of two attributes for each item in a sample. You can recognize two-variable situations when you see:
- Ordered pairs
- Scatter plots
- Two-column tables of values

3.1 One- and Two-Variable Data

Example 1
Identifying Situations Involving One- and Two-Variable Data

State whether each situation involves one-variable or two-variable data. Justify your answers.
a) Noah researches annual hours of sunshine in Canadian cities.
b) A study compared the length of time children spend playing video games and the time they spend reading.

Solution

a) Noah's research could be analysed using mean, median, or mode. So, the situation involves one-variable data.
b) The study involves two pieces of information for each child, time spent on video games and time spent reading. These data could be represented in a scatter plot.
So, the study involves two-variable data.

Example 2
Deciding Which Type of Graph to Draw

For a class project, Dylan surveyed students about their part-time jobs.

Student	Hours Spent at Part-Time Job During the week (h)	Hours Spent at Part-Time Job On the weekend (h)
Adil	0.0	18.0
Anya	5.0	12.5
Ellen	8.0	12.0
Fiona	17.0	8.0
Aaron	0.0	16.5
Leila	10.0	16.0
Mason	9.5	8.0
Petra	15.0	6.0

a) What type of graph would be best to show how many hours each student worked on the weekend? Justify your choice. Does the graph display one-variable or two-variable data?
b) What type of graph would best show a possible relationship between weekday and weekend hours? Justify your choice.

132 CHAPTER 3: Two-Variable Data

> **Solution**
>
> a) A bar graph would display all the information together so that the reader can easily compare the number of hours for each student. One piece of data would be displayed for each student.
> So, the graph would display one-variable data.
>
> b) A scatter plot could show a possible relationship between weekday hours and weekend hours. Since each point on the scatter plot would display two pieces of information about a student, the graph would reflect two-variable data.

Practice

A **1.** a) Does each graph illustrate one-variable or two-variable data?

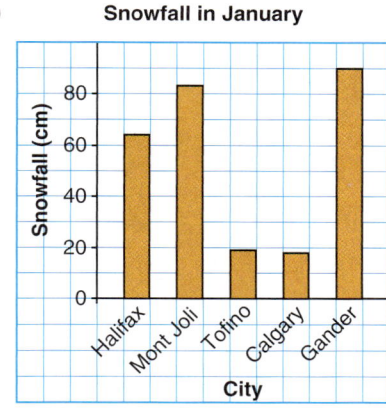

i) Snowfall in January

ii) Destination of Apprenticing Students
- Motive power 22%
- Industrial 24%
- Service 19%
- Construction 35%

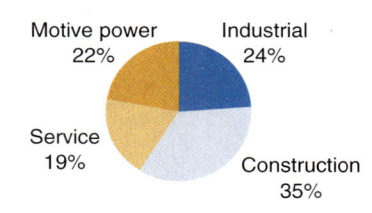

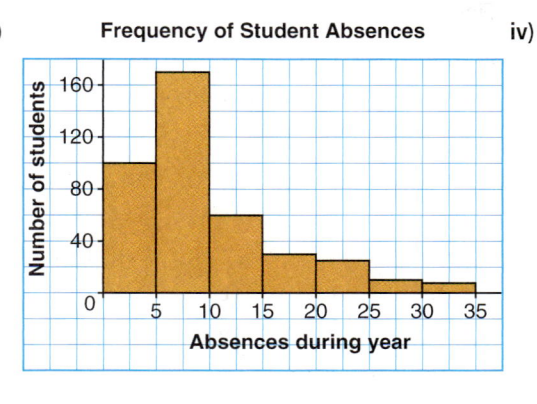

iii) Frequency of Student Absences

iv) Pant Measurements

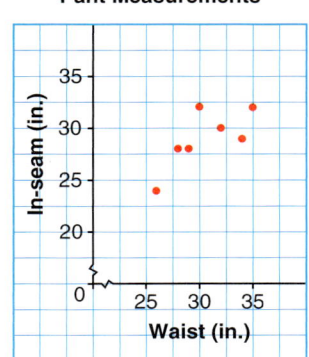

b) Choose one graph from part a. How did you decide whether the graph showed one- or two-variable data?

3.1 One- and Two-Variable Data **133**

2. a) Does each table illustrate one-variable or two-variable data?

i)

Household size	Number of TVs in household
1	2
2	3
3	3
4	3
5	4

ii)

Student	Dollars
Anne	15
Lars	25
Mason	5
Thom	20
Riaz	25
Loni	10

iii)

Candle burn time (h)	10	15	25	40
Cost ($)	7.49	10.99	15.49	19.99

b) Choose one table from part a. How did you decide whether the table showed one- or two-variable data?

■ For help with question 3, see Example 1.

3. Identify the two variables in each situation.
   a) The more purchases made with a credit card, the more reward points earned.
   b) Anthropology students read an article that claimed that people with greater brain mass have higher IQs.
   c) Across Ontario, the mosquito population remained low due to the lower than average rainfall.

4. a) State whether each situation involves one-variable or two-variable data.
   i) Marcus calculates the median mark of the class on an exam.
   ii) An article discusses the possible link between prolonged cell phone use and the increased probability of brain cancer.
   iii) A classroom survey shows that 70% of the students plan to attend university, 15% plan to attend college, 10% are going directly into the workplace, and 5% are uncertain.
   b) Choose one situation from part a. Explain how you decided whether the situation involved one- or two-variable data.

> The median is the middle value when data are arranged in numerical order. If there are two middle values, their mean is the median.

B

For help with question 5, see Example 2.

5. What type of graph would you use to display the data in each table? Justify your choices.

a)

Number of Sit-Ups Students Can Do in 1 min					
Number of sit-ups	0–9	10–19	20–29	30–39	40–49
Frequency	0	3	5	8	4

b)

Land Area of Selected Provinces/Territories							
Province or territory	Alberta	Manitoba	Ontario	Quebec	Nova Scotia	PEI	Nunavut
Land area (1000 km²)	642	554	918	1365	53	6	1936

c)

Ages of Selected Students by Grade							
Grade	9	11	12	9	10	10	11
Age	15	18	18	14	15	16	16

6. A company manufactures fuses. The quality control department frequently tests sample lots to determine how many fuses are defective. This chart shows the results of testing over several days.

Quality Control Results								
Sample size	50	50	100	150	150	200	200	200
Defective fuses	2	1	3	3	4	3	5	4

a) Suppose the supervisor wants a graph showing the frequency of each sample size. What type of graph would be best? Why?

b) What kind of graph would you use to display all the data in the table? Justify your choice.

c) If you were in charge of quality control for this company, which of the samples would concern you most? Least? Explain.

7. Cheyenne recorded the foot lengths of 10 students in her math class. Ayub recorded the heights of 10 students in his English class. They created a scatter plot with the variables foot length and height to determine if there is a relationship between them.

a) Why are the data they collected not two-variable data?

b) What kind of analysis could they do with their data?

3.1 One- and Two-Variable Data **135**

8. A teacher surveyed her students about how long they had studied for a test.
   a) Which graph displays two-variable data? Justify your answer.

   How Long Did Students Study?

   Study Time and Test Scores

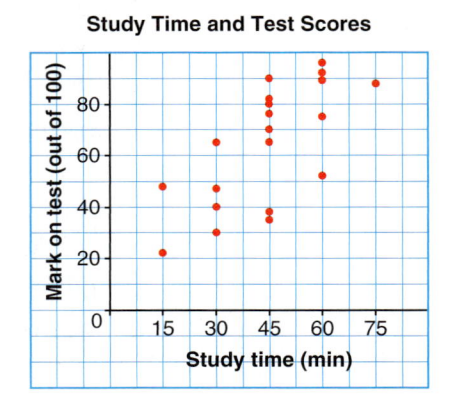

   b) Which graph provides information about a possible relationship? What variables does the relationship involve?
   c) For each graph, write a question someone could answer using the data in the graph. Answer your questions.

9. **Literacy in Math** Select a graph from this lesson. Describe one piece of information you can learn from the graph.

10. **Assessment Focus** A school basketball team had these data for a game.

Player	Minutes played	Points scored 1st quarter	2nd quarter	3rd quarter	4th quarter	Total
Willans	23	2	6	4	4	16
Ohira	11	3	1	3	0	7
Jasquith	19	5	2	2	6	15
Meyers	14	0	0	1	2	3
Salinski	28	5	5	4	4	18
Tobin	26	6	4	6	4	20
Ramanathan	8	0	2	2	0	4
Olander-Hinns	10	2	0	0	2	4
Leenders	5	2	0	0	0	2
Wardhaugh	16	4	2	2	2	10
Total		29	22	24	24	

CHAPTER 3: Two-Variable Data

a) Suppose you calculated the mean number of minutes played by a team member in the game. Is this one-variable or two-variable statistics? Explain.

b) Suppose you want to create a graph to display the number of points scored in each quarter. Would the graph display one- or two-variable data? What type of graph would you create? Explain your thinking.

c) Suppose you want to create a graph to display the number of minutes a player played in the game and the number of points the player scored. Would the graph display one- or two-variable data? What type of graph would you create? Explain your thinking.

C **11.** a) Construct the graphs you described in parts b and c of question 10.

b) Write a question about the graphs that would involve one-variable analysis.

c) Write a question about the graphs that would involve two-variable analysis.

d) Trade graphs and questions with a classmate. Answer your classmate's questions. Check each other's work.

In Your Own Words

Describe a situation where you would need to collect, analyse, and graph two-variable data.

3.1 One- and Two-Variable Data **137**

3.2 Using Scatter Plots to Identify Relationships

Ergonomic designs suit the way humans think, see, and move. They are based on the measurement and study of human body dimensions. Ergonomics is studied in college programs such as industrial design, fitness, and lifestyle management.

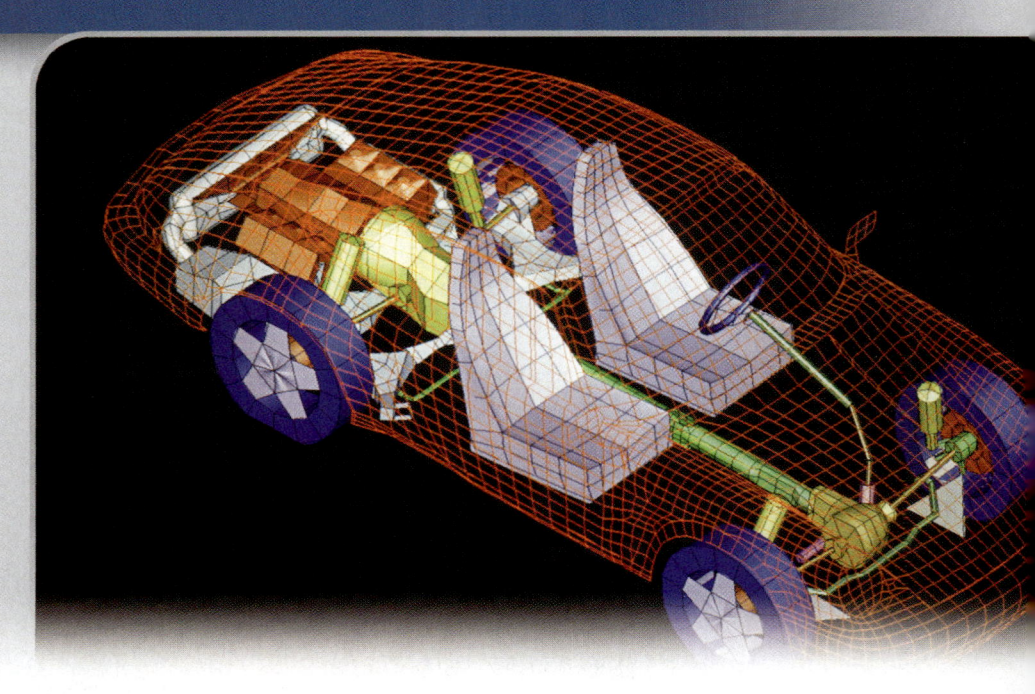

Investigate

Creating a Scatter Plot

Materials
- grid paper
- measuring tape at least 3 ft. or 1 m long

Work with a partner to collect data from at least 10 students.

- For each student, measure and record the distance:
  - From the elbow to the outstretched tip of the middle finger
  - From the knee to the ankle
- Graph your data set, and describe the results.
- Do you think the two body measurements are related? Why or why not?

Elbow-to-fingertip length	Knee-to-ankle length

Reflect

- Compare graphs with other pairs. Do your graphs appear to show the same relationship between the measurements?
- Identify at least one reason the relationship you identified may not be true in general. What could you do to find out?

138 CHAPTER 3: Two-Variable Data

Connect the Ideas

Scatter plots represent two-variable data as points. Scatter plots may reveal a relationship between the two variables.

Independent variable and dependent variable

In two-variable situations, one variable may be **dependent** on another: its value changes according to the value of the **independent** variable. For example, the value of a car depends on its age.

Typically, we plot the independent variable on the horizontal axis, and the dependent variable on the vertical axis.

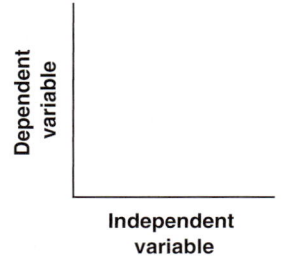

Example 1

Interpreting a Scatter Plot

Jay researched estimates for a job painting his house.
The scatter plot shows Jay's results. 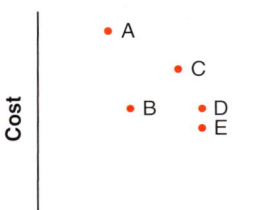
a) Which is the dependent variable? Justify your choice.
b) Which two companies will take the longest? Which of these is cheaper?
c) Which two companies charge the same amount?
d) Why might you pick company E? Company B?

Solution

a) Employees are paid for the time they work, so the cost depends on the time required for the job. The dependent variable is cost.
b) Company D and Company E take the longest time for the job. Of Companies D and E, Company E is cheaper.
c) Company B and Company D charge the same amount.
d) You might pick Company E if you are not in a hurry. They take a long time, but are the cheapest.
Company B is the second fastest and charges a low price. So, you might pick Company B if you want the job done fairly quickly, but economically.

Types of correlations

A **correlation** is a relationship between two variables. Graphing two-variable data on a scatter plot may show a correlation between the variables.

A *positive correlation* describes a situation in which both variables increase together.
On a scatter plot, the points go up to the right.

Positive Correlation

A *negative correlation* describes a situation in which one variable decreases as the other variable increases.
On a scatter plot, the points go down to the right.

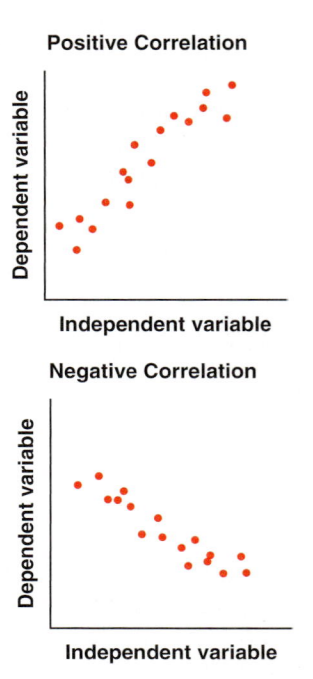

Negative Correlation

Example 2

Analysing Data Using a Scatter Plot

Davis conducted an experiment comparing a person's leg length and how long it takes the same person to walk 100 m. He gathered these data showing (leg length in centimetres, time taken in seconds).
(80, 66), (73, 74), (60, 83), (64, 62), (63, 75), (78, 76), (83, 64), (54, 81), (73, 70), (78, 76)

a) Graph the data.
b) Does the graph suggest a relationship between leg length and time taken to walk 100 m? If so, describe the relationship.
c) Use the scatter plot to estimate the time it would take a person with leg length 85 cm to walk 100 m. Explain.
d) How might Davis make the results of his experiment more reliable?

140 CHAPTER 3: Two-Variable Data

Solution

a) Time taken may depend on leg length, so use the vertical axis to show time taken, and the horizontal axis to show leg length.

b) The points go down to the right. This suggests a negative correlation: the time taken to walk 100 m tends to decrease as leg length increases.

Leg Length and Time Taken to Walk 100 m

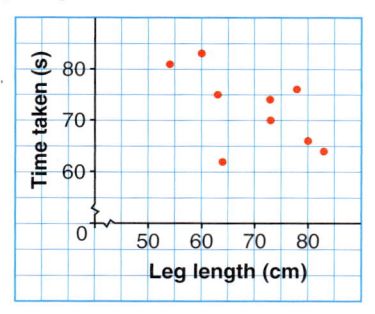

c) A point with a first coordinate of 85 might have a second coordinate of about 63. So, someone with legs 85 cm long might take about 63 s to walk 100 m.

d) More data would make Davis' results more reliable. He could recruit people with a wider range of leg lengths while keeping other factors, such as age, gender, and body type, very similar.

Cause-and-effect relationships

Observing a relationship between two variables does not mean that one variable causes a change in the other. Other factors could be involved, or the correlation could be a coincidence.

Demonstrating cause and effect conclusively is a challenging task that requires careful analysis and specialized statistical tools.

Example 3

Considering Possible Cause and Effect

State whether the claim in each situation is reasonable.

a) A scientific study showed a negative correlation between aerobic exercise and blood pressure. It claimed that the increase in aerobic activity was the cause of the decrease in blood pressure.

b) Mila discovered a positive correlation between gasoline price and average monthly temperature. She concluded that temperature determines the price of gasoline.

c) Since the 1950s, the concentration of carbon dioxide in the atmosphere has been increasing. Crime rates in many countries have also increased over this time period. Does more carbon dioxide in the atmosphere cause people to commit crimes?

3.2 Using Scatter Plots to Identify Relationships

Solution

a) It is reasonable to think there may be a cause-and-effect relationship. There are many factors that affect blood pressure, however. Since this is a scientific study, we might reasonably expect that the researchers made efforts to neutralize the other factors, for example by studying subjects in a very close age and fitness range.

b) It is not reasonable to say there is a cause-and-effect relationship between temperature and gasoline prices. A more likely explanation for the correlation is that higher temperatures occur in the summer, when more people are travelling out of town for weekends and vacations. This increased demand could cause the price increase.

c) It is not reasonable to say there is a cause-and-effect relationship. It is much more likely that carbon dioxide levels and crime rates are each determined by many other factors, such as increasing populations.

Practice

A

For help with questions 1 or 3, see Example 1.

1. a) What does each point show?
   b) Which child wears the smallest shoe? The biggest shoe?
   c) Which two children are the same age?
   d) Which two children wear the same shoe size?

   Age and Shoe Size of Six Children

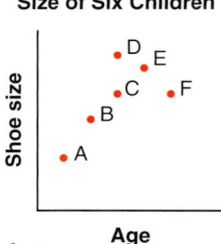

2. Identify the dependent variable in the scatter plot in question 1. If you do not think there is a dependent variable, tell why.

3. a) Which company has the lowest moving cost?
   b) Which two companies use the largest trucks?
   c) Which two companies use the smallest trucks?
   d) Based on the scatter plot, how many different truck sizes are there?
   e) Which company would you use for a small load? For a very large load?

   Moving Company Estimates

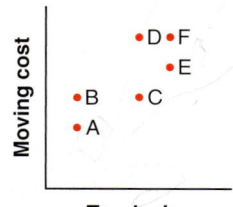

4. Identify the dependent variable in the scatter plot in question 3. If you do not think there is a dependent variable, explain why.

142 CHAPTER 3: Two-Variable Data

5. Which building is represented by each point in the scatter plot?

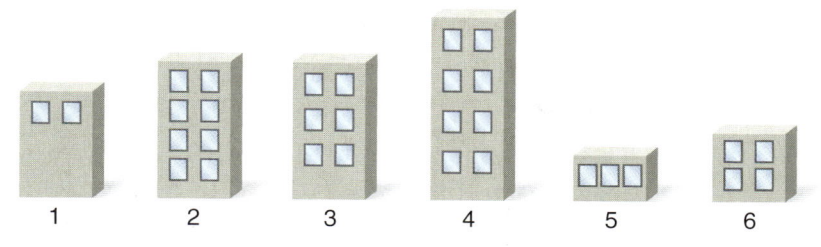

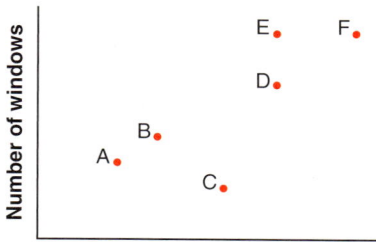

6. For each situation, state whether you think the two variables would have a positive correlation, a negative correlation, or no correlation.
   a) Cost of a restaurant bill and the amount left as a tip
   b) Blood pressure reading and IQ
   c) Number of applicants for a job and probability that you will get the job
   d) Speed of current and time taken to travel upstream
   e) Number of kilometres driven and price of gas per litre

B

7. Create a scatter plot of the data in each table.

a)

Number of people in household	5	4	5	6	3
Average electricity consumption per month (kWh)	1152	928	953	1067	893

b)

Number of daylight hours	10.54	10.57	10.60	10.63	10.66
Time spent brushing teeth (s)	47	52	38	40	42

8. For each part, state whether you think the two variables will have any correlation. If you think a correlation exists, describe it briefly.
   a) Summer temperatures and sales of bottled water
   b) Price of gasoline and number of people who go to movies
   c) Price of gasoline and number of day trips people take
   d) Cost of tuition and number of students who apply to college

3.2 Using Scatter Plots to Identify Relationships

9. For each variable, describe a variable that could be correlated with it.
   a) The amount of time you sleep and …
   b) The size of an in-ground pool and …
   c) The amount of traffic on the road and …

10. Each graph shows the time taken for two race cars to travel around an oval track for a few practice laps.

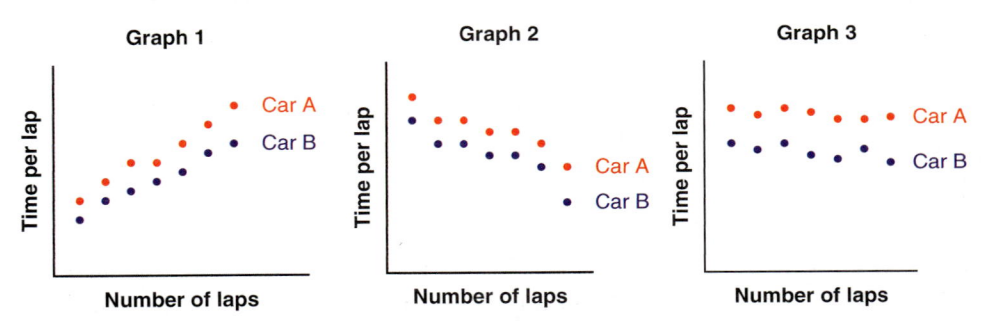

   Graph 1 — Time per lap vs Number of laps; Car A, Car B
   Graph 2 — Time per lap vs Number of laps; Car A, Car B
   Graph 3 — Time per lap vs Number of laps; Car A, Car B

   a) How do these graphs show that there are two cars?
   b) How many laps are shown in each graph? What assumption are you making?
   c) Which graph shows both cars maintaining a fairly constant lap time?
   d) In each graph, which car has the greater average speed?
   e) Suppose it starts to rain, making the track slippery and forcing the cars to slow down. Which graph would best show this? Explain.

11. Each graph shows the time taken for two cars to travel around an oval track for the first few laps of a race.

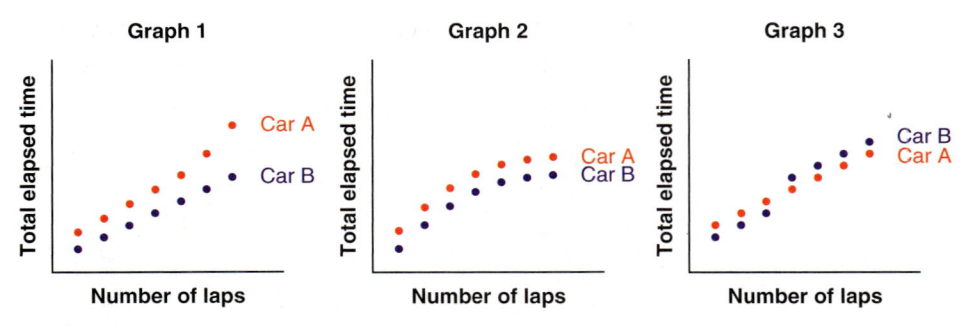

   Graph 1 — Total elapsed time vs Number of laps; Car A, Car B
   Graph 2 — Total elapsed time vs Number of laps; Car A, Car B
   Graph 3 — Total elapsed time vs Number of laps; Car B, Car A

   a) In Graph 1, which car completes each lap faster?
   b) Which graph shows both cars increasing their speed during the first few laps?
   c) Which graph shows Car B making a brief pit stop to repair a loose tire and being passed by Car A?
   d) Which graph shows Car A having engine trouble and therefore taking longer and longer to complete each lap? Justify your choice.

144   CHAPTER 3: Two-Variable Data

For help with question 12, see Example 2.

12. This table shows the hourly cost of heating a pool with a given surface area.

Surface area (sq. ft.)	Hourly cost (cents)
100	24
200	50
400	102
500	127
800	206
1200	310
1500	390

a) Using the table, describe what happens to the hourly cost as the surface area increases. What does this suggest about the trends you might see in a scatter plot of these data?
b) Create a scatter plot for these data. How does it compare to your prediction in part a?
c) Does the scatter plot reveal a correlation between the two variables? If so, describe it.
d) About how much would it cost to heat a 20 feet × 40 feet pool for 24 h?

13. Assessment Focus In Canada, fuel consumption ratings for vehicles are expressed in L/100 km, or the number of litres of fuel used to travel 100 km.

Fuel Consumption for a Vehicle Driven at Different Speeds on a Test Track								
Speed (km/h)	60	70	80	90	100	110	120	130
Fuel consumption (L/100 km)	5.6	5.9	6.2	6.7	7.4	8.1	8.7	9.4

a) Which is the independent variable? Justify your choice.
b) What happens to the fuel consumption as the speed increases? Predict the general shape of a scatter plot of the data. Justify your prediction.
c) Create a scatter plot of the data. How does the shape of the scatter plot compare to your prediction from part b?
d) Describe the relationship between the variables.
e) Use the plotted data to estimate the fuel consumption for this car at a speed of 85 km/h. Explain your thinking.

3.2 Using Scatter Plots to Identify Relationships **145**

14. **Literacy in Math** Create a Frayer model about correlations. Include descriptions of positive and negative correlations under *Facts/Characteristics*.

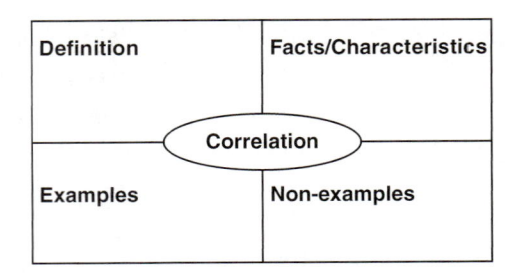

Definition	Facts/Characteristics
Examples	Non-examples

Correlation

■ For help with question 15, see Example 2.

15. For each situation, decide whether you think it's reasonable to conclude a cause-and-effect relationship.
    a) Scientific studies found that as exposure to second-hand smoke increased, so did the risk of lung cancer. Based on these studies, a panel concluded that exposure to second-hand smoke increases the risk of lung cancer.
    b) Jovanna found a strong correlation between students' science and English marks. She concluded that a student's success in science causes increased marks in English.
    c) In Ontario, there seems to be a negative correlation between the consumption of hot chocolate and the number of motorcycle accidents. So, drinking more hot chocolate causes a decrease in the number of motorcycle accidents.
    d) Zach researched a negative correlation between the price of a concert ticket and the distance of the seat from the stage. He concluded that the distance from the stage determines the price of a ticket.

16. Vivek is an amateur meteorologist. He collected data on temperature and relative humidity, each hour for 12 h.

Time	Temperature (°C)	Relative humidity (%)
1:00 p.m.	30	51
2:00 p.m.	31	48
3:00 p.m.	32	46
4:00 p.m.	31	46
5:00 p.m.	28	62
6:00 p.m.	30	49
7:00 p.m.	30	47
8:00 p.m.	29	52
9:00 p.m.	27	58
10:00 p.m.	24	69
11:00 p.m.	23	73
12:00 a.m.	21	82

a) Which is the independent variable? Why do you think so?
b) Create a scatter plot of Vivek's data.
c) Does there appear to be a correlation? If so, describe it. Otherwise, explain why you think there is no trend.

17. Use the scatter plot you drew in question 16.
a) Vivek heard that the temperature will reach a low of 18°C overnight. Estimate the relative humidity overnight.
b) Vivek has found it is very likely to rain when the relative humidity reaches 90%. Is it very likely to rain overnight? Explain.

In Your Own Words

Describe a situation where a set of data can be displayed in a scatter plot. How could the scatter plot help you interpret the data?

3.2 Using Scatter Plots to Identify Relationships

3.3 Line of Best Fit

In recent years, natural disasters seem to be more intense and occur more frequently. Will this trend continue? Climatologists try to answer questions like this by identifying trends in historical data and using them to make predictions about future events.

Investigate

Sketching a Line of Best Fit

Materials
- grid paper
- coloured pencils

Work with a partner.

Sabah researches the cost of houses in a new area. She is looking for
- A two-storey, detached house
- Asking prices of $300 000 or less
- 2000 square feet of living space

Size (sq. ft.)	Price ($)
1700	271 900
1850	289 900
1600	277 900
1650	289 900
1700	279 000
1800	294 900
1550	269 900

- Graph the data. Use axes that show house sizes up to 2500 square feet and prices of up to $400 000.
- Describe any trend you see in the graph. With a coloured pencil, draw the line that you think comes closest to matching the trend.
- Use the line to estimate how much a 2000 square foot house might cost in Sabah's area.

You've drawn the **line of best fit**.

148 CHAPTER 3: Two-Variable Data

Sabah notices a new *For Sale* sign on a house in the area.
At 2300 square feet and $384 500, it does not match her criteria.
However, she decides to add the data to her collection anyway.

- Add the point (2300, 384 500) to your scatter plot.
- With a different colour, draw the new line of best fit.
- Use this line to estimate the price for a 2000 square foot house in Sabah's neighbourhood. Compare this estimate to your first estimate.

Reflect

- How does the arrangement of plotted data affect the way you draw a line of best fit? How does it affect your confidence in using the line to make predictions?
- A real estate agent tells Sabah that 2000 square foot houses often come up for sale in this area, usually at prices from $295 000 to $305 000. How close were your estimates to these prices? Which line of best fit helped you make a closer prediction?

Connect the Ideas

Line of best fit

A **line of best fit** is a line drawn through data points to best represent a linear relationship between two variables. Other names for a line of best fit are *regression line* or *trend line*.

Drawing the line of best fit involves more than just finding a pathway through the middle of the data. The line of best fit is the line that is closest to each point. The more varied the position of points, the more difficult it is to draw the line of best fit.

Outliers

In a scatter plot, a point that lies far away from the main cluster of points is an **outlier**. An outlier may be caused by inaccurate measurements, or it may be an unusual, but still valid, result.

Outliers and the line of best fit

The line of best fit should reflect all valid points from a data set. This includes outliers.

If most of the plotted points are clustered along a linear path, even a single outlier can affect the path of the line of best fit.

Example 1

Exploring the Effect of Outliers on a Line of Best Fit

Which line is the line of best fit? Justify your choice.

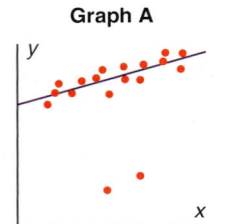

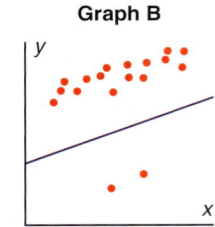

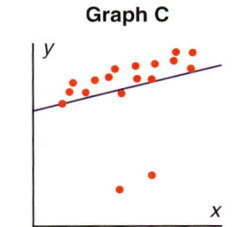

Graph A   Graph B   Graph C

Solution

The line in Graph C seems most likely to be the line of best fit.
- The line in Graph A passes through the middle of the main cluster of data. It suggests the two outliers are not present or are invalid.
- The line in Graph B follows a path exactly halfway between the main cluster of data and the outliers. It suggests that the outliers and the main cluster of data are equally important.
- The line in Graph C passes just below the middle of the main cluster of data. Its path is affected by the outliers, but it is affected more by the main cluster of data. It is the most reasonable choice.

Interpolating and extrapolating from a line of best fit

A line of best fit can be used to estimate or predict values.
- Estimating values that lie among the known values on the graph is **interpolation**.
- Predicting values that lie beyond the known values is **extrapolation**.

Example 2

Using a Line of Best Fit to Make Predictions

These are the pre-exam term marks and exam marks for some students in a Grade 12 math course.

Term mark (%)	84	76	70	95	92	61	25	55	51	73	62
Exam mark (%)	80	72	68	96	90	58	29	60	53	77	67

a) Graph the data and draw the line of best fit.
b) Determine the equation of the line of best fit.
c) Use the data to predict the exam mark of a student with a pre-exam term mark of 98%.
d) Use the data to predict the exam mark of a student with a pre-exam term mark of 10%.

150   CHAPTER 3: Two-Variable Data

Solution

a) Plot the points. Place a ruler on the graph so that it comes as close as possible to all of the points. Draw a line along the ruler's edge.

b) The equation of a line is $y = mx + b$, where m is the slope and b is the vertical intercept. Choose two points on the line to find the slope: (40, 43) and (70, 70)

Slope: $\frac{rise}{run} = \frac{70 - 43}{70 - 40}$

$= \frac{27}{30}$, or 0.9

Term Marks and Exam Marks

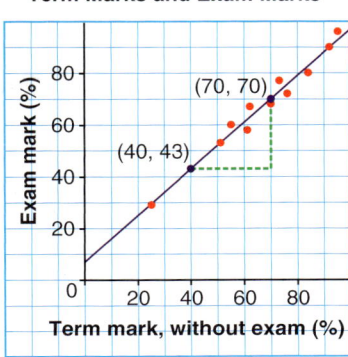

From the graph, the vertical intercept of the line is about 7.
The equation of the line of best fit is approximately $y = 0.9x + 7$.

c) Extrapolating from the graph, a student with a pre-exam term mark of 98% would get an exam mark of about 95%.

d) Substitute $x = 10$ into the equation of the line of best fit.
$y = 0.9x + 7$
$= 0.9 \times 10 + 7$
$= 16$

The predicted exam mark is 16%.

> The point where a line crosses the vertical axis is called the *vertical intercept*.

> *x* and *y* represent the marks as percents, so do not convert them to decimals.

How confident can we be of predictions made from scatter plots?

Data spread and reliability

A model with data spread over a small interval is less reliable than a model based on data spread over a larger interval.
The farther we get from the main cluster of data, the less confidence we should have on predictions made from that model.

> How does this relate to the predictions in *Example 2*?

Sample size and reliability

The more data we use, the more reliable the prediction should be.

Non-linear data

Not all relationships between variables are linear. Over a small interval, a linear model may provide a reasonable fit for non-linear data, but it will not be reliable at the extremes.

> You will learn more about other types of models in Chapter 5.

3.3 Line of Best Fit

Example 3

Deciding if a Correlation Is Linear

Reanna created a scatter plot using data about the world record times for women's 500-m speed skating. If the world record changed more than once in a year, she used the best time for that year.

World Record for Women's 500-m Speed Skating

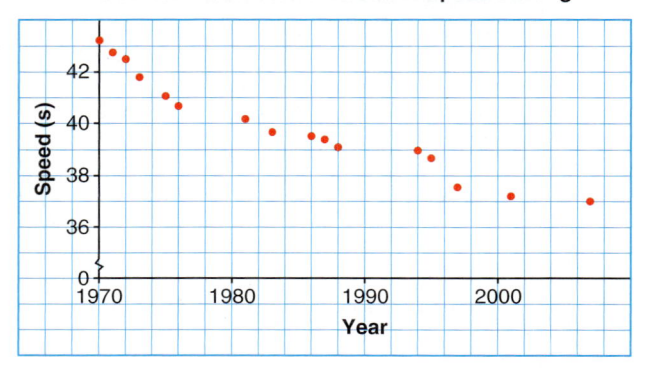

(scatter plot: Speed (s) vs. Year, points decreasing from about 42.5 s in 1970 to about 37 s near 2000)

Describe the relationship between the two variables. Does it appear to be linear? If not, describe how Reanna could better model the data.

Solution

As the years increase, the time decreases; there is a strong negative correlation between the variables.

Although the relationship appears to be linear in some places, the overall relationship is non-linear. The world record times decreased most rapidly in the four-year period from 1970 to 1976. A model that decreases steeply at first and then more slowly would better represent the data.

Assessing a linear model

A linear model may be unreliable in these situations:
- The model is based on too few pieces of data.
- The model is based on data that are clustered together.
- There does not appear to be any correlation.
- There are outliers.
- The data do not appear to be linear.

Practice

A

For help with questions 1 to 4, see Example 1.

1. For each scatter plot, select the line of best fit. Justify each choice.

   a) Graph A, Graph B, Graph C

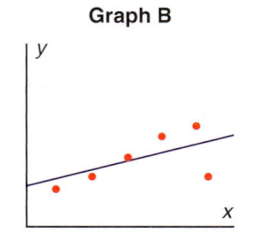

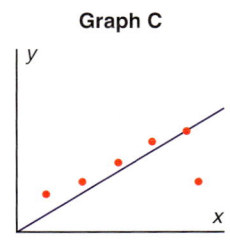

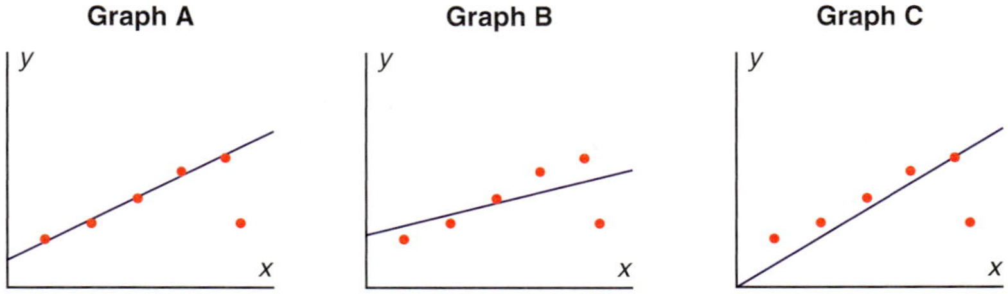

   b) Graph D, Graph E, Graph F

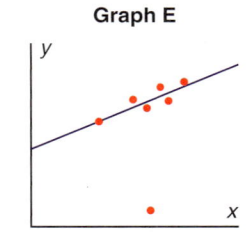

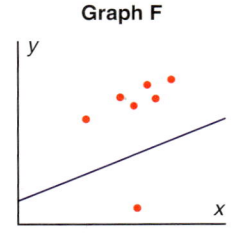

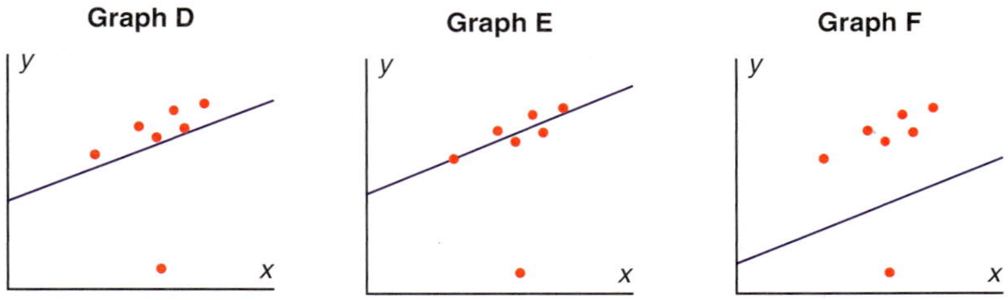

2. Points A and B are both outliers. Which outlier would have the greater effect on the path of the line of best fit? Justify your choice.

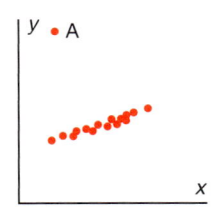

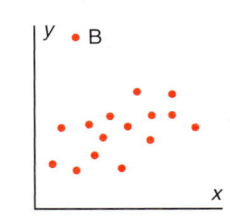

3. Use this graph to identify whether you would use interpolation or extrapolation to predict each value.
   a) Volume of water in a hot tub that requires 10 g chlorine.
   b) Mass of chlorine needed for a 500 L hot tub.
   c) Mass of chlorine needed for a 35 000 L hot tub.
   d) Volume of water in a hot tub that requires 18 g chlorine.

   Hot Tub Maintenance

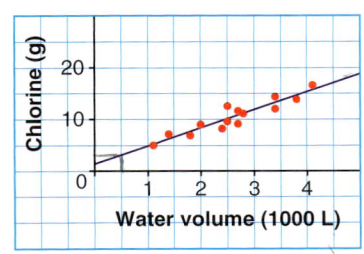

3.3 Line of Best Fit   **153**

For help with question 4, see Example 2.

Use the *Course Study Guide* at the end of the book to recall how to calculate slope.

B 4. a) On a copy of these graphs, sketch the line of best fit for each scatter plot.
b) Determine the equation of each line of best fit.

i) Retail Cost of Evergreens

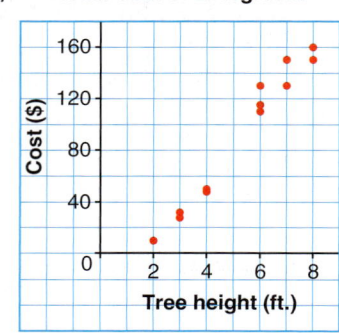

ii) Eruptions of Cerro Negro Volcano since 1900

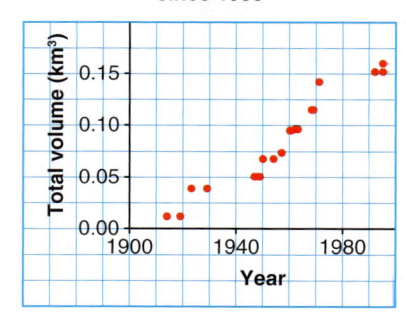

5. The line drawn in this graph passes through the points so that half the points lie above the line and half the points lie below the line.
a) Why is this line not the line of best fit?
b) Describe how the line of best fit would look for these data.

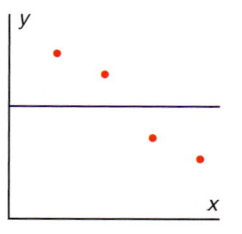

6. Describe the problems with drawing the line of best fit for these data.

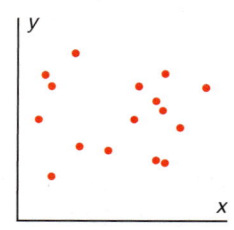

7. a) For each scatter plot, describe the relationship between *x* and *y*.
b) Would you model each relationship with a linear or non-linear model? Justify your answers.

i)

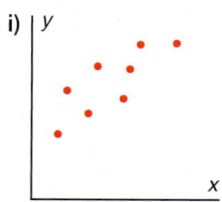

ii)

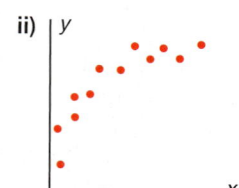

iii)

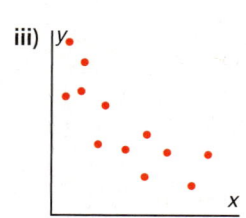

iv)

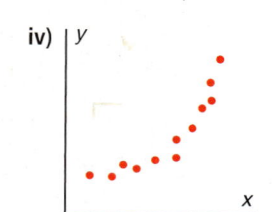

154 CHAPTER 3: Two-Variable Data

For help with question 8, see Example 3.

8. a) Use the line of best fit for this scatter plot to make each prediction.
   i) Number of games played by a player who averages 10 shifts per game
   ii) Number of games played by a player who averages 5 shifts per game
   b) Which of your predictions in part a do you think is more reliable?

Play Time for Some Toronto Maple Leaf Forwards

9. A power utility company warns that if the demand for electricity is greater than 27 000 MW, there will be some power outage. The table shows daily high temperature and peak electricity demand for a 10-day period one summer.
   a) Create a graph and draw the line of best fit for these data.
   b) Determine an equation of the line of best fit.
   c) Estimate the peak electricity demand for a daily high temperature of 28°C.
   d) Predict the likelihood of a power outage when the daily high temperature is 37°C.
   e) The power company can take one of their generators offline for maintenance when the demand is below 17 000 MW. On what days would the peak electricity demand fall below 17 000 MW?

Daily high temperature (°C)	Peak electricity demand (MW)
24	19 503
23	18 832
24	19 150
27	20 613
29	21 544
30	22 237
26	20 082
29	21 819
32	23 488
34	24 950

3.3 Line of Best Fit   **155**

10. **Assessment Focus** Use the table of data about life expectancy of Canadian males.

Life Expectancy at Birth of a Canadian Male

Birth year	1920	1930	1940	1950	1960	1970	1980	1990
Life expectancy at birth (years)	59	60	63	66	68	69	72	75

a) Create a graph and draw the line of best fit for these data.
b) Find an equation of the line of best fit.
c) Estimate the life expectancy of a male born in 1975.
d) Predict the life expectancy of a male born in 2000.
e) Predict the birth year of males with a life expectancy of 80 years.
f) Write a question someone could answer using your graph. Prepare an answer for the question.

11. Use the table of data about life expectancy of Canadian females.

Life Expectancy at Birth of a Canadian Female

Birth year	1920	1930	1940	1950	1960	1970	1980	1990
Life expectancy at birth (years)	61	62	66	71	74	76	79	81

a) How do the life expectancies for male and female Canadians compare?
b) Repeat question 10 using these data.

12. **Literacy in Math** Another name for a line of best fit is a *trend line*. Explain why this is an appropriate name.

13. How would you describe the relationship between fuel consumption for city and highway driving? Does it appear to be linear? If not, describe a better model for the data.

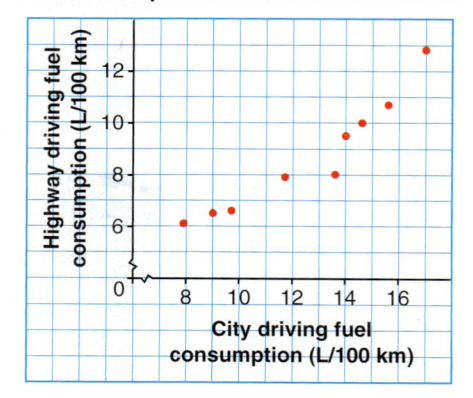

156 CHAPTER 3: Two-Variable Data

14. For each situation, explain why the data analysis is not reasonable. There may be more than one reason for each graph.

   a) A technician graphed data and concluded he would get a 100% reduction in heat loss by using $4\frac{1}{2}$ inches of insulation.

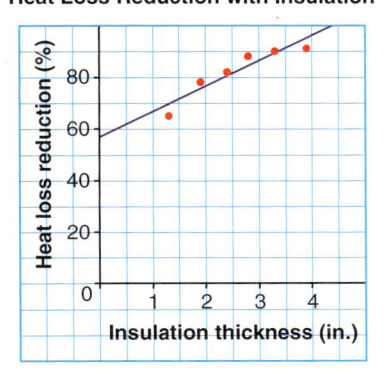

   b) A carnival weight-guesser created this scatter plot using data from his last eight customers. He plans to ask customers how tall they are and use the linear model to estimate their weights.

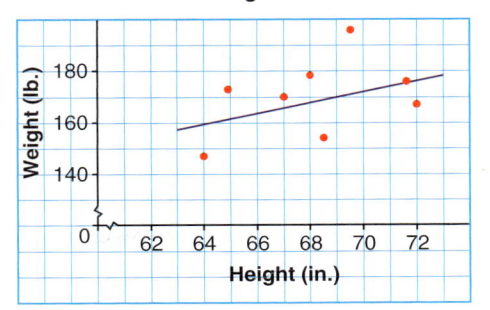

Heat Loss Reduction with Insulation

Carnival Weight-Guesser

15. You have seen that outliers affect the placement of the line of best fit.

   a) Do you think that the degree to which an outlier influences the line of best fit depends on the size of the data sample? Explain.

   b) What other factors can you think of that might affect the degree to which an outlier affects the line of best fit?

> **In Your Own Words**
>
> Suppose a classmate missed the lesson on line of best fit. Write instructions for her or him about how to draw a line of best fit and why it is useful.

3.3 Line of Best Fit   **157**

Mid-Chapter Review

3.1 **1.** Does the situation illustrate one-variable or two-variable data? Explain.

a) **Students Competing in a Math Contest**

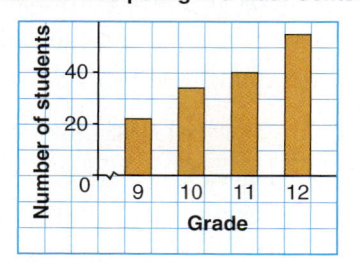

(bar graph: Number of students vs Grade 9, 10, 11, 12)

b) Students in a science class measured the height of their plants each week for 12 weeks.

3.2 **2.** State whether you think the variables in each situation would have a negative correlation, a positive correlation, or no correlation.

a) Driving speed and time to travel 100 km
b) Size of a house and its interior temperature
c) One's age and the number of colds one's had
d) Cost of gasoline and fuel efficiency of a vehicle

3.3 **3.** Which linear model better represents the data? Explain your choice.

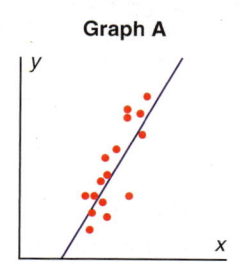

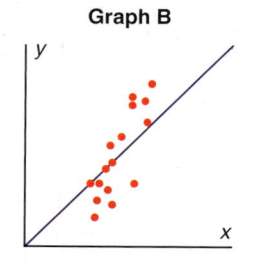

Graph A   Graph B

3.2 **4.** This table compares the parking facilities
3.3 of several large companies.

Available Land and Parking Capacity for Various Companies	
Acres of land	Parking spaces
2.0	145
1.5	160
4.0	500
1.0	95
5.0	600
4.0	425
2.0	550
3.0	280

a) Create a scatter plot of the data.
b) Describe any trends you see.

5. Use the table of data on tire pressure.
a) Graph the data; draw a line of best fit.
b) Describe the correlation.
c) Predict the pressure at:
   i) 70°F   ii) 40°F

Tire Pressure and Temperature	
Outside temperature (°F)	Tire pressure (psi)
58	35
79	38
63	36
61	36
85	39
55	34
74	37
88	40

CHAPTER 3: Two-Variable Data

3.4 Analysing Data Using a Graphing Calculator

Smog is created when air pollutants react with sunshine and heat. The Ontario Ministry of the Environment measures the amount of smog in the air and issues a smog advisory when high smog levels are likely.

Inquire: Analysing Scatter Plots with the TI-83/TI-84

Materials
- TI-83 or TI-84 graphing calculator

This table shows the number of smog advisories issued by the Ontario Ministry of the Environment from 1995 to 2006.

Number of Smog Advisories and Smog Days in Ontario

Year	1995	1996	1997	1998	1999	2000	2001	2002	2003	2004	2005	2006
Number of advisories	6	3	3	3	5	3	7	10	7	8	15	6
Total number of days	14	5	6	8	9	4	23	27	19	20	53	17

Drawing a scatter plot

Steps	Display	Notes
Press [STAT] 1 to access the Stat List Editor. To clear any data from **L1**, use the arrow keys to move the cursor to the **L1** heading. Press [CLEAR] [ENTER]. Repeat for **L2** and **L3**. Enter the data from the smog advisory table.	(L1/L2/L3 table with years 1995–2001) L3(7)=23	Enter the years in **L1**, the number of advisories in **L2**, and the number of days in **L3**. You can press either [ENTER] or ▼ after each number.
Press [2nd] [Y=] 1 to access the Stat Plots menu for Plot 1. With the cursor over ON, press [ENTER]. Press ▼ [ENTER] to select ⌐ (scatter plot). Press ▼ [2nd] 1 [ENTER] to set **Xlist** to **L1**, [2nd] 2 [ENTER] to set **Ylist** to **L2**, and [ENTER] to set **Mark** to □ (box).	Plot1 Plot2 Plot3 On Off Type: ... Xlist:L1 Ylist:L2 Mark: □ + ·	Make sure **Plot2** and **Plot3** are turned off. Use the arrow keys to place the cursor over **Plot2**. Press [ENTER] ▶ [ENTER]. Repeat for **Plot3**.
Press [Y=] to see the list of equations stored in the calculator. Clear them or turn them off. Then press [ZOOM] 9 to graph the data from L1 and L2 on an appropriate scale.	(scatter plot display)	When an equation is turned on, the equals sign is highlighted. To turn off an equation, place the cursor over = and press [ENTER].

1. a) What does the scatter plot suggest about how the number of smog advisories is changing over time?
  b) Describe any correlation between the variables.
  c) Do you think it would be appropriate to model the relationship with a line of best fit? Justify your answer.

160 CHAPTER 3: Two-Variable Data

■ **Creating a line of best fit**

Steps	Display	Notes
Press STAT ▶ 4 to select linear regression. Then press 2nd 1 , 2nd 2 , to have the regression performed on the data in lists **L1** and **L2**. Press VARS ▶ 1 1 to have the equation of the line of best fit stored in **Y1**. Press ENTER to perform the regression.	LinReg y=ax+b a=.6363636364 b=-1266.712121 r²=.4176136364 r=.6462303276	If the values of r^2 and r do not appear on your screen, turn diagnostic mode on. Press 2nd 0, scroll down to **DiagnosticOn** and press ENTER ENTER.

> **Correlation coefficient**
> The correlation coefficient, r, is a number between -1 and 1.
> It describes the strength of the linear correlation.
> Generally, the closer r is to 1 or -1, the stronger the linear correlation and the more closely the data approximate a line.

2. a) The screen shows that the equation of the line is $y = ax + b$ and gives the values of a and b to many decimal places. What are these values? Round a to one decimal place and b to the nearest whole number. What does a tell you about the line of best fit?

b) What is the correlation coefficient, r?

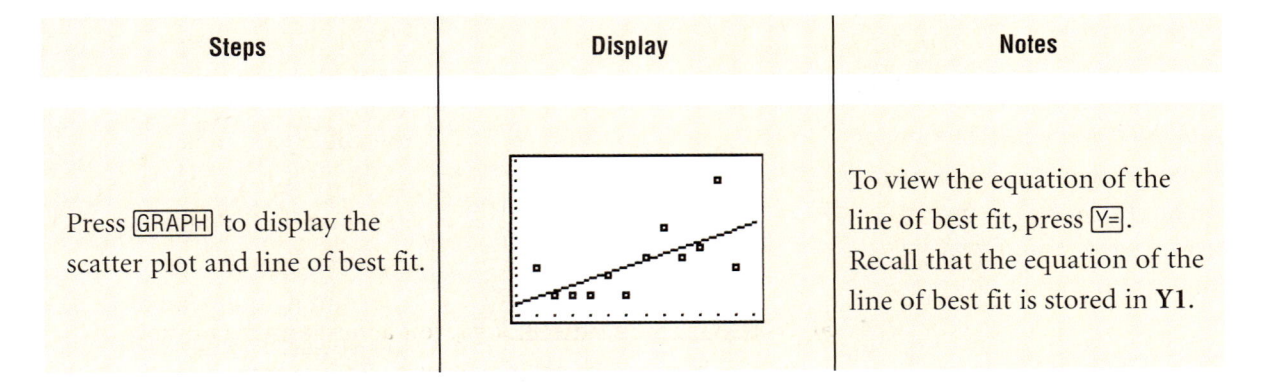

Steps	Display	Notes
Press GRAPH to display the scatter plot and line of best fit.		To view the equation of the line of best fit, press Y=. Recall that the equation of the line of best fit is stored in **Y1**.

3. Describe the line of best fit. Is it what you expected? Explain why or why not.

3.4 Analysing Data Using a Graphing Calculator **161**

■ Extrapolating with the line of best fit

Steps	Display	Notes
Use the TRACE feature to determine the value of **Y** when **X** is 2007. Press TRACE. Then press ▼ to place the cursor on the line of best fit. Press: 2007 ENTER The corresponding **Y**-value is displayed at the bottom right of the screen.	Y1=.6363636364X+ -12_ X=2007 Y=10.469697	The TRACE feature will not work for **X**-values that are beyond the window settings. You could extrapolate algebraically by substituting 2007 for x in the equation of the line of best fit and determining the value of y. Press: CLEAR VARS ▶ 1 1 (2007) ENTER

4. a) What was the predicted number of smog advisories for 2007?
   b) As of October 22, 2007, there were 13 smog advisories in Ontario. Is the total number for the year likely to be much greater than this? Explain.
   c) Was the prediction from the line of best fit close to the actual number?

5. Follow these steps to predict the number of smog advisory days in 2010.

Steps	Display	Notes
Press WINDOW to access the **WINDOW** menu. The cursor should be on the **Xmin** setting. Press ▼ and input 2012 for the **Xmax** value. Press TRACE. Then press ▼ to place the cursor on the line of best fit. Press: 2010 ENTER. What is the corresponding **Y**-value?	Y1=.6363636364X+ -12_ X=2010 Y=12.378788	You could extrapolate algebraically by substituting 2010 for x in the equation of the line of best fit and determining the value of y. Press: CLEAR VARS ▶ 1 1 (2010) ENTER

6. a) Which point appears to be an outlier?

b) The summer of 2005 was one of the hottest and most humid summers in Ontario. Toronto recorded 41 days with temperatures greater than 30°C. How might this have resulted in the unusual data for 2005?

■ **Removing an outlier**

Steps	Display	Notes
Press STAT 1 to access the lists you have stored. In **L1**, use the ▼ key to move down to the entry for 2005. Then press DEL ▶ to delete it and move to **L2**. Press DEL ▶ DEL to delete the corresponding entries in **L2** and **L3**.		To re-enter 2005 in list **L1**, place the cursor over 2006 and press 2nd DEL 2005 ENTER. Then press ▲ ▶ to move to list **L2**. Repeat the process to enter the data for lists **L2** and **L3**.

7. Repeat the process from *Creating a line of best fit*.
  a) What are these values of *a* and *b*?
  Round *a* to one decimal place and *b* to the nearest whole number. What does *a* tell you about the line of best fit?
  b) What is the correlation coefficient, *r*?
  c) How does the correlation coefficient compare to the coefficient from question 2? What does this suggest about the new model?

8. Repeat the process from *Extrapolating with the line of best fit*.
  a) What are the new predicted number of smog advisories for 2007 and 2010?
  b) Which line of best fit gave a more accurate prediction for 2007?
  c) Would you say that including outliers always weakens a model? Justify your answer.

9. Suppose an environmental group wants to show Ontarians that air quality in Ontario is getting worse. Use the data about the number of days with a smog advisory you stored in L3. Be sure to re-enter the data for 2005.
Analyse the data.
- Create a scatter plot relating the year and the number of days with a smog advisory.
- Describe the relationship as it appears on your scatter plot.
- Record the *r*-value for the correlation and compare it to the previous correlation coefficients you recorded in questions 2 and 7.
- Create a line of best fit and describe whether it provides a good model for the data.
- Write a conclusion stating whether the graph supports the environmental group's case.

10. The Ministry of the Environment says air quality in the province has been improving since 1988, but it expects the number of smog advisories to increase because it is doing a better job of monitoring air quality. What does this show about drawing conclusions about cause and effect based only on numerical data?

Reflect

- Suppose you used linear models to represent two different sets of data. Describe how you could decide which set of data is better represented by a linear model.
- What are advantages of using a graphing calculator to analyse data?

GAME

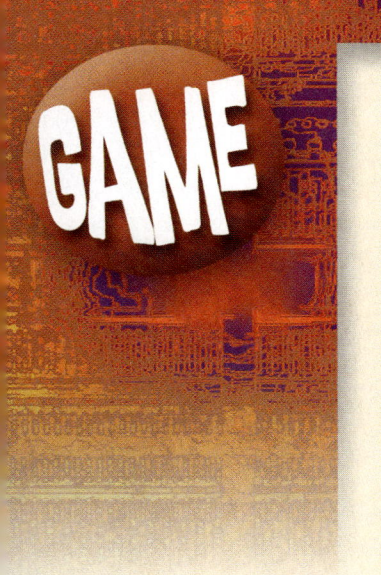

Give It to Me Straight

Materials

- TI-83 or TI-84 graphing calculator
- dice
- grid paper

Play the game in pairs. Each player will need a calculator.

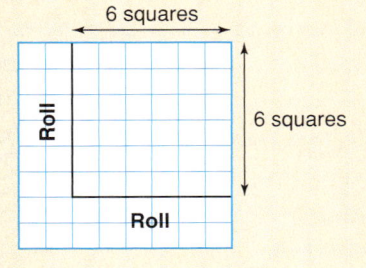

- On grid paper, draw and label axes as shown.
  On the graphing calculator, press [STAT] [ENTER] to access the Stat List Editor. Clear any data from L1 and L2.
- One player rolls the dice.
- The other player uses the numbers that are rolled as the coordinates of a point. For example, if 3 and 5 are rolled, the player can create (3, 5) or (5, 3). The player enters the coordinates in lists L1 and L2 of the graphing calculator and plots the point on the grid.
- Players take turns rolling the dice and choosing coordinates until each player has created 6 points. The goal is to get a higher correlation coefficient.
- Each player predicts the correlation coefficient for her or his data, and records the prediction as a decimal to the nearest hundredth.
- Players use the graphing calculators to determine the r-value for their data.
  Press [STAT] [▶] 4 [ENTER] to perform linear regression.
- The player whose data have an r-value closer to 1 or −1 gets 1 point.
- The player whose prediction of the r-value is closer to the actual value also gets 1 point.
- After five rounds, the player with more points is the winner.
  If there is a tie, play an extra round. The player with the higher r-value is the winner.

If the values of r^2 and r are not showing on the screen, the diagnostic mode is turned off. To turn it on, press [2nd] 0, scroll down the list to **DiagnosticOn** and press [ENTER] [ENTER].

Reflect

- Explain how you decided which number to use as the x-value and which to use as the y-value in a coordinate pair.
- What strategy did you use to predict the correlation coefficient?

3.5 Analysing Data Using a Spreadsheet

Do you like to sleep in? Maybe you should move to Resolute Bay, Nunavut where the sun doesn't rise for three months. But you will have to leave by April, when the town begins a three-month stretch of daylight!

Inquire

Analysing Scatter Plots Using Software

Materials
- *Microsoft Excel*
- daylighthours.xls
- snowrain.xls
- access to the Internet and E-STAT

Work with a partner.

Part A: Using *Microsoft Excel*

■ Open the file *daylighthours.xls*.

1. The spreadsheet shows the latitudes of different locations and the number of daylight hours on August 15. Describe the relationship between latitude and hours of daylight.

	A	B
1		Daylight Hours on August 15
2	Latitude (°N)	Length of daylight time (h)
3	0	12.1
4	10	12.5
5	20	12.8
6	30	13.3
7	40	13.8
8	50	14.5
9	60	15.8
10	70	18.5
11	80	24.0
12	90	24.0

CHAPTER 3: Two-Variable Data

- Highlight cells **A3** to **B12**.
  Click the **Chart Wizard** icon.
  Under *Chart type*: select **XY** (**Scatter**).
  Click **Next**. Accept the data range by clicking **Next** again.
  The title screen should appear. If it does not, click on the **Titles** tab.
  Enter the data headings as the chart titles.

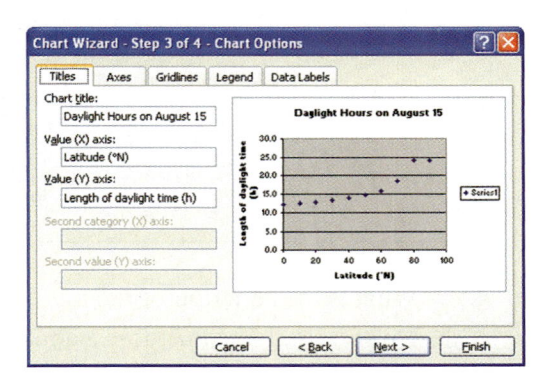

> You may find it easier to type "degrees North" than "°N".

- Click on the **Legend** tab. Deselect **Show Legend**.
  Click **Finish** to embed the graph into your spreadsheet.
- Right click the horizontal axis of the graph and select **Format Axis…**.
  Click on the **Scale** tab.
  Set **Minimum** to 0 and **Maximum** to 90.
- Right click the vertical axis and select **Format Axis…**.
  Click on the **Scale** tab.
  Set **Minimum** to 0 and **Maximum** to 24.

> The latitude of the equator is 0°N.
> The latitude of the North Pole is 90°N.

> There are 24 h in 1 day.

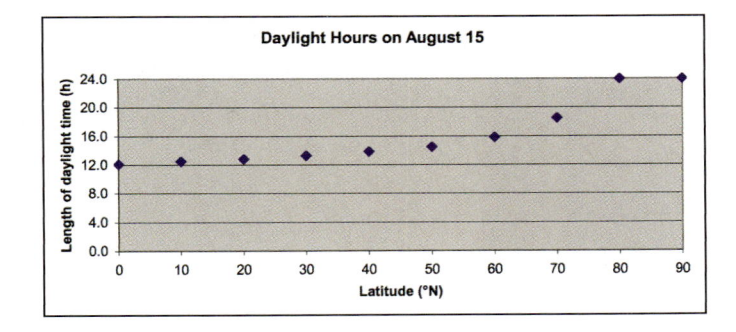

2. Describe the correlation between the variables. Compare it with your description of the relationship from question 1.

3. Do you think a linear model would represent the data well? Explain your thinking.

3.5 Analysing Data Using a Spreadsheet   **167**

> The **Chart** menu will not appear unless a graph is selected.

- From the **Chart** menu, select **Add Trendline**.
  Select **Linear** from the list of Trend/regression types.
  Click on the **Options** tab, then select **Display equation on chart**.
  Click **OK**. Your graph should look similar to the one shown here.

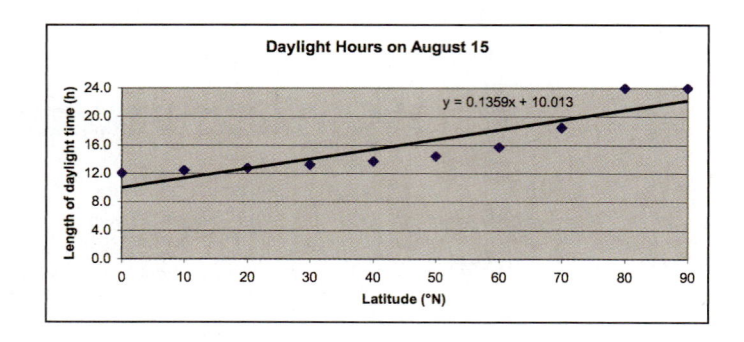

Daylight Hours on August 15

$y = 0.1359x + 10.013$

(Length of daylight time (h) vs. Latitude (°N))

4. a) What is the equation of the line of best fit?
  b) Do you think the line does a good job of representing these data? Justify your answer.
  c) Would the linear model provide reliable estimates of daylight hours? Justify your answer.

5. Environment Canada calculates "weather normals" that represent typical weather data for different locations. The current normals are based on data collected from 1971 to 2000.

	A	B	C
1	City or town	Average annual snowfall (cm)	Average annual rainfall (mm)
2	Belleville	155.7	735
3	Cameron Falls	237.5	576
4	Chalk River	195.4	669
5	Cobourg	106.0	765
6	Dresden	84.6	759
7	Hamilton	161.8	764
8	Kapuskasing	313.0	544
9	Renfrew	195.5	616
10	Sarnia	125.0	732
11	Sault Ste. Marie	302.9	634
12	Timmins	313.4	558
13	Toronto	133.1	709

Open the file *snowrain.xls*.

a) Do the two variables appear to be related? If so, describe the relationship. If not, explain why not.
b) Create a scatter plot for the data. Describe any correlation you see. Does the graph support your answer to part a?
c) Add a line of best fit to the graph. How well do you think it represents the data? Justify your answer.
d) Petawawa receives an average of 228.5 cm of snow each year. Based on the line of best fit, what would you expect the average annual rainfall to be in Petawawa? How close was the prediction to the actual average of 615.9 mm?

168 CHAPTER 3: Two-Variable Data

Part B: Using E-STAT

- Go to the Statistics Canada Web site.
  Click **English**.
  Select **Learning resources** from the menu on the left.
  Click on **E-STAT** in the yellow box on the right.
  Click on **Accept and enter**.
  If you are working from home, you will need to enter the user name and password assigned to your school.

> *CANSIM* provides data taken over time. *Census databases* offer information about entire populations taken once every 5 years.

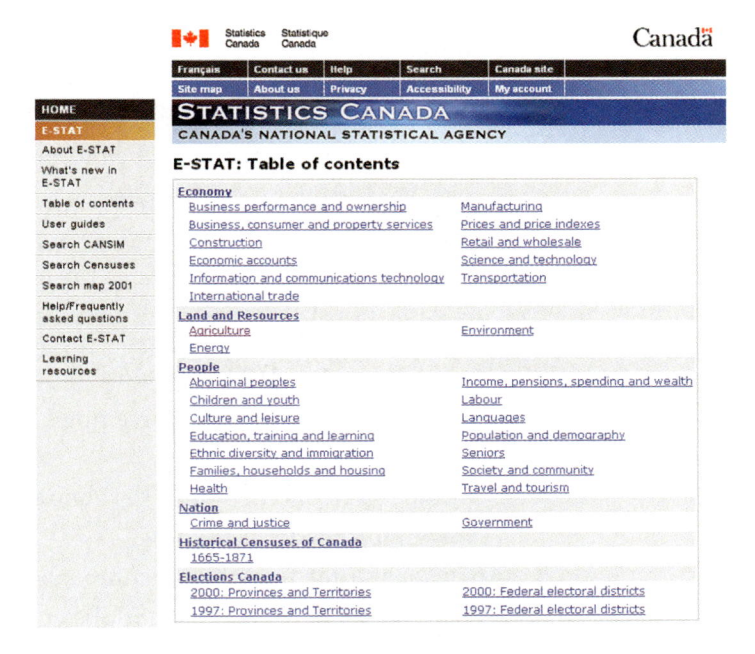

- The E-STAT table of contents will be displayed.
  In the *Land and Resources* section, click on **Agriculture**.
  From the list of CANSIM data, click on **Food and nutrition**.
  Click on table **002-0011**. This table contains data about the per capita consumption of various food and beverage items throughout Canada.

> *Per capita* means per person.

- Under *Food Categories*, select **Food available**.
  Under *Commodity* you will be selecting two items, but there are too many choices to see all at once. Click on **View checklist and footnotes** to display all the items more conveniently.
  Scroll down and select **Ice cream (litres per year)** and **Yogurt (litres per year)**.
  Scroll up to the top of the page and click **Return to picklist**.
  Set the Reference period from **1996** to **2006**.
  Click on **Retrieve as individual Time Series**.

3.5 Analysing Data Using a Spreadsheet **169**

- Create a scatter plot. Select **Scatter graph with line of best fit (linear regression)** and click **Retrieve Now.** Click on **Modify Graphic.** After **Origin:** select **Start axis at 0.** Click **Replot.**

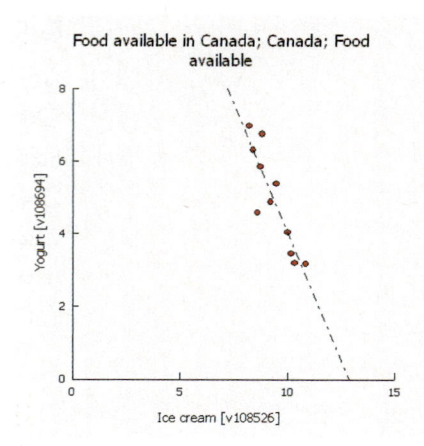

6. a) What two variables are being compared in this graph?
   b) What data are not included in the graph?
   c) Describe the correlation displayed by the scatter plot.
   d) What does the direction of the line of best fit tell us about the relationship between ice cream consumption and yogurt consumption? Why do you think this relationship occurs?

- Click on the back button three times to return to the *Output specification screen*.
  In the *Screen output – table* box, under *HTML, Table*, select **Time as rows** and click **Retrieve now**.
  Click and drag to highlight entire table, including the headings.
  With the table highlighted and the cursor on the highlighted table, right click and **Copy**.
- Open a new *Microsoft Excel* spreadsheet.
  Right click on cell A1 and select **Paste**.
  The data from E-STAT will appear.

Widen the columns to reveal the table headings.

	A Annual	B v108526 - Canada; Food available; Ice cream (litres per year)	C v108694 - Canada; Food available; Yogurt (litres per year)
2	1996	10.87	3.17
3	1997	10.35	3.19
4	1998	10.18	3.46
5	1999	10.02	4.05
6	2000	8.62	4.59
7	2001	9.22	4.88
8	2002	9.49	5.39
9	2003	8.76	5.85
10	2004	8.4	6.31
11	2005	8.84	6.76
12	2006	8.22	6.98

- Highlight the columns containing the ice cream and yogurt data. Click on the **Chart Wizard** icon in the toolbar.
  Follow the steps you learned in the first part of this *Inquire* to create a scatter plot. Enter appropriate titles and embed the graph into your spreadsheet.
  Add a trend line using the **Chart** menu.

7. Compare this scatter plot to the one you created in E-STAT. Which one do you prefer? Justify your answer.

8. Return to E-STAT table 002-0011 on food consumption in Canada.
   a) Retrieve data about the consumption, in litres, of standard (3.25%) milk and partly skimmed (1%) milk between 1996 and 2006.
   b) Would you expect the consumption of these two products to be related? Justify your answer.
   c) Graph the data as a scatter plot with a line of best fit. Describe the shape and direction of the correlation. What does the scatter plot tell you about the consumption of these two kinds of milk?

Reflect

- How does creating a scatter plot help you identify and describe trends in data that might not be obvious from looking at a table?
- Choose a data set from Lesson 3.2 or 3.3. Describe how to use a spreadsheet to graph the data set. What would be an advantage of using spreadsheet software to construct the scatter plot and draw the line of best fit?

3.5 Analysing Data Using a Spreadsheet

3.6 Analysing Data Using *Fathom*

In Ontario, every driver is required to have liability insurance on his or her vehicle. Liability insurance covers the potentially enormous costs if a person's actions cause damage to property or injury to a third party.

Inquire: Creating and Analysing Scatter Plots Using *Fathom*

Materials
- *Fathom*
- access to the Internet and E-STAT

Work with a partner.

Part A: Working with Given Data

Lena is researching the cost of the liability portion of car insurance to determine whether it is related to the car's value.
She has collected this data from various insurance brokers.

Value of car ($)	Liability insurance cost ($/year)	Collision insurance cost ($/year)
5 000	247	138
10 000	233	163
10 000	275	154
15 000	291	192
15 000	277	185
20 000	243	257
20 000	315	201
25 000	254	233
30 000	288	252
35 000	302	261

172 CHAPTER 3: Two-Variable Data

■ **Entering data in a table**
  • Start *Fathom* and begin a new blank document.
  • Click on the **Table** icon, then click in the document to start a table.
  • In the table, click **<new>** and enter the attribute **Value_of_car**. Press **Enter**.
  • A new cell should appear. Click **<new>** and enter the attribute **Liability_ins_cost**. Press **Enter**.
  • Repeat the process to enter the attribute **Collision_ins_cost**.
  • Enter the data into the table.
  • Double click on **Collection 1** in the upper left corner. Rename it **Car Insurance Costs**. Your table should look something like this.

Fathom allows only letters, digits, and underscores in attribute names.

Do not enter spaces between digits or the numbers will be treated as text.

To easily see all the information in a table, click and drag the edges of the table object and of the attribute cells.

	Value_of_car	Liability_ins_cost	Collision_ins_cost	<new>
1	5000	247	138	
2	10000	233	163	
3	10000	275	154	
4	15000	291	192	
5	15000	277	185	
6	20000	243	257	
7	20000	315	201	
8	25000	254	233	
9	30000	288	252	
10	35000	302	261	

■ **Creating a scatter plot**
You will begin by investigating whether there is a relationship between the liability insurance cost and the value of the car.
  • Click on the **Graph** icon, then click in the document to start a graph.
  • Drag the **Value_of_Car** table attribute to the horizontal axis of the graph. You should see some data plotted. Ignore them for now.
  • Drag the **Liability_ins_cost** table attribute to the vertical axis of the graph. The data should be plotted accurately now.

1. Does there appear to be a relationship between the two variables? If so, explain why and describe it. If not, explain why not.

3.6 Analysing Data Using *Fathom* **173**

> The **Graph** menu is in the toolbar above the **Graph** icon. It will not appear unless a graph is selected.

> *Least-squares line* is another term for the *line of best fit*.

■ **Drawing a line of best fit**
 • Click on the graph, then click on the **Graph** menu.
 • Select **Least-Squares Line**.
  The line of best fit should appear on the graph.
  The equation of the line appears below the graph.

2. a) Describe the line of best fit.
   b) Do you predict the correlation coefficient will be positive or negative? Will it be closer to 1 or closer to 0?
   Justify your predictions.

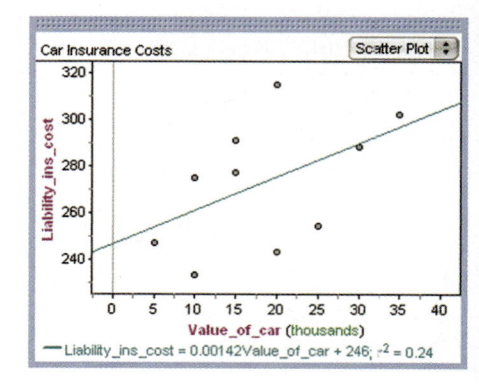

■ **Determining the correlation coefficient**
 • Click the **Model** icon, then click in the document to start a statistical model of the data.
 • Click on **Empty model**. A pop-up menu should display statistical calculations that can be performed.
   Select **Simple Regression**.
 • Drag the **Value_of_car** table attribute onto **Predictor attribute** in the model object.
 • Drag the **Liability_ins_cost** table attribute onto **Response attribute** in the model object.
   Simplify the information that appears by right-clicking in the model object, then deselecting **Verbose**.

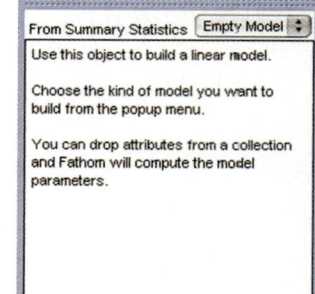

> The *predictor attribute* is the independent variable. The *response attribute* is the dependent variable.

3. The correlation coefficient, *r*, is displayed in the model object.
   a) Record the value of *r* to 2 decimal places.
   b) Is the correlation is strong or weak? Justify your answer.
   c) How does it compare to your prediction in question 2 part b?

174 CHAPTER 3: Two-Variable Data

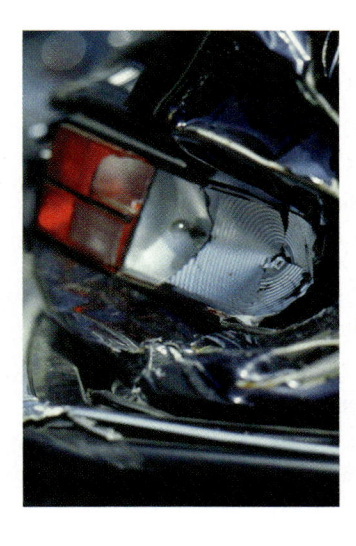

4. The equation of the line of best fit also appears in the model object.
  a) Write the equation. Use *x* to represent **Value_of_car**, and *y* to represent **Liability_ins_cost**.
  b) How confident would you be in interpolating or extrapolating liability insurance costs using this equation? Explain your thinking.

■ **Analysing another set of data**

Collision insurance only pays for damages to your vehicle in an accident. Collision insurance is not required by law in Ontario. You will investigate whether collision insurance cost is related to the value of the car.

5. Without deleting your first graph, drop a new graph in the document. You will graph car values and collision insurance costs.
  a) Which data should you place on the horizontal axis? Which data should you place on the vertical axis? Justify your answers.
  b) Plot the data as you described in part a.
  c) Does there appear to be a relationship between the variables? If so, tell why and describe it. If not, tell why not.

6. Display the line of best fit on the graph.
  a) Describe the line.
  b) Predict whether the correlation coefficient will be closest to 0, 0.25, 0.5, 0.75, or 1. Explain your thinking.

7. Drop the **Collision_ins_cost** table attribute onto **Response attribute** in the model object.
  a) What is the value of the correlation coefficient? Record your answer to 2 decimal places.
  b) How does it compare to your prediction in question 6 part b?

8. a) Write the equation of the line of best fit. Use *x* to represent **Value_of_car**, and *y* to represent **Collision_ins_cost**.
  b) How confident would you be in interpolating or extrapolating collision insurance costs using this equation? Explain your thinking.
  c) Use the equation in part a. Predict the costs of collision insurance for cars with each value.
  Car P: $12 000   **Car Q:** $40 000

3.6 Analysing Data Using *Fathom* **175**

9. When Lena was getting a collision insurance quote from a broker, she mistakenly stated that the vehicle would be used for business rather than for pleasure. As a result, one piece of data is invalid.
   a) Which point do you think it is? What is the name for this kind of point? Delete the point by right-clicking on it and selecting **Delete Case**.
   b) Describe how the line of best fit changes.
   c) What is the new *r*-value?
   d) What is the new equation of the line of best fit?
   e) What change occurred in the table when you deleted the point?

10. What assumptions do you need to make about Lena's data collection method in order to consider her data reliable?

Part B: Working with Data Imported from E-STAT

■ **Developing a conjecture**

In today's busy world, it can be difficult to juggle work and family. For example, parents may need to take time off work to care for sick children.

11. How might the number of absences due to family responsibilities for working mothers with children under the age of 5 be related to the number of children under 5 in Canada? Describe the correlation, if any, you might expect between these variables. This is your **conjecture** about the relationship.

■ **Finding E-STAT data**
   • Go to the Statistics Canada Web Site.
     Click **English**.
     Select **Learning resources** from the menu on the left.
     Click on **E-STAT** in the yellow box on the right.
     Click on **Accept and enter**.
     If you are working from home, you will need to enter the user name and password assigned to your school.

176 CHAPTER 3: Two-Variable Data

- The table of contents window will open.
  In the *People* section, click on **Labour**. From the list of CANSIM data, click on **Labour mobility, turnover and work absences**. Click on table **279-0033**.
  Under *Sex*, select **Females**.
  Under *Presence of children*, select **Preschoolers, under 5 years**.
  Under *Absence rates*, select **Days lost per worker in a year, personal or family responsibility**. Set the *Reference Period* from **1997** to **2006**.
  Click on **Retrieve as individual Time Series**.

You have selected data about the number of work days lost due to personal or family responsibilities for women with children under 5 from 1997 to 2006. Now you need to find the population of children under 5 over the same period of time.

- Scroll to the bottom of the screen and click on **Add more series**. Click on **Browse by subject**.
  The CANSIM table of contents appears.
  Click on **Population and demography.**

3.6 Analysing Data Using *Fathom*   **177**

- Click on **Population estimates and projections**.
  Then click on table **051-0001**.
  Under *Geography*, select **Canada**.
  Under *Sex*, select **Both sexes**.
  Under *Age group*, select **0 to 4 years**.
  Set the *Reference Period* from **1997** to **2006**.
  Click on **Retrieve as individual Time Series**.
  The titles of both data series you requested should be displayed.

Select an output format for the data.
- In the *Screen output – table* section, under *Plain text*: select **Table, time as rows**.
  Click on **Retrieve now**.
  The table should appear in text format.

■ **Importing and using E-STAT data in *Fathom***
- Highlight the data in the table, but do not include the column headings or any other information. Right click and **Copy** the highlighted material.
- Open a new *Fathom* document.
  Click the **Collection** icon, then click in the document to start a collection of data.
  Right click and **Paste Cases** into the collection.
- Click on the collection object, then drop a table into the document.
  The E-STAT data should immediately fill the table.
  Change the table attributes to **Year, Population_under_5, and Family_ absences**.

> To rename a table attribute, double click the name.

	Year	Population_under_5	Family_absences	<new>
1	1997	1917294	4.1	
2	1998	1872747	3.5	
3	1999	1828982	4.1	
4	2000	1791178	4.1	
5	2001	1759196	4.5	
6	2002	1730473	5.2	
7	2003	1710647	4.8	
8	2004	1705488	4.5	
9	2005	1702406	5.1	
10	2006	1712848	6.2	

Collection 1

12. Drop a graph into the document.
- Make a scatter plot with population as the independent variable and absences as the dependent variable.
- Describe the relationship and whether your conjecture from question 11 was correct.

13. In the last 10 to 15 years, many companies have improved their policies about personal days for employees. Drop a new graph into the document.
- Make a scatter plot with year as the independent variable and absences as the dependent variable.
- Describe the relationship.

14. Display the line of best fit for each graph.
- Use a **Model** object to view the correlation coefficient for each line.
- Which relationship shows a stronger correlation?
- What does this suggest about the data?

15. What other factors could be affecting the number of days parents with young children take off for personal and family matters?

> ### Reflect
>
> - Suppose you want to create a scatter plot relating two variables for a table of data in *Fathom*. Explain how to decide which variable to plot along the horizontal axis.
> - Why is it useful to know the equation of the line of best fit? Include an example.
> - Researchers often make a conjecture before they begin to investigate a possible correlation between variables. What are advantages and disadvantages of making a conjecture before collecting and analysing data?

3.7 Conducting an Experiment to Collect Two-Variable Data

College students in fishery and aquaculture programs look for correlations between variables such as water temperature and egg hatching, water depth and water quality, or types of food and fish growth.

Inquire | Designing and Conducting Experiments

Materials
- *Fathom, Microsoft Excel*, TI-83 or TI-84 graphing calculator, or grid paper
- materials for the experiment

Work with a partner or in a group.

■ **Planning a question or conjecture**

When you pose a question or conjecture for an experiment to collect two-variable data, you should follow these steps.
- Find a topic that interests you.
- Identify the two variables for the topic.
- Pose a question or make a conjecture about a possible relationship between the variables.

1. Identify the variables that are being studied in each experiment.
   a) Kasey wants to prove that a person's height is positively related to the number of basketball free throws he or she can successfully make in 10 attempts.
   b) Ramon wants to determine if there is a relationship between a person's height and walking speed.

2. Write a question or conjecture for each experiment in question 1.

CHAPTER 3: Two-Variable Data

- **Developing a method**

    When you develop a method for measuring or collecting data, think about these issues.
    - Reduce the effects of other variables by keeping all other factors in the experiment the same, as much as possible.
    - If you cannot think of a way to measure each variable, plan a different experiment.

3. Refer to Ramon's experiment in question 1.
    a) What method could Ramon use to measure height?
    b) What method could he use to measure walking speed?
    c) What can be different about the people he includes in his experiment?
    d) Explain why the people should be about the same in these ways: age, body type, and fitness level.

- **Determining what materials you will need**

    As you develop a method, plan what you will need for the experiment. Keep the availability and the cost of materials in mind.

4. Refer to question 3. Describe materials for Ramon's experiment.

- **Writing a plan**

    A plan should include these items.

    - A statement outlining the question you want to answer or the conjecture you want to prove
    - Details of your method, including the expected length of observation time or the number of observations or trials
    - A strategy to limit the influence of other variables
    - A list of the materials
    - Any tables or checklists needed for recording your findings

    You should have at least one person read your plan. Ask the reader if he or she understands your plan, if anything has been left out, and if the plan seems possible.

5. Safety/health issues, expense issues, privacy issues, and sensitivity to people's feelings are reasons an experiment may not be appropriate. Explain how one or more of these issues might apply to each experiment.
   a) Faria wanted to conduct a blind taste test where people of various ages had to identify what they had just eaten as quickly as possible.
   b) Davian's experiment was aimed at finding a connection between hours spent doing homework and the mark earned in a course.
   c) Travis likes to work out. His experiment involved looking for a correlation between the circumference of a person's bicep and the maximum weight he or she can bench-press.

■ **Avoiding problems when collecting experimental data**
   • Except for the variables you are measuring, keep the characteristics of your sample as similar as possible.
   • Collect a large sample of data that includes a range of values.
   • Measure or count as accurately as possible.

6. Explain the problem with how each person collected data.
   a) Alisha investigated how height is related to successful basketball free throws. To find people of different heights, she asked a student from each of Grades 2 to 12.
   b) For a ball-rolling experiment, Dexter needed to mark a distance of 20 feet. He estimated that one of his paces is 4 feet long, then took five paces, and marked the distance.

c) For an experiment comparing leg length to jump height, Alex collected data from his parents and three sisters. He plotted all five points, and concluded there was no correlation.

d) Madison wanted to determine if there is a relationship between a person's age and the ability to hear high-pitched frequencies. She collected data from many students in her grade.

7. The student-run store at Bernadette's school sells coffee, tea, and hot chocolate. She wants to determine if there is a relationship between the outside temperature and the number of hot drinks the store sells before school.

   a) What factors other than temperature might affect the sales? How can Bernadette work around these factors?
   b) For how many days should Bernadette collect data? Explain your reasoning.
   c) Write a plan for Bernadette's experiment. Include tables or checklists she would need.

■ **Planning and conducting your own experiment**

8. Choose one of these topics or think of your own topic.

   > Is there a correlation between these variables?
   > - The circumference of a straw and how long it takes to drink a specific volume of liquid
   > - Hand span and the number of small items a person can pick up at once
   > - The number of students in the cafeteria during a period and the number of students in the library
   > - The number of sit-ups and number of push-ups that a person can do in 1 min
   > - Age and a person's ability to remember objects that are displayed for 1 min, then hidden
   > - Height and a person's lung capacity
   > - The type size of a page of text and how long it takes to read the text

   - Identify the variables for your topic. Develop a question or conjecture about a possible relationship between the variables.
   - Write a plan for an experiment to test your question or conjecture. Include the elements outlined in this lesson.
   - Get approval from your teacher.
   - Gather the materials.
   - Conduct the experiment.

3.7 Conducting an Experiment to Collect Two-Variable Data   **183**

■ **Displaying and analysing your data**

> If computers or graphing calculators are available, use them to plot your data.

9. Make decisions about how you will display your data in a scatter plot.
- Create a scatter plot for your data.
- Is there a correlation? If so, describe it.
  If not, tell how the scatter plot shows this.
- If appropriate, include a line of best fit and its equation.
  If a linear model is not appropriate, explain why.
- Write a conclusion about your question or conjecture.
  Tell how your scatter plot supports this.

Reflect

- What was the most challenging part of planning or conducting your experiment? How did you deal with this challenge?
- What would you do differently if you were to repeat the experiment?
- If you modelled your data with a line of best fit, how confident are you in the model? Justify your answer.

Study Guide

Scatter Plots and Correlations

A correlation indicates a relationship between two variables.
- In a positive correlation, points on the scatter plot go up to the right.
- In a negative correlation, points go down to the right.

A correlation does not necessarily indicate a cause-and-effect relationship.

A linear correlation is weak if the points are spread out.
A linear correlation is strong if the points appear to lie along a line.
If there is no trend, there is no correlation.

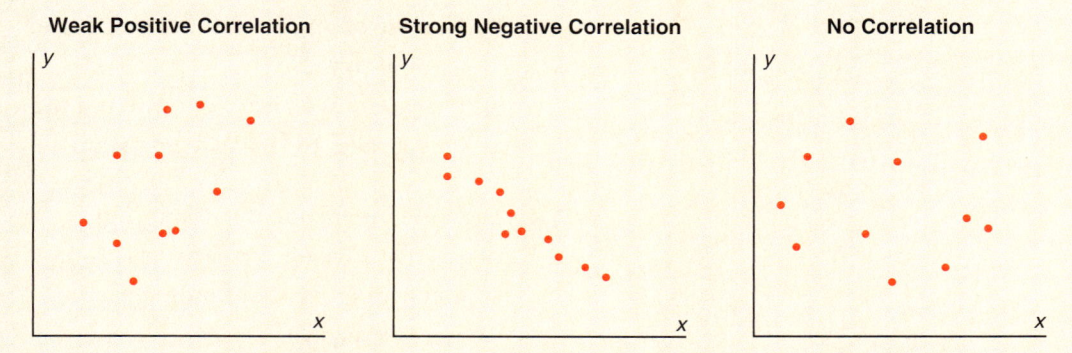

Weak Positive Correlation **Strong Negative Correlation** **No Correlation**

A point that does not follow the trend shown by the rest of the data is an outlier.

Line of Best Fit

Other names for the line of best fit are *regression line* or *trend line*.

To determine an equation of the line of best fit, find its slope, m, and y-intercept b:
$y = mx + b$

Predictions for data values between known points are called **interpolation**.
Predictions for data values beyond known points are called **extrapolation**.

These factors can reduce the reliability of a linear model:
- A weak linear correlation
- A non-linear correlation
- Data that are too clustered
- Outliers in the data
- A data sample that is too small
- A data sample that is biased

Chapter Review

3.1

1. State whether each situation involves one-variable or two-variable data.

a) **Driving Status of Students in Course**

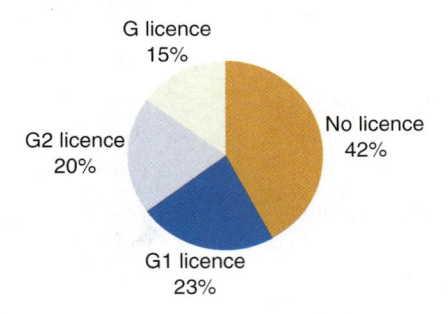

- G licence 15%
- No licence 42%
- G1 licence 23%
- G2 licence 20%

b) **Monthly Precipitation**

Rainfall (mm)	Snowfall (cm)
19	32
21	26
35	20
56	7
69	0

c) According to data Hayden collected, his classmates watch an average of 11.4 h of television per week.

2. Each situation below involves two-variable data. Identify the two variables.
a) In science class, students learn that air pressure decreases as height above Earth's surface increases.
b) Assuming that rainfall stays within the seasonal range, higher amounts of rainfall increase crop yield.
c) The time taken to cook a turkey increases as the turkey's mass increases.

3.2

3. Predict whether the variables in each situation would have a negative correlation, a positive correlation, or no correlation.
a) The number of people playing a board game and the likelihood that you will win
b) The mass of a car and its selling price
c) The number of people in a household and the monthly household water use
d) The size of a pizza and the number of toppings on it
e) The speed of a river's current and the time it takes to travel downstream

4. State the independent variable and dependent variable for each situation.
a) Aquariums that hold more water are more expensive.
b) The more rain we have, the less time people spend watering their lawns.

5. Identify which cylinder is represented by each point in the scatter plot.

Cylindrical Containers

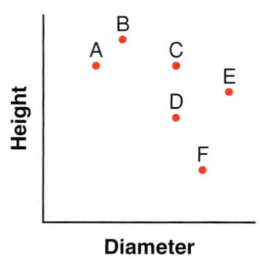

186   CHAPTER 3: Two-Variable Data

6. This table compares the number of jigsaw puzzle pieces and the manufacturer's recommended minimum age.

Number of pieces in puzzle	Recommended minimum age
100	5
500	7
48	3
65	4
300	9
150	6
550	12
300	6
200	8
100	4

a) Graph the data and sketch the line of best fit.
b) What conclusion can you make from the graph?

7. Explain why the following data analysis is not reasonable. There may be more than one reason.

A car manufacturer tested one of their vehicle models and concluded that the linear correlation between speed and fuel consumption is strong.

Effect of Vehicle Speed on Fuel Consumption

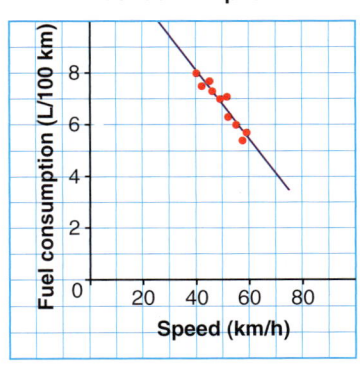

8. Describe each correlation.
a)
b)
c)

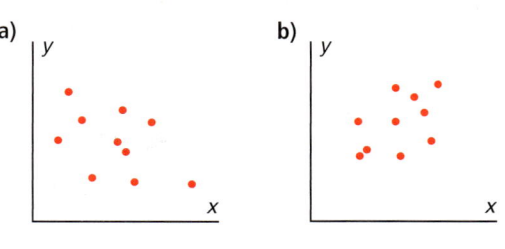

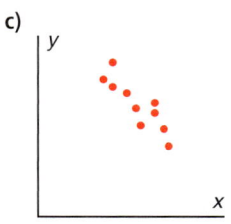

9. The Body Mass Index (BMI) relates a person's weight and height. A clinician gathered age and BMI data for a group of people aged 8 to 17 years.

Age (years)	BMI
8	15.5
8	17.0
10	16.0
10	18.0
12	20.0
13	19.0
13	20.0
14	20.0
15	21.0
15	20.0
16	22.5
16	21.0
17	22.0
17	21.0

a) Graph the data. Identify any outliers, if they occur.
b) Sketch the line of best fit and determine its equation.
c) Describe any correlation you see between the variables. What conclusion could someone make based on the scatter plot?
d) How many people were included in the sample? Do you think these data are enough to justify a conclusion about people in this age group? Explain.

Chapter Review **187**

10. Refer to question 6.
   a) Create a scatter plot of the data.
   b) Use technology to determine the equation of the regression line.
   c) How many pieces would be in a puzzle with a recommended minimum age of 10?
   d) What would be the recommended minimum age for a jigsaw puzzle with 400 pieces?

11. Open the file *daytona.xls*, which gives data on average speeds of the winning drivers of the Daytona 500.

	A	B	C
1	Year	Winning driver	Average speed (mph)
2	1960	Junior Johnson	125
3	1965	Fred Lorenzen	142
4	1970	Pete Hamilton	150
5	1975	Benny Parsons	154
6	1980	Buddy Baker	178
7	1985	Bill Elliot	172
8	1990	Derrike Cope	166
9	1995	Sterling Marlin	142
10	2000	Dale Jarrett	156
11	2005	Jeff Gordon	135

   a) Create a scatter plot for the data. Describe the correlation.
   b) Create a line of best fit. Use it to predict the speed of the winning driver in 2010.
   c) How reliable is the prediction you made in part b? Explain.
   d) Do you think it is acceptable to include only every 5th year? Justify your answer.

12. The table shows the rate of injury among young workers compared to the actual number of injury claims that were submitted between 1996 and 2005.

Number of claims	Injury rate (%)
11 657	5.0
11 612	5.0
11 006	4.8
9980	4.4
11 040	4.4
10 595	4.1
8630	3.5
8962	3.4
9460	3.5
10 280	3.4

   a) Open *Fathom*. Create a table and scatter plot for these data.
   b) Draw a least-squares line and determine the value of the correlation coefficient.
   c) What does the *r*-value suggest about the type of correlation that exists between the variables?
   d) How might you explain that in 2005, the injury rate decreased but the number of injury claims increased?

13. Esin wants to determine if there is a correlation between a person's weight and the number of chin-ups the person can perform in 1 min.
   a) Write a plan for an experiment Esin could perform to collect data.
   b) Esin wants volunteers with a variety of weights. How could he keep other physical factors from affecting his results?
   c) What ethical issues will Esin have to consider?

Practice Test

Multiple Choice: Choose the best answer for questions 1 and 2. Justify each choice.

1. Which is another name for the line of best fit?
   A. Response line   B. Explanatory line   C. Trend line   D. Predictor line

2. Which phrase best describes the correlation between the number of people attending a movie and the number of empty seats remaining in the theatre?
   A. Strong and positive   B. Strong and negative
   C. Weak and positive   D. Weak and negative

Show your work for questions 3 to 6.

3. **Knowledge and Understanding** This table shows data for ten players from the Toronto Blue Jays for the 2007 season.
   a) Create a scatter plot of the data.
   b) Does there appear to be a correlation between the number of times at bat and the number of hits? If so, describe it. If not, explain why not.
   c) Draw a line of best fit. Use it to estimate the number of hits a player might have after 100 times at bat.

At bats	Hits
643	191
385	101
608	177
584	143
531	147
327	82
357	103
425	102
290	69
331	80

4. **Communication** Rosa wants to use data she has collected to investigate the correlation between people's leg strength and arm strength, then use the data to make predictions. Describe the steps she should follow.

5. **Thinking** Zaki heard that if you drop a piece of buttered bread, it lands with the buttered side down. Zaki decided to test this theory.
   a) Briefly describe an experiment Zaki could conduct.
   b) What are some issues Zaki should consider to ensure that his experimental results are valid?
   c) Is Zaki's experiment about a correlation between two variables? Explain.

6. **Application** Use this scatter plot.
   a) Describe the correlation.
   b) Would it be reasonable to draw a line of best fit for these data? Justify your answer.
   c) Identify a point that might be considered an outlier. What might the outlier suggest about this particular school district?

Schools in Nine Ontario School Districts

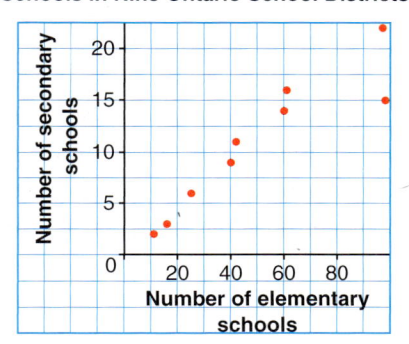

Chapter Problem: Temperatures around the Globe

Materials
- *Microsoft Excel* or *Fathom* or grid paper
- jantemps.xls or jantemps.ftm

Latitude and *longitude* describe the location of places on Earth.
- Latitude describes the location in degrees north or south of the Equator.
- Longitude describes the location in degrees east or west of the Prime Meridian.

Ottawa
Latitude: 45.3°N
Longitude: 75.7°W

1. Predict whether there is a relationship between the mean January temperature of a North American city and the city's position north of the Equator. Explain.

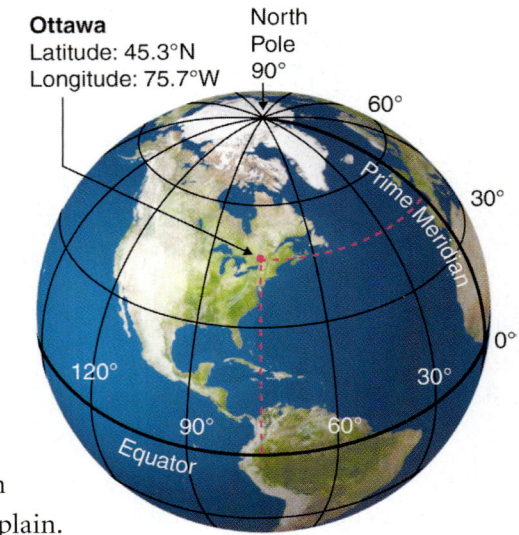

2. Open *jantemps.xls* or *jantemps.ftm*. Create a scatter plot for January temperature and latitude or use grid paper to plot the data shown in this table. Describe the correlation.

   Plot latitude on the horizontal axis.

	City	JanuaryTemp_deg_F	Latitude_deg_N	Longitude_deg_W
1	Key West, FL	65	25.0	82.0
2	New Orleans, LA	45	30.8	90.2
3	Atlanta, GA	37	33.9	85.0
4	San Francisco, CA	42	38.4	123.0
5	St. Louis, MO	24	39.3	90.5
6	Denver, CO	15	40.7	105.3
7	New York, NY	27	40.8	74.6
8	Toronto, ON	23	43.6	79.6
9	Ottawa, ON	12	45.3	75.7
10	St. John's, NF	22	47.9	52.5
11	Vancouver, BC	37	49.2	123.2
12	Winnipeg, MB	-1	49.9	97.2
13	Edmonton, AB	9	53.5	113.5
14	Whitehorse, YT	-1	60.6	135.6
15	Yellowknife, NT	-18	62.5	114.5

3. Draw a line of best fit and determine the equation of the line.

4. Predict whether there is a relationship between the mean January temperature of a city in North America and the city's position west of the Prime Meridian. Check your prediction.

5. Make your own prediction about the data. Check your prediction.

4 Statistical Literacy

What You'll Learn

To interpret statistical data and to assess whether conclusions based on statistical analysis are valid

And Why

Statistical data are often included in news articles, informational pamphlets, and advertisements. As a media consumer, you need to know how to analyse data to avoid being misled.

Key Words

- percentile
- quartile
- poll
- margin of error
- representative sample
- bias
- valid conclusion
- index
- base value
- inflation

CHAPTER 4

Activate Prior Knowledge

Ratios

Prior Knowledge for 4.1

A **ratio** compares two or more quantities. Ratios that make the same comparison are **equivalent**.

Example

A marketing company conducted a survey at a mall. Researchers asked 40 teens whether they owned a car. Eight teens owned a car.
a) What is the ratio of teens who owned a car to the total number surveyed?
b) Write an equivalent ratio with each second term:
  i) 20
  ii) 100

Solution

a) 8 teens owned cars out of 40 teens surveyed.
The ratio is 8:40.

b) Divide or multiply each term of the ratio in part a by the same number.
  i) 8:40 = ?:20
  Since 40 ÷ 2 = 20,
  divide each term by 2.
  8:40 = (8 ÷ 2):(40 ÷ 2)
  = 4:20

  ii) 8:40 = ?:100
  Since 40 × 2.5 = 100,
  multiply each term by 2.5.
  8:40 = (8 × 2.5):(40 × 2.5)
  = 20:100

CHECK ✓

1. Thirty people tasted a new brand of cheese. Twenty of them liked the taste.
   a) What is the ratio of those who liked the taste to the total number who tried it?
   b) Write an equivalent ratio with each second term:
     i) 6
     ii) 100

2. There are 700 students enrolled in a high school. The ratio of girls to boys is 4:3.
   a) How many boys and how many girls go to this school?
   b) The average class size is 28 students. Suppose this class is representative of all the students in the school. How many students in the class are girls? How many are boys?

3. In question 1, what is the meaning of the equivalent ratio with a second term of 100?

Measures of Central Tendency and Range

Prior Knowledge for 4.1

The **mean, median,** and **mode** are measures of central tendency for a data set. They represent a typical value for the set.
The **range** is a measure of spread. It tells you the difference between the greatest and least numbers in the set.

> Mean, median, and mode are defined in the glossary at the back of this book.

Example

Eight students received these marks on a test: 85, 76, 91, 65, 68, 72, 78, 43
a) Calculate the mean, median, and mode mark.
b) Which measure of central tendency best describes these data? Explain.
c) Determine the range of the data set.

Solution

a) For the mean:
$(85 + 76 + 91 + 65 + 68 + 72 + 78 + 43) \div 8 = 578 \div 8$, or 72.25
The mean is 72.25.
For the median, list the numbers in order.
43, 65, 68, **72, 76,** 78, 85, 91
Since there are 2 middle numbers, add them and then divide by 2.
$(72 + 76) \div 2 = 148 \div 2$, or 74
The median is 74.
The mode is the number that occurs most often.
Each number occurs once. So, there is no mode for this data set.

b) The mark 43 is much less than the other marks. It reduces the mean, but has no effect on the median. There is no mode. So, the median best describes the data.

c) The range is the difference between the greatest and least marks:
$91 - 43 = 48$

CHECK ✓

1. The heights, in metres, of trees in a woodlot are as follows:

| 18.0 | 21.3 | 17.1 | 23.5 | 19.8 | 17.9 | 17.0 | 21.5 | 19.2 | 19.0 | 20.6 | 19.5 |

a) Calculate the mean, median, and mode tree height.
b) Determine the range of this data set.

2. Which measure of central tendency best describes the data in question 1? Explain.

Activate Prior Knowledge   **193**

Percent Increase and Decrease

Prior Knowledge for 4.5

Percent increase and **decrease** are often used to describe how quantities have changed over time. They describe the change in a quantity as a percent of its original value.

Example

In August, a sports store priced a shoe at $125.99.
The shoes were so popular that the store increased the price to $134.99 in September. Shoe purchases declined. So, the store reduced the price to $119.99 in October.

a) What was the percent increase in price from August to September?
b) What was the percent decrease in price from September to October?
c) What was the percent decrease in price from August to October?

Solution

For each calculation, write the difference in the prices as a fraction of the earlier price. Then multiply by 100 to determine the percent increase or decrease.

a) $\dfrac{\$134.99 - \$125.99}{\$125.99} \times 100 \doteq 7.1$ The price increased by about 7.1%.

b) $\dfrac{\$119.99 - \$134.99}{\$134.99} \times 100 \doteq -11.1$ The price decreased by about 11.1%.

c) $\dfrac{\$119.99 - \$125.99}{\$125.99} \times 100 \doteq -4.8$ The price decreased by about 4.8%.

CHECK ✓

1. The population of a city was 18 500 last year.
   This year the population is 21 300.
   What is the percent increase in population?

2. Ms. Voisin was trying to sell her house for $325 000.
   When it had not sold after several weeks, she lowered the asking price to $298 000.
   What is the percent decrease in price?
   Use this information to write an advertisement headline for the house.

Transitions

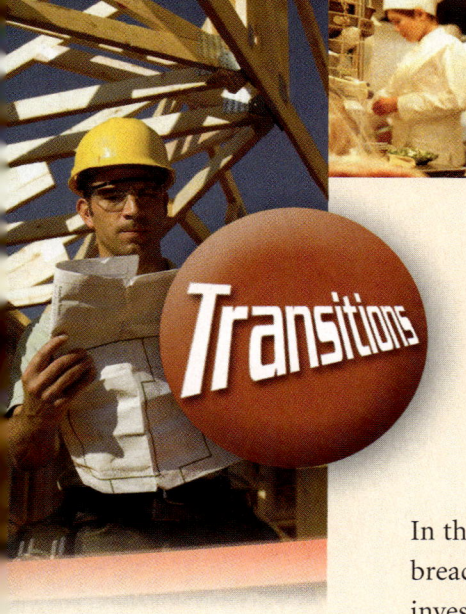

Ethical Conduct

Colleges and workplaces have strict rules about ethical conduct. Breaking these rules can have serious consequences. College students can be suspended; tradespeople may lose their right to practise; or employees can lose their jobs.

In these scenarios, the student, apprentice, or employee has committed a breach of ethics. Use your Internet skills and general knowledge to investigate rules and penalties for such conduct.

- Troy, a second year fashion student, hands in essays with paragraphs copied from the Internet. He does not reference the sites.
- Nora quotes $1200 to insulate an attic with 10 inches of cellulose fibre. She uses only 8 inches, but says the job is done and collects $1200.
- Joshua uses his 15% discount to purchase appliances for friends.
- Jane got an A on an essay in a business management course. She lends the paper to friends who take similar courses with different professors and at different schools.
- Stefan is an apprentice plumber who does jobs on weekends without permits. On these jobs, he does not make sure everything is "to code."
- Sharon, a receptionist at a doctor's office, discusses some of the patients and their medical problems with friends.
- Jon takes home printer paper from work to use for his evening courses.

1. Describe ethical problems in these scenarios.
2. As you create work of your own and work with others in this chapter, make notes about how you can follow ethical standards.
3. Imagine being an apprentice, employee, or college student. How could you apply what you have learned about ethics?

Search words for student ethics
- ☐ Academic misconduct
- ☐ Academic dishonesty
- ☐ Academic integrity
- ☐ Plagiarism
- ☐ Student resposibilities
- ☐ Student regulations
- ☐ College policies

• Complete questions 2 and 3 after you complete Chapter 4.

Search words for workplace ethics
- ☐ Ethical conduct
- ☐ Ontario employment standards
- ☐ Ontario plumber ethics
- ☐ Ontario paramedic ethics
- ☐ Workplace fraud
- ☐ Workplace racism
- ☐ Corporate bullying
- ☐ Employee code of conduct

4.1 Interpreting Statistics

A nurse uses growth charts to monitor a child's development. Growth charts are graphs showing what percent of children in the population are at or below particular heights or weights for different ages.

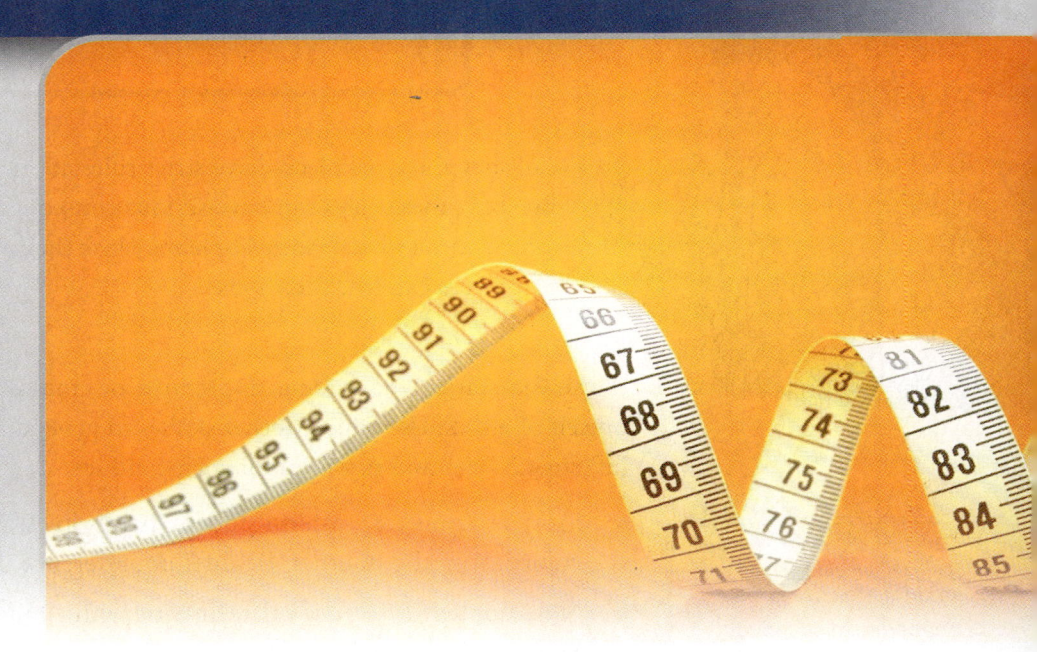

Investigate — Organizing and Describing Data

Materials
- scientific calculator

Work with a partner.

The 16 students in Jesse's math class measured their heights to the nearest centimetre.

> 160, 178, 167, 180, 168, 157, 164, 179,
> 163, 182, 176, 170, 172, 165, 175, 167

- Determine the measures of central tendency and the range for this set of data.
- What percent of the class is shorter than each measure of central tendency?
- Rylan is taller than 65% of the class. How many students are shorter than he is? What is Rylan's height?

196 CHAPTER 4: Statistical Literacy

> **Reflect**
>
> - Why might people want to know how they compare in height to classmates or to other people their age?
> - How might information about typical heights and weights at different ages be of use to a clothing manufacturer?

Connect the Ideas

Television, radio, newspapers, and Web sites often report statistical data. To understand these reports, you need to be familiar with the statistical language they use.

Percentiles

A **percentile** tells approximately what percent of the data are *less than* a particular data value. Percentiles are a good way to rank data when you have a lot of data or want to keep data private.

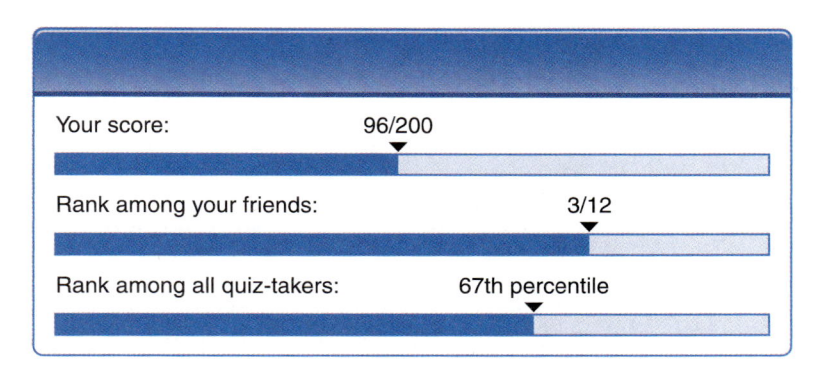

```
Your score:                    96/200
Rank among your friends:                  3/12
Rank among all quiz-takers:        67th percentile
```

Quartiles

A **quartile** is any of three numbers that separate a sorted data set into four equal parts.

- The second quartile is the median.
  It cuts the data set in half.
  So, it is the same as the 50th percentile.
- The first, or lowest, quartile is the median of the data values less than the second quartile.
  It separates the lowest 25% of the data set.
  So, it is the same as the 25th percentile.
- The third, or upper, quartile is the median of the data values greater than the second quartile.
  It separates the highest 25% of the data set.
  So, it is the same as the 75th percentile.

Example 1 — Working with Percentiles

Here are the hourly pay rates, in dollars, for 17 high-school students with part-time jobs.

11.50	10.20	8.00	8.25	9.00	9.15
9.75	7.50	8.00	12.50	13.00	11.25
10.75	9.50	9.25	9.45	7.75	

a) What are the quartiles for this data set?
b) Damien's pay is in the 85th percentile for this group. What does the percentile mean? What is Damien's hourly pay rate?

Solution

a) Start by finding the median. Order the data.
7.50, 7.75, 8.00, 8.00, 8.25, 9.00, 9.15, 9.25, **9.45,**
9.50, 9.75, 10.20, 10.75, 11.25, 11.50, 12.50, 13.00
The second quartile is the median: $9.45 per hour

Look at the ordered data that are less than the second quartile.
The lower quartile is the median of these data:
$$\frac{\$8.00 + \$8.25}{2} \doteq \$8.13 \text{ per hour}$$

Look at the ordered data that are greater than the second quartile.
The upper quartile is the median of these data:
$$\frac{\$10.75 + \$11.25}{2} = \$11.00 \text{ per hour}$$

7.50, 7.75, 8.00, 8.00, 8.25, 9.00, 9.15, 9.25, **9.45**, 9.50, 9.75, 10.20, 10.75, 11.25, 11.50, 12.50, 13.00

- Lower quartile: 8.13
- Second quartile: 9.45
- Upper quartile: 11.00

b) The 85th percentile means approximately 85% of the students in the group earn less money per hour than Damien.
$17 \times 0.85 = 14.45$
Round down to the nearest whole number to determine the number of students who earn less money per hour than Damien: 14
So, Damien is the 15th student in the ordered list.
He earns $11.50 per hour.

Data reliability

When you read statistical data, you need to think about the reliability of the source. Data from a government agency are usually more reliable than data from someone who is trying to sell a product or promote a point of view.

Example 2

Comparing Data Sources

In each case, a research topic and two sources of information are described. Decide which data source is more likely to provide reliable data.

	Topic	Source 1	Source 2
a)	The benefits or adverse effects of drinking milk	A pamphlet from an animal rights group that opposes dairy farming	Canada's Food Guide produced by Health Canada
b)	The effect of logging on the population of a species of bird	A pamphlet from a wildlife protection organization	A forestry company advertisement
c)	Possible complications of flu shots	A Ministry of Health Web site	A Web site run by a group that opposes immunizations

Solution

a) The animal rights group is promoting a particular point of view and may not be objective. The Food Guide was developed in consultation with thousands of dietitians, scientists, physicians, and public health personnel from across Canada. It will represent a more balanced view.

b) Both the wildlife organization and the forestry company are promoting particular points of view. It would be best to look for additional sources.

c) While the Ministry of Health Web site promotes the use of immunizations, it also describes possible complications. A group opposing immunizations is more likely to present only one side of the issue.

Polls

Polling companies conduct interviews with randomly selected Canadians to determine their opinions about a variety of topics. These surveys are called **polls**.

The results of polls are often reported in the media, particularly during elections. Poll results usually state a **margin of error** that describes how reliable the data are. As a media consumer, you need to know how to interpret these results.

Example 3

Interpreting Poll Results

The results of a poll conducted by EKOS in 2005 are shown.

"Canada should increase its humanitarian aid to poor countries even if it means less spending in other important areas"

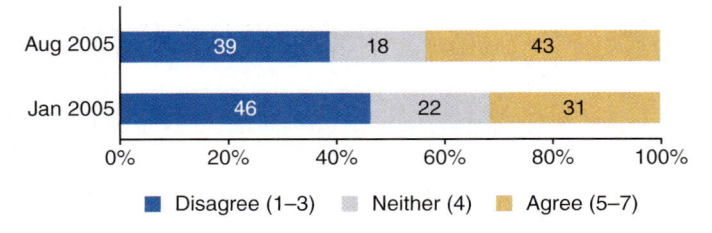

	Disagree (1–3)	Neither (4)	Agree (5–7)
Aug 2005	39	18	43
Jan 2005	46	22	31

a) What question were people asked?
b) How did the favourable responses compare in January and August?
c) A line below the graph stated "The results are valid within a margin of error of plus or minus 2.5 percentage points, 19 times out of 20." What does this mean?

Solution

a) People were asked whether they agreed with the statement: "Canada should increase its humanitarian aid to poor countries even if it means less spending in other important areas."
b) In January, 31% of the people polled agreed with the statement. In August, the percent agreeing had increased to 43% of those polled.
c) If you had taken another poll, there is a 19 out of 20, or 95% chance that the results would be within 2.5 percent of these results. That is, in August, somewhere between 40.5% and 45.5% of the people polled would agree with the statement.

Practice

A

For help with questions 1 to 3, see Example 1.

1. Determine the quartiles for each data set.
   a) 10, 8, 12, 15, 9, 9, 11, 12, 12, 8, 14, 11
   b) 170, 162, 150, 165, 180, 165, 154, 163, 168, 164, 172

2. Identify the statistical information in each excerpt from a Statistics Canada news release about a survey on college and university graduates' debt 5 years after graduation.

a) Two out of five graduates from the class of 2000 who had left school owing money to government student loans had completely repaid their debt five years after graduation.

b) Of all graduates from a Canadian college or university in 2000, 56% had no debt from government student loan programs while 44% owed money to such programs.

c) Slightly less than half of the graduates who still owed money on their student loans reported having difficulty repaying these loans, compared to one out of five among graduates who had paid off their loans by 2005.

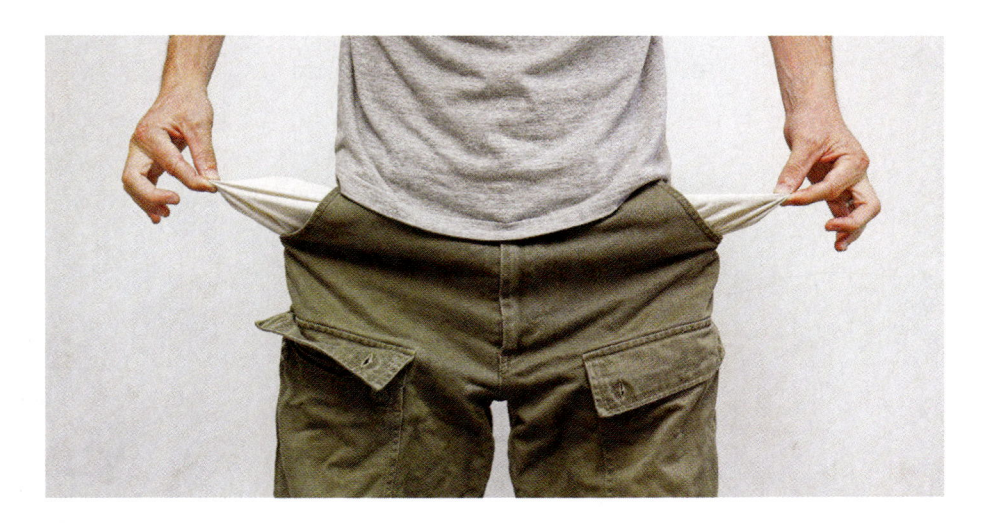

3. The Grade 9 students in a high school participated in a national standardized math test. The principal reported on the school Web site that three of the students placed in the upper quartile. Which sentence best describes the meaning of this statement?
   i) Three students received a mark of at least 75%.
   ii) Three students did better than 75% of all those who wrote the test.

4.1 Interpreting Statistics

B

■ For help with question 4, see Example 2.

4. In each case, a research topic and two sources of information are described. Decide which data source is more likely to provide reliable data. Justify your answers.

	Topic	Source 1	Source 2
a)	The sound quality of a particular stereo	An advertisement in a magazine	A review in a consumer magazine
b)	The possible side effects of a medication	A health information Web site run by a hospital	A blog written by someone who has taken the medication
c)	Job prospects in a particular field	A PDF document written by the Ontario Ministry of Training, Colleges, and Universities	A brochure you receive advertising mail-order courses in this field

5. **Literacy in Math** The headline on the press release for the poll results in *Example 3* read: "Support for Foreign Aid Rises Compared with Other Priorities." Would you say this headline is accurate? Justify your answer.

6. In December 2006, UNU World Institute for Development of Economics Research released a study on the global distribution of personal wealth in 2000. It included these statements:
   - The richest 1% of adults alone owned 40% of global assets.
   - The richest 10% of adults accounted for 85% of the world total.
   - The bottom half of the world adult population owned barely 1% of global wealth.

   Demonstrate the meaning of these statements visually by colouring squares on graph paper or using another method of your choice.

202 CHAPTER 4: Statistical Literacy

7. Sunny works as an assistant to a real estate agent.
   She prepares a price comparison for a client thinking of selling a house.
   Sunny investigates the list prices and sale prices of comparable houses that have recently sold in nearby neighbourhoods.

	House 1	House 2	House 3
List price ($)	324 500	379 000	299 900
Sale price ($)	315 000	370 000	295 000

   a) Determine the mean and median list and sale prices and the price ranges.
   b) Which measure would you use for estimating what the client's house might sell for? Explain your choice.

8. Use the growth chart for girls aged 10 to 20.
   a) Determine the percentile ranking for each girl.
      i) Tameika is 14 years old and 165 cm tall.
      ii) Audra is 16 years old and 152 cm tall.
      iii) Sabrina is 19 years old and 174 cm tall.
   b) Asayo is 17 years old and 165 cm tall.
      Between what 2 quartiles is her percentile ranking?

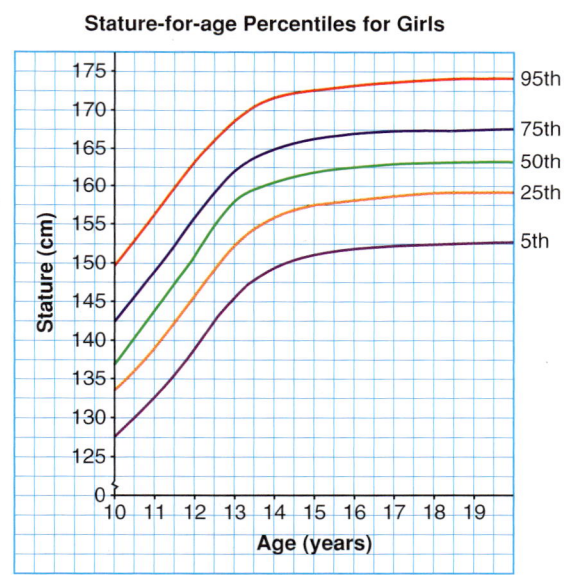

9. Here are the exam marks for a class of 20 math students.

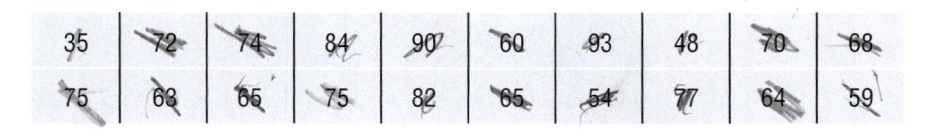

   | 35 | 72 | 74 | 84 | 90 | 60 | 93 | 48 | 70 | 68 |
   | 75 | 63 | 65 | 75 | 82 | 65 | 54 | 77 | 64 | 59 |

   a) Determine the mean, median, and mode.
      Which measure of central tendency best represents the data?
   b) What are the quartiles for this data set?
   c) Vince's mark is in the 37th percentile for this group.
      Explain what the percentile means. What is Vince's mark?

4.1 Interpreting Statistics

For help with question 10, see Example 3.

10. In November 2007, Ipsos Reid conducted a poll of 1314 randomly selected Ontarians. The report stated:

> …eight in ten (80%) Ontarians support "legislation that would ban smoking in cars and other private vehicles where a child or adolescent under 16 years of age is present". Moreover, a majority of non-smokers (86%) and smokers (66%) would support this legislation…

a) What question were people asked?
b) How did the responses of smokers and non-smokers compare?
c) The poll results are considered accurate to within ±2.7 percentage points, 19 times out of 20. Explain what this statement means.

11. Assessment Focus In November 2007, Harris/Decima polled just over 1000 Canadians about their toy-buying habits. The response to one question is shown.

Please say whether you are certain to, likely to, unlikely to or certain not to…
Avoid toys made in China because of concerns about health or safety risks.

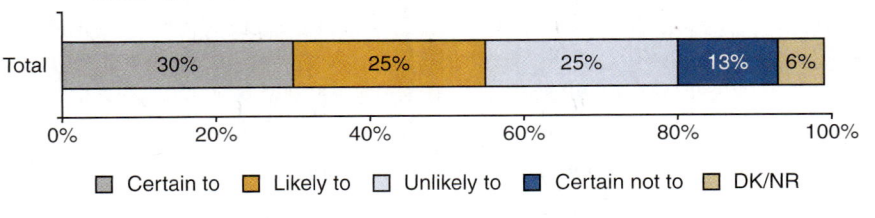

Total: 30% | 25% | 25% | 13% | 6%

☐ Certain to ☐ Likely to ☐ Unlikely to ■ Certain not to ☐ DK/NR

a) What was the question?
b) What percent of respondents said they were certain or likely to avoid toys made in China?
c) The poll's margin of error is 3.1%, 19 times out of 20. Explain what this statement means.
d) Write a headline that could be used in a news story about this poll.

12. Cong reads on a Web site that determining the "85th percentile speed" is part of the process of setting speed limits.
a) Which sentence describes the meaning of an 85th percentile speed? Justify your choice.
  i) The speed below which 85% of motorists are travelling.
  ii) The speed at which 85 out of every 100 cars on the road are driving.
b) What percent of drivers travel faster than the 85th percentile speed?
c) How could an 85% percentile speed be determined for a particular road?
d) Do you think it is reasonable to use this information to help set speed limits? Explain your thinking.

C **13.** The real estate agent in question 7 knows that each of the three houses has upgrades and features that the client's house lacks. She prepares an itemized list for each house and asks Sunny to adjust the prices.

	House 1	House 2	House 3
List price ($)	324 500	379 000	299 900
Sale price ($)	315 000	370 000	295 000
Adjustment ($)	−40 000	−63 000	−23 000

a) Determine the adjusted list and sale price for each house.
b) Determine the mean and median adjusted list and sale prices and the adjusted price ranges. How do they compare to the measures in question 7?
c) Which measure would you use for estimating what the client's house might sell for? Explain your choice.

14. Transportation engineers are considering changing the speed limit on a rural road. Every day for 1 week, a technician records speeds of vehicles using the road in kilometres per hour. Here is a representative sample of data:

| 76 | 74 | 78 | 75 | 69 | 68 | 87 | 90 | 73 | 70 |
| 68 | 72 | 85 | 78 | 72 | 70 | 75 | 75 | 76 | 65 |

a) What are the quartiles for this data set?
b) Determine the 85th percentile speed for these data.
c) The current speed limit on this road is 70 km/h. Based on these data, would you recommend changing it? What other factors should be considered?

In Your Own Words

Vivian thinks the first quartile of a data set is always a piece of data in the set. Alessandro thinks the first quartile is never in the set. Is either person correct? Justify your answer.

4.2 Statistics in the Media

Whether a journalist reports for television, radio, newspapers, magazines, or the Internet, part of her or his job is to describe statistical data in a way that people can understand. How good a job do journalists do?

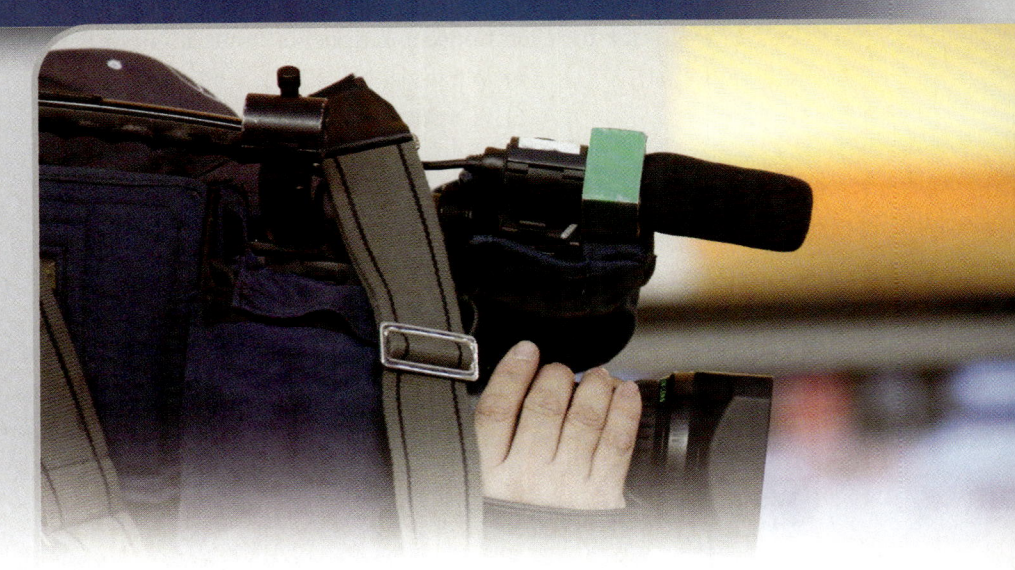

Inquire

Finding Statistics in the Media

Materials
- newspapers or magazines
- computer with Internet access

If you come across a statistical term you do not know, look it up in a dictionary or the glossary at the back of this book.

Work with a partner.

1. With your partner, create a list of words and phrases related to statistics, based on the materials you have available for this activity.

2. Select two examples from your work in question 1:
   - State where you found the data.
     If a source for the data is given, state that as well.
   - Describe the context in which the data were used.
   - Decide how reliable you think the data are.
     Explain what factors influenced your decision.

3. Find some data that are presented with graphs or other visual presentations. Write to explain how the image helped you understand the data. If you think it did not help, describe the problems with the presentation and suggest how it could be improved.

4. Find an advertisement or an article that is using data to promote a viewpoint.
   a) Decide whether or not you agree with the viewpoint. Explain your decision.
   b) If you agree with the viewpoint, find more data from another source that supports it. If you disagree with it, find data from another source that contradicts it.

5. Many media organizations use their Web sites to pose a "question of the day." Find a few of these informal surveys on the Internet. For each example you find, answer these questions.

a) What was the survey question?

b) How many people responded? What were their opinions?

c) Is the question related to any information presented on the Web site? If so, might the information affect how people answer? Explain your thinking.

d) Are people able to respond to the survey more than once? How can you find out? How might this affect the results?

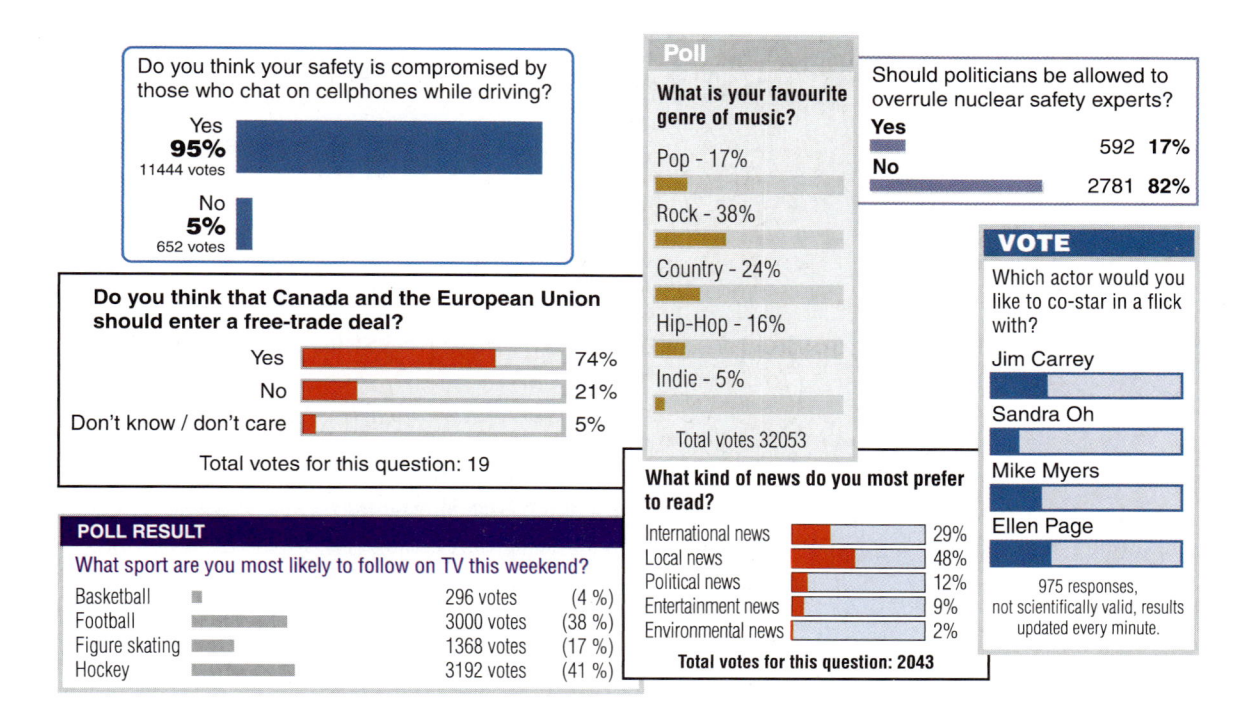

Do you think your safety is compromised by those who chat on cellphones while driving?
- Yes **95%** 11444 votes
- No **5%** 652 votes

Do you think that Canada and the European Union should enter a free-trade deal?
- Yes — 74%
- No — 21%
- Don't know / don't care — 5%

Total votes for this question: 19

POLL RESULT
What sport are you most likely to follow on TV this weekend?
Basketball		296 votes	(4 %)
Football		3000 votes	(38 %)
Figure skating		1368 votes	(17 %)
Hockey		3192 votes	(41 %)

Poll
What is your favourite genre of music?
- Pop - 17%
- Rock - 38%
- Country - 24%
- Hip-Hop - 16%
- Indie - 5%

Total votes 32053

What kind of news do you most prefer to read?
- International news — 29%
- Local news — 48%
- Political news — 12%
- Entertainment news — 9%
- Environmental news — 2%

Total votes for this question: 2043

Should politicians be allowed to overrule nuclear safety experts?
- Yes — 592 **17%**
- No — 2781 **82%**

VOTE
Which actor would you like to co-star in a flick with?
- Jim Carrey
- Sandra Oh
- Mike Myers
- Ellen Page

975 responses, not scientifically valid, results updated every minute.

Reflect

- Share your list from question 1 with the class or other pairs. Create a class list of common statistical terms and expressions.
- Describe three reasons why it is important to be able to recognize and understand common statistical words used in the media.
- Choose one statistic that you think was used effectively in a news article or advertisement. Explain what the statistic means and describe its importance to the article or advertisement in which you found it.

4.3 Surveys and Questionnaires

Government agencies, news organizations, and marketing companies often conduct surveys. The data collected can be factual, such as the number of brothers and sisters a person has, or subjective, such as a person's opinion about the use of lawn pesticides.

Investigate: Assessing the Validity of Survey Results

Work with a partner.

Choose to analyse one of these case studies:
- Case Study 1: *What Drinks Should Be Sold in the School Vending Machine?*
- Case Study 2: *Are Part-time Jobs Related to Sleep Deprivation?*

- Decide whether the survey is valid. List reasons for your decision. Think about:
  - Sample size; population size
  - The method of selecting respondents
  - The survey questions

Reflect

- Find a pair that analysed the other case study.
- Discuss your findings, inviting the students in the other pair to add any thoughts they have.
- Listen to their analysis of the other case study and discuss their decision.

Case Study 1
What Drinks Should Be Sold in the School Vending Machine?

A high school vending machine can sell four different bottled drinks. The student council surveys the students to determine what drinks should be sold.
- There are 1200 students in the school.
- The student council prepares this interview script and tally sheet:

> *Please help us select four bottled drinks for the vending machine by answering these questions.*
>
> *What grade are you in?*
>
> *Which drink would you be most likely to buy from a school vending machine? Please choose one.*

Vending Machine Survey

Student's grade	Cola	Diet cola	Sport drink	Iced tea	Cranberry juice	Apple juice

- Several council members gather outside the cafeteria during lunch one day. They interview students walking by and record the students' choices on tally sheets. They ask each student to participate only once.
- The council members stop when they have 100 responses.
- They tabulate the results and determine the four most popular drinks.

4.3 Surveys and Questionnaires

Case Study 2
Are Part-time Jobs Related to Sleep Deprivation?

A reporter wants to see if there is a correlation between the number of hours a high school student spends working at a part-time job and the number of hours of sleep the student gets.
He designs this questionnaire.

1. ☐ Male ☐ Female

2. ☐ Grade 9 ☐ Grade 10
   ☐ Grade 11 ☐ Grade 12

3. Do you have a part-time job?
   ☐ No ☐ Yes
   Number of hours you
   work in a typical week: _____

4. Doctors say teens require 8.0 h to 9.5 h of sleep each night. About how many hours of sleep do you get on a typical night?

   ☐ Less than 6 h
   ☐ Between 6 h and 7 h
   ☐ Between 7 h and 8 h
   ☐ Between 8 h and 9 h
   ☐ Between 9 h and 10 h
   ☐ 10 h or more

He puts the questionnaire in his newspaper and invites students or their parents to e-mail or call in their responses.
He collects 200 responses, then analyses the data.

Connect the Ideas

Sample size

Sample size can affect survey results. If the sample is too small, the survey results may not be reliable. If it is too great, the survey may be costly and difficult to administer.

Representative samples

A sample needs to be typical of the entire population. This is called a **representative sample**. If the sample is not representative, it is **biased** and the survey results are invalid.

Sampling techniques

Some sampling techniques are **random**, which means each member of the population has the same chance of being selected. A non-random technique may not yield a representative sample.

These sampling techniques are described in the glossary.

Random techniques	Non-random techniques
• Simple random sampling	• Convenience sampling
• Stratified sampling	• Judgement sampling
• Cluster sampling	• Voluntary sampling
• Systematic sampling	

Example 1

Assessing the Sample

A town has a population of 20 000 people. The town council conducts a vote at a public meeting about constructing a new ice-hockey rink.
- 50 people attend the meeting.
- 40 of the people at the meeting vote in favour of the hockey rink.
- The council decides to build the hockey rink since 80% of the people support the idea.

a) What percent of the people at the meeting voted for the rink?
b) What percent of the people in the town attended the meeting?
c) Is the sample representative? Justify your answer.

4.3 Surveys and Questionnaires   **211**

Solution

a) Since $\frac{40}{50} \times 100 = 80$, 80% of the people at the meeting voted for the rink.

b) Since $\frac{50}{20\,000} \times 100 = 0.25$, only 0.25% of the residents of the town attended the meeting.

c) The sample is not representative, for several reasons.
- The sample size is too small.
- It is a voluntary sample; only people who attended the meeting could vote.
- It is probably a biased sample; people who chose to attend the meeting probably have an opinion about the arena.

Biased questions

Biased questions restrict people's choices unnecessarily or use words that could influence people to answer in a certain way.
For results to be valid, survey questions must be unbiased.

Survey techniques

Another factor to consider is how the survey is conducted.
This is particularly important if any of the questions are about sensitive subjects. People may be more likely to answer honestly if they can reply anonymously in writing rather than responding to an interviewer in person or over the phone.

Example 2

Assessing the Question

People were asked this survey question in phone interviews:
"We harm the planet when we use pesticides on our lawns. Should the government ban all residential pesticide use?"
Will the survey results be valid? Justify your answer.
If you feel the survey is not valid, how could it be improved?

Solution

The results will not be valid. The first statement biases the question. Respondents may agree even if they wish to disagree. To determine peoples' true opinions on this issue, the group conducting the survey should omit the first sentence and use a written survey so people can respond anonymously.

Example 3

Assessing the Entire Survey Process

About 4000 people visited a large sports equipment store during its annual sale. The store surveyed 100 customers after they paid for their purchases. An employee recorded their answers.

---

1. Good sports equipment can greatly improve performance. How much do you spend on equipment each year?
   - ___ $200 or less
   - ___ $200–$400
   - ___ $400–$600
   - ___ $600–$800
   - ___ $800–$1000
   - ___ More than $1000

2. How much do you earn per year?
   - ___ Less than $10 000
   - ___ $10 000–$20 000
   - ___ $20 000–$40 000
   - ___ $40 000–$60 000
   - ___ $60 000–$80 000
   - ___ More than $80 000

Why are the survey results invalid? How could the survey be improved?

Solution

To assess the survey, ask yourself these questions.

- **Is the sample size large enough?**

    The store sampled 100 people out of 4000 people.
    $$\frac{100}{4000} \times 100 = 2.5$$
    A sample of 2.5% of the customers is too small.

- **Is the sample representative?**

    The store only surveyed people who made a purchase.
    The sample does not represent people who visited the store and did not purchase anything.
    As people leave the store, every 10th person could be asked to answer the survey questions. This way everyone has a chance of being asked and the store has a greater sample size.

- **Are the survey questions unbiased?**

    The first question contains a statement that may encourage people to exaggerate the amount of money they spend on sports equipment. This sentence should be omitted.

- **Was the collection method appropriate?**

    Having an employee record the answers may be intimidating. People may not wish to share information about their spending habits or their salaries.
    The survey should be conducted as an anonymous written survey.

4.3 Surveys and Questionnaires

Practice

A

■ For help with questions 1, 3, and 9, see Example 1.

1. Three schools each survey 300 students about whether they want a longer lunch. What percent of students in each survey want a longer lunch?

School	Number of students who want a longer lunch
a) 1	60
b) 2	270
c) 3	175

2. For each population, determine how many people should be surveyed to include 10% of the population.
   a) 350 people   b) 930 people   c) 1180 people   d) 10 360 people

3. The student council at a school surveys 50 students. What percent of each population is this? Choose one population and explain whether you think it is a large enough sample.
   a) 450 students   b) 750 students   c) 1200 students

4. Would you conduct a survey on each topic using personal interviews or written forms?
   a) Household income and spending on travel
   b) Time spent on homework and student marks
   c) Preferences for different brands of shampoo
   d) Favourite colours and gender
   e) Whether people have encountered discrimination in their lives

■ For help with questions 5 and 10, see Example 2.

5. Identify whether each survey question is biased or unbiased.
   a) Old gasoline powered lawn mowers pollute more than cars. People should be forced to replace them with more efficient mowers.
   ❑ Agree ❑ Disagree

   b) We will offer yoga classes one weeknight each week. Which night would you prefer?
   ❑ Thursday
   ❑ Monday
   ❑ Wednesday
   ❑ Tuesday
   ❑ Friday

   c) Should owners of hybrid vehicles be given an energy efficiency rebate from the government?
   ❑ Yes ❑ No

   d) Speed kills! Speed limits on our highways should be reduced to 90 km/h.
   ❑ Agree ❑ Disagree

B   **6.** An Internet survey asks people's opinions about a new software package. Which question is unbiased? How is the other question biased?

i)
> This software is used by some of the biggest names in business. If you have tried this software, what did you think about it?
> ___ Excellent   ___ Good   ___ Fair   ___ Poor

ii)
> Have you tried this software? ___ Yes ___ No
> If you have tried this software, what did you think about it?
> ___ Excellent   ___ Good   ___ Fair   ___ Poor

7. Choose one part of question 4. Explain how you decided whether to recommend a personal interview or written form.

8. For each part of question 5 that involved a biased question, explain how the question is biased and suggest how it could be improved.

9. An urban music radio station asks its listeners to e-mail or text an answer to this question:
"Do you think students in our city should wear school uniforms?"
95% of respondents say "No". The radio station announces that city schools should not introduce school uniforms since 95% of city residents are against the idea. Is the sample representative? If not, how could it be improved?

10. A newspaper columnist wants to find out what people think of a proposed by-law that would limit the height of fences they can build in their yards. He writes this survey question in his weekly column. Will the survey results be valid? Justify your answer.
If you feel the survey is not valid, how could it be improved?

> Once again the government is trying to control us. This time they are interfering with our backyards.
> Do you agree with the proposed law to limit the height of a fence residents can put up in their yards to 2.44 m?
> ❑ No   ❑ Yes

11. Describe how the town council in *Example 1* could conduct a valid survey to collect people's opinions about the arena.

■ For help with question 12, see Example 3.

12. The owner of a coffee shop plans to collect data to see if there is a relationship between the number of cups of coffee a person drinks per day and how happy he or she feels.

> Survey Plans
> - Set up a stand outside the coffee shop on Saturday morning.
> - Hand out a free cup of coffee to each person who participates in the survey.
> - Survey 100 people.
> - Ask these questions and record each person's answers.
>   1. How many cups of coffee do you drink in a day?
>   ___1-2 cups   ___3-4 cups   ___5-6 cups   ___7-8 cups
>   2. How happy are you? Rate yourself using a scale of 1 to 10, with 1 being very unhappy and 10 being very happy. ___

a) Why will the survey results be invalid?
b) What changes could you recommend to improve the survey?

13. **Assessment Focus** A salesperson for a new cellular phone service provider randomly selected the names of 20 small business owners from an association list of 500 members. She phoned each of them and asked them this question.

   > Studies have shown that you can lose business if your cell phone service is not up-to-date. On a scale of 1 to 4, where 1 is low importance and 4 is high importance, how important is it to you to have up-to-date cell phone service for your business?

   - All responded with a rating of 3 or 4.
   - The salesperson prepared a business proposal for her boss, stating: "100% of the local small business owners I surveyed want up-to-date cell phone service. We should launch a large sales campaign very soon."

   a) Are the survey results valid? Justify your answer.
   b) What changes would you make to improve the survey?

14. **Literacy in Math** Create a concept map describing a good survey. Copy and complete this map, adding explanations of each component as well as any additional features you feel are missing, or create your own map.

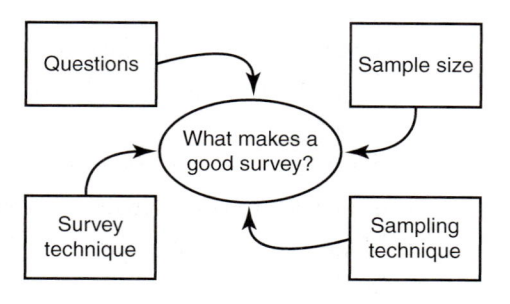

CHAPTER 4: Statistical Literacy

C **15.** An employee with a social service agency in Ottawa wants to know if there is a relationship between the number of years immigrants live in Canada and their sense of "fitting in." He selects the first 25 names from a list of immigrant families in the Ottawa area. He calls each family, states who he works for, and asks these questions:

> - How long have you lived in Canada?
> - On a scale of 1 to 4, with 4 being very well, how well are you fitting in to life in Canada?

Suggest one way to improve each of these components of the survey.
a) Sample size
b) Survey questions
c) Sampling technique
d) Survey technique

16. Full-time employees of the town of Sunderton belong to a union. Part-time employees who work on 10-month contracts from September to June do not belong to the union. The union held a vote for part-time employees to choose whether to join the union.
- Part-time employees received a letter in early July saying that the vote would be held in the town office on the second Wednesday in July.
- On the day of the vote, there was a power outage. A sign on the door of the town office stated the vote was postponed and told voters to check the office bulletin board for a new vote date.
- The vote was held four weeks later.
- 15% of the part-time employees voted. The majority voted to join the union.
- In September, the union announced that the part-time employees had voted to join the union.

a) In what ways is a vote similar to a survey? How is it different?
b) Based on what you know about planning a survey, describe at least two problems with the way the vote was held. How might these problems have affected the outcome?

In Your Own Words

Explain why it is important to consider respondents' privacy when planning a survey. Describe an example where privacy concerns could affect survey results.

4.4 Conducting a Survey to Collect Two-Variable Data

A marketing research assistant helps to prepare and conduct surveys. The data collected are then analysed to help clients make decisions about the design, advertising, and pricing of their products or services.

Inquire: Collecting and Using Data to Answer a Question

Materials
- computer with statistical software (optional)

In this lesson, you will collect and analyse data about your own research topic.

Work with a partner or in a small group.

1. **Choosing a topic**

   Choose a topic that relates two variables. Both variables must be measurable. Here are some examples.
   - The number of minutes per day students spend talking to friends on the phone and the number of minutes per day they spend on homework
   - The number of hours per week people spend using a computer and how physically fit they are on a scale of 1 to 10
   - A driver's age and how many kilometres he or she drives in a typical week
   - The number of minutes people spend listening to music per day and how happy they feel on a scale of 1 to 10

   Write your research question in this format:
   Is there a correlation between _____ and _____?

For more information about two-variable data and correlations, look back at Chapter 3.

218   CHAPTER 4: Statistical Literacy

> The population may be the students in your school or your grade, the people in your neighbourhood, or another group.

2. **Selecting a sample**

   To help select a sample, answer these questions.
   - What is the population for your investigation?
   - What is the size of the population?
   - What should be the size of your sample? Justify your choice.
   - What sampling technique will you use to select your sample?
   - How can you ensure your sample will be representative of the population? Explain.

3. **Designing your questionnaire**

   Write your survey questions. Make sure they are unbiased.
   As you begin writing, ask yourself these questions.
   - Will you conduct your survey by interviewing respondents and recording their answers, or by handing out forms for respondents to complete privately?
   - Do you want to include background information about age, gender, or grade on your survey?

   Research question:
   Is there a correlation between the amount of time students talk on the phone and the amount of time they spend doing homework?

   Questionnaire:
   Phone Habits and Homework

   1. Are you male or female?
      ____M   ____F

   2. About how much time each day do you talk on the phone to your friends?
      _____ min

   3. About how much time each day do you spend on your homework?
      _____ min

   ---

4. **Collecting your data**

   Collect your data using the sampling technique you described in question 2.

5. Organizing your data

Organize your data in a table.

> You could use a computer with spreadsheet software or a graphing calculator if it is available.

Survey data:

	A	B	C
1	Gender	Phone (min/day)	Homework (min/day)
2	M	5	120
3	F	10	80
4	M	10	90
5	M	14	85
6	F	14	80
7	M	20	75
8	F	20	80
9	F	20	85
10	M	25	90

The background information is shown in the first column.

The answers to the survey questions are in the second and third columns.

6. Drawing a line of best fit

- Create a scatter plot.
- Does there appear to be a correlation? Justify your answer.
- If there is a correlation, describe it, and construct a line of best fit. Determine the equation of the line.

Create a scatter plot:
- Independent variable: phone time
- Dependent variable: homework time

Is There a Correlation Between Time Spent on the Phone and Time Spent Doing Homework?

$y = -0.7x + 95.8$

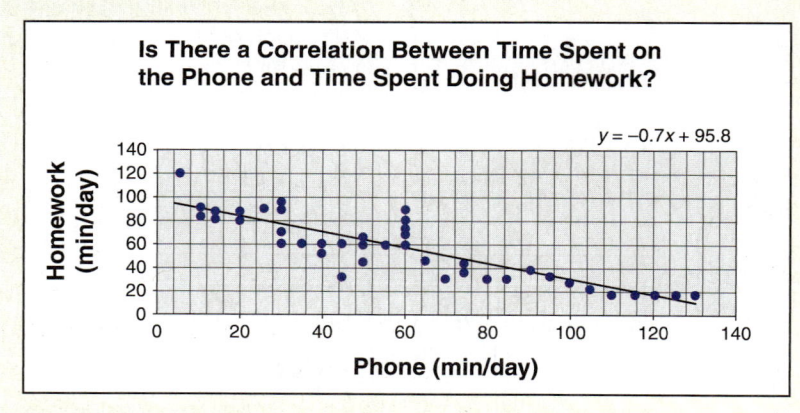

The data lie along a line with negative slope, so there is a strong, negative correlation.

7. **Analysing your scatter plot and drawing a conclusion**
   - Write a conclusion that summarizes your results and answers your research question.

   > **Conclusion:**
   > As time students spend on the phone each day increases, the time they spend doing homework decreases.
   > There is a strong, negative correlation.
   >
   > While we know it is not the only variable affecting time spent on homework, we believe that spending too much time on the phone will cause students to neglect their homework.

8. **Extending your investigation**
   - If you collected additional data for each subject, you can extend your investigation.
   - If you think there is a cause-and-effect relationship, you might want to do additional research about other variables that might affect the variables you have been examining.

   > **Possibilities for further research:**
   > - Create, then compare separate scatter plots for males and females using the background information already collected.
   > - Go back and ask students about other factors that may have affected their homework time, such as hours worked at part-time jobs.

Reflect

- How could you improve your investigation if you were to repeat it? Justify your answer.
- What related questions could you investigate? Would you have to collect more data? If so, describe your next steps for collecting that data.

Mid-Chapter Review

4.1

1. Determine the quartiles of these marks.

73	45	79	88	64	70
96	68	72	94	56	81

2. Explain the meaning of each statistic.
   a) The median salary at an advertising firm is $85 000.
   b) In a blind taste test, 95% preferred the new cereal over all the others.
   c) A local athlete scores in the top quartile in a province-wide fitness challenge.
   d) Last year, Ontario's electricity consumption was 11 996 kWh per capita.
   e) 1 in 4 Canadian adolescents is considered overweight.

4.2

3. Four candidates ran for mayor in a town.
   - 35% of the town voted.
   - The distribution of votes was:

   Elson: 30%   Singh: 25%
   Jinah: 23%   Watkins: 22%

   The headline in the paper the next day read: "The People Have Spoken: Elson To Be Mayor"
   Explain why the headline is not appropriate.

4.3

4. For each topic, would you conduct a survey using personal interviews or written forms? Justify your answers.
   a) Age and exercise frequency
   b) Adult literacy
   c) Income and education level
   d) Favourite leisure activities and time spent on them

5. A politician wants to know if seniors in her town would use a seniors' centre. She designs a questionnaire and has her assistant call every 10th phone number in the local phonebook.
   a) What is the population for the survey?
   b) In what way is the sample not representative of the population? How should the sample be changed?

6. A newspaper had the following headline: "People Against New Breed-Specific Dog Ban." The reporter who wrote the story tells you he visited a leash free park and talked to 100 dog owners. His question:

   > A new by-law bans owning a Pit Bull, Staffordshire Terrier, or Bull Terrier dog breed, or any hybrid or similar crossbreed. Do you support the by-law?

   a) Why are the survey results invalid?
   b) What changes would you make to improve the survey?

4.4

7. a) Write unbiased survey questions you could use to determine if there is a correlation between the amount of time high-school students spend e-mailing each other and the amount of time they spend talking to each other on the phone.
   b) Describe how you would select your sample and conduct your survey.

222   CHAPTER 4: Statistical Literacy

4.5 The Use and Misuse of Statistics

Statistics can both lead and mislead. We are often presented with conclusions that are based on statistical analysis. However, it is our responsibility to determine whether the conclusions are valid.

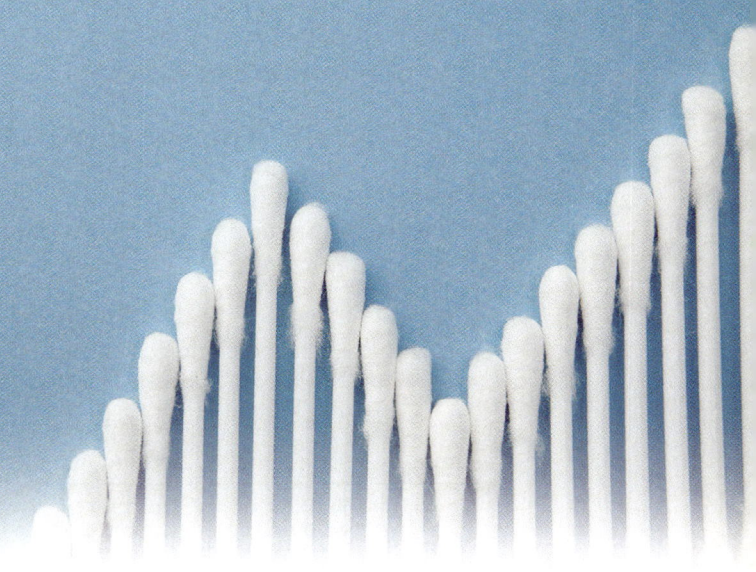

Investigate

Assessing the Validity of Conclusions

Work with a partner.

Refer to Case Study: *Women in the Workforce.*
- Do you think the conclusion is valid or invalid?
  Include answers to these questions in your response.
  - Are the data reliable?
  - Is there a possible bias in the person analysing the data?
  - Is the sample size reasonable?
  - Is the correlation strong?
  - Is there any evidence to support a cause-and-effect relationship?
  - Does the graph represent the data appropriately?

Reflect

- List three questions you should ask yourself before you accept a conclusion drawn from statistical data. Choose one question and describe how the answer would affect your decision about whether a conclusion is valid.
- Why might someone want to mislead others using statistics?

Case Study
Women in the Workforce

The members of a high school debating club are preparing for a debate.

Debate position:
It would be better for families if women stayed at home rather than joined the workforce.

Preparation:
- Analyse data about women in the workforce and single-parent families in Canada.
- Use data collected by Statistics Canada.

Data:

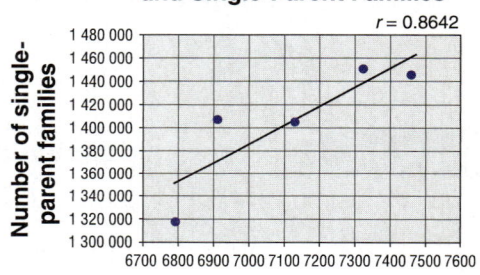

Year	Number of women in workforce (thousands)	Number of single-parent families
2000	6790.4	1 317 760
2001	6910.3	1 406 390
2002	7126.0	1 404 250
2003	7324.2	1 451 150
2004	7466.4	1 444 150

Women in the Workforce and Single-Parent Families

$r = 0.8642$

Conclusion:
These data support our position.
There is a strong positive correlation between the two variables.

As the number of women in the workforce increases, the number of single-parent families increases. If we want to reduce the number of single-parent families in Canada, women should stay at home and not go out to work.

Connect the Ideas

Assessing statistical data

A **valid conclusion** is one that is supported by unbiased data that has been interpreted appropriately.

When you read a conclusion someone has made based on statistics, you must decide whether the conclusion is valid. To do this, ask yourself:
- Is there any bias in the data collection—in the way the sample was selected, the questions were phrased, or the survey was conducted?
- If the data involve measurements, were they accurate?
- Are any graphs drawn accurately or do they mislead the viewer?

Example 1

Assessing Graphs

The graphs in each pair show the same data. Choose the graph that displays the data more accurately. Justify your choice.

a) Canada's population by age according to the 2001 census

i) Ages of Canadians, 2001 Census

- 85 and older 6%
- 65 to 84 11%
- 45 to 64 20%
- 25 to 44 24%
- 15 to 24 20%
- 0 to 14 19%

ii) Ages of Canadians, 2001 Census

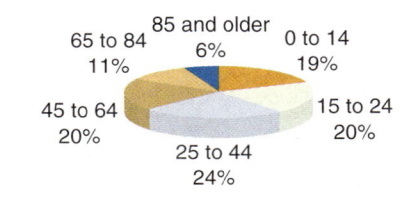

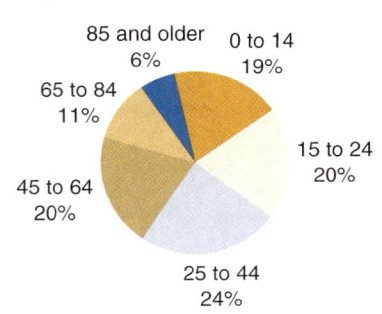

- 85 and older 6%
- 65 to 84 11%
- 45 to 64 20%
- 25 to 44 24%
- 15 to 24 20%
- 0 to 14 19%

b) A company's profits over a 5-year period

i) Company Profits

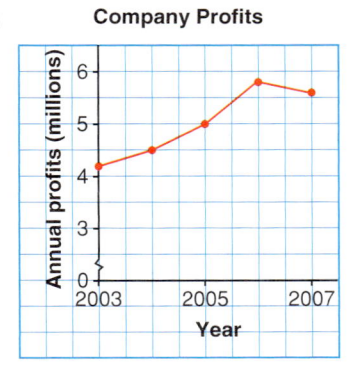

ii) Company Profits

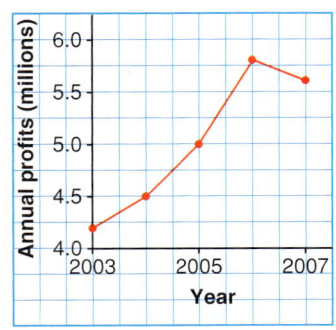

4.5 The Use and Misuse of Statistics

Solution

a) The graph in part ii displays the data more accurately.
- The graph in part i is a three-dimensional graph.
- Using three dimensions makes some pieces of the graph appear larger than they should compared to the other pieces.

b) The graph in part i displays the data more accurately.
- Part of the vertical axis has been omitted in each graph. This makes the differences between the values more striking, particularly in the second graph where more has been deleted.
- The graph in part i uses a symbol to alert the viewer that part of the axis is missing, while the other graph does not.

Example 2 — Assessing How Data Were Collected and Graphed

Four Grade 9 students collected data on school lunch preferences.

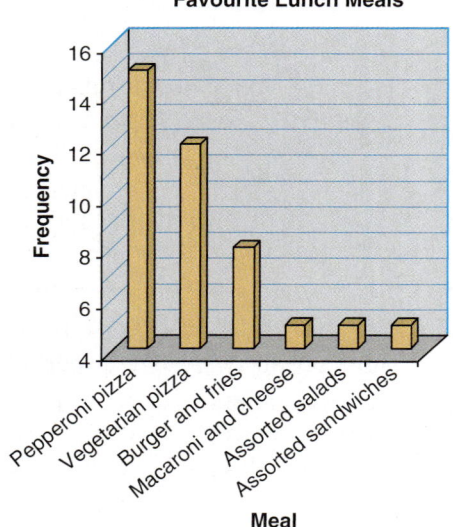

Favourite Lunch Meals

(Bar graph with Frequency on vertical axis from 4 to 16, and Meal on horizontal axis showing: Pepperoni pizza, Vegetarian pizza, Burger and fries, Macaroni and cheese, Assorted salads, Assorted sandwiches)

They concluded:

> We asked students to tell us their favourite lunch meals and displayed the results in this bar graph. We conclude that the school cafeteria should serve more pizza since it is clearly the favourite lunch of students.

Is this conclusion valid?

CHAPTER 4: Statistical Literacy

Solution

- **Was the sample size appropriate?**

  By adding the frequencies, you can see that 50 students were surveyed. Depending on the size of the school, this may not be enough data.

- **Was the sample representative?**

  You cannot tell from the information given. Perhaps the researchers surveyed only their friends or only Grade 9 students. You need more information about the sampling technique to judge this.

- **Was the survey question biased?**

  It appears that students were simply asked their favourite lunch meal. This is an unbiased question since it does not try to influence the answer.

- **How was the survey conducted?**

  It appears the survey was conducted orally. This could bias the results because some students might be self-conscious about their eating habits.

- **Is the graph constructed accurately?**

  In general, a three-dimensional bar graph tends to distort the relative quantities being displayed.

  Starting the vertical axis at 4 also distorts the relative quantities.

The conclusion is *not* valid, although it may be true. You need more information about the sample selection before recommending any change in lunch choices at the cafeteria.

Example 3

Assessing Assumptions about Cause and Effect

A group of Grade 12 students performed a linear regression on data they collected from Statistics Canada about the number of seniors and the number of weapons crimes in Canada.

> The *r*-value is a measure of how strong a correlation is. The closer to 1 or –1, the stronger the correlation.

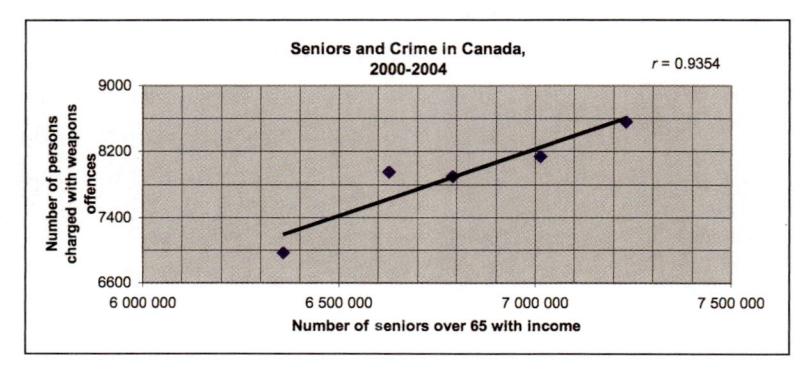

[Scatter plot: Seniors and Crime in Canada, 2000-2004, r = 0.9354. Vertical axis: Number of persons charged with weapons offences (6600 to 9000). Horizontal axis: Number of seniors over 65 with income (6 000 000 to 7 500 000).]

They concluded:

> There is a strong positive correlation between the two variables. As the number of seniors increases, weapons charges increase. Therefore, criminals in Canada are becoming bolder because of our ageing population.

Is this conclusion valid?

Solution

To assess the validity of the conclusion, ask yourself these questions.

- **Was there bias in the data collection?**
  The students gathered the data from Statistics Canada, which is a reliable source of data. However, only 5 years of data were included.
- **Is the graph constructed accurately?**
  Yes. Because the numbers are large, it would be impractical to start the scale on the vertical axis at zero.
- **Is the correlation strong?**
  Yes. The points are close to the regression line.
- **Does the analysis support a cause-and-effect relationship?**
  Not necessarily. Both variables may be increasing because the population of Canada is increasing.

The conclusion is *not* valid, although it may be true.

You need more data and you need to eliminate any other variables before you can fairly draw the conclusion made by these students.

> The *r*-value 0.9354 is very close to 1. So, the correlation is strong.

Practice

A

For help with questions 1 and 3, see Example 1.

1. The graphs in each pair show the same data. Choose the graph that displays the data more accurately.

a) Favourite cola drinks of 95 shoppers in a city mall

i) **Favourite Cola Drinks**
- Extra caffeine 3%
- Sugar-free 35%
- Caffeine-free 16%
- Regular 25%
- Sugar- and caffeine-free 21%

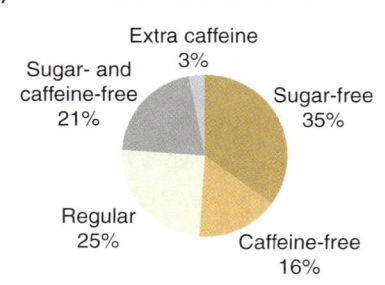

ii) **Favourite Cola Drinks** (bar graph)

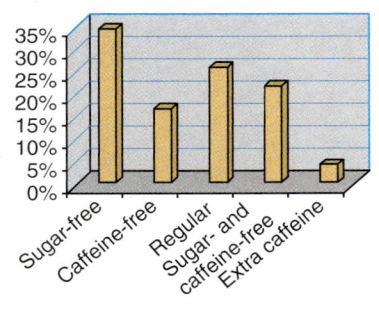

b) Temperature change over time

i) Average Daily High Temperatures

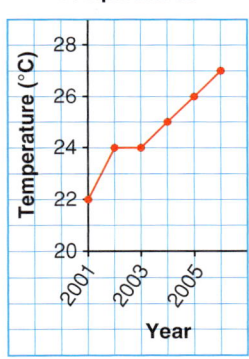

ii) Average Daily High Temperatures

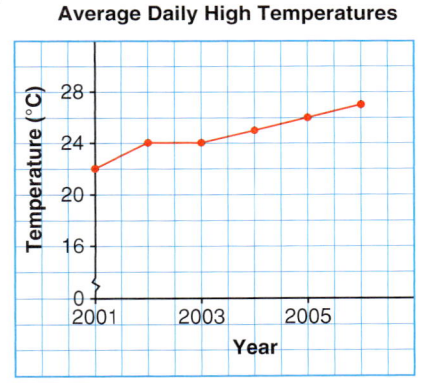

2. For each part in question 1, describe both graphs. What features misrepresent the data in the graph that represents the data *less* accurately?

3. For each survey, who do you think would be less biased in collecting data?
  a) A survey on recycling rates for plastic bottles
     i) A bottled water manufacturer   ii) A town's public works department
  b) A survey on people's opinions about health care
     i) A college student doing a project   ii) A group of doctors

4. A soft-drink company wants to test consumers' reaction to a new soft drink. Which group should collect the data? Why?
  i) The sales and marketing department of the company
  ii) An outside agency specializing in statistical surveys

4.5 The Use and Misuse of Statistics   **229**

5. Decide whether you would expect there to be a correlation between each pair of variables.
   a) The numbers of students and teachers in a school
   b) The number of years a person has worked for a company and the number of vacation days he or she receives each year
   c) The population of a town and the amount of precipitation the town receives each year
   d) A person's height and her or his mark in mathematics

B 6. For each part of question 6 in which you felt there would be a correlation, describe the correlation. Explain whether you think there might be a cause-and-effect relationship between the variables, and why.

7. **Literacy in Math** What additional information would you need before deciding whether each statistical analysis is valid?
   a) The host of a TV infomercial demonstrates a cleaning product. Then a man in a laboratory coat says, "Studies have shown that this product eliminates more bacteria from household surfaces than the leading brands."
   b) You research athletic shoes on the Internet before you purchase a new pair. On one site, a pop-up advertisement displays results from an online survey in a bar graph. The graph shows that people prefer shoes made by Robur to those made by several other brands.

8. Which of the following statements best describes the information in the scatter plot? Justify your choice.
   i) There is no correlation between the number of advertisements shown per month and the monthly cereal sales.
   ii) There is a strong positive correlation between the number of advertisements shown per month and monthly cereal sales.
   iii) As the number of advertisements shown per month increases, cereal sales increase.

Breakfast Cereal Advertisement Effectiveness

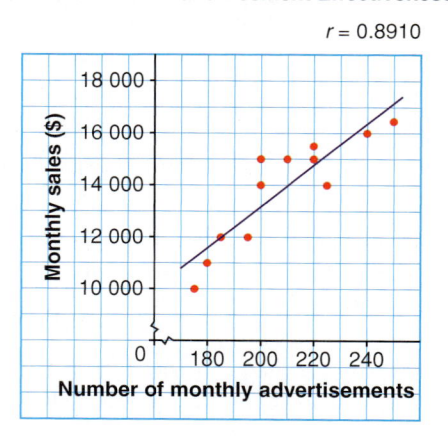

$r = 0.8910$

For help with question 9, see Example 2.

9. A reporter from a TV news show asks 5 people on the street this question: "In light of the many recent home invasions, do you think police are doing all they can to keep us safe?"
   Four of those interviewed say the police are not keeping us safe.
   On the news that evening, the reporter announces, "4 out of 5 citizens are worried about personal safety," and then shows the interviews. What is wrong with this statistical analysis?

10. The prom committee researched four possible locations for the prom.
    - They wanted the graduating students to make the final selection.
    - They provided homeroom teachers with copies of a questionnaire to hand out to all graduating students.
    - They received completed questionnaires from 85% of the graduating students.

    The questionnaire, a graph of the results, and their conclusion are shown. Is the conclusion valid? Justify your answer.

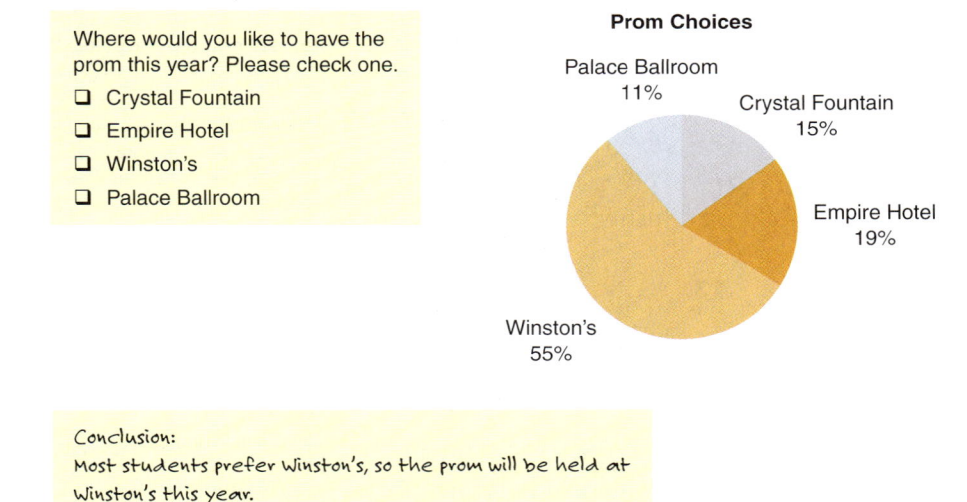

    Where would you like to have the prom this year? Please check one.
    ❑ Crystal Fountain
    ❑ Empire Hotel
    ❑ Winston's
    ❑ Palace Ballroom

    Prom Choices
    Palace Ballroom 11%
    Crystal Fountain 15%
    Empire Hotel 19%
    Winston's 55%

    Conclusion:
    Most students prefer Winston's, so the prom will be held at Winston's this year.

For help with question 11, see Example 3.

11. The headline in a newspaper reads:

    Hockey Contributes to Increase in Crime

    - The article describes a study that compared the number of young people playing on minor hockey teams and the number of arrests over a 5-year period.
    - Both variables increased over time.
    - A linear regression performed on the data had an r-value of 0.82.

    Describe the errors that make this statistical analysis questionable.

4.5 The Use and Misuse of Statistics

12. **Assessment Focus** A stress management clinic in a city of 250 000 people wanted to find out whether there is a relationship between the number of hours worked in a week and job stress level.
    - They hired an outside agency to collect the data.
    - The agency randomly selected 2500 adults who work in the city.
    - They asked people to tell them the number of hours they work per week and to rate their level of job stress from 1 to 10.

    Hours Worked and Stress Level
    $r = 0.9176$

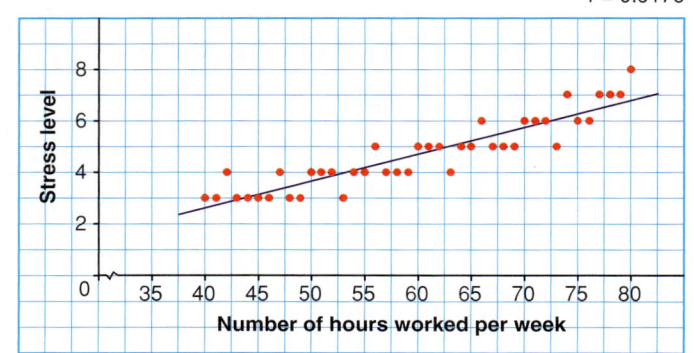

    - The clinic concluded:
    Is this conclusion valid?
    Justify your answer.

    > There is a strong positive correlation between the number of hours people work per week and their stress level on the job. We believe that an increase in working hours is likely to cause an increase in stress level.

13. A city plans to widen a road from two lanes to four lanes. Some residents of nearby neighbourhoods are concerned that traffic noise will increase. They want the city to construct noise barrier walls when they widen the road. They hire you to do a study. You outline what you will need to do.
    - Design and conduct a survey.
    - Graph and analyse the survey results.
    - Search for data from similar situations in other neighbourhoods or cities.
    - Hire a consultant to test current noise levels and develop projections for the future.
    - Make a recommendation.
    a) Describe how you could design the study so that the recommendation is to construct a wall.
    b) Describe how you could design the study so it is unbiased.

    In Your Own Words

    Explain how someone could be misled by a statistical analysis. Include an example in your explanation.

4.6 Understanding Indices

Prices for everyday items such as gasoline change over time or depending on the geographic location.

Investigate

Creating a Gasoline Price Index

Materials
- scientific calculator

Work with a partner.

- Graph the data in this table. Describe the change in gasoline prices over time.
- Create a new table from this one. Express each price as a percent of the price in January 2006. Graph the data in the table.
- Compare the graphs.

	Gasoline Prices (¢/L)	
	January	95.0
	March	93.3
	May	104.6
2006	July	109.7
	September	89.7
	November	86.5
	January	87.1
2007	March	102.4
	May	111.5

Reflect

- For each graph, identify the information it shows that the other graph does not.
- Use the graph of prices as a percent of the price in January 2006. Create a question about gasoline price increase or decrease that can be answered using the graph. Answer the question. Exchange questions with your partner. Check your partner's solution.

Connect the Ideas

Price indices

Price indices help citizens, businesses, and industries follow and predict trends in prices. A price index describes the price of an item compared to a **base value** measured at a particular time or in a particular place.

Statistics Canada tracks price changes using several different indices. The most important is the Consumer Price Index (CPI).

To determine the CPI, Statistics Canada collects thousands of price quotations from across the country for a basket of about 600 popular consumer goods and services. These items range from French fries and bus fares to tuition and Internet service.

Example 1

Reading the Consumer Price Index

Use this CPI graph to answer these questions.

Consumer Price Index (CPI)

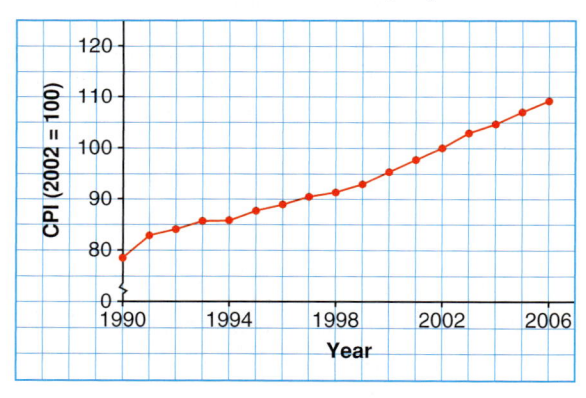

[Graph: CPI (2002 = 100) on y-axis ranging from 0 to 120, Year on x-axis from 1990 to 2006. Line rises from about 78 in 1990 to about 109 in 2006.]

a) What is the base year for the CPI?
b) In what year was the cost of the basket of goods about 90% of the base cost?
c) What was the CPI in 1990? What does this mean?
d) Describe the change in the CPI from 1990 to 1991. What do you notice about the line segment representing this period?
e) Describe the overall trend in the CPI and its significance.

CHAPTER 4: Statistical Literacy

Solution

a) Look for the year with a CPI of 100.
   The base year is 2002.
b) When the cost of a basket of goods is 90% of the cost in the base year, the CPI will be 90. The CPI was 90 in 1997.
c) In 1990, the CPI was about 78.
   So, prices in 1990 were about 78% of the prices in 2002.
d) The CPI increased from about 78 to about 83. This is an increase of 5% of the base value in one year. This is the greatest one-year increase. The line segment representing this increase is the steepest on the graph.
e) The CPI increases over the years shown.
   So, Canadians spend more money each year to buy the same basket of products and services.

Example 2

Solving Problems Using an Index

Use the graph in *Example 1*.
a) Calculate the average annual rate of inflation from 1990 to 2006.
b) Use your answer to part a to predict the CPI for 2010. Justify your prediction.

Solution

a) From 1990 to 2006, the CPI rose from about 78 to about 109.
   So, the CPI increased by $109 - 78$, or 31.
   This represents a 31% increase in prices in 16 years.
   $31\% \div 16 \doteq 1.9\%$
   Therefore, the average annual rate of inflation during this time was about 1.9%.
b) From 2006 to 2010 is a 4-year period. If the trend observed in part a continues, you would expect inflation to increase by 1.9% each year.
   $4 \times 1.9\% = 7.6\%$
   Add this increase to the CPI for 2006: $110 + 7.6 = 117.6$
   So, the CPI for 2010 would be about 118.
   It is reasonable to assume that the trend seen in the 14 years from 1990 to 2006 would continue for the next 4 years.

4.6 Understanding Indices **235**

Other price indices

Some price indices do not show a change over time. Instead, they compare prices among different geographical regions.

Example 3

Using an Index to Compare Cities

The 2006 UBS *Prices and Earnings* report includes a comparison of clothing prices in 71 cities. The base price is the price in New York.
a) Which cities in this table have index values less than 100? What does this tell you?
b) How do clothing prices in Zurich and Hong Kong compare to clothing prices in New York?

City	Clothing Price Index (New York = 100)
Zurich	115.6
Oslo	114.4
Dublin	97.5
New York	100.0
Toronto	73.8
Tokyo	148.1
Rome	87.5
Hong Kong	75.0
Delhi	43.8

Solution

a) Dublin, Toronto, Rome, Hong Kong, and Delhi have index values less than 100. This means clothing prices in these cities are cheaper than in New York.
b) Zurich's index is 115.6.
$115.6 - 100 = 15.6$
Clothing prices are 15.6% higher in Zurich than in New York. For every $100 spent on clothing in New York, you would spend $115.60 spent in Zurich for similar items.
Hong Kong's index is 75.0.
$75.0 - 100 = -25.0$
Clothing prices are 25% lower in Hong Kong than in New York. For every $100 spent on clothing in New York, you would spend only $75 in Hong Kong for similar items.

Other types of indices

Some indices do not use a base value. Instead, they use formulas to produce a number that describes something about a person, place, or thing. These numbers can then be compared.

Practice

A

For help with question 1, see Example 1.

1. a) What is this price index measuring?
   b) What is the base year for the index?
   c) Estimate the index value for each year.
      i) 1994   ii) 2002

Farm Product Price Index (FPPI) for Fruit

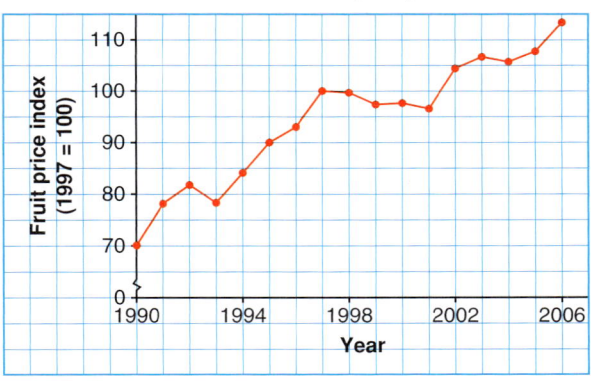

Fruit price index (1997 = 100) vs Year

2. For each price, calculate the percent price increase from a base value of $124. Round each answer to the nearest percent.
   a) $186   b) $155   c) $248   d) $131

3. For each price, calculate the percent price decrease from a base value of $124. Round each answer to the nearest percent.
   a) $92   b) $62   c) $115   d) $25

4. Order these top 10 happiest countries from most to least happy.

The world map of happiness is based on a **subjective well-being** (SWB) index. The greater the index value, the happier the population.

Country	SWB Index
Austria	260
The Bahamas	257
Bhutan	253
Brunei	253
Canada	253
Denmark	273
Finland	257
Iceland	260
Sweden	257
Switzerland	273

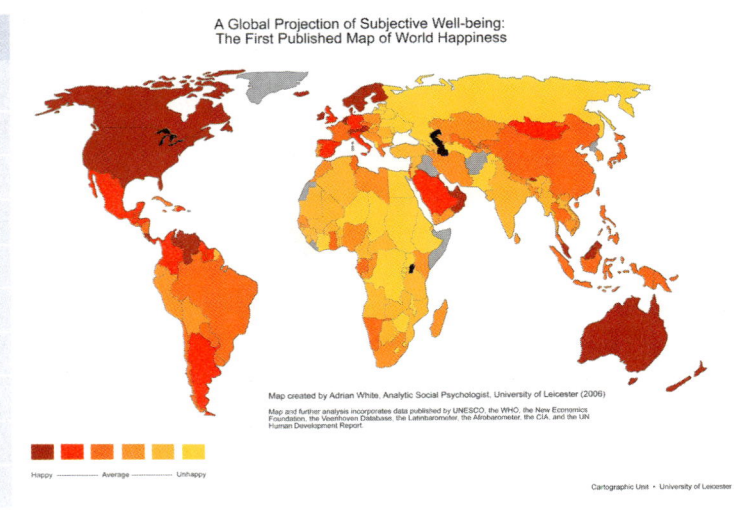

A Global Projection of Subjective Well-being: The First Published Map of World Happiness

4.6 Understanding Indices **237**

■ For help with question 5, see Example 2.

5. Use the Consumer Price Index graph in *Example 1* to answer these questions.
   a) What was the CPI in January 1996? What does this value mean?
   b) What was the CPI in January 2001? What does this value mean?
   c) Describe the change in the CPI from January 1996 to January 2001.
   d) Calculate the average annual inflation rate from January 1996 to January 2001.

B

6. Use the graph of FPPI for fruit in question 1:
   a) Describe the general trend in the graph.
   b) Explain what this trend means.

7. Meteorologists and forestry technicians use the Canadian Forest Fire Weather Index (FWI) to predict the intensity of potential forest fires. Severe fires have FWI values greater than 30.

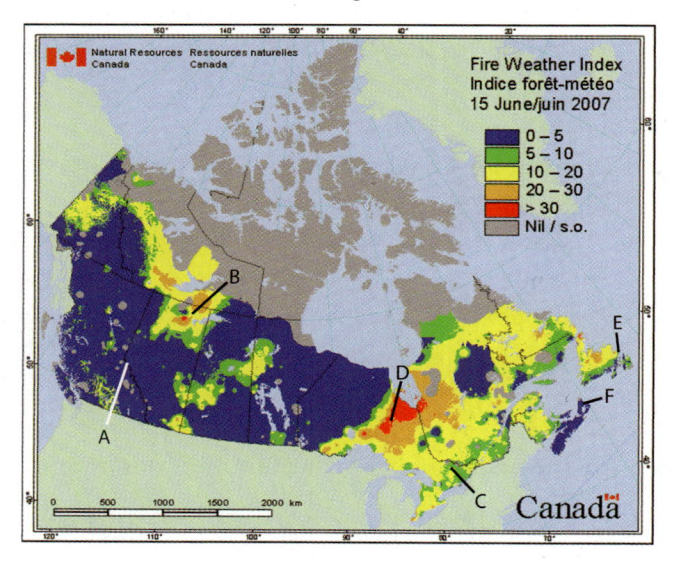

For each pair of forested regions, identify the location that likely had more intense forest fires on June 15, 2007.
   a) Jasper National Park (A) or Wood Buffalo National Park (B)
   b) Algonquin Provincial Park (C) or Cochrane District, Ontario (D)
   c) Terra Nova National Park (E) or Cape Breton Highlands National Park (F)

8. Literacy in Math Choose a person that you could talk to about prices from long ago. Ask what the person remembers about salaries and prices of a few common items, such as a bottle of pop, a haircut, or a new car. Compare the salaries and prices then and now. Do you think things are more affordable today or in the past? Explain. How does this relate to the idea of an index?

Use this graph to answer questions 9, 10, and 11. It shows an index for government spending on education in Canada from 1986 to 2003.

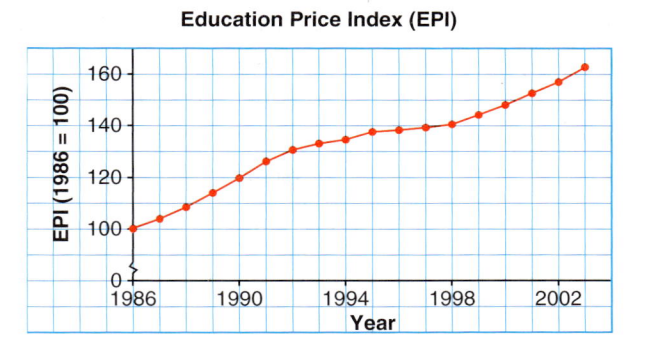

Education Price Index (EPI)

9. a) What is the base year for this index? Explain how you know.
   b) Estimate the EPI for each year.
      i) 1992 ii) 1998 iii) 2003
   c) By what percent did spending on education rise during each time period?
      i) Base year to 1992 ii) 1992 to 1998 iii) 1998 to 2003
   d) Compare your answers to part c. Which period had the greatest increase? Which period had the least increase? How does this relate to the line segments on the graph? Explain your thinking.

10. a) Calculate the overall change in the EPI from 1986 to 2003. What was the average rate of change per year for this 17-year period?
    b) Predict the EPI for 2010 if the rate of change you determined in part a continues. Explain your method.

11. **Assessment Focus**

    This graph shows an index for government spending on instructional supplies in Canada from 1986 to 2003. Instructional supplies are part of the education price in Canada.

    Education Price Index (EPI) for Instructional Supplies

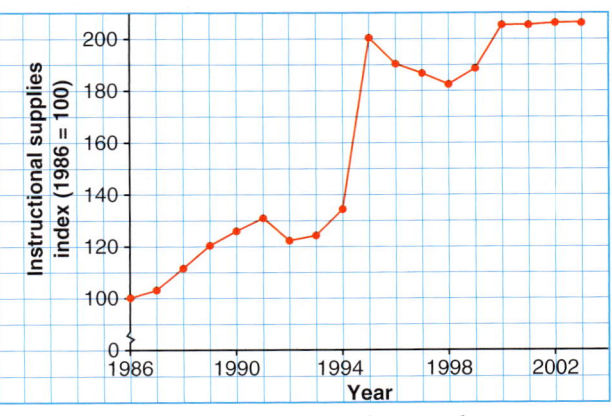

    a) How is this graph the same as the education price index graph?
    b) How is this graph different from the education price index graph?
    c) Calculate the overall change in the instructional supplies index from 1986 to 2003. Then determine the average annual rate of change for the 17-year period.
    d) Compare your answer from part c with your answer in question 10 part a. Give a possible explanation for any differences.

4.6 Understanding Indices   **239**

■ For help with question 12, see Example 3.

12. The 2006 UBS *Prices and Earnings* report compares the cost of a basket of food in 71 cities. The base cost is the cost in New York.
Data for 10 cities is given.
a) Which cities have index values greater than 100? What does this tell you about food prices in these cities?
b) How do food prices in Oslo and Delhi compare to food prices in New York?
c) Name a pair of cities that have similar food prices. Justify your answer.
d) Write a question someone could answer using these data. Answer the question.

City	Food Price Index (New York = 100)
Zurich	115.6
Oslo	112.1
Dublin	86.6
New York	100.0
Copenhagen	99.5
Toronto	80.8
Tokyo	130.3
Rome	87.8
Hong Kong	86.6
Delhi	35.1

C

13. Use the data from the food index in question 12.
a) Recalculate the index values using Toronto as the base value. That is, Toronto = 100. Explain your method.
b) Which cities have new index values less than 100? What does this mean?
c) Suppose you were to recalculate the original index using Tokyo as the base value. How would the index values change? Justify your answer.

14. The S&P/TSX Composite Index compares the current value of certain stocks traded at the Toronto Stock Exchange relative to their value at a previous time. Research this index. Prepare a brief report that includes:
- A description of what the index measures
- The base value and the year in which the base value is fixed
- A line graph showing the changes in the index over time using at least 12 pieces of data

In Your Own Words

Explain the difference between a graph showing *Wage Rates* and a graph showing a *Wage Rate Index*.

4.7 Indices and E-STAT

Statistics Canada provides many data sets involving indices, including data about housing prices across Canada. In this lesson, you will research indices on the Statistics Canada Web site using E-STAT.

Inquire

Researching Indices

Materials
- computer with Internet access
- E-STAT user name and password

Work with a partner.

- Go to the Statistics Canada Web site.
    Click **English**.
    Select **Learning Resources** from the menu on the left.
    Click on **E-STAT** in the golden box on the right.
    Click on **Accept and enter**.
    If you are working from home, you will need to get a user name and password from your teacher.

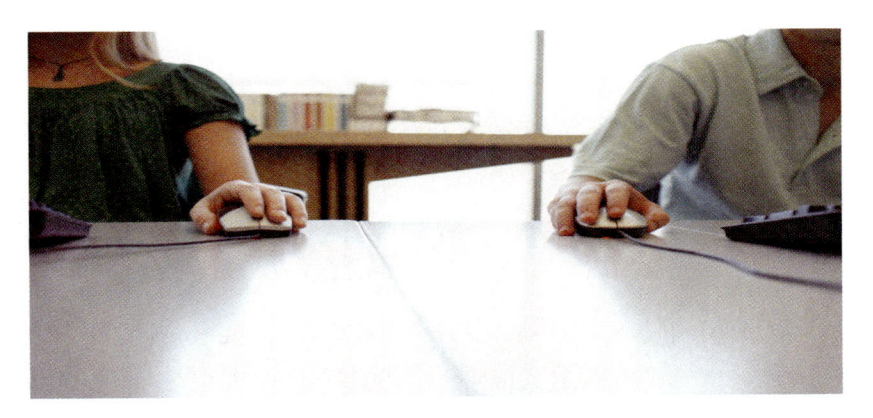

- The E-STAT table of contents will be displayed.

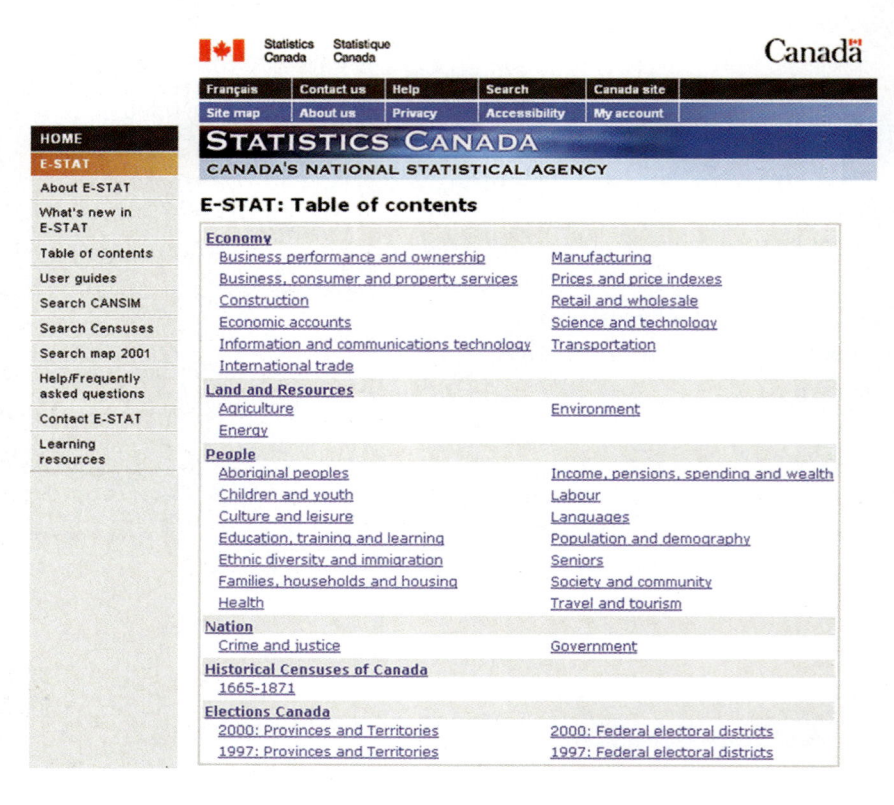

E-STAT: Table of contents

Economy
- Business performance and ownership
- Business, consumer and property services
- Construction
- Economic accounts
- Information and communications technology
- International trade
- Manufacturing
- Prices and price indexes
- Retail and wholesale
- Science and technology
- Transportation

Land and Resources
- Agriculture
- Energy
- Environment

People
- Aboriginal peoples
- Children and youth
- Culture and leisure
- Education, training and learning
- Ethnic diversity and immigration
- Families, households and housing
- Health
- Income, pensions, spending and wealth
- Labour
- Languages
- Population and demography
- Seniors
- Society and community
- Travel and tourism

Nation
- Crime and justice
- Government

Historical Censuses of Canada
- 1665-1871

Elections Canada
- 2000: Provinces and Territories
- 1997: Provinces and Territories
- 2000: Federal electoral districts
- 1997: Federal electoral districts

> Statistics Canada uses the word *indexes* instead of *indices*.

1. a) List the categories of data sets.
   b) List the topics under the category *Land and Resources*.
   c) Which category contains the topic *Prices and price indexes*?

- Select **Prices and price indexes**, and then **Construction price indexes**.

242 CHAPTER 4: Statistical Literacy

- You are going to explore the New Housing Price Index.
  Click on table **327-0005**.
  This table contains data sets for many Canadian cities.
  - Scroll through the list of cities in the box labelled *Geography*.
  - Hold down the **Ctrl** key and select:
    Montreal, Quebec [24462]
    Toronto and Oshawa, Ontario [35535, 35532]
    Vancouver, British Columbia [59933]
  - In the next box, click on **Total (house and land)**.
  - In the next box, click on **1997 = 100**.
  - Set the *Reference period* to begin in **Jan 1997** and end in **Jan 2007**.
  - Click on **Retrieve as individual Time Series**.
    A summary of the data you have requested is shown.
  - Select the output format. Under *Screen output – table*: and *HTML, table*: select **Time as Rows**.
  - Click on **Retrieve Now**.

Table 327-0005
New housing price indexes

Monthly	v21148172 - Montréal, Quebec [24462]: Total (house and land); 1997=100 (index)	v21148181 - Toronto and Oshawa, Ontario [35535, 35532]: Total (house and land); 1997=100 (index)	v21148238 - Vancouver, British Columbia [59933]; Total (house and land); 1997=100 (index)
Jan 1997	100.4	98.3	101.3
Feb 1997	100.4	98.6	101.4
Mar 1997	100.4	98.7	101.1
Apr 1997	100.1	99.4	101.3
May 1997	99.4	99.8	101.3
Jun 1997	99.7	99.8	100.7
Jul 1997	99.7	100.3	100.1
Aug 1997	100.0	100.2	99.5
Sep 1997	100.1	100.4	99.2
Oct 1997	100.1	100.5	98.2

4.7 Indices and E-STAT

2. This data set is part of the New Housing Price Index (NHPI).
   a) In which year was the base value set?
   b) How often is the NHPI calculated?
   c) For each city, how many pieces of data are in this set?
   d) Scroll through the data set for one city. What do you notice about the index values?

- To view the data as a graph, click **Back** on the tool bar.
  - Under *Screen output – Graph*, select **Line graph**. Then click on **Retrieve now**.
  - Click on **Modify Graphic** if you wish to revise the title for your graph or add gridlines. When you are finished, click **Replot**.
  - If you wish to print the graph, click on **Printer-friendly format** at the bottom of the page.

3. Use the graph to answer these questions.
   a) How are the lines the same?
   b) How are the lines different?
   c) What most surprises you about the information in the graph?

4. Use the table to answer these questions.
   a) For each city, calculate the percent rate of increase in new housing prices from Jan 1997 to Jan 2007.
   b) Suppose you bought a new house in Toronto in 1997 for $100 000. What would be the price for an equivalent new house in 2007?

c) Repeat part b for Montreal and Vancouver.

d) For each city, suppose the percent rate of increase you calculated in part a remains the same for the next 10-year period. What would be the price of an equivalent new house in each city in 2017?

5. Suppose you were offered the same job in all three cities.
  a) If you were to make your decision based on the NHPI, which city would you select? Justify your answer using data from your graph.
  b) In addition to the NHPI, what other information on housing would you need to know to help you make your decision?

6. Return to the E-STAT table of contents.
Select another price index and conduct similar research to investigate the index for three cities in Canada.

7. List three things you found interesting about the E-STAT data sets.

> **Reflect**
> - What features made the Statistics Canada Web site and E-STAT data sets easy to use? Explain.
> - What did you find challenging about using the E-STAT data?
> - Describe a real-life question or problem that one of the data sets from E-STAT could help solve.

4.7 Indices and E-STAT

4.8 Statistical Literacy and Occupations

Data management skills are important in industries such as manufacturing, health sciences, hospitality services, financial services, and resource management. As a result, many college programs include courses that involve organizing, displaying, and analysing data.

Inquire: Researching College Programs and Occupations

Materials
- computer with Internet access

Work with a partner.

■ Record the data management skills you have learned in this course. If you need some hints, look through the *Chapter Reviews* in Chapter 3 and Chapter 4.

> *Extrapolate from a graph*
>
> *Assess validity of a survey*
>
> *Interpret statistical information*

■ On the Internet, search for Web sites of Ontario colleges. Investigate the post-secondary programs offered at a few colleges. Look for courses that involve data management.

You may be able to search for courses using keywords.

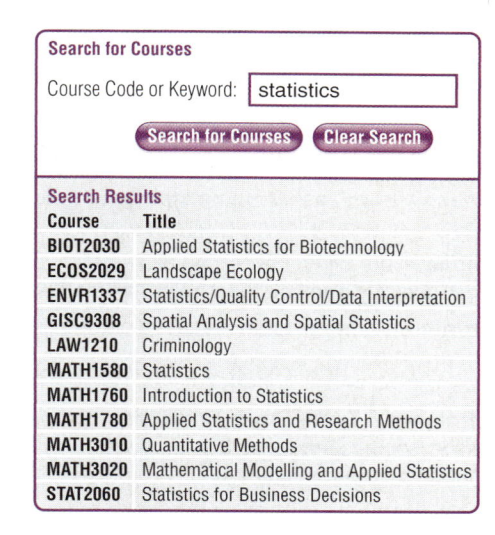

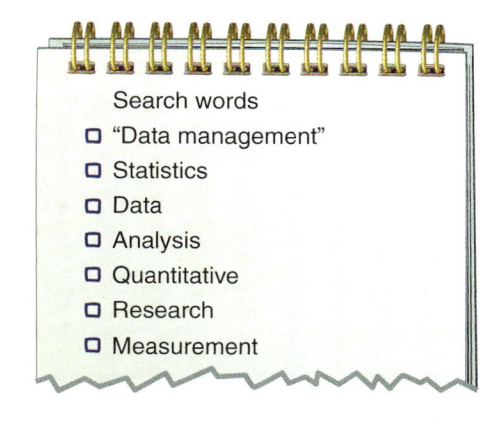

Search for Courses

Course Code or Keyword: statistics

[Search for Courses] [Clear Search]

Search Results

Course	Title
BIOT2030	Applied Statistics for Biotechnology
ECOS2029	Landscape Ecology
ENVR1337	Statistics/Quality Control/Data Interpretation
GISC9308	Spatial Analysis and Spatial Statistics
LAW1210	Criminology
MATH1580	Statistics
MATH1760	Introduction to Statistics
MATH1780	Applied Statistics and Research Methods
MATH3010	Quantitative Methods
MATH3020	Mathematical Modelling and Applied Statistics
STAT2060	Statistics for Business Decisions

Search words
- ❑ "Data management"
- ❑ Statistics
- ❑ Data
- ❑ Analysis
- ❑ Quantitative
- ❑ Research
- ❑ Measurement

Or, you may have to search for programs likely to involve data management, then look through the course listings for those programs.

Search words
- ❑ Advertising
- ❑ Business administration
- ❑ Computer programming
- ❑ Environmental technology
- ❑ Health information management

■ Select three courses that interest you and describe each of them. Include the following information in your description.
- Course name
- College that offers the course
- Programs that require or recommend the course
- Brief description of the data management skills used in the course

4.8 Statistical Literacy and Occupations **247**

- On the Internet, search for occupations that require data management skills. You might try job search Web sites, or career centres on college Web sites.

> **Some Occupations Requiring Data Management Skills**
> - **Business** – legal assistant, marketing consultant, forecasting and inventory planner, customs broker
> - **Sciences** – forensic technologist, automation technician, network security analyst, laboratory assistant
> - **Environment** – earth sciences technician, conservation officer, forestry technician, eco-tourism business operator
> - **Government & Community Services** – property assessor, social service worker, archivist, international trade specialist
> - **Health & Medicine** – health information manager, registered practical nurse, lab technician

- Select three occupations that interest you and describe each of them. Include the following information in your description.
  - Job title
  - A description of the data management skills used in the job
  - Educational background required for the job
  - Typical salary range for the job
  - Whether there is a demand for people who can perform this job and future prospects for job availability
  - Source of your information

- How might you use data management skills in the next few years, either at school or in everyday life?

Reflect

- Describe one thing that surprised you or might surprise someone else about the information you found.
- What would you say it means to be "statistically literate?"

GAME

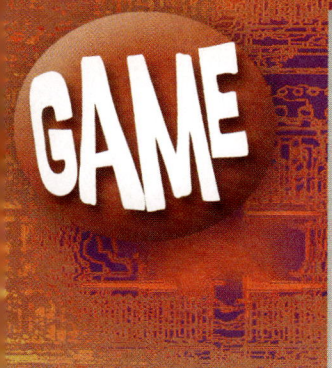

Concept Clues

Materials
- 5 *Concept* cards (see *Preparation* section below)
- stopwatch, watch with second hand, or 1-min sand timer

> You can use your notes, textbook, and any supplementary material in the classroom to help you create your *Concept* card.

Preparation

Each student should create at least one *Concept* card. Two sample cards are shown. Give your completed card(s) to the teacher.

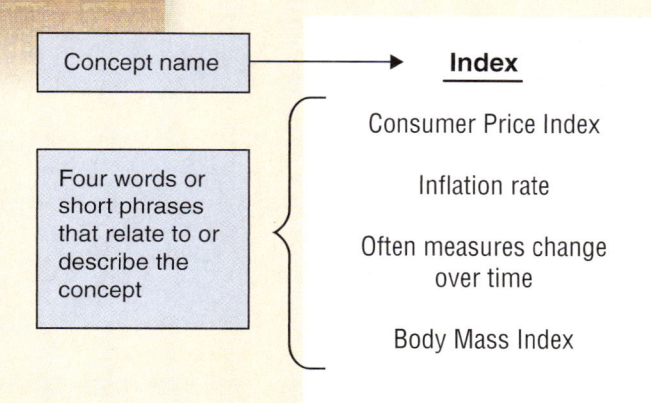

Concept name → **Index**

Four words or short phrases that relate to or describe the concept:
- Consumer Price Index
- Inflation rate
- Often measures change over time
- Body Mass Index

Sources of Bias
- Sample is not representative
- Question unfairly restricts choices
- Question wording influences response
- Personal interview used to collect sensitive information

How to Play

- Play in groups of 3 to 5.
- Each group should get five *Concept* cards from the teacher. Shuffle these and place them face down in a pile.
- The first player reads aloud the concept on a *Concept* card and starts timing. The other group members try to identify each of the other four words or phrases on the card. The player with the card may give clues, but cannot say any of the words or phrases.
- Play ends after 1 min. The group gets 1 point for each word or phrase identified. If all four were identified, the group earns a bonus point.
- After all groups finish their cards, the group with the most points wins.

Reflect
- Explain whether creating a *Concept* card helped you when you played the game.

Study Guide

Assessing the Validity of a Survey

When you read or hear about survey results, ask questions to help you decide whether the survey is valid.
- What is the sample size?
- Is the sample representative?
- Are the survey questions unbiased?
- Is the survey technique suitable?

Assessing the Validity of Data Analysis

When you read or hear a conclusion based on data analysis, ask questions to help you decide whether or not the conclusion is valid.
- What is the source of the data?
- How many pieces of data were used?
- Are graphs constructed accurately?
- Is the correlation strong?
- Is the relationship likely to be cause and effect?

Indices

An index is a single number calculated from several pieces of data.
Many indices describe a cost or quantity in comparison to a cost or quantity at a particular time or in a particular place. This is called the base for the index.

Price indices are often presented as broken line graphs with these features:
- The horizontal axis is time.
- The vertical axis values are relative to a base value.
- The graph shows changes in value over time.
- The index may be used to predict future trends.

Some other indices, such as the Body Mass Index, do not have a base value.
They use a formula to determine a number that describes something about a person, place, or thing. This number allows comparisons to be made.
These indices are often presented as tables or bar graphs.

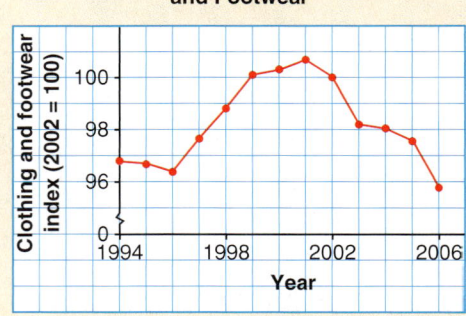

$$BMI = \frac{\text{mass in kilograms}}{(\text{height in metres})^2}$$

250 CHAPTER 4: Statistical Literacy

Chapter Review

4.1

1. A group of people were discussing marks of high-school students. Irfon overheard this statement: "All of our students should have marks above average."
   a) Explain the error in the statement.
   b) What do you think the speaker actually intended by the statement?

2. Here are the batting averages of 18 Blue Jays players for the 2007 season.

.297	.178	.245	.233	.240	.262
.289	.242	.251	.238	.236	.208
.238	.277	.216	.204	.291	.167

   a) Order the data from least to greatest.
   b) What are the quartiles for this data set?
   c) Frank Thomas's batting average is in the 80th percentile for this group. Explain what this percentile means. What was Frank Thomas's batting average for the 2007 season?

3. Rahim researches whether it is better to drink bottled water or tap water. Explain which of these organizations' Web sites are likely to have reliable information.
   a) A think tank studying environmental issues
   b) An association of water bottling companies
   c) A company selling water purifying equipment for the home
   d) The city of Kingston
   e) A scientific organization concerned with water management issues

4.2

4. Livia found this graph on the Web site of the Certified General Accountants of Ontario. It is based on data collected in an e-mail survey in 2006. Almost 6000 people responded to the survey—a response rate of 38%.

Annual Salary of Certified General Accountants in Ontario

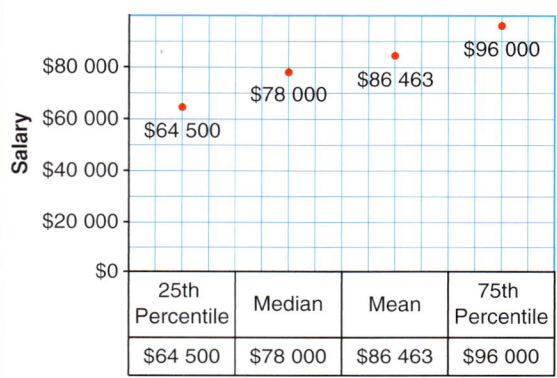

	25th Percentile	Median	Mean	75th Percentile
	$64 500	$78 000	$86 463	$96 000

   a) What was the 25th percentile salary? What does this value mean?
   b) What was the 75th percentile salary? What does this value mean?
   c) How confident are you in the results of the survey? Justify your answer.

4.3

5. A school has 1240 students. How many should be surveyed to sample each portion of the population?
   a) 5%   b) 10%   c) 15%   d) 30%

6. A TV station wants to know if the community would like more local news included in the evening news. Viewers are asked to e-mail or text their opinions.
   a) Why is this sample not representative of the community?
   b) Describe how you could make the sample more representative.

Chapter Review **251**

7. A town's recreation director conducts a survey about building a swimming pool in the community centre.
- He designs a questionnaire and places it at the community centre.
- People using the centre pick up a copy, fill it in, and drop it in a box.

a) The sample is not representative of the people who live in the town. Explain why not.

b) Suggest a way to make the sample more representative.

8. The mayor of a town of 15 000 wants to find out if there is a relationship between the amount of time people spend commuting to work and the amount of time they spend with family and friends. She hires you to conduct a survey.

a) What sample size would you use?

b) How would you select your sample?

c) How would you make sure your sample is representative?

9. Is each statement true or false? Justify your answers.

a) Three-dimensional graphs display data more accurately than two-dimensional graphs.

b) When the population of a survey is the students at a school, the sample should include students from each grade.

c) People may be reluctant to provide salary information or other sensitive information in a personal interview.

d) A graph of two-variable data that has a strong positive correlation means there is a cause-and-effect relationship between the variables.

10. For each survey, decide who you think would be more likely to collect reliable data.

a) A survey on pesticide restrictions
 i) A company that makes pesticides
 ii) A college student doing a project

b) A survey on college students' opinion of a career in the entertainment industry
 i) A radio show
 ii) A college's student life office

11. A PhD student in health economics studied U.S. fuel prices and health data. In a research paper, he wrote:

> A $1 (U.S.) increase in gas prices would, after 7 years, reduce U.S. obesity by approximately 9%, saving 11 000 lives and $11 billion per year.

a) List at least two questions you could ask him about his research before deciding whether you agree with his conclusions.

b) Why might gas prices have an effect on obesity?

12. A flyer advertising a children's soccer camp is delivered to Jan's home. It says: "Parents prefer soccer to any other sport!" Jan asks about the statement. He is told that at a soccer tournament, 50 parents were asked: "Which sport do you prefer your children to play: soccer, tennis, or judo?" All of the parents chose soccer.

a) Record as many reasons as you can that would cause you to question the above conclusion.

b) Compare your list with a partner.

c) Select the most convincing reasons and write a response to the soccer camp, explaining why you think their statement is misleading.

4.6 **13.** Here are the 2008 Environmental Performance Index (EPI) results for North and Central American countries. The EPI was designed to help countries evaluate their environmental policies. The index uses a formula to calculate a score out of 100 for each country, with 100 being the best possible score.

Country	EPI
Canada	86.6
United States	81.0
Mexico	79.8
Belize	71.7
Guatemala	76.7
Honduras	75.4
El Salvador	77.2
Nicaragua	73.4
Costa Rica	90.5
Panama	83.1

a) Which country has a higher EPI score than Canada's?
b) Which countries have a lower EPI score than Canada's?

Use this graph for questions 14 and 15.

Computer Price Index for Consumers, Jul 2001 to Jan 2007

14. a) What is the price index measuring?
b) What was the base year? What does this mean?
c) What was the index value in January 2002? What does this value mean?
d) Describe the general trend in the data.

15. a) When was the computer cost about 40% of the cost in 2001? How can you tell this from the graph?
b) Determine the change in the Computer Price Index between January 2002 and January 2007. Calculate the average annual rate of decrease in the Computer Price Index during this 5-year period.
c) What does this information mean to consumers?

4.7 **16.** Go to the Statistics Canada Web site. Follow the steps from Lesson 4.7 to use data from E-STAT, but select **Agriculture price indexes** rather than **Construction price indexes**. Use the annual Farm Product Price Index from table **002-0022**. Retrieve data about the prices of all the listed commodities from 1997 to 2006.
a) Which two commodities showed the greatest increase in price from 1997 to 2006?
b) Which two commodities showed the least increase in price from 1997 to 2006?

4.8 **17.** Name an occupation that involves data management. Describe the educational background required for the job and how data management skills are used.

Chapter Review **253**

Practice Test

Choose the best answer for questions 1 and 2. Justify your choice.

1. "An extremely cold day occurs when the temperature is below the 10th percentile of historical temperatures for that day."
   What does this phrase mean?
   A. The temperature is below −10°C.
   B. 10% of the days in a year are extremely cold.
   C. February 2 is extremely cold about 10% of the time.
   D. In 100 years, February 2 will be extremely cold 10 times.

2. Which of the following statements is false?
   A. A price index can be used to compare changes in prices over time.
   B. A graph of the average monthly price of gas is an example of an index.
   C. Price index values are often calculated relative to a base value at a certain point in time or in a certain place.
   D. Index values can be greater than 100.

Show your work for questions 3 to 6.

3. **Knowledge** These tables show the number of paid admissions to movie theatres in Canada and the population of Canada for 5 years.

| Paid Admissions to Movie Theatres in Canada ||
Year	Paid Admissions
1996	89 024 000
1997	96 805 000
1998	109 688 000
1999	117 352 000
2000	117 574 000

| Population of Canada ||
Year	Population
1996	29 610 757
1997	29 907 172
1998	30 157 082
1999	30 403 878
2000	30 689 035

   a) In 1996, the number of movie admissions per capita was 89 024 000 ÷ 29 610 757, or about 3.0. What does this mean?
   b) For each year, determine the per capita rate of movie theatre admissions.
   c) From your results in part b, can you say that Canadians are going to the movies more often? Explain.
   d) Why is the per capita rate more meaningful than the raw numbers?

254 CHAPTER 4: Statistical Literacy

4. **Application** A high-school soccer coach wonders if there is a relationship between the number of hours students work at part-time jobs each week and the number of hours they spend on school sports and clubs.
   a) Create unbiased survey questions that could be used to research this topic. Explain why your survey questions are unbiased.
   b) The school has 1000 students in Grades 9 through 12. Suggest a sample size and sampling technique that could be used for this study.
   c) Would you recommend a written questionnaire or interview for this study? Justify your answer.

5. **Thinking** Ari and Bianca plan to open a computer store in their town of 25 000 people. They need a bank loan to begin the business. As part of their loan application, they prepare a business plan. It includes this information.
   - We surveyed 60 adults in town to determine the level of support for our business venture.
   - We selected our sample by randomly choosing numbers out of the town phone book.
   - We called, asked to speak to an adult, then posed our question.

   > It is nearly impossible to function today without a computer. Our town has no computer store. Do you agree that it would be a good idea to have a computer store in town to serve all your computer needs?
   > ❏ Strongly agree
   > ❏ Somewhat agree
   > ❏ Somewhat disagree
   > ❏ Strongly disagree

   - Our conclusion: Since 70% of the people would like a computer store in town, we will have a good customer base for our business.

   If you were the bank manager, would you find their survey persuasive? Justify your answer.

   It Would Be a Good Idea to Have a Computer Store in Town
   Strongly disagree 10%
   Somewhat disagree 20%
   Somewhat agree 15%
   Strongly agree 55%

6. **Communication** Use this graph.
   a) Your friend has never heard of a price index. Explain to your friend what the graph is showing and how it can be used.
   b) In which year did the price of new houses in Calgary have the greatest increase? Explain how you know.
   c) What was the average annual percent price change for new houses in Calgary from 1997 to 2006?

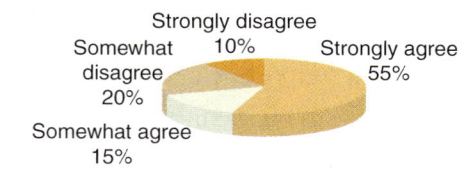

   New Housing Price Index for Calgary

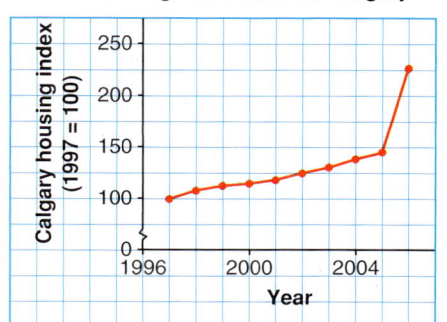

Chapter Problem — To Ban or Not to Ban?

Materials
- *Microsoft Excel*
- *frenchfries.xls*

Suppose your school was considering banning French fries from its cafeteria. Do you think students would agree or disagree with this plan?

You will prepare a report on this issue for the school administration. For the report, you will collect, analyse, and display data. Your report should be fair and unbiased.

1. Prepare a survey about students' opinions on banning French fries from the cafeteria. Include questions about grade, gender, and how often students purchase food from the cafeteria. Decide on a sample size. Plan a sampling technique.

2. Conduct your survey or use the sample data provided in the file *frenchfries.xls*.

3. Analyse the data, using a computer if one is available. Create tables and graphs.

4. Find supporting data in the media or on the Internet.

5. Prepare your report.

	A	B	C	D	E	F	G
1	Collected in Gym				Collected in Cafeteria		
2	Grade	Gender	French Fries should be banned from the school cafeteria.		Grade	Gender	French Fries should be banned from the school cafeteria.
3	9	M	Agree strongly		9	M	Disagree strongly
4	9	M	Agree strongly		9	M	Disagree strongly
5	9	F	Disagree strongly		11	F	Disagree

256 CHAPTER 4: Statistical Literacy

PROJECTS

A For the Birds!

B Making Headlines with Statistics

Your teacher may give you an expanded version of either project.

What You'll Apply

Use trigonometry and geometry to design a birdhouse and analyse the design for Project A, or plan and conduct research involving two-variable data, represent the data, and draw a conclusion for Project B.

And Why

Designing a birdhouse requires making decisions about how to follow building requirements and perform calculations for optimization. Developing and applying research plans, then representing and analysing the data, offer an opportunity to extend data management to a real-life situation.

PROJECT A

For the Birds!

Suppose you plan to set up a small business building and selling birdhouses. Because of financial constraints and limited carpentry skills, you will design and market a very simple and cost-effective birdhouse.

In groups or as a class, discuss these questions.
1. How can a business make a profit selling a simple and inexpensive product?
2. What are some ways to increase the profit made on each birdhouse?
3. What factors should you consider when designing the birdhouse?
4. What role might technology play in the design process?
5. What mathematics might be used in the design, construction, and marketing of the birdhouse?

CAREERS

- Carpenter
- Technical designer
- Architectural technologist
- Marketing specialist
- Industrial designer

Math Focus

Calculate side and angle measures in oblique and right triangles, solve problems involving the volume of a triangular prism, determine the optimal dimensions of a three-dimensional figure for a given constraint. (Chapter 1 Trigonometry, Chapter 2 Geometry)

PROJECT B

Making Headlines with Statistics

Often, news reports begin with a phrase such as "A new study has shown…" or "Researchers say…" How do you decide whether the report is reasonable? In groups or as a class, discuss these questions.

1. a) Here's a headline of a news article.

   New Study Links Student Achievement to Hours of Sleep

   Which statement do you think is implied by this headline? Why do you think so?
   - A student's academic achievement depends on the number of hours of sleep the student gets.
   - The number of hours of sleep a student gets depends on her or his academic achievement.

   b) Is either cause-and-effect statement necessarily true?

   c) Predict whether the correlation between the variables would be positive or negative.

   d) List three questions to ask the reporter who wrote the news story that may help you decide whether the study she quoted is valid.

2. Researchers found a strong positive correlation between the number of fast-food restaurants in cities and the number of schools in the area.

   a) Is this headline misleading? Explain your thinking.

   Latest Research Shows Number of Fast Food Restaurants Related to Number of Schools

   b) Does this correlation mean that increasing the number of schools in an area causes an increase in the number of fast-food restaurants? Justify your answer.

   c) How could this relationship be used to predict the number of fast-food restaurants in a city?

260

CAREERS

- Market researcher
- Survey interviewer
- Quality control technician
- Forestry and wildlife manager
- Risk assessor

Math Focus

Collect, analyse, and summarize two-variable data, and interpret and draw conclusions from the data; demonstrate an understanding of applications of data management (Chapter 3 Two-Variable Data, Chapter 4 Statistical Literacy)

261

Chapters 1–4 Cumulative Review

CHAPTER 1

1. Use primary trigonometric ratios to determine each measure.
   a) Side c
   b) $\angle N$

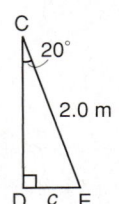

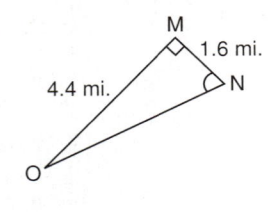

2. Sketch and solve each triangle.
   a) $\triangle ABC$ with $\angle A = 15°$, $\angle C = 90°$, and $c = 8$ cm
   b) $\triangle CDE$ with $\angle E = 90°$, $e = 14.0$ yards, and $c = 9.2$ yards
   c) $\triangle XYZ$ with $\angle Y = 67°$, $\angle Z = 90°$, and $y = 21$ m
   d) $\triangle PQR$ with $\angle P = 90°$, $\angle R = 51°$, and $q = 150$ mm
   e) $\triangle GHI$ with $\angle I = 90°$, $g = 1.5$ m, and $h = 1.2$ m

3. Determine the measure of obtuse $\angle D$ for each ratio.
   a) $\sin D = 0.45$
   b) $\cos D = -0.21$
   c) $\tan D = -0.43$
   d) $\sin D = 0.60$
   e) $\cos D = -0.99$
   f) $\tan D = -0.84$

4. Decide whether you use the Sine Law or the Cosine Law to solve each triangle. Then, solve each triangle.

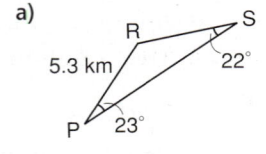

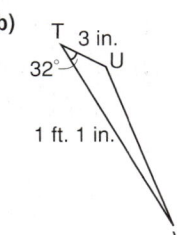

5. Determine z.

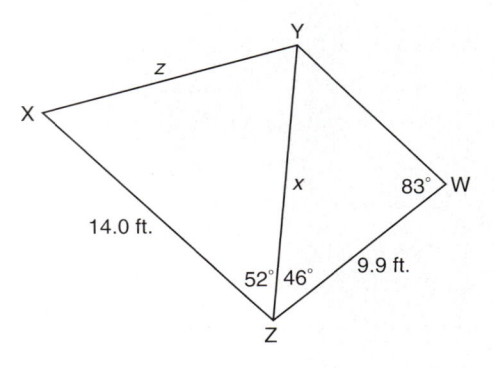

6. Two ships sail out from a harbour at the same time. One sails on a bearing of 015° and travels a distance of 32 miles. The other ship sails 47 miles on a bearing of 165°.
   a) How far apart are the ships?
   b) What is the bearing from the first ship to the second ship?

CHAPTER 2

7. a) Determine the area of this composite figure. All curves are quarter circles or semicircles.

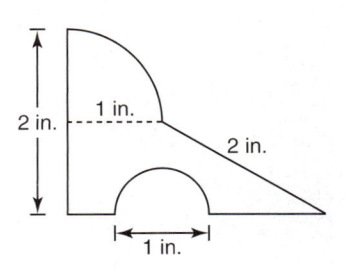

   b) Determine the surface area and volume of this composite object.

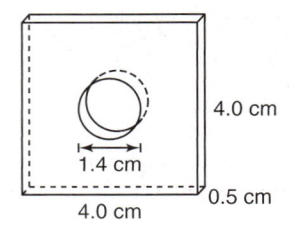

262 Chapters 1–4

CHAPTER 2

8. For each perimeter, what are the dimensions of the rectangle with the maximum area? What is the area?
   a) 28 m
   b) 44 inches
   c) 10 cm
   d) 94 feet

9. Gizelle is designing an art project for children at her day care. She will have them use 10 paper clips to create a border for an art project. She is debating whether to use a rectangular or triangular border.
   a) Each paper clip is 2 inches long. What are the side lengths of the rectangles and triangles she can construct?
   b) What is the greatest rectangular or triangular area she can enclose? What shape does it have?

10. For each volume, what are the dimensions of the rectangular prism with the minimum surface area? What is the surface area?
   a) 64 cubic feet
   b) 729 m^3
   c) 225 cm^3
   d) 3000 cubic inches

11. Evan is creating a rectangular planter with square ends. He will use 200 ceramic tiles to create a design on the sides and bottom of the planter. Each tile is a square with side length 1 inch.
   a) What are possible dimensions for the planter?
   b) What is the maximum volume of soil the planter can contain?

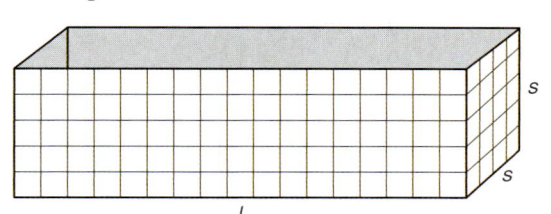

CHAPTER 3

12. Avery created this graph using data from the Statistics Canada Web site.

Incidence of Lung Cancer in Ontario by Gender, 2006

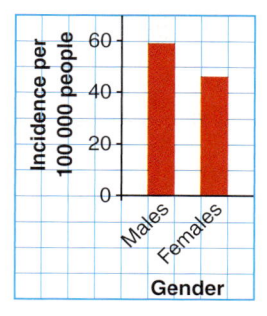

   a) What type of graph is it?
   b) Avery concluded that since the graph is comparing lung cancer incidence rate and gender, it is displaying two-variable data. Is Avery correct? Justify your answer.

13. For each scenario, state whether you think the two variables have a positive correlation, a negative correlation, or no correlation.
   a) Number of air conditioners sold and average daily summer temperature
   b) Hours spent sleeping and hours spent awake
   c) Number of applicants for a job and probability that you will get the job
   d) Number of kilometres driven and total fuel cost of trip

14. For each of these variables, describe a variable that could be correlated with it.
   a) The price of oil and …
   b) The age of a car and …
   c) The number of cigarettes a person smokes per day and …

Cumulative Review **263**

CHAPTER 3

15. Select the line of best fit for the data. Justify your choice.

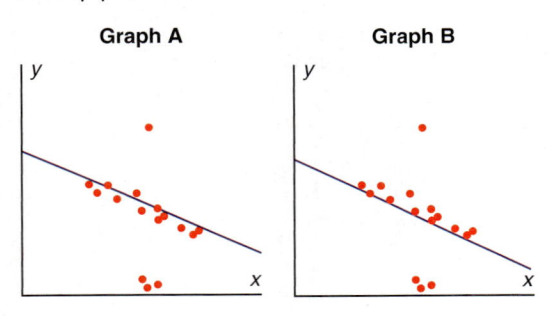

Graph A

Graph B

16. a) For each scatter plot, describe the relationship between x and y.
  b) Would you model each relationship with a linear or non-linear model? Justify your answers.
  i)
  ii)

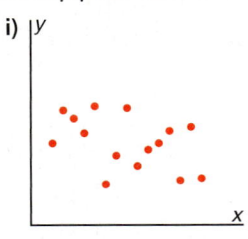

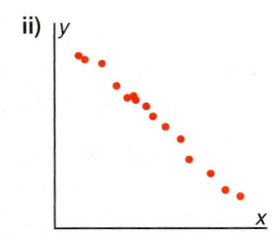

CHAPTER 4

17. Determine the quartiles for each data set.
  a) 7, 6, 1, 6, 2, 1, 10, 10, 4, 1, 3, 8
  b) 107, 109, 102, 113, 102, 110, 108, 104, 116, 108, 109

18. For each population, determine how many people should be surveyed to include 15% of the population.
  a) 20 people
  b) 120 people
  c) 360 people
  d) 11 500 people

19. Carmelo and his friends oppose switching to year-round schooling. They survey about 100 students in their high school to find out their opinions on year-round schooling.
  • Carmelo asks each person in his math class her or his opinion and records the answers.
  • 3 of his friends each choose one of their classes and ask everyone in that class.
  a) Explain why the sampling technique the students use is not random.
  b) Describe a random sampling technique for this survey.
  c) Which sampling technique from parts a and b do you think would produce a more representative sample? Explain your thinking.
  d) How might the survey technique Carmelo uses affect his results? What changes would you suggest to improve the survey? Explain.

20. A box of doggie dental chews contains a flyer with this graph. What additional information would you need before deciding whether the claim is valid?

DOGS PREFER THE TASTE OF YummyBone over other dental chews

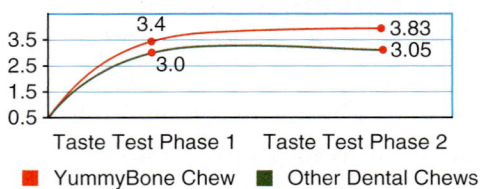

Taste Test Phase 1: YummyBone Chew 3.4, Other Dental Chews 3.0
Taste Test Phase 2: YummyBone Chew 3.83, Other Dental Chews 3.05

264 Chapters 1–4

5 Graphical Models

What You'll Learn
To use graphs to model real-world situations and data, and to compare, interpret, and analyse graphs

And Why
Recognizing trends and patterns in real-world graphs helps us better understand the situations being modelled, and allows us to make predictions about future behaviour.

Key Words
- trends
- rate of change
- first differences
- linear regression
- quadratic regression
- exponential regression

CHAPTER 5

Activate Prior Knowledge

Linear, Quadratic, and Exponential Graphs

Prior Knowledge for 5.3, 5.4, and 5.5

The graph of a **linear relation** is a straight line.

Cost of Magician

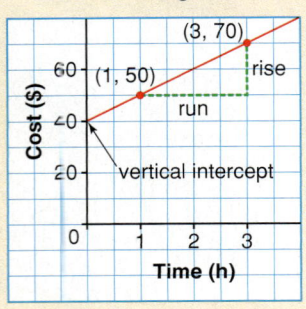

- Slope = $\frac{\text{rise}}{\text{run}}$

The graph of a **quadratic relation** is a **parabola**.

Sweatshirt Profits

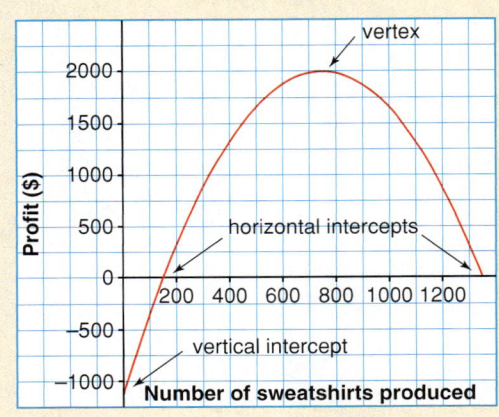

The graph of an **exponential relation** is an exponential curve.

Exponential Growth

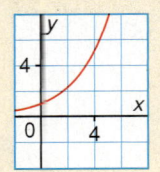

Exponential Decay

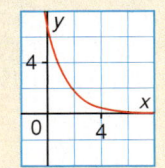

Example

This exponential graph shows the mass of pain medication left in a person's bloodstream after a tablet is swallowed.

a) How much medication was initially taken?

b) Determine the ratio of the mass of medication after 1 h to the mass of medication after 0 h. What does this value represent?

Pain Medication in the Body

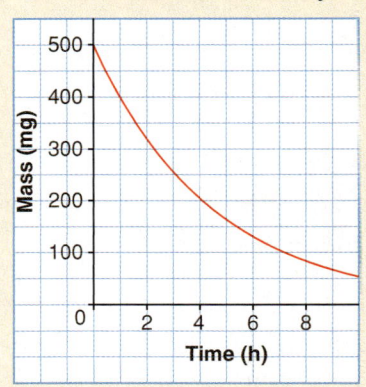

266   CHAPTER 5: Graphical Models

Solution

a) The vertical intercept represents the mass of medication initially taken, 500 mg.

b) Decay factor = $\dfrac{\text{Mass of medication after 1 h}}{\text{Mass of medication after 0 h}}$

= $\dfrac{400 \text{ mg}}{500 \text{ mg}}$

= 0.8

So, each hour, the mass of medication remaining in the body decreases by a factor of 0.8.

> We can find the growth or decay factor by calculating the ratio of any two successive *y*-values. In this graph, the masses after 0 h and 1 h are easiest to read.

CHECK ✓

1. Determine the slope of each line. What does each slope represent?

   a) Distance Travelled by a Car

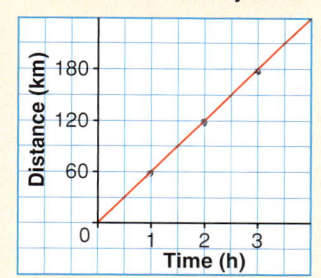

   b) Depreciation of the Value of a Computer

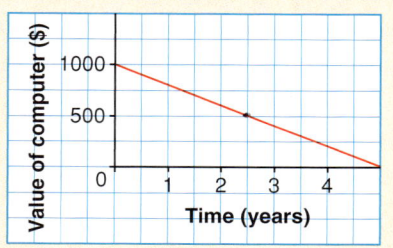

2. Determine the coordinates of the vertex of each parabola. What do these coordinates represent in each situation?

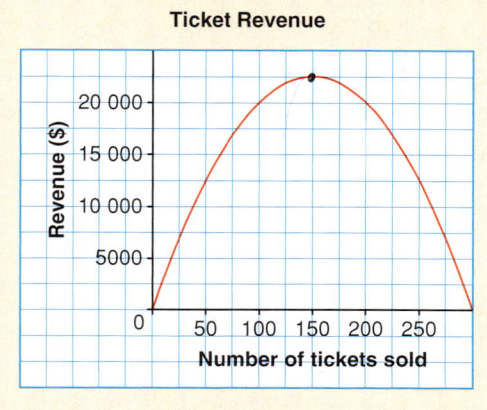

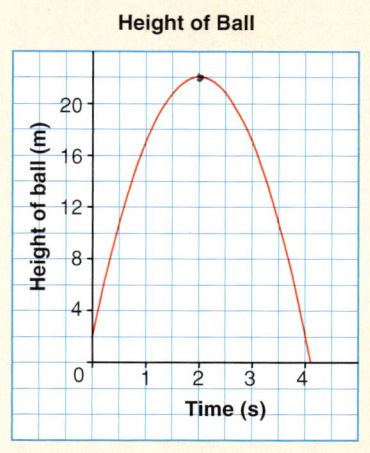

   a) Ticket Revenue

   b) Height of Ball

3. An exponential pattern of change results when an initial value is repeatedly multiplied by a constant factor. What role does each play in the graph?

Activate Prior Knowledge   **267**

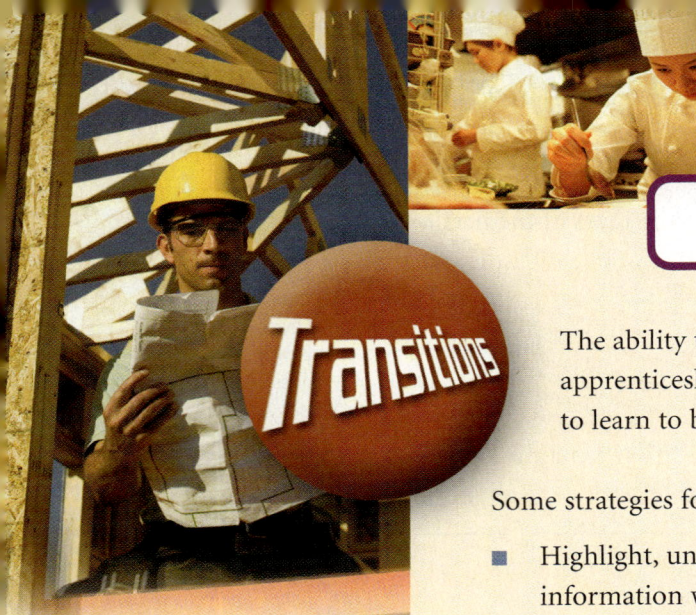

Teaching Yourself

The ability to teach yourself is key to success in school, apprenticeships, and occupations. One of the best ways to do this is to learn to be an active reader.

Some strategies for active reading can be used in any subject.

- Highlight, underline, and make notes as you read. Identifying key information will make it easier to review.
- Use the glossary or a dictionary, or search on the Internet, to read about unfamiliar terms.
- Connect new concepts to ones you already know.
- Put the information into your own words, or create a graphic organizer such as a concept map, to show how pieces of information fit together.
- When possible, read information several times during a few days to allow time for reflection and to improve understanding.

Some additional strategies are needed for mathematics.

- Put yourself in the problem. For example, in the *Investigate* for Lesson 5.1, think about understanding the situation. Focus on the change in the global average temperature and carbon dioxide levels.
- Make a model, sketch, or diagram to help clarify your ideas.
- Do the math! If there is an example, you might rewrite the solution without looking at the book, then check.
- After you understand the example or question, try another question from the book, or make up a question yourself.

• Complete questions 1 and 2 as you complete the lessons.

1. Apply active reading as you work with graphing calculators and use regression to model data in the *Investigates* for Lessons 5.3 to 5.6 and the *Inquire* in Lesson 5.7.

2. Imagine being an apprentice, an employee, or a college student. How do you think you might use active reading in this role?

5.1 Trends in Graphs

Environmental scientists study issues such as global warming, air and water pollution, and resource management. Their work includes identifying trends in data and using these trends to predict future change.

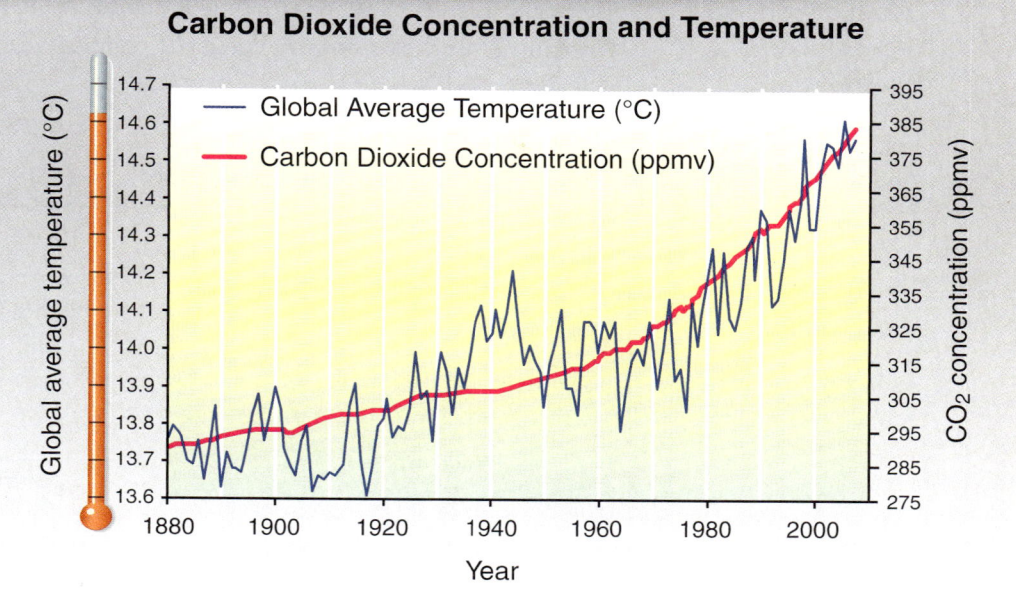

Carbon Dioxide Concentration and Temperature

— Global Average Temperature (°C)
— Carbon Dioxide Concentration (ppmv)

(Graph: Global average temperature (°C) on left y-axis from 13.6 to 14.7; CO_2 concentration (ppmv) on right y-axis from 275 to 395; Year on x-axis from 1880 to 2000.)

Investigate

Analysing a Climate Change Graph

Work with a partner.

The graph above shows the change in global average temperatures and carbon dioxide (CO_2) levels in Earth's atmosphere from 1880 to 2007.

- What information can you read from the graph about global average temperatures? About CO_2 levels?
- What can you predict about global average temperatures and CO_2 levels over the next 20 years? How confident can you be about your predictions?

Reflect

- What factors did you consider in making your predictions?
- What factors may affect the reliability of your predictions?
- Does the graph show that the increase in CO_2 levels causes global warming? Justify your answer.

Connect the Ideas

Using graphs to visualize relationships

A graph is a visual representation of the relationship between two quantities. It shows how one quantity changes with respect to the other.

Example 1 — Describing Relationships in Graphs

Describe the relationship shown in each graph.

a) Jack's Babysitting Earnings
(Earnings vs. Hours worked — straight line rising)

b) Amount of a Compound Interest Investment
(Amount vs. Time — curve rising more rapidly)

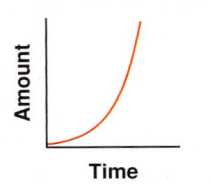

c) Temperature of a Cooling Cup of Coffee
(Temperature vs. Time — curve decreasing, leveling off)

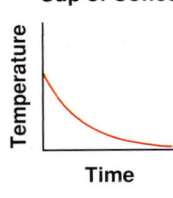

d) Fertilizing a Field
(Crop yield vs. Fertilizer used — curve rising then falling)

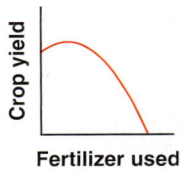

Solution

a) Jack's Babysitting Earnings

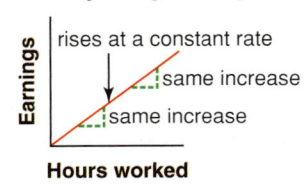

- rises at a constant rate
- same increase
- same increase

Pairs of points with equal horizontal distances have equal vertical distances. As the number of hours Jack works increases, his earnings increase by a constant amount.

b) Amount of a Compound Interest Investment

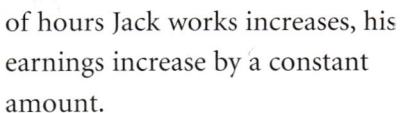

- rising more rapidly
- rising slowly
- greater increase
- smaller increase

The vertical distances between pairs of points with equal horizontal distances are increasing. The amount of the compound interest investment increases over time, slowly at first and then more rapidly.

270   CHAPTER 5: Graphical Models

c)

Temperature of a Cooling Cup of Coffee

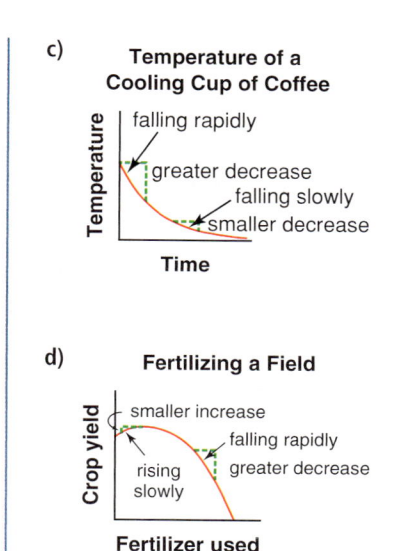

(graph: Temperature vs Time, showing falling rapidly → greater decrease, falling slowly → smaller decrease)

The vertical distances between pairs of points with equal horizontal distances are decreasing. The coffee temperature decreases over time, rapidly at first, then more slowly, and finally levelling off at room temperature.

d)

Fertilizing a Field

(graph: Crop yield vs Fertilizer used, showing rising slowly → smaller increase, falling rapidly → greater decrease)

The vertical distances between pairs of points with equal horizontal distances are decreasing, then increasing. As fertilizer use increases, the crop yield increases, reaches a maximum, and then decreases.

Trends in graphs

Trends, or patterns of change, in a graph are often used to justify decisions and make predictions.

Example 2

Describing the Trends in a Graph

This graph shows the number of births in Ontario from 1945 to 2005. Describe the trends in the graph.

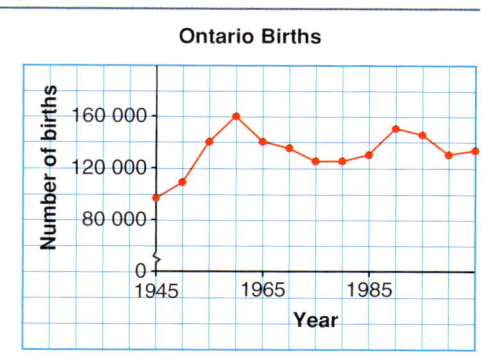

Ontario Births

(graph: Number of births vs Year, 1945–1985, with values around 80 000 to 160 000)

Solution

Trends occur in 3 broad groups: increasing, decreasing, and constant (no change). Divide the graph into intervals of time when the number of births is increasing, constant, or decreasing.

- From 1945 to 1960, the number of births is increasing rapidly. There is a maximum number of births in 1960.
- From 1960 to 1975, the number of births is decreasing, rapidly at first, then slowly, then more rapidly again.
- From 1975 to 1980, the number of births is constant.
- From 1980 to 1990, the number of births is increasing, slowly at first, then rapidly. There is another maximum of births in 1990.
- From 1990 to 2000, the number of births is decreasing rapidly.
- From 2000 to 2005, the number of births is increasing slowly.

5.1 Trends in Graphs **271**

Example 3 — Using Trends to Make Predictions and Justify Decisions

a) Use the graph to predict the number of Canadians in each age group in 2010.

b) What decisions might the Canadian government make in response to the trends in the graph?

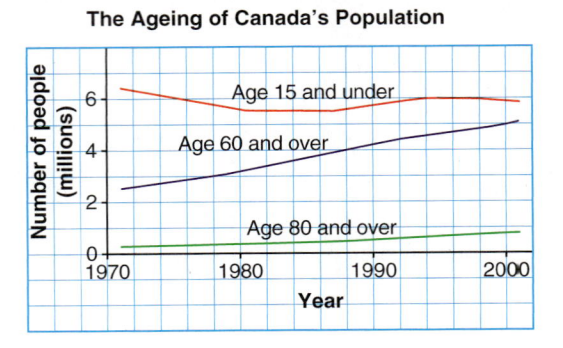

The Ageing of Canada's Population

Solution

a) Continue the trends to 2010.

The Ageing of Canada's Population

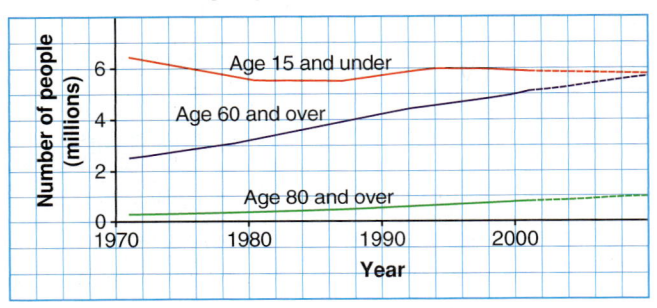

Record the projected number of Canadians in each age group.
- Under age 15: About 5.8 million
- Age 60 and over: About 5.7 million
- Age 80 and over: About 1.0 million

b) The trends suggest that a declining number of younger Canadians will have to support an increasingly larger number of elderly Canadians. Some decisions the Canadian government may make in response are:
- Increasing immigration levels to prevent future labour shortages
- Increasing the retirement age
- Strengthening health care and social security to better address the needs of older Canadians

Example 3 illustrates that often the best prediction we can make is to continue the trend. However, only a short-term prediction is reliable because we cannot be certain that the trend will continue, or, there may be several reasonable ways to continue the trend.

CHAPTER 5: Graphical Models

Practice

In questions 1 to 4, choose the graph that best represents the given description. Justify your choice.

A

For help with question 1 to 4, see Example 1.

1. The number of bacteria in a laboratory colony increases over time, slowly at first and then more rapidly.

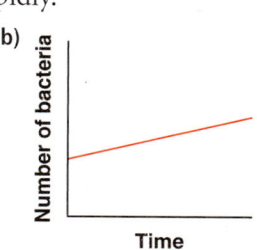

   a) [Graph: Number of bacteria vs Time — flat horizontal line]
   b) [Graph: Number of bacteria vs Time — gently increasing line]
   c) [Graph: Number of bacteria vs Time — exponentially increasing curve]

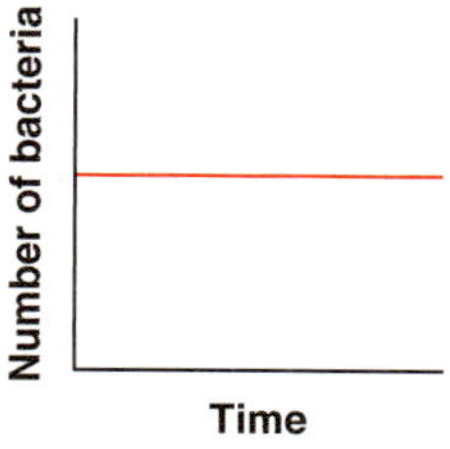

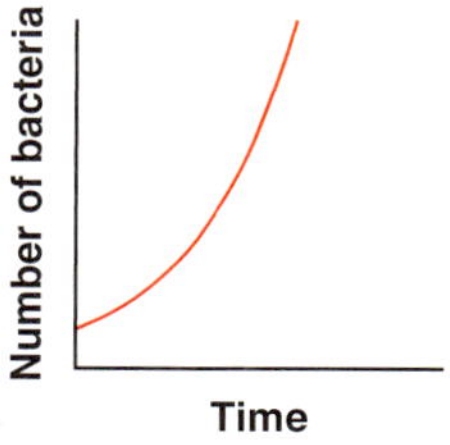

2. The fuel used increases steadily as the distance driven increases.

   a) [Graph: Fuel used vs Distance driven — straight line increasing]
   b) [Graph: Fuel used vs Distance driven — exponentially increasing curve]
   c) [Graph: Fuel used vs Distance driven — decreasing line]

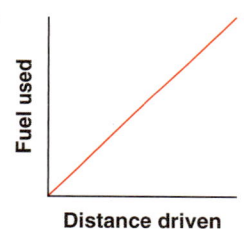

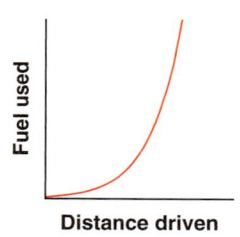

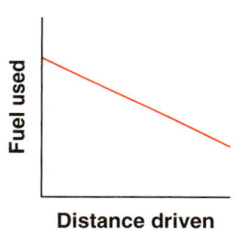

3. As the price increases, the revenue earned increases, reaches a maximum, then decreases.

   a) [Graph: Revenue vs Sale price — inverted parabola]
   b) [Graph: Revenue vs Sale price — increases, dips, then increases]
   c) [Graph: Revenue vs Sale price — decreasing curve]
   d) [Graph: Revenue vs Sale price — V-shape, decreases then increases]

5.1 Trends in Graphs **273**

4. The radioactive substance decayed over time, rapidly at first, then more slowly.

   a) [graph: Mass vs Time] b) [graph: Mass vs Time] c) [graph: Mass vs Time] d) [graph: Mass vs Time]

5. Match each graph with the statement that best describes it. Which words gave clues about the shape of the graph?

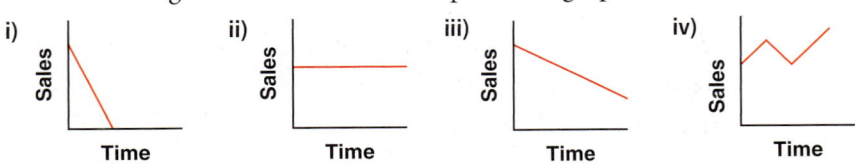

   i) [graph: Sales vs Time] ii) [graph: Sales vs Time] iii) [graph: Sales vs Time] iv) [graph: Sales vs Time]

   a) Sales have fallen dramatically over the last year.
   b) Sales have fallen steadily over the last year.
   c) Sales have remained constant over the last year.
   d) Sales have fluctuated over the last year.

B

For help with question 6, see Example 2.

6. Describe the trends in each graph.

   a) Percent of Immigrants in the Population of Canada

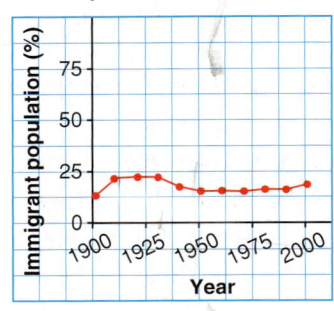

   b) Number of Births in Each Month in 2004

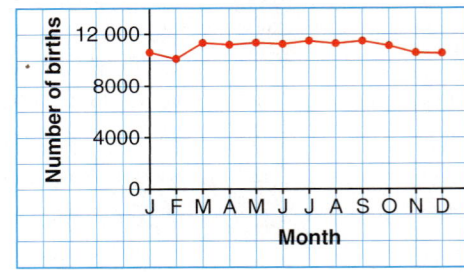

   c) Value of $1Can in US Dollars

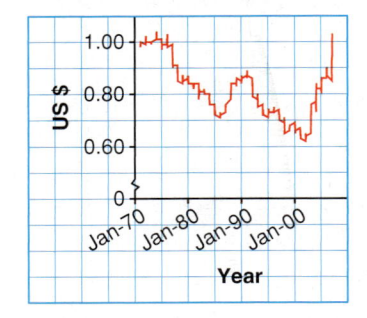

   d) Maximum Safe Heart Rate During Exercise

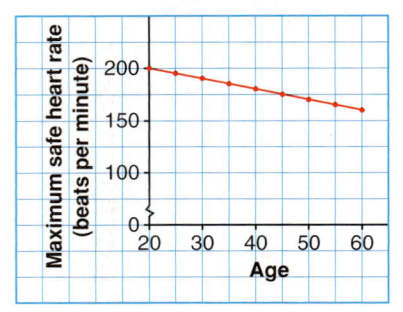

274   CHAPTER 5: Graphical Models

7. Wind power is becoming a practical source of renewable energy. This graph shows how the power generated by a wind turbine with radius 5 m changes with the wind speed.

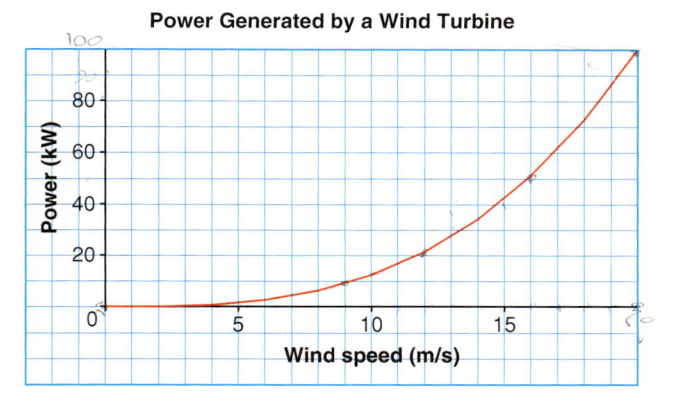

Power Generated by a Wind Turbine

(Graph: Power (kW) vs. Wind speed (m/s), curve rising from 0 through approximately (10, 12), (15, 45), to (20, 100))

a) Describe the general trends in the graph.
b) What is the vertical intercept? What does it represent in this situation?
c) How much power is generated when the wind speed is 20 m/s?
d) Does the power generated double when the wind speed doubles? Explain.

8. Each Tuesday, the Independent Electricity System Operator publishes a summary of the power generated and used in Ontario in the previous week.

Week ending August 26, 2007	
Ontario Peak Demand	**(MW)**
4 p.m. to 5 p.m. August 24, 2007	23,497
Wholesale Prices	**(¢ per kWh)**
Average Weekday Prices (8 a.m. to 8 p.m.)	6.41
Average Prices (other times)	4.72
(Prices weighted by Ontario Demand)	

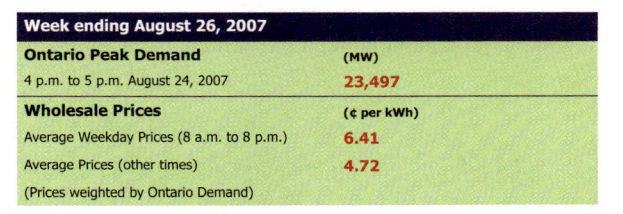

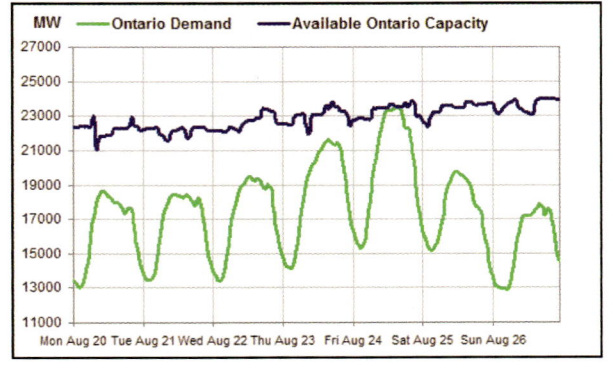

(Graph: Ontario Demand and Available Ontario Capacity, MW vs. days Mon Aug 20 – Sun Aug 26)

a) Describe the trend of Ontario electric power capacity for the week of August 20–26, 2007.
b) Explain the daily pattern of the Ontario electric demand.
c) How does peak daily demand change over the week? Why might this happen?
d) When does demand increase to or above capacity? How do you know?
e) What happens to electric service when demand surpasses capacity?

5.1 Trends in Graphs   **275**

■ For help with question 9, see Example 3.

9. You design a triangular garden for a park. One side of the garden has length 60 m and another side has length 80 m. This graph shows how the area of the garden is related to the length of the third side.

   Garden Area
   (Graph: Area (m²) vs. Length of third side (m); curve peaks around 2000 m², x-axis from 0 to 120)

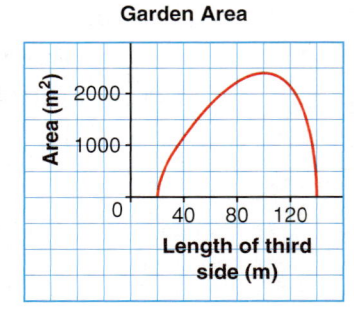

   a) Describe the relationship between the length of the third side and the area of the garden.
   b) How could you use the graph to decide what the length of the third side should be in each situation? Justify your answers.
      i) You want the area of the garden to be 1500 m².
      ii) You want the garden to have the maximum possible area.

10. A small company makes and sells T-shirts. This graph shows how the cost, sales, and profit vary as the number of T-shirts produced increases.

    T-shirt Costs, Sales, and Profits
    (Graph: Cost ($); Sales ($); Profit ($) vs. Number of T-shirts produced)

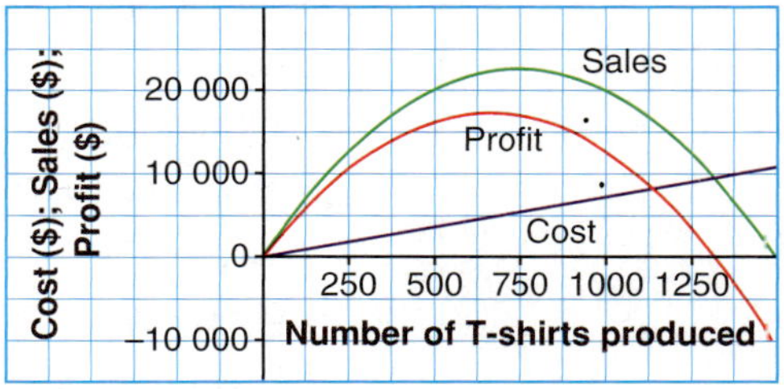

    a) Describe each relationship. Explain your reasoning.
       i) The relationship between the cost and the number of T-shirts produced.
       ii) The relationship between the profit and the number of T-shirts produced.
    b) Describe at least 2 decisions that the sales manager of the company might make based on the trends in the graph. Justify your answers.

11. **Assessment Focus**

    U.S. Mortality by Age
    (Graph: Number of deaths per 100 000 in age group vs. Age)

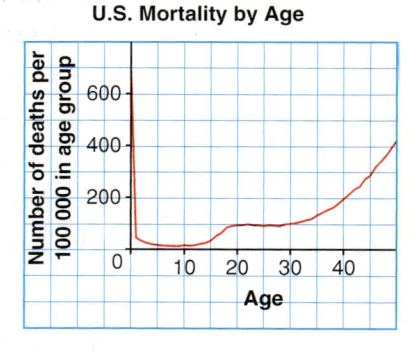

    a) Describe the trends in this graph.
    b) What is the minimum point on the graph? What do its coordinates represent?
    c) Describe the trend in mortality from age 15 to age 20. Why do you think this occurs? What decisions may be made in response to this trend? Explain.
    d) Predict the mortality rate at age 70. Justify your prediction.
    e) The actual mortality rate at age 70 is 1754 deaths per 100 000 seventy-year-olds. Compare your predicted result to this result. What might account for differences between the actual and predicted values?

276   CHAPTER 5: Graphical Models

12. **Literacy in Math** Use a matrix or another graphic organizer to summarize the important features of a graph. Some possible headings for a matrix are given below.

	Intercepts	Maximum or minimum points	Trends
Definition			
How to recognize			
Example			

C 13. This graph shows the fuel economy of Eva's car at various speeds.
   a) Describe the relationship between speed and fuel economy.
   b) Use the graph and the current cost of fuel. How much money would Eva save on a 1200 km trip by driving at the speed that produces the greatest fuel economy instead of the speed limit of 100 km/h? Justify your answer.
   c) How much longer will the trip in part b take at the speed with the greatest fuel economy?

14. a) Describe the general trend in this graph.
    b) Predict when the world population will reach 9 billion under each scenario.
       i) The current rate of growth continues
       ii) The growth rate decreases
       iii) The growth rate increases
    Explain how you made your predictions.

In Your Own Words

Suppose you collect data about the mass of a puppy over 15 weeks and plan to present the data in a graph. What trends would you expect to see in the graph? Explain.

5.1 Trends in Graphs   **277**

5.2 Rate of Change

A child's height and growth rate are important indicators of the child's overall health.

Investigate

Analysing Patterns of Growth in Girls and Boys

Work with a partner.

> The World Health Organization is an agency of the United Nations.

This graph presents data from the World Health Organization on average heights of boys and girls.

- What information can you read from the graph? Record your ideas.

Comparing the Growth of Boys and Girls

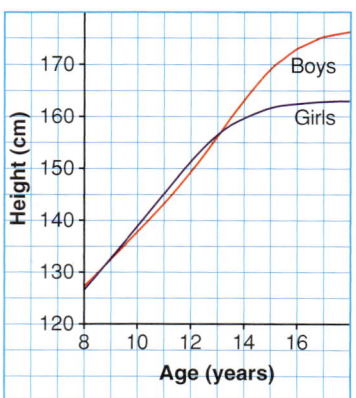

Reflect

- How does a graph communicate information about change?
- Compare the information you read from the graph with another pair of students. What additional information did they find?

278   CHAPTER 5: Graphical Models

Connect the Ideas

Rate of change

We can determine an average rate of change using a table or graph.

Table

Independent variable	Dependent variable
x_1	y_1
x_2	y_2

Graph

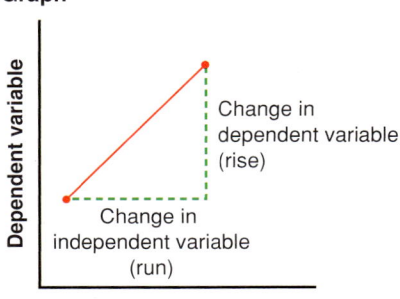

The average rate of change between two points is the slope of the line segment joining the points.

Average rate of change

$$\text{Average rate of change} = \frac{y_2 - y_1}{x_2 - x_1} \text{ or } \frac{\text{Rise}}{\text{Run}}$$

Example 1 — Calculating and Interpreting Rates of Change

Calculate the average rate of change between each pair of points. Explain what the rate of change represents.

a)

Time (min)	Height of airplane (m)
0	2000
4	1400

b) **Vehicle Rental Cost**

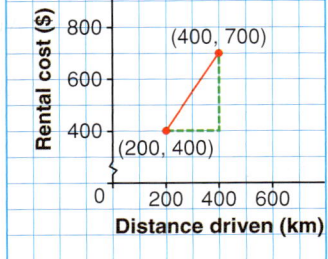

Points shown: (200, 400) and (400, 700)

Solution

a) Average rate of change
$= \dfrac{\text{Change in height}}{\text{Change in time}}$

$= \dfrac{1400 \text{ m} - 2000 \text{ m}}{4 \text{ min} - 0 \text{ min}}$

$= \dfrac{-600 \text{ m}}{4 \text{ min}}$ or -150 m/min

The height of the airplane is decreasing by an average of 150 m each minute.

b) Average rate of change
$= \dfrac{\text{Change in cost}}{\text{Change in distance driven}}$

$= \dfrac{\$700 - \$400}{400 \text{ km} - 200 \text{ km}}$

$= \dfrac{\$300}{200 \text{ km}}$ or $\$1.50$/km

The cost increases by an average of $1.50 for each kilometre driven.

5.2 Rate of Change

Example 2   Comparing Rates of Change

The reaction distance is the distance the car travels from the time the driver decides to stop the car until the driver applies the brakes. The braking distance is the distance the car travels from the time the brakes are applied until the car stops.

The distance required to stop a car depends on the speed at which the car is travelling.

These tables show the reaction distance and braking distance needed to stop a car on dry pavement for given speeds.

Speed (km/h)	0	10	20	30	40	50
Reaction distance (m)	0	2	4	6	8	10

Speed (km/h)	0	10	20	30	40	50
Braking distance (m)	0.0	0.5	2.0	4.5	8.0	12.5

The sum of the reaction distance and braking distance is the stopping distance.

a) Calculate the average rate of change between consecutive points in each table. Describe the rates of change revealed in each table.
b) Graph the data in the tables. Describe how the graph reflects the rates of change across the data.

Solution

a)

Speed (km/h)	Reaction distance (m)	$\dfrac{\text{Change in distance}}{\text{Change in speed}}$
0	0	
10	2	$\dfrac{2-0}{10-0} = \dfrac{2}{10} = 0.2$
20	4	$\dfrac{4-2}{20-10} = \dfrac{2}{10} = 0.2$
30	6	$\dfrac{6-4}{30-20} = \dfrac{2}{10} = 0.2$
40	8	$\dfrac{8-6}{40-30} = \dfrac{2}{10} = 0.2$
50	10	$\dfrac{10-8}{50-40} = \dfrac{2}{10} = 0.2$

The rates of change are constant: 0.2. So, the reaction distance increases by 0.2 m for every 1-km/h increase in speed.

280   CHAPTER 5: Graphical Models

Speed (km/h)	Stopping distance (m)	$\dfrac{\text{Change in distance}}{\text{Change in speed}}$
0	0.0	
10	0.5	$\dfrac{0.5 - 0.0}{10 - 0} = \dfrac{0.5}{10} = 0.05$
20	2.0	$\dfrac{2.0 - 0.5}{20 - 10} = \dfrac{1.5}{10} = 0.15$
30	4.5	$\dfrac{4.5 - 2.0}{30 - 20} = \dfrac{2.5}{10} = 0.25$
40	8.0	$\dfrac{8.0 - 4.5}{40 - 30} = \dfrac{3.5}{10} = 0.35$
50	12.5	$\dfrac{12.5 - 8.0}{50 - 40} = \dfrac{4.5}{10} = 0.45$

The rates of change are increasing.

b) **Reaction Distance and Speed**

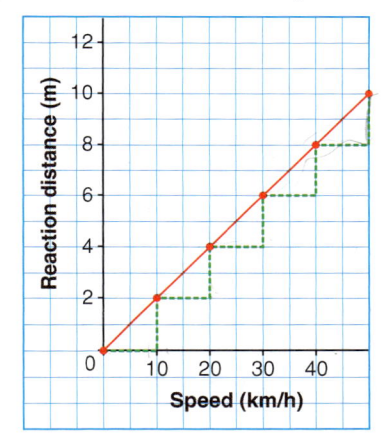

Since the rate of change is the same in each 10 km/h interval, the points on the graph lie on a straight line.

Braking Distance and Speed

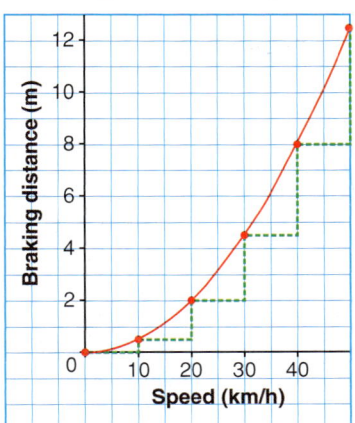

Since the average rate of change is different for each 10 km/h interval, the points on the graph do not lie on a straight line.

> Since a linear relation has a constant rate of change, we refer to the *rate of change* of a linear relation and the *average rate of change* of a non-linear relation.

5.2 Rate of Change

Example 3 Identifying Rates of Change in a Table and Graph

This table shows the change in height of a tomato plant from germination until the tomatoes ripen.

Time (weeks)	0	2	4	6	8	10	12	14	16	18
Height (cm)	0	5	10	20	40	58	75	86	90	90

a) Determine when the rate of change in the height is:
   i) Zero   ii) Constant   iii) Changing
b) When is the rate of change in height the greatest?
c) Describe the growth of the plant.

Solution

a) Create a table of first differences. Graph the data from the table.

Growth of Tomato Plant

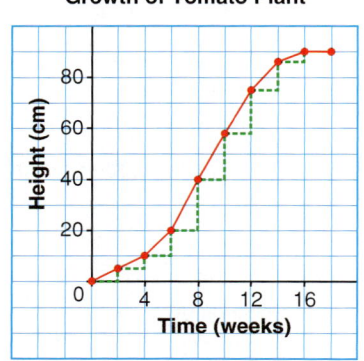

Time (weeks)	Height (cm)	First differences
0	0	
2	5	$5 - 0 = 5$
4	10	$10 - 5 = 5$
6	20	$20 - 10 = 10$
8	40	$40 - 20 = 20$
10	58	$58 - 40 = 18$
12	75	$75 - 58 = 17$
14	86	$86 - 75 = 11$
16	90	$90 - 86 = 4$
18	90	$90 - 90 = 0$

> On the graph, the vertical distances between consecutive points correspond to the first differences.

i) In the table, look for a first difference of 0. The rate of change is zero from week 16 to 18.

ii) On the graph, look for points lying along a line. The rate of change is constant from week 0 to 4.

> The vertical distances between consecutive points on a line are constant.

iii) The rate of change is changing when the vertical distances between consecutive points change. The rate of change is changing from week 4 to 16.

CHAPTER 5: Graphical Models

b) The rate of change is greatest from week 6 to 8.
This can be seen in the graph, where the steepest line segment is between these two weeks.

c) The plant grows at a constant rate for the first 4 weeks. The rate of growth increases over the next 4 weeks, and then decreases as the plant reaches maturity. After 16 weeks, the plant has reached its full height, and the rate of growth is 0.

Identifying when a rate of change is zero, constant, or changing

Given a table or graph, we can identify when the rate of change is zero, constant, or changing without actually calculating the rates of change.
- In a table, we look at the first differences.
- In a graph, we decide whether the points lie on a line.

Identifying Rates of Change in a Table or Graph

Rate of change	Table	Example of graph
Zero	The first differences are 0.	(horizontal line)
Constant	The first differences are equal.	(straight increasing line)
Changing	The first differences are changing.	(curve)

5.2 Rate of Change

Practice

A

For help with questions 1 to 5, see Example 1.

1. For each table, name the variables.

a)
Hours worked	Earnings ($)
4	32
20	160

b)
Pages printed	Cost ($)
1000	56
5000	145

c)
Distance driven (km)	Fuel used (L)
45	3
60	12

2. State the units of the rate of change for each table in question 1. What does the rate of change represent?

3. Refer to the tables in question 1. Determine the average rate of change between each pair of points in the table.

4. For each graph, name the variables.

a) **Graph A** — Temperature (°C) vs Depth (m); points (1, 30) and (3, 50).

b) **Graph B** — Distance (m) vs Time (s); points (2, 8) and (4, 32).

c) **Graph C** — Interest earned ($) vs Year; points (2, 50) and (4, 50).

5. State the units of the rate of change in each graph in question 4. What does each rate of change represent?

6. Refer to the graphs in question 4. Determine the average rate of change between the indicated points on the graph.

B

7. To save energy, an office building is only heated during business hours.

a) When is the temperature:
  i) Constant?
  ii) Decreasing?
  iii) Increasing?

b) Calculate the rate of change during each time period from part a.

c) Describe the connection between your answers in parts a and b.

Office Building Temperature — Temperature (°C) vs Time.

284 CHAPTER 5: Graphical Models

8. a) Describe the trends in this graph.

Community College Enrollment in Ontario

b) From 1980 to 1998, the number of students attending a community college changed from 260 761 students to 403 516 students. What was the average rate of change in attendance during this time?

c) The graph is approximately horizontal from 1983 to 1989. What does this tell you about the rate of change in attendance during this time?

■ For help with question 9, see Example 2.

9. Bipin is a financial advisor. He uses these tables to help his clients understand the difference between simple interest and compound interest.

Simple Interest

Year	0	5	10	15
Amount ($)	500	700	900	1100

Compound Interest

Year	0	5	10	15
Amount ($)	500	735	1079	1586

a) Calculate the average annual rate of change for consecutive pairs of data in each table.

b) Describe the rates of change in each table. What do these values indicate about each type of interest?

c) Graph the data in the tables. How is the rate of change reflected in the graph?

5.2 Rate of Change   **285**

10. **Assessment Focus**

 A tanker runs aground, creating a circular oil spill.

 a) For each graph, calculate the average rate of change:
   i) From 0 min to 5 min
   ii) From 10 min to 15 min
  What do the rates of change represent in this situation?
 b) Describe the change in the radius of the spill.
 c) Describe the change in the area of the spill.

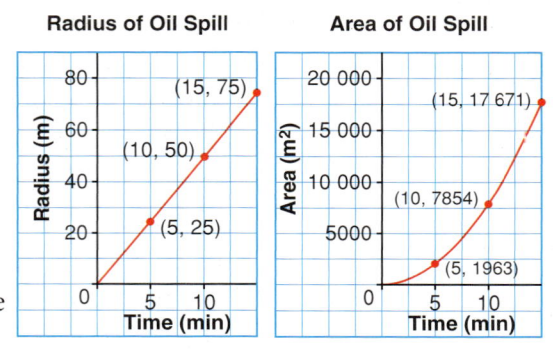

Radius of Oil Spill — points (5, 25), (10, 50), (15, 75); Radius (m) vs Time (min)

Area of Oil Spill — points (5, 1963), (10, 7854), (15, 17 671); Area (m²) vs Time (min)

■ For help with questions 11 and 12, see Example 3.

11. Use this table to analyse changing power needs in Canada.

Total Amount of Electric Power Generated, Canada

Year	1950	1955	1960	1965	1970	1975	1980	1985	1990	1995	2000	2005
Amount (MWh)	64	79	108	149	209	285	360	433	490	530	564	578

 a) Create a table of first differences. What do they represent in this situation?
 b) When did the electric energy generated increase most rapidly?
 c) The first differences are approximately equal from 1975 to 1980 and from 1980 to 1985. What does this mean about the electric energy generated in those years?
 d) Describe how the electric energy generated has changed from 1950 to 2005.

12. The height of a sunflower was recorded every week over a growing season. Describe the growth of the sunflower.

Height of a Sunflower — Height (cm) vs Week

286   CHAPTER 5: Graphical Models

13. Children learn new words at varying speeds. This graph shows a typical learning curve for children from ages 10 months to 72 months.

Vocabulary Growth Chart

a) When are new words learned most quickly?
b) What is the average rate of learning at that time?
c) What is the average rate of learning from 60 to 72 months of age?
d) Suppose a child started elementary school at 72 months. Predict and sketch a continuation of the graph for another 72 months.
e) Explain the shape you chose for part d.

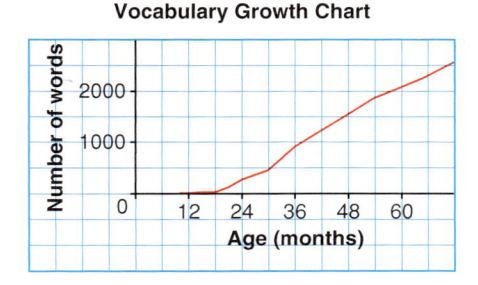

14. Literacy in Math You have described change and rates of change in Lessons 5.1 and 5.2. Use a graphic organizer of your choice to summarize some of the key ideas and vocabulary you have used in these lessons.

C 15. Refer to the graph for the growth of a sunflower in question 12.
a) Sketch a graph for a sunflower that takes the same time to mature but has half the final height. Explain your reasoning.
b) Sketch a graph for a sunflower that takes longer to mature but has the same final height. Explain your reasoning.

16. The temperature in a sunroom of a house changes throughout the day. Use the following information to draw a graph of the temperature over time.
- The early morning temperature is 12°C.
- After the sun reaches the sunroom at 8:00 A.M., the temperature rises 4°C/h until 12:00 P.M.
- The temperature is approximately constant from 12:00 P.M. in the afternoon until 3:00 P.M.
- After that it drops about 2°C/h until it reaches 12°C.

In Your Own Words

Draw or find a graph that shows how a quantity changes over time. Use words and numbers to describe the change and rates of change shown in the graph. What do these tell you about how the quantity is changing?

5.3 Linear Models

The chirping rate of a cricket increases as the temperature rises and decreases as the temperature falls. Biologists call crickets "nature's thermometers" because the chirp rates can be used to predict the temperature.

Investigate | Chirping Rates of a Cricket

Materials
- TI-83 or TI-84 graphing calculator

Work with a partner.

A biologist collects data about the temperature and chirp rate of a cricket.

- What trends do you see in the data? How well does a line model the trends in the data? Justify your answers.

- Is it reasonable to use a linear model to predict the chirp rate when the temperature is:
  a) 70°F?
  b) 0°F?
  c) 120°F?
  Explain.

Temperature (°F)	Chirp rate (chirps/min)
61	87
66	102
67	109
73	136
74	154
76	150
77	154
78	160

288   CHAPTER 5: Graphical Models

> **Reflect**
>
> What assumptions do we make when we create a line of best fit and use it to make predictions? Why might these assumptions not be valid?

Connect the Ideas

Mathematical models

Tables, graphs, and equations are examples of **mathematical models**. Mathematical models allow us to represent the relationship between real-world quantities, analyse current behaviour, and predict future behaviour.

> Mathematical models can be numerical (tables), graphical (graphs), or algebraic (equations).

Linear models

> **Linear models**
> A linear model represents quantities that increase or decrease by a constant amount over equal intervals.
> - In a table of values, the first differences are equal.
> - The graph is a straight line.
> - The equation of the line can be written in the form $y = mx + b$, where m is the slope and b is the vertical intercept.
> - The rate of change is constant.

Example 1 — Identifying Linear Models

Which models represent linear relations? Justify your answers.

a)

Time (s)	0	1	2	3
Height (m)	60	55	40	15

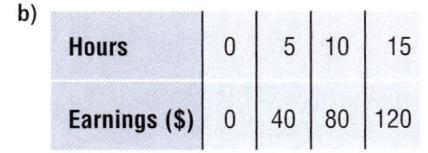

b)

Hours	0	5	10	15
Earnings ($)	0	40	80	120

c) Depreciation in Value of Printer

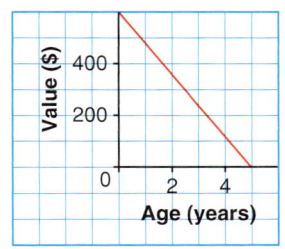

d) Surface Area of a Cube

e) $y = 2x + 5$

f) $y = x^2 + 5$

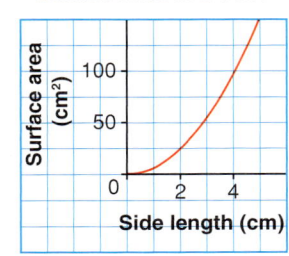

5.3 Linear Models **289**

Solution

a)

Time (s)	Height (m)	First differences
0	60	
		55 − 60 = −5
1	55	
		40 − 55 = −15
2	40	
		15 − 40 = −25
3	15	

The first differences are not equal. The relationship is non-linear.

c) The graph is a line. The relationship is linear.

e) $y = 2x + 5$ is of the form $y = mx + b$. The relationship is linear.

b)

Hours	Earnings ($)	First differences
0	0	
		40 − 0 = 40
5	40	
		80 − 40 = 40
10	80	
		120 − 80 = 40
15	120	

The first differences are equal. The relationship is linear.

d) The graph is a curve. The relationship is non-linear.

f) $y = x^2 + 5$ cannot be expressed in the form $y = mx + b$. The relationship is non-linear.

Analysing the graph of a linear relation

In real-world graphs of linear relations:
- The vertical intercept represents the initial value of the dependent variable.
- The slope represents the rate of change in the dependent variable with respect to the independent variable.

We can compare the graphs of pairs of relations to investigate how the initial value and rate of change are reflected in the graph.

Example 2 — Comparing Pairs of Linear Relations

Materials
- TI-83 or TI-84 graphing calculator

A cup of coffee is reheated in a microwave. The temperature, C degrees Celsius, of the coffee after t seconds can be modelled by linear equations.

500-W power setting: $C = 0.5t + 20$
1000-W power setting: $C = t + 20$

a) Explain what the numbers in the equations represent.
b) Graph both equations on the same screen of a graphing calculator. How are the numbers in the equation reflected in the graph?

CHAPTER 5: Graphical Models

Solution

a) • The coefficients of t, 0.5 and 1, represent the rate of change of the temperature with respect to time. At the 500-W setting, the temperature increases at a rate of 0.5°C/s. At the 1000-W setting, the temperature increases at a rate of 1°C/s.
• The constant 20 represents the temperature of the coffee at $t = 0$ s. This is the temperature of the coffee when it was placed in the microwave.

b) Use a graphing calculator to generate the graphs of $y = 0.5x + 20$ and $y = x + 20$ on the same screen.

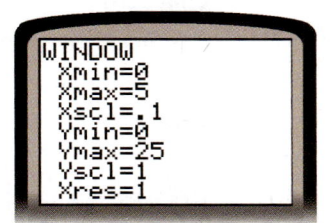

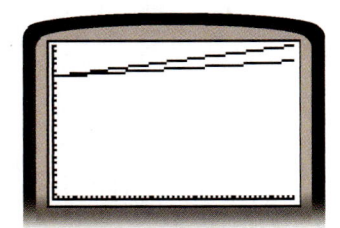

Both graphs have the same vertical intercept, but the line representing the 1000-W setting is steeper than the line representing the 500-W setting. This is because both cups of coffee have an initial temperature of 20°C, and the rate of change in temperature is greater at the 1000-W setting.

Fitting a linear model to data

We can use linear regression to model data that appear to be linearly related. The regression line can then be used to analyse the data and make predictions. The closer the regression line is to the data points, the more reliable the predictions are likely to be.

Example 3

Materials
• TI-83 or TI-84 graphing calculator

Fitting a Linear Model to Data

This table shows the median age of Canada's population from 1975 to 2000.

Year	1975	1980	1985	1990	1995	2000
Median age (years)	27.4	29.1	31.0	32.9	34.8	36.8

a) Create a scatter plot of the data and describe any trends.
b) Determine the equation of the regression line. Graph the regression line on the scatter plot.
c) Predict the median age of Canada's population in 2020.

5.3 Linear Models **291**

Solution

a) Create a scatter plot of the data.

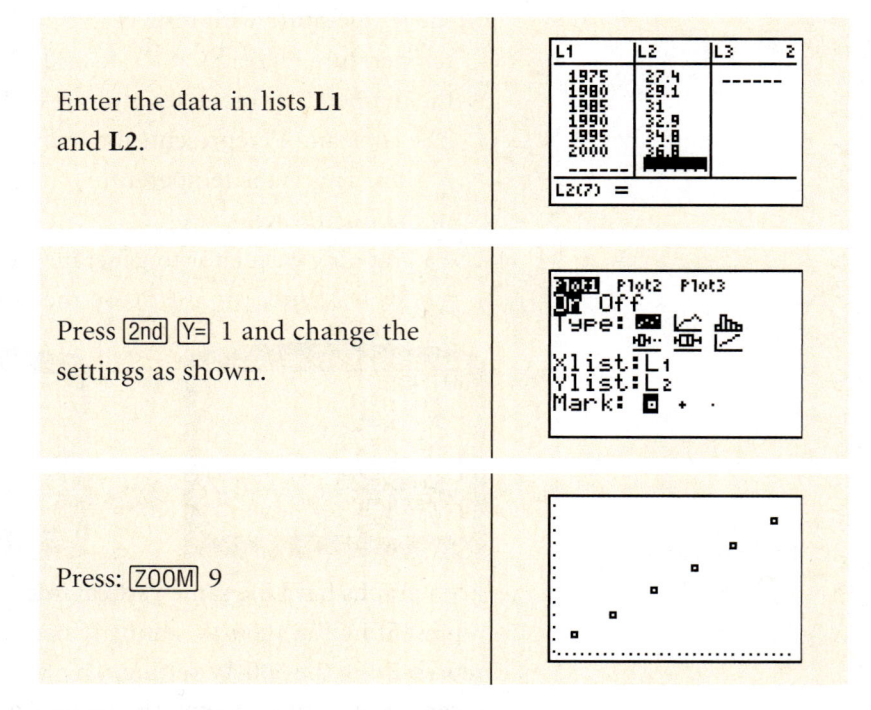

Enter the data in lists **L1** and **L2**.

Press [2nd] [Y=] 1 and change the settings as shown.

Press: [ZOOM] 9

The data appear to lie on a straight line. This suggests that the median age has been increasing at a constant rate.

b) Use the **LinReg(ax+b)** command.

Press [STAT] [▶] 4 [2nd] 1 [,] [2nd] 2 [,] [VARS] [▶] 1 1 [ENTER] to perform linear regression on the data and store the regression equation as **Y1**.

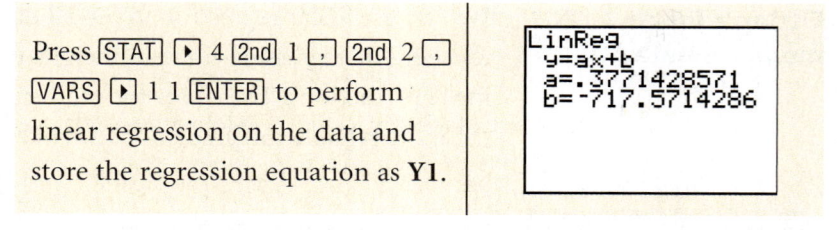

> Since the points lie close to the line, the regression line is a good fit for the data.

The regression equation is given in the form $y = ax + b$, where a is the slope of the line and b is the y-intercept.

So, the regression equation is: $y \doteq 0.3771x - 717.5714$

Graph the data and the regression curve.

Press: [GRAPH]

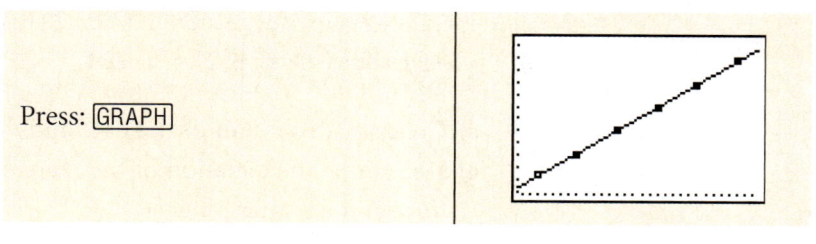

292 CHAPTER 5: Graphical Models

c) Here are two methods for predicting the median age in 2010.

Method 1: Using the regression line

Use the **TRACE** feature to estimate the median age of Canada's population in 2020.

Press WINDOW and set
Xmax = 2020.
Press: TRACE
Press ▼ to place the cursor on the graph of the regression equation.
Press: 2020 ENTER

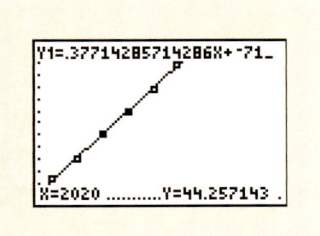

At $x = 2020$, $y \doteq 44.2$

Method 2: Using the regression equation

> You could also use the command **Y1(2020)** on your graphing calculator.
> Press: VARS ▶ 1 1 (
> 2020) ENTER

Substitute $x = 2020$ into the regression equation.
$y \doteq 0.3771x - 717.5714$
$\doteq 0.3771(2020) - 717.5714$
$\doteq 44.2$

If the current trends continue, the median age of Canada's population in 2020 will be about 44 years.

Practice

A

▪ For help with questions 1 to 3, see Example 1.

1. Which tables of values model a linear relation? How do you know?

a)

r	0	5	2	3	4	5
C	0.0	31.4	62.8	94.2	125.7	157.1

b)

t	0	1	2	3	4	5
h	282.5	272.7	243.3	194.3	125.7	37.5

2. Which equations model a linear relation? How do you know?
   a) $y = -2x$ b) $y = x^2 + 1$ c) $y = 5 - 2x$

5.3 Linear Models

3. Which graphs model a linear relation? How do you know?

a) Number of tickets vs Cost ($)

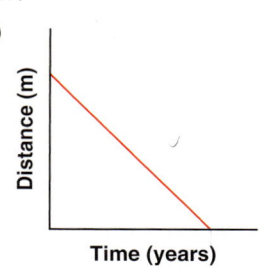

b) Population vs Year

c) Distance (m) vs Time (years)

For help with questions 4 and 5, see Example 2.

4. Match each graph with the statement that best describes it.
   a) Same initial value, different rates of change
   b) Different initial values, same rate of change
   c) Different initial values, different rates of change

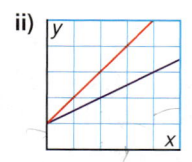

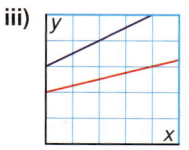

   i)   ii)   iii)

5. For each part, sketch a line that is different from this line.
   a) Different initial value
   b) Greater rate of change
   c) Same initial value and a zero rate of change

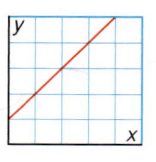

B

6. This graph shows how the distance a car travels changes over time.
   a) Calculate the rate of change. Does it matter which points you use? Explain.
   b) What does the rate of change represent? How do you know?

   Distance Travelled by a Car

7. An energy auditor uses a temperature probe to check the insulation in a wall of a house.
   a) Describe the rates of change revealed in the graph.
   b) Calculate the rate of change in temperature with respect to distance into the wall.
   c) Write an equation that describes the temperature-distance relation.

   Insulation Temperature at Various Distances from Interior Wall

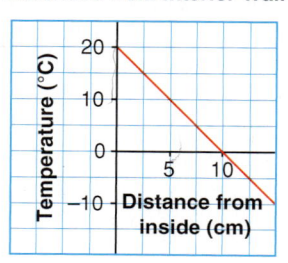

8. Two springs are stretched to determine which one is stiffer. The force, F newtons, needed to stretch the springs a distance of x cm is given by these formulas.

   Spring 1: $F = 0.1x$ **Spring 2:** $F = 0.5x$

   a) Describe what the graph would look like if you graphed both equations on the same grid. What would be the same about both graphs? What would be different? How do you know?
   b) Use graphing technology to verify your answer to part a.
   c) Determine the rate of change in force with respect to distance for each spring. Explain your method.
   d) The rate of change in force for a spring is called the *spring constant*. If a spring is easy to stretch, does it have a high or low spring constant? Explain.

9. Two friends go for a walk during their lunch break. This graph shows how their distance from work changes over time.
   a) Compare their initial positions and speeds.
   b) Describe how their motions differ.

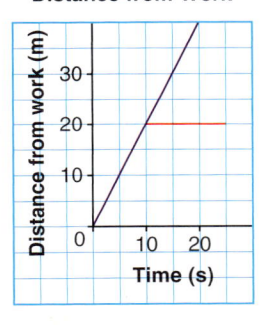

10. **Assessment Focus** The graduation committee is making arrangements for the prom. This graph shows the total cost of the prom based on the number of people who attend it.
    a) What is the vertical intercept? What does it represent?
    b) Calculate the rate of change in the cost with respect to the number of people. What does this rate of change represent?
    c) The committee wants to sell tickets at $20 per person. On a copy of the graph, sketch a line that shows the total sales.
    d) How many tickets would have to be sold to break even?
    e) Suppose the ticket price is $25 per person. How would this change the sales graph? How would this change the break-even point? Justify your answers.

To *break even* means to neither make nor lose money.

11. This graph shows the positions of four cars over time.

    Motion of Four Cars

    - Car A and Car B start from the same position, but travel at different speeds.
    - Car C follows Car B at a constant distance behind.
    - Car D travels in the opposite direction, passing each of the other cars in turn.

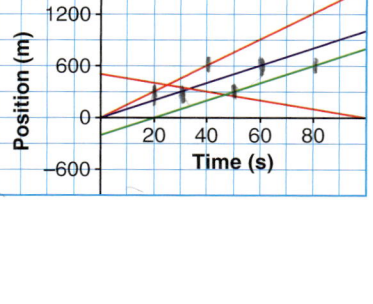

    a) On a copy of the graph, label each line with the car it represents. Justify each choice.
    b) Determine the speed of each car.

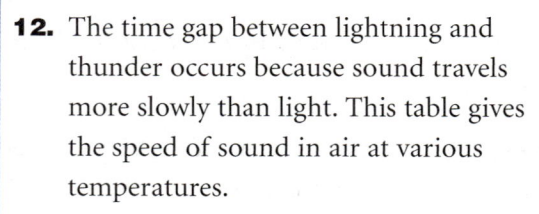

■ For help with questions 12 to 14, see Example 3.

12. The time gap between lightning and thunder occurs because sound travels more slowly than light. This table gives the speed of sound in air at various temperatures.

Temperature (°C)	0	10	20	30
Velocity (m/s)	332	338	344	350

   a) Draw a scatter plot of the data and determine the equation of the line of best fit.
   b) Predict the speed of sound at these air temperatures.
      i) 15°C   ii) 50°C   iii) −3°C

13. Mass-for-age data were collected for two baby girls.

Age (months)	21	24	27	30	33	36
Alea's mass (lb.)	21.5	22.4	23.3	24.1	24.9	25.7
Yanxia's mass (lb.)	30.5	32.0	33.5	35.0	36.5	38.0

   a) Do the babies' masses grow linearly? Explain.
   b) Which baby is growing more quickly? Explain.
   c) Find a linear regression equation that models the data for each baby.
   d) How well does the line of best fit model the data? Explain.
   e) Determine the rate of change in mass for each baby.

296   CHAPTER 5: Graphical Models

14. The table shows the distance travelled by a transport truck after each number of hours.

Time (h)	2	3	5	6
Distance travelled (m)	195	302	508	599

    a) Find a linear regression equation that models the data.
    b) Estimate the distance travelled in 4 hours. Why should this estimate be accurate?
    c) Estimate the distance travelled in 24 hours. Why might this estimate be inaccurate?

15. **Literacy in Math** Make a Frayer Model for linear relations. Include *rate of change* in your model.

C 16. When a bedroom air conditioner is turned on High, the temperature in the bedroom drops as shown in this graph.
    a) How fast is the temperature decreasing?
    b) Sketch a graph for the situation when the initial temperature of the bedroom is greater. Explain your reasoning.
    c) Sketch a graph for the situation where the air conditioner is on Low. Explain your reasoning.

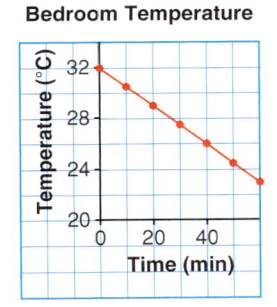

17. Draw a graph to show how the price of gasoline changes with time.
    - On January 1, the price is 95.0¢/L.
    - The price increases 3.0¢/L each month for 4 months.
    - The price decreases 1.5¢/L each month for 2 months.
    - The price jumps to 115.0¢/L overnight.
    - The price doesn't change for the rest of the year.

In Your Own Words

Why can we say that rate of change is the same as slope for linear graphs? Why is this not true for other graphs? Use examples to illustrate your answer.

5.3 Linear Models   **297**

5.4 Quadratic Models

Aram is an accident reconstructionist for the Ontario Provincial Police. He investigates car accidents in which serious injury or death occurred. Aram uses mathematical models to estimate the speed at which the vehicle was moving when it crashed.

Investigate: Investigating Relationships in Formulas

Materials
- TI-83 or TI-84 graphing calculator

Work with a partner.

The distance a vehicle travels in icy conditions, after the driver applies the brakes, can be modelled by the formula $d = 0.75sm^2$, where
- d metres is the distance travelled
- m tonnes is the vehicle's mass
- s kilometres per hour is the vehicle's speed

■ Choose a reasonable speed s for the vehicle.
 Substitute into the formula $d = 0.75sm^2$ and simplify.
 Graph the resulting formula.

■ Choose a reasonable mass m for the vehicle.
 Substitute into the formula $d = 0.75sm^2$ and simplify.
 Graph the resulting formula.

Reflect

What would be the effect of choosing a different value for s? For m?

298 CHAPTER 5: Graphical Models

Connect the Ideas

Quadratic models

In Lesson 5.3, we used linear models to represent a variety of real-world situations involving constant rates of change. In situations that do not involve constant rates of change, we must use a non-linear model. One possible non-linear model is a quadratic model.

> **Quadratic models**
> Quadratic models have these characteristics.
> - In a table of values, the second differences are equal.
> - The graph is a curve called a *parabola*.
> - The equation can be written in the form $y = ax^2 + bx + c, a \neq 0$.

A quadratic relation must have an x^2 term, so $a \neq 0$.

Example 1

Identifying Quadratic Models

Which models represent quadratic relations? Justify your answers.

a)

h	0	1	2	3
p	250	238	202	142

b)

r	0	1	2	3
q	32	48	72	108

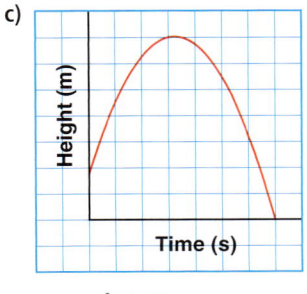

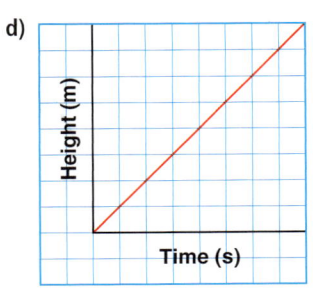

c) [graph: Height (m) vs Time (s), parabola curve]

d) [graph: Height (m) vs Time (s), straight line]

e) $y = x^2 + 7$

f) $y = 3x + 2$

Solution

a)

h	p	First differences	Second differences
0	250		
		$238 - 250 = -12$	
1	238		$-36 - (-12) = -24$
		$202 - 238 = -36$	
2	202		$-60 - (-36) = -24$
		$142 - 202 = -60$	
3	142		

Since the x-values increase by 1, the first differences are the rates of change.

The second differences are equal. The relationship is quadratic.

5.4 Quadratic Models **299**

b)

r	Q	First differences	Second differences
0	32		
		48 − 32 = 16	
1	48		24 − 16 = 8
		72 − 48 = 24	
2	72		36 − 24 = 12
		108 − 72 = 36	
3	108		

The second differences are not equal. The relationship is not quadratic.

> In a quadratic model, the first differences and rates of change are not constant.
>
> In a linear model, the first differences and rates of change are constant.

c) The graph appears to be a parabola.
   The relationship appears to be quadratic.
d) The graph is not a parabola.
   The relationship is not quadratic.
e) Rewrite the equation as $y = 1x^2 + 0x + 7$. Since $a \neq 0$, the relationship is quadratic.
f) The equation has no x^2 term. So, the relationship is not quadratic.

Relationships in formulas

In *Investigate*, the formula $d = 0.75sm^2$ involved three variables: the dependent variable d, and two independent variables, s and m.

To investigate the relationship between d and s, we set m constant to get a formula involving only s and d. Similarly, to investigate the relationship between d and m, we set s constant to obtain a formula involving only d and m. We can use this strategy to investigate relationships in other formulas with more than two variables.

Example 2 — Investigating Relationships in Formulas

Materials
- TI-83 or TI-84 graphing calculator

The formula $d = \frac{1}{2}at^2$ gives the distance travelled by a car as it accelerates from a stopped position; d metres is the distance travelled, a metres per second squared is the acceleration, and t seconds is the time elapsed.

a) Investigate the relationship between d and a when $t = 2$ s.
b) Investigate the relationship between d and t when $a = 2$ m/s².
c) What types of relationships were obtained in parts a and b? Why does it make sense that these relationships were obtained?

Solution

a) Substitute $t = 2$ in the formula.

$$d = \frac{1}{2}at^2$$
$$d = \frac{1}{2}a(2)^2$$
$$d = 2a$$

Enter the formula $d = 2a$ on a graphing calculator as $y = 2x$.

b) Substitute $a = 2$ in the formula.

$$d = \frac{1}{2}at^2$$
$$d = \frac{1}{2}(2)t^2$$
$$d = t^2$$

Enter the formula $d = t^2$ on a graphing calculator as $y = x^2$.

> Press ZOOM 6 to view the equations using the standard window settings.

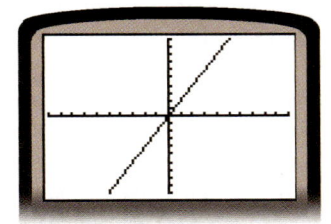

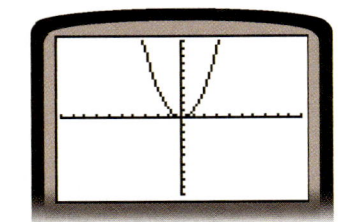

c) Setting t constant produces the linear relation $d = 2a$. This makes sense since a is raised to the exponent 1 in the formula $d = \frac{1}{2}at^2$.
Setting a constant produces the quadratic relation $d = t^2$. This makes sense since t is raised to the exponent 2 in the formula $d = \frac{1}{2}at^2$.

Quadratic regression

In Lesson 5.3, we used linear regression to find the line of best fit. Similarly, we can use quadratic regression to find the parabola of best fit.

Example 3

Fitting a Quadratic Model to Data

Materials
- TI-83 or TI-84 graphing calculator

A fountain of sparks from a Canada Day rocket follows an arc in the air. This table shows the height of the sparks at various horizontal distances from the launching point.

Distance (m)	5	10	15	20	25	30
Height (m)	43	75	97	108	109	100

a) Determine the equation of the parabola of best fit.
b) Determine the maximum of the regression curve. What does it represent?

5.4 Quadratic Models   **301**

Solution

a) Use the **QuadReg** command.

Enter the data in lists **L1** and **L2**.

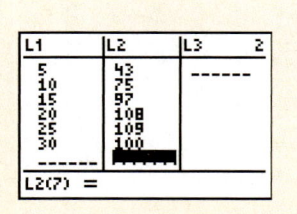

Press [STAT] [▶] 5 [2nd] 1 [,]
[2nd] 2 [,] [VARS] [▶] 1 1 [ENTER]
to perform quadratic regression on the data and store the regression equation as **Y1**.

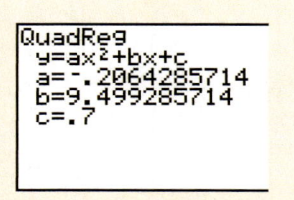

So, the regression equation is: $y \doteq -0.2064x^2 + 9.499x + 0.7$, where x represents the horizontal distance from the launching point and y represents the height of the sparks.

b) Graph the data and the regression curve.

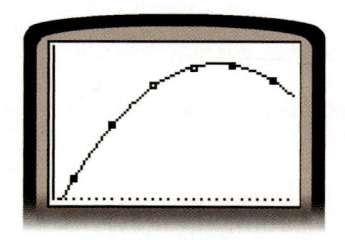

Use the **maximum** feature.

Press: [2nd] [TRACE] 4
Move the cursor to the left of the maximum and press [ENTER], then to the right of the maximum and press [ENTER], then close to the maximum and press [ENTER].

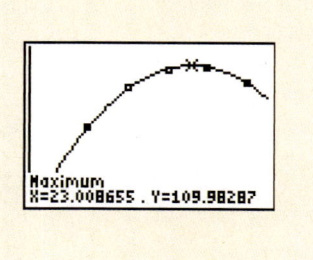

The maximum of the regression curve is about 110.
It represents the maximum height of the sparks from the rocket: about 110 m

Practice

A

For help with questions 1 to 3, see Example 1.

1. Which table of values models a quadratic relation? How do you know?

a)
t	0	1	2	3	4
c	0.5	2	8	32	128

b)
t	0	1	2	3	4
c	0.5	5.5	20.5	45.5	80.5

2. Which graphs might model a quadratic relation? Why do you think so?

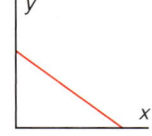

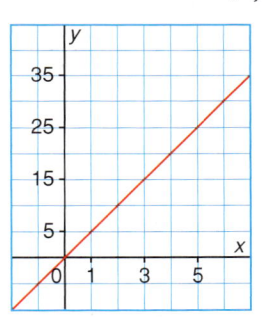

a) [parabola opening up]   b) [line decreasing]   c) [curve]

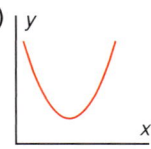

3. Which equations model a quadratic relation? How do you know?

a) $y = -2x$ b) $y = x^2 + 1$ c) $y = 5 - 2x$

4. Match each equation with a graph.

a) $y = x + 3$ b) $y = x^2 - 4x + 3$ c) $y = 5x$
d) $y = -2x^2 + 2$ e) $y = -2x + 10$ f) $y = 4x^2 - 4x$

i) [decreasing line]   ii) [increasing line]   iii) [downward parabola]

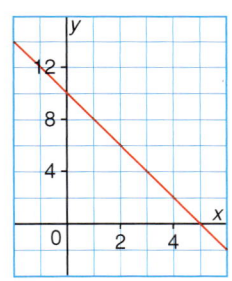

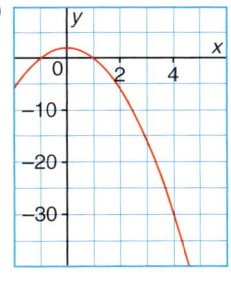

iv) [upward curve]   v) [upward parabola]   vi) [increasing line]

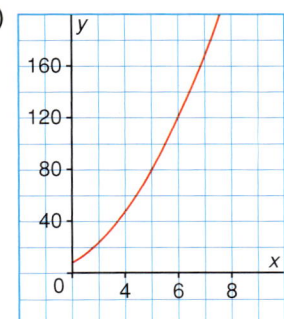

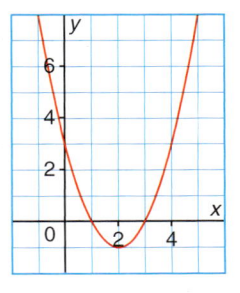

5.4 Quadratic Models   **303**

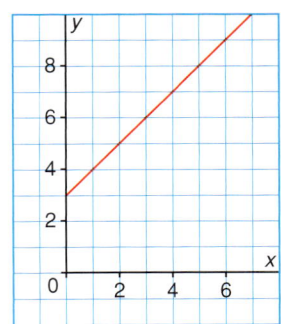

5. Tell whether each statement is true or false.
  a) For a quadratic relation, the second differences are equal.
  b) No quadratic relation can be represented by a parabola.
  c) If the first differences are equal, the relation is quadratic.
  d) All relations with equal first differences are linear.
  e) Some quadratic relations, but not all, can be represented by an equation of the form $y = mx + b$.

B

6. An orange is tossed straight up in the air.

Height of Orange

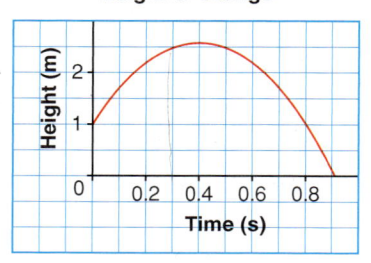

  a) When is the height of the orange increasing? When is it decreasing?
  b) When is the height of the orange changing rapidly? When is it changing slowly?

■ For help with questions 7 and 8, see Example 2.

7. A T-shirt manufacturer wants to maximize her revenue.
  a) Describe the trends in the graph.
  b) As the number of shirts produced increases, why does the revenue increase and then decrease?
  c) Explain how the manufacturer can use the graph to decide how to maximize her revenue.

Shirt Revenue

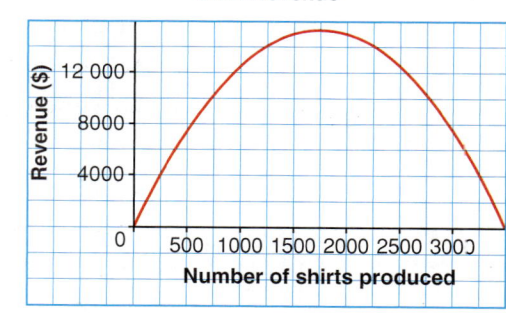

8. The formula for the volume of a cylinder with radius r and height h is: $V = \pi r^2 h$
  a) Which variable(s) in the formula $V = \pi r^2 h$ should you set constant to generate a linear relationship? Explain why you made that choice.
  b) Which variable(s) in the formula $V = \pi r^2 h$ should you set constant to generate a quadratic relationship? Explain why you made that choice.
  c) Verify your answers to parts a and b by graphing $V = \pi r^2 h$ when $r = 5$ cm and when $h = 5$ cm. Were you correct? Explain.

304 CHAPTER 5: Graphical Models

9. The volume V of a pyramid with height h and square base of side length s is given by the formula: $V = \frac{1}{3}s^2h$
   a) Which variable should you set constant to generate a linear relationship?
   b) Which variable should you set constant to generate a quadratic relationship?
   c) How can you check that your answers in parts a and b are correct? Explain.

■ For help with questions 10 to 12, see Example 3.

10. **Assessment Focus** This graph shows how the fuel consumptions of two vehicles change as their speed increases.
    a) Describe the trends in each graph.
    b) When does the fuel efficiency for each vehicle increase or decrease rapidly? When does it increase or decrease at a constant rate? How do you know?
    c) Which car is more fuel efficient at each speed?
       i) 55 km/h
       ii) 110 km/h
    d) Nyarai thinks the graph for car B appears to be quadratic. Do you agree or disagree? Justify your answer.

Fuel Efficiencies of Two Cars

11. A juggler tosses balls from one hand to the other. These two equations represent the height of a ball, B metres above its release point, after t seconds.
    Right-hand toss: $B = 8.4t - 9.8t^2$
    Left-hand toss: $B = 3.4t - 9.8t^2$
    a) Graph the equations on the same screen of a graphing calculator.
    b) What is the maximum height of a ball tossed by each hand?
    c) Assume that the ball is caught at the same height from which it was released. How long is a ball tossed by each hand in the air?
    d) Which hand releases the ball at a faster speed? Justify your answer.

$$\text{Average speed} = \frac{\text{Distance travelled}}{\text{Time taken}}$$

5.4 Quadratic Models **305**

12. The management of a hockey arena plans to increase ticket prices to obtain more revenue. A survey was conducted to estimate the revenue generated for different ticket prices. These data were obtained.

Ticket price ($)	Projected revenue ($)
10	16 000
15	19 500
20	20 300
25	14 750

a) Why might a quadratic model be a good fit for the data?
b) Perform a quadratic regression on the data.
c) Determine the maximum value of the regression equation. What does it represent?
d) Predict the revenue that would be obtained if the tickets cost $30.

13. a) Why might a quadratic model be a good fit for the data in this table?
b) Perform a quadratic regression on the data.
c) Determine the minimum value of the regression equation. What does it represent?
d) Predict the number of males registered in 2004. How does this compare with the actual value of 241 995?

Males Registered in Apprenticeship Programs in Canada

Year	Number of males
1991	184 705
1992	172 740
1993	160 020
1994	153 275
1995	151 945
1996	152 840
1997	157 875
1998	161 595
1999	170 710
2000	181 610
2001	195 220
2002	209 650

14. Transit fares are growing as shown in the table.

Year	1955	1960	1965	1970	1975	1980	1985	1990	1995	2000	2005
Fare ($)	0.10	0.15	0.20	0.30	0.30	0.60	0.95	1.20	2.00	2.00	2.25

a) Determine the equation of the parabola of best fit.
b) Estimate the fare for each year.
   i) 1987   **ii)** 2020
c) Which of your estimates in part b do you think is more accurate? Explain.

15. Literacy in Math Make a list of terms related to quadratic relations. Write a definition for each term. Use a sketch when helpful.

16. A grandfather clock that is running too fast or too slow can be fixed by adjusting the length of the pendulum. Use the formula $l = \frac{1}{4\pi^2}gT^2$, where:
- l metres represents the length of the pendulum
- T seconds represents the time to make one complete swing
- g m/s^2 is the acceleration due to gravity

On Earth, the value of g is 9.8 m/s^2.

a) Sketch a graph of length versus time.
b) Describe the relationship between length and time. Justify your answer in two different ways.
c) Imagine that the clock is taken to the Moon, where $g = 1.6$ m/s^2. How would the graph in part a change?
d) The clock is moved from planet to planet but always set so $T = 2$ s. Describe the relation between l and g.

17. Sketch a distance-time graph for the following race between the Tortoise and the Hare. Justify your graph.
- The Tortoise walks at a speed of 0.5 m/s for the entire 1-km race.
- The Hare runs fast and then more slowly, stopping after 50 s at the 100-m mark.
- The Hare then takes at nap.
- The Hare wakes up when the Tortoise is 10 m from the finish line.
- The Hare starts to run faster and faster but loses the race by 200 m.

> **In Your Own Words**
>
> How are the steps involved in quadratic regression similar to steps for linear regression? How are they different? Explain.

5.4 Quadratic Models

Mid-Chapter Review

5.1

1. Describe the relationship illustrated by each graph.

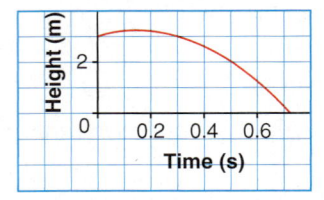

a) **Diver Height**

b) **Museum Admission Cost**

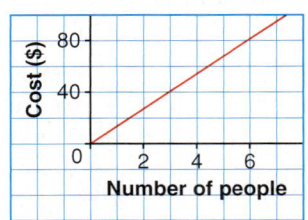

5.2

2. A computer store gives volume discounts.

Computer Volume Discount

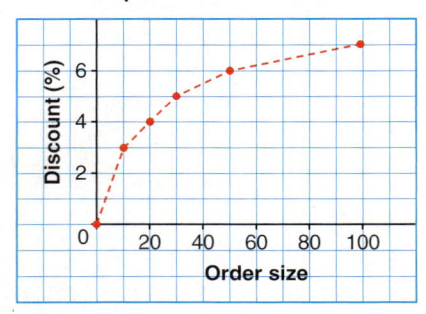

a) Describe the trends in the graph.
b) Predict the number of computers you would have to buy to get an 8% discount. Justify your prediction.

5.3

3. Determine when the rate of change in a hockey player's plus/minus score is zero, constant, or changing.

Hockey Player's Plus/Minus Scores

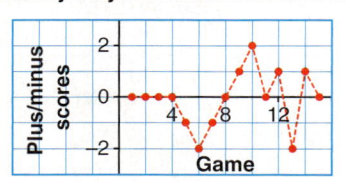

4. The median age, A, in each province n years after 1991 is:
Prince Edward Island: $A = 0.34n + 24$
Alberta: $A = 0.43n + 24$

a) Graph the equations on the same grid. How do the graphs compare?
b) How would the graph for Alberta change if the median age in 1991 was 30?

5.4

5. The volume V of a cone with height h and radius r is given by the formula $V = \frac{1}{3}\pi r^2 h$.

a) Which variable should you set constant to generate a linear relationship?
b) Which variable should you set constant to generate a quadratic relationship?

6. a) Why might a quadratic model be a good fit for the data in this table?

Year	Population of Kingston
2001	152 652
2002	154 439
2003	155 676
2004	156 123
2005	155 685
2006	154 971

b) Perform quadratic regression.
c) Do you think the trend modelled by the regression equation will continue? Explain.

308 CHAPTER 5: Graphical Models

GAME

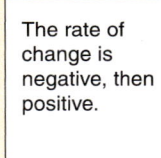

Curves of Concentration

Materials
- 6 *graph cards*
- 6 *rate of change cards*
- 6 *feature cards*

Play with a partner.

- Shuffle the cards. Spread them out face down so no cards overlap.
- One player turns over three cards.
  If the cards make a matching set, the player keeps them.
  If not, the player turns them over again.
- A matching set of cards contains each type of card.
  The feature and rate of change cards accurately describe the graph card. For example, this is a matching set:

Graph card	Rate of change card	Feature card
(parabola with minimum, y-values 4, 8, 12; x from −2 to 2)	The rate of change is negative, then positive.	The graph has a minimum.

- Players take turns turning over the cards.
- The game ends when all cards have been picked up.
  The player with more cards wins.

Reflect

- Explain your strategy for deciding whether cards match.
- Compare strategies with your partner. What have you learned from your partner's strategy?

GAME: Curves of Concentration **309**

5.5 Exponential Models

You may have heard statements like these in the media:
- The economy is projected to grow by 3.4% this year.
- College tuition fees are expected to rise by nearly 5% next year.

Each of these situations involves a rate of change expressed as a percent.

Investigate

Comparing Constant Growth and Constant Percent Growth

Materials
- TI-83 or TI-84 graphing calculator

Work with a partner.

Suppose you are offered a choice of jobs.
- Job A pays $10/h with a $1/h raise every year.
- Job B pays $10/h with a 10% raise every year.

■ How would your wages grow under each job over 5 years? Organize your work in a table like this.

■ Would you prefer to have Job A or Job B? Explain your reasoning.

Year	Job A ($)	Job B ($)
0	10	10
1		
2		
3		
4		
5		

Reflect

■ Why do you think constant growth is also called additive growth?

■ Why do you think constant percent growth is also called multiplicative growth?

■ Why do your wages grow faster under Job B than under Job A?

Connect the Ideas

Linear and exponential models
- Linear models represent quantities that change at a *constant rate*; that is, a fixed amount is *added* to the quantity at regular intervals.
- Exponential models represent quantities that change at a *constant percent rate*; that is, the quantity is *multiplied* by a fixed amount at regular intervals.

Example 1

Recognizing Linear and Exponential Models

Last year, a school had a population of 1000 students. This year, 1100 students attend the school, an increase of 100 students or 10%.
a) Determine the population after 3 more years under each scenario.
   - **Scenario A:** The population increases by 100 students each year.
   - **Scenario B:** The population increases by 10% each year.
b) What type of growth does each scenario illustrate?

Solution

a) Use a table to record the growth in the student population.

	Student Population	
Year	Scenario A: increase of 100	Scenario B: increase by 10%
0	1100	1100
1	1100 + 100 = 1200	1100 × 1.10 = 1210
2	1200 + 100 = 1300	1210 × 1.10 = 1331
3	1300 + 100 = 1400	1331 × 1.10 ≐ 1464

A 10% increase means that you have 100% + 10% = 110% of the original amount. To find 110% of a number, multiply by 1.10.

After 3 years, the student population is 1400 students under Scenario A and 1464 students under Scenario B.

b) Under Scenario A, we repeatedly added 100. This is linear growth. Under Scenario B, we repeatedly multiplied by 1.10. This is exponential growth.

Exponential models

Exponential models
Exponential models have these characteristics.
- In a table of values, the growth/decay factors are equal.
- The graph resembles an exponential curve.
- The equation can be written in the form $y = ab^x$, where a is the initial value and b is the growth/decay factor.

Example 2 — Identifying Exponential Models

Which models represent exponential relations?
Justify your answers.

a)

t	0	1	2	3
A	35	25	15	5

b)

d	0	1	2	3
P	51.2	64	80	100

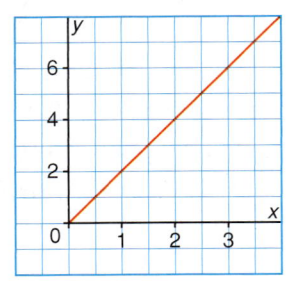

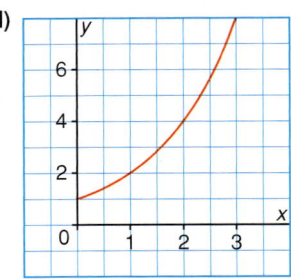

c) [Graph of a straight line through origin rising to about y = 6 at x = 3]

d) [Graph of an exponential curve rising from about y = 1 at x = 0 to beyond y = 6 at x = 3]

e) $y = 10(2)^x$

f) $y = 10x^2$

Solution

a)

t	A	Decay factor
0	35	
		$\frac{25}{35} \doteq 0.71$
1	25	
		$\frac{15}{25} = 0.60$
2	15	
		$\frac{5}{15} \doteq 0.33$
3	5	

b)

d	P	Growth factor
0	51.2	
		$\frac{64.0}{51.2} = 1.25$
1	64.0	
		$\frac{80.0}{64.0} = 1.25$
2	80.0	
		$\frac{100.0}{80.0} = 1.25$
3	100.0	

The decay factor is not constant. The relationship is not exponential.

The growth factor is constant. The relationship is exponential.

c) The graph is a line. The relationship is not exponential.

d) The graph appears to be an exponential curve. The relationship appears to be exponential.

e) The equation is of the form $y = ab^x$. The relationship is exponential.

f) The equation is not of the form $y = ab^x$. The relationship is not exponential.

312 CHAPTER 5: Graphical Models

Example 3

Materials
- TI-83 or TI-84 graphing calculator

Comparing Pairs of Exponential Relations

Two colonies of bacteria each start with 100 bacteria.
- The population of Colony A doubles every hour.
- The population of Colony B triples every hour.

These equations represent the population, P bacteria, of the two colonies after t hours.

Colony A: $P = 100(2)^t$
Colony B: $P = 100(3)^t$

a) Graph the equations on the same screen. How do the graphs compare? Explain.
b) How would the graph for Colony A change if there were 200 bacteria initially?

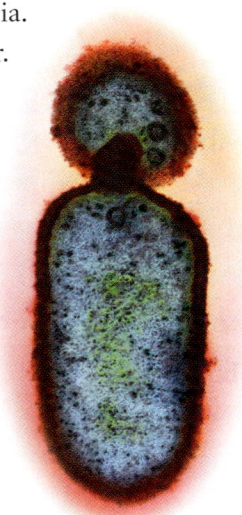

Solution

a) Use a graphing calculator to generate the graphs of $y = 100(2)^x$ and $y = 100(3)^x$ on the same screen.

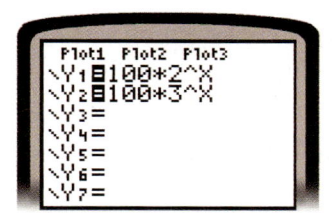

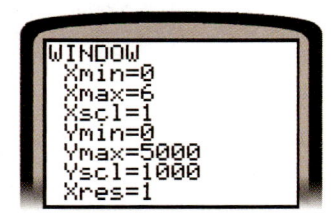

Both graphs have the same vertical intercept because each colony has the same initial number of bacteria, 100. The growth factor for Colony B is greater, so the curve for Colony B increases much faster than the curve for Colony A.

b) Generate the graphs of $y = 100(2)^x$ and $y = 200(2)^x$ on the same screen.

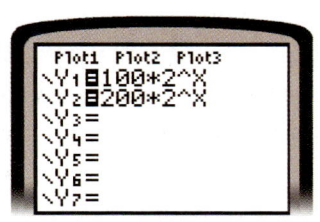

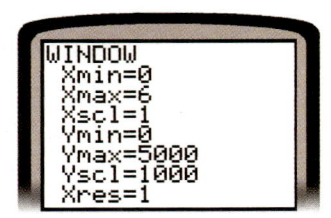

The vertical intercept would be 200 instead of 100.
Because the initial value would be greater, the graph would increase slightly faster than the original graph.

5.5 Exponential Models **313**

Exponential regression

We can use exponential regression to fit an exponential model to data.

Example 4

Materials
- TI-83 or TI-84 graphing calculator

Fitting an Exponential Model to Data

a) Determine the exponential relation $y = ab^x$ that best fits the data in this table, where x is the number of years since 1921 and y is the population of British Columbia in millions.

Year	1921	1931	1941	1951	1961	1971	1981	1991	2001
B.C. Population (millions)	0.52	0.69	0.82	1.17	1.63	2.18	2.82	3.37	4.08

b) What do the values of a and b represent in this situation? Explain.

c) Estimate the population of British Columbia in 1985.

Solution

a) Use the **ExpReg** command.

Since x represents years since 1921, enter these values in list **L1**: 0, 10, 20, 30, 40, 50, 60, 70, 80. Enter the population data in list **L2**.

Press [STAT] ▶ 0 [2nd] 1 [,] [2nd] 2 [,] [VARS] ▶ 1 1 [ENTER] to perform exponential regression on the data and store the regression equation as **Y1**.

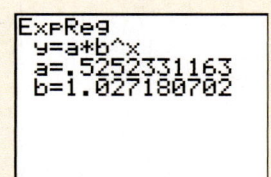

So, the regression equation is: $y \doteq 0.5252(1.02718)^x$.

b) The initial value a represents the approximate population, about 530 000 people, of British Columbia in 1921. The growth factor b represents a growth rate in the population of about 2.7% per year.

c) The number of years between 1985 and 1921 is: $1985 - 1921 = 64$
Substitute $x = 64$ into the regression equation.
$y = 0.5252(1.02718)^x$
$= 0.5252(1.02718)^{64}$, or about 2.9
The population was about 2.9 million people in 1985.

You could also use the **TRACE** feature or the command **Y1(64)** on your graphing calculator.

Practice

A

For help with question 1, see Example 1.

For help with questions 2 to 5, see Example 2.

1. A population grows by each percent per year. By what factor is each year's population multiplied?
   a) 3%   b) 5%   c) 12%

2. Which situations represent linear growth? Exponential growth? Justify your answers.
   a) A hairdresser increases the price of a haircut by $0.50 every year.
   b) Gua's investment doubles every 20 years.
   c) The players in each round of a tennis tournament are the winners from each pair in the previous round.
   d) Daryl makes $10/h working as a line chef.

3. Which tables of values model an exponential relation? How do you know?

a)
t	0	1	2	3	4	5
A	400	420	441	463	486.2	510.5

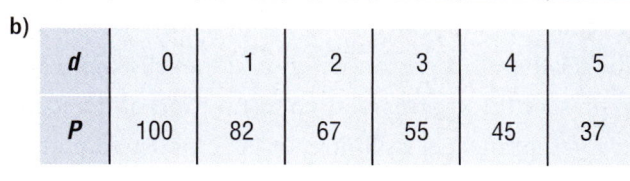

b)
d	0	1	2	3	4	5
P	100	82	67	55	45	37

4. Which graphs model an exponential relation? How do you know?

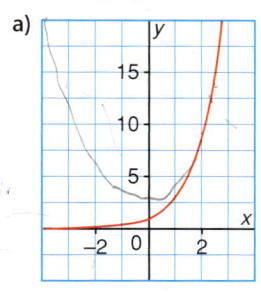

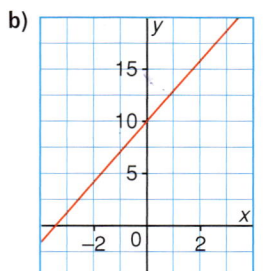

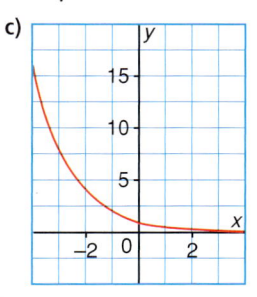

5. Which equations model an exponential relation? How do you know?
   a) $y = 2 + 4x$ b) $y = 2 + 4x^2$ c) $y = 2(4)^x$

5.5 Exponential Models **315**

B

■ For help with questions 6 and 9, see Example 3.

6. For the given curve, sketch a different curve with:
   a) A different starting value
   b) A greater rate of change
   c) A lesser rate of change

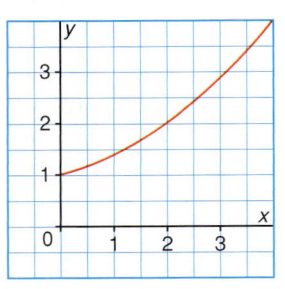

7. A city council wants to discourage illegal parking. It has two plans.
   - **Plan A:** A $10 fine for the first offence. The fine increases by $20 for each subsequent offence.
   - **Plan B:** A $10 fine for the first offence. The fine doubles for each subsequent offence.
   a) Determine the cost of a third fine under each plan.
   b) What type of growth does each plan illustrate?

8. The half-life of penicillin in a patient with kidney disease is often longer than in a person without kidney disease. These equations represent the approximate concentration of penicillin, C micrograms per millilitre, in the bloodstream of each patient after t hours.
   Patient with kidney disease: $C = 250(0.63)^t$
   Patient without kidney disease: $C = 250(0.37)^t$
   a) Graph the equations on the same screen. How do the graphs compare? Explain.
   b) How would the graph for the patient with kidney disease change if the initial concentration of penicillin was 1000 µg/mL?

9. A formula to model the radioactive decay of Chromium-51 is $A = A_0(0.975)^t$, where A_0 is initial amount of Chromium-51 and A is the amount of Chromium-51 remaining after time t.
   a) Which variable should you set constant to generate a linear relationship?
   b) Which variable should you set constant to generate an exponential relationship?
   c) How can you check that your answers to parts a and b are correct? Explain.

10. A general formula for population growth is $N = N_0(1 + r)^t$, where N_0 is the initial population, r is the growth rate as a decimal, and t is the time that has passed. Describe the shape of these graphs.
   a) The graph of N against N_0 for fixed r and t
   b) The graph of N against t for fixed r and N_0
   Justify your answers.

■ For help with questions 11 and 12, see Example 4.

11. This table shows the growth in cell phone subscribers for a particular company.

Year	2000	2001	2002	2003	2004
Number of subscribers (thousands)	15.9	33.8	43.9	55.3	86.1

   a) Determine the exponential relation $y = ab^x$ that best fits the data, where x is the number of years since 2000 and y is the number of cell phone subscribers in thousands.
   b) What do the values of a and b represent in this situation? Explain.

12. A camera stores charge in a capacitor and then uses it to create light when a flash is needed. This table shows the charge left in the capacitor after the flash begins.

Time (s)	Charge (µC)
0.00	100
0.02	82
0.04	67
0.06	55
0.08	45
0.10	37

   a) Create a scatter plot of the data and describe any trends.
   b) Determine the exponential relation $y = ab^x$ that best fits the data, where y is the charge left in the capacitor, in microcoulombs, after x seconds. Graph the regression curve on the scatter plot.
   c) Estimate the charge left in the capacitor after each length of time.
      i) 0.05 s
      ii) 0.20 s

5.5 Exponential Models

13. **Assessment Focus** The volume of dough for cinnamon buns is measured at 10-min intervals.

Time (min)	0	10	20	30	40	50	60
Volume (L)	1.5	1.7	1.9	2.2	2.5	2.8	3.2

    a) Create a scatter plot of the data and describe any trends.
    b) Determine the exponential relation $y = ab^x$ that best fits the data, where y is the volume of dough, in litres, after x minutes.
       Graph the regression curve on the scatter plot.
    c) Estimate the volume of dough after each length of time.
       i) 45 min   ii) 90 min
    d) Which estimate is likely to be more accurate? Explain.
    e) How long do you think the trend in the data will continue? Give reasons for why it might change.

14. **Literacy in Math** Explain how you can determine whether a relationship is exponential. Include an example of a relationship that is exponential and an example of a relationship that is not exponential.

15. The formula for the number of contestants C in a competition that began with B contestants and eliminates n contestants in each round r is $C = B\left(\frac{1}{n}\right)^r$. Is it possible to generate each type of relationship using this formula? If so, explain which variables you would set constant. If not, explain why not.
    i) Linear
    ii) Quadratic
    iii) Exponential

> **In Your Own Words**
>
> Use an example to explain why an exponential relation has a constant growth or decay factor but a changing rate of change.

318 CHAPTER 5: Graphical Models

5.6 Selecting a Regression Model for Data

Diane is a quality control technician at a tire manufacturing company. She helps collect and analyse data to test how well the tires perform under different situations. Part of Diane's analysis involves finding a suitable model for the data.

Investigate

Comparing Regression Models

Materials
- TI-83 or TI-84 graphing calculator

Work in a group of 3.

This table shows how the air pressure in a car tire changes in the first 40 s after the tire is punctured.

- Have each person fit a different regression model to the data: linear, exponential, or quadratic.
- Compare the regression models. Use the model that you think best fits the data to predict the tire pressure:
  i) After 12 s ii) After 45 s

Time (s)	Tire pressure (kPa)
0	207
5	186
10	145
15	110
20	90
25	62
30	48
35	41
40	28

Reflect

How can you decide whether a linear, quadratic, or exponential relation provides the best fit for a set of data?

Connect the Ideas

Selecting a regression model

In Lessons 5.3 to 5.5, we were told whether to use linear, quadratic, or exponential regression to model data. In practical applications, we may not know what model to use. Examining first and second differences and growth or decay factors can help us make the best choice.

Example 1

Identifying Relationships in Data

Electrical appliances such as a VCR or a digital clock contain a capacitor for power during brief electrical outages. The table shows how the voltage in a capacitor decreases over time after a power outage.

Time (s)	0	2	4	6	8	10
Voltage (V)	9.0	7.0	5.2	3.9	3.0	2.3

What type of relationship seems to exist between voltage and time? Justify your answer.

Solution

Calculate the first differences, second differences, and decay factors.

Time (s)	Voltage (V)	First differences	Second differences	Decay factors
0	9.0			
		$7.0 - 9.0 = -2.0$		$\frac{7.0}{9.0} \doteq 0.8$
2	7.0		$-1.8 - (-2.0) = 0.2$	
		$5.2 - 7.0 = -1.8$		$\frac{5.2}{7.0} \doteq 0.7$
4	5.2		$-1.3 - (-1.8) = 0.5$	
		$3.9 - 5.2 = -1.3$		$\frac{3.9}{5.2} \doteq 0.8$
6	3.9		$-0.9 - (-1.3) = 0.4$	
		$3.0 - 3.9 = -0.9$		$\frac{3.0}{3.9} \doteq 0.8$
8	3.0		$-0.7 - (-0.9) = 0.2$	
		$2.3 - 3.0 = -0.7$		$\frac{2.3}{3.0} \doteq 0.8$
10	2.3			

The first and second differences are not equal, but the decay factors are approximately equal. There appears to be an exponential relationship between voltage and time.

Example 2

Materials
- TI-83 or TI-84 graphing calculator

Selecting a Model to Represent Data

In a science experiment, students punched a hole near the bottom of a 2-L pop bottle. They filled the bottle with water and measured how the water level changed over time. The results are shown in the table below.

Time (s)	0	25	50	75	100
Water level (cm)	30.0	22.3	16.1	11.2	7.8

a) Perform linear, quadratic, and exponential regressions on the data.
b) Which model best represents the data? Justify your answer.
c) Determine the height of the hole in the bottle. Justify your answer.

Solution

a) Perform each regression. Store the regression equations as **Y1**, **Y2**, and **Y3**.

Enter data in lists **L1** and **L2**.

Press [STAT] ▶ 4 [2nd] 1 , [2nd] 2 , [VARS] ▶ 1 1 [ENTER] to perform linear regression on the data and store the regression equation as **Y1**.

LinReg
y=ax+b
a=-.222
b=28.58

Press [STAT] ▶ 5 [2nd] 1 , [2nd] 2 , [VARS] ▶ 1 2 [ENTER] to perform quadratic regression on the data and store the regression equation as **Y2**.

QuadReg
y=ax²+bx+c
a=.0011314286
b=-.3351428571
c=29.99428571

Press [STAT] ▶ 0 [2nd] 1 , [2nd] 2 , [VARS] ▶ 1 3 [ENTER] to perform exponential regression on the data and store the regression equation as **Y3**.

ExpReg
y=a*b^x
a=30.79930756
b=.9865598555

5.6 Selecting a Regression Model for Data   **321**

b) Set up a scatter plot of the data. Press ZOOM 9 to view the scatter plot and the graphs of the regression equations.
Graph the data. Press WINDOW and set **Xmax** = 200 to show the minimum of the quadratic regression curve.

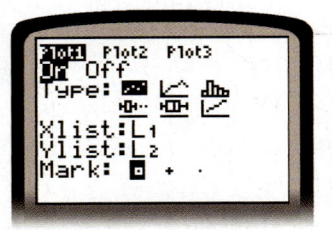

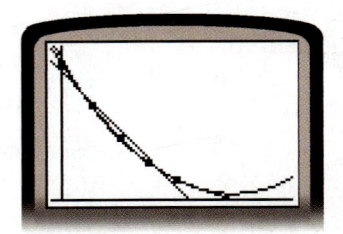

The linear model is clearly the worst fit for the data. Both the quadratic and exponential graphs look like a close fit for the data. Water below the hole cannot drain out of the bottle. So, the model that best represents the data will stop decreasing at some point. The graphs of the linear and exponential equations constantly decrease, while the graph of the quadratic equation has a minimum. So, a quadratic model best represents the data.

c) Use the quadratic model. Graph the data and the regression parabola. Water will stop flowing out of the bottle when the water level is below the hole. So, the minimum of the regression parabola represents the height of the hole.

> The quadratic relation is an accurate model of the data up to $t = 148$ s. After that, the water level will stay at 5.2 cm but the quadratic relation starts to increase. The quadratic model is not valid for later times.

Press: Y=
Choose to display only the graph of the quadratic regression equation.
In row \Y1, highlight = and press ENTER to hide the linear graph.
Repeat for row \Y3 to hide the exponential graph.

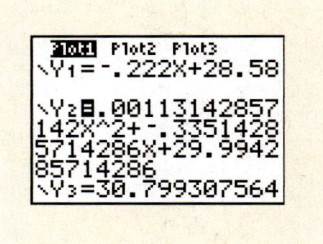

Use the **minimum** feature.
Press: 2nd TRACE 3
Move the cursor to the left of the minimum and press ENTER, then to the right of the minimum and press ENTER, then close to the minimum and press ENTER.

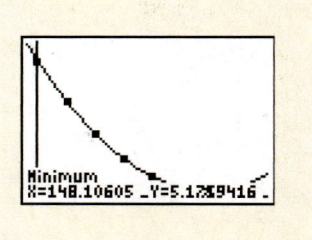

The minimum of the regression parabola is about 5.2. The hole is about 5.2 cm above the bottom of the bottle.

Practice

A

For help with questions 1 to 3, see Example 1.

1. Use first differences to verify that a linear relation best represents each set of data.

a)
r	0	5	10	15	20
C	0	31	63	94	126

b)
x	0	1	2	3
y	10	12.5	15	17.5

2. Use second differences to verify that a quadratic relation best represents each set of data.

a)
v	0	1	2	3	4	5
E	0	0.5	2	4.5	8	12.5

b)
x	1	3	5	7
y	23	55	103	167

3. Use the growth/decay factors to verify that an exponential relation best represents each set of data.

a)
x	0	1	2	3
y	0.3	1.5	7.5	37.5

b)
x	10	15	20	25
y	54	36	24	26

4. What type of regression should Jamil use for each set of data?

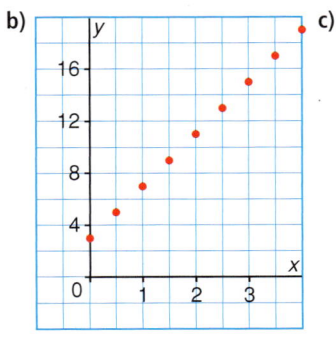

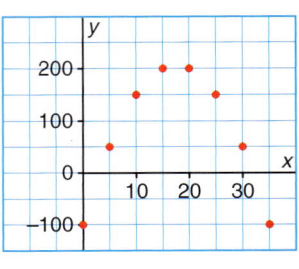

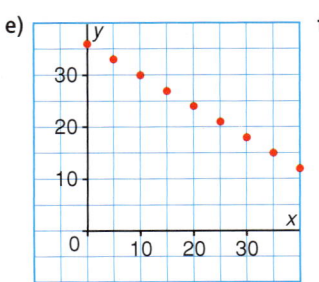

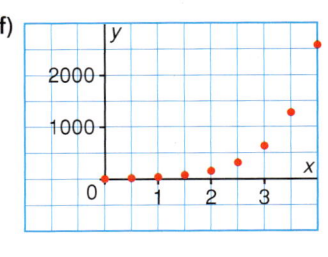

a)

b)

c)

d)

e)

f)

5.6 Selecting a Regression Model for Data

5. Toria fits a linear, quadratic, and exponential relation to a given set of data. Which relation do you think best models the data? Justify your answer.

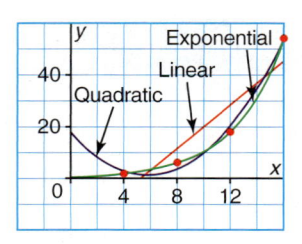

6. Tell whether each set of data models a linear, quadratic, or exponential relation.

a)
x	0	1	2	3	4
y	0.5	2	8	32	128

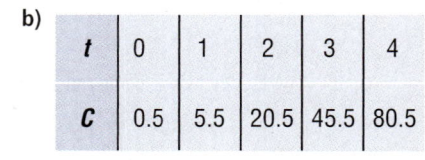

b)
t	0	1	2	3	4
c	0.5	5.5	20.5	45.5	80.5

c)
p	0	5	10	15	20
q	9.6	11.3	13.0	14.7	16.4

d)
v	3	6	9	12	15
w	11 250	2250	450	90	18

B 7. A company creates snow tubes with fixed radius B. Depending on how much air is pumped into the snow tube, radius A and volume change as shown in the table.

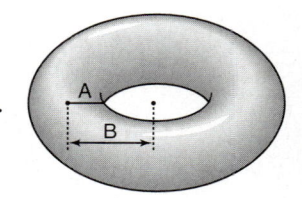

Radius A (cm)	15	16	17	18
Volume (cm³)	73 282	83 378	94 126	105 526

a) What type of relationship seems to exist between radius A and volume? Justify your answer.

b) A snow tube is an example of a torus. The formula for the volume of this torus is $V = 2\pi^2 A^2 B$. Does this formula agree with the type of relationship you found in part a? Explain.

8. Rahul inherits money on his 25th birthday. He deposits it in an account that earns a constant rate of interest compounded annually. This table shows how much money he has after each year.

a) What type of relationship seems to exist between the amount and the number of years? Justify your answer.

Number of years	Amount
0	$6000.00
1	$6318.00
2	$6652.85
3	$7005.45
4	$7376.74

b) The formula for compound interest is $A = P(1 + i)^n$, where A is the amount, P is the principal invested, i is the interest rate, and n is the number of years. Does this formula agree with the type of relationship you found in part a? Explain.

9. A toy manufacturer makes two sets of plastic cylindrical cups. The cups in set A have the same height, but varying radii. The cups in set B have the same radius, but varying heights. These tables show the volumes of each set of cups.

Set A

Radius (cm)	Volume (cm³)
1	23
2	92
3	207
4	368
5	575

Set B

Height (cm)	Volume (cm³)
5	161
6	193
7	225
8	257
9	290

a) What type of relationship seems to exist between height and volume? Justify your answer.
b) What type of relationship seems to exist between radius and volume? Justify your answer.
c) The formula for the volume of a cylinder is $V = \pi r^2 h$, where V is the volume, r is the radius, and h is the height. How can you use this formula to verify your answers to parts a and b? Explain.

10. A diabetic on an insulin pump gives herself a bolus of insulin when she eats a meal or a snack. This table shows the units of insulin, in hundredths of a cubic centimetre, in the bolus depending on the number of carbohydrates she eats.

Carbohydrates (g)	50	60	70	80	90
Units of insulin	3.7	4.0	5.3	5.7	6.0

a) Perform linear, quadratic, and exponential regressions on the data.
b) Which model best represents the data? Justify your answer.

11. Verify your answers to question 6 by performing linear, quadratic, and exponential regressions on the data.

12. These data show the mass of algae in a pond during summer changes over time.

Time (s)	5	10	15	20	25	30
Mass (kg)	4.4	5.2	6.6	8.4	10.8	13.6

    a) Perform linear, quadratic, and exponential regressions on the data.
    b) Which model best represents the data? Justify your answer.

13. **Assessment Focus** These data show the speed of a rocket after it is launched.

Time (s)	5	10	15	20	25	30
Speed (m/s)	14	23	38	61	98	157

    a) Create a scatter plot of the data.
    b) Perform linear, quadratic, and exponential regressions on the data.
    c) Which model best represents the data? Justify your answer.
    d) How could you use the scatter plot to tell that a linear model would not be a good fit? Explain.

14. A 2-L bottle of pop is placed in a cooler filled with ice. The table shows how the temperature of the pop changes with time.

Time (h)	0	1	2	3	4
Temperature (°C)	26	11	4.8	2.1	1.0

    a) Perform linear, quadratic, and exponential regressions on the data.
    b) Which model best represents the data? Justify your answer.
    c) Use the model you chose in part b to estimate the temperature of the pop after each length of time.
       i) 6 h   ii) 24 h
    d) Which estimate from part c do you think is more accurate?
    e) Sketch a possible graph for temperature versus time for values of time from 0 h to 48 h. Explain your assumptions and the main features of the graph.

15. **Literacy in Math** Use a matrix or graphic organizer of your choice to compare linear, quadratic, and exponential relations. Include these categories: basic trends, first differences, second differences, growth/decay factors, and rate of change.

16. This table shows the number of hours of daylight in Dryden every 15 days.

Date	May 2	May 17	June 1	June 16	July 1	July 1	July 31
Hours of daylight	14.75	15.50	16.05	16.32	16.25	15.87	15.23

a) Create a scatter plot of the data.
b) Fit the data to an appropriate relation.
c) Explain how you chose which type of relation to use.
d) Estimate the number of hours of daylight on August 15.
e) Explain why it would be invalid to use your relation much past September 1.
f) Sketch a graph of the actual hours of daylight for a full year.

17. These data show the prices of different lengths of kitchen cabinet door handles.

Length (mm)	190	290	390	490	590	690	790	890	990
Price ($)	23.20	30.00	46.80	52.20	59.40	70.00	85.20	95.60	106.00

a) Create a scatter plot of the data. Describe the trend in the data.
b) Perform linear, quadratic, and exponential regressions on the data.
c) Which model best represents the data? Justify your answer.

In Your Own Words

Suppose you are given a set of data and asked to make a prediction based on the data. Describe how you would do this. What factors may affect the accuracy of the prediction?

5.7 Applying Trends in Data

Energy and the environment appear regularly in media headlines as society struggles to understand the issues and decide upon appropriate action. You can use what you have learned in this chapter to examine and interpret some of the data underlying these stories.

Inquire: Analysing Trends in Energy Usage and Glacier Melting

Materials
- TI-83 or TI-84 graphing calculator

Work in a small group.

Part A: World Energy Consumption

Scientists have recorded annual world energy consumption over the last 140 years.

Global Energy Consumption

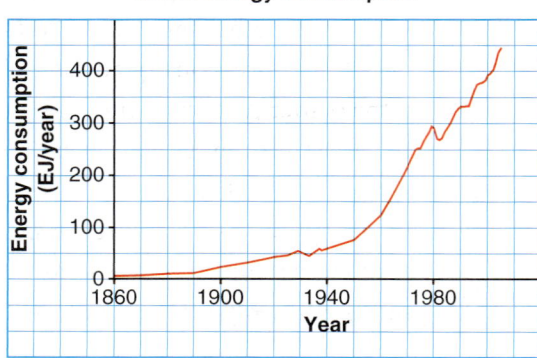

(Graph: Energy consumption (EJ/year) vs. Year, from 1860 to beyond 1980, showing a rising curve from near 0 to over 400 EJ/year)

328 CHAPTER 5: Graphical Models

1. **Reading values from the graph**
   a) Create a table like this.

Year	Number of years since 1860	World energy consumption (EJ)
1860	0	
1870	10	
1990	130	
2000	140	

   b) Read the world energy consumption values from the graph and enter them in the table.

2. **Creating an exponential model**
   a) Describe the features of the graph that suggest that the growth in world energy consumption is approximately exponential.
   b) Create a scatter plot of the data, and fit an exponential regression model to the data.
   Record the regression equation.
   c) Display the regression line and the scatter plot on the same screen. How well does the regression line model the data? Explain.
   d) Use the model to predict the world energy consumption in 2015 and 2050.

3. **Creating a linear model**
   a) Perform two linear regressions on the data: one for 1860 to 1950 and the other for 1950 to 2000.
   b) Record the regression equations and draw the regression lines on the scatter plot from question 2.
   c) Compare the fit of the two linear models to the fit of the exponential model.
   d) Use the appropriate linear regression equation to predict the world energy consumption in 2015 and 2050. How do these predictions compare to those from the exponential model?

Part B: Glacier Melting

4. Choosing the best model to represent a set of data

This table shows the approximate area of the glaciers on Mount Kilimanjaro since 1880.

Year	1880	1912	1953	1976	1989	2003
Area (km^2)	20	12.1	6.7	4.2	3.3	2.5

a) Create a scatter plot of the data.
b) Perform linear, quadratic, and exponential regressions on the data.
c) Which model do you think best estimates the trend in the data? Do you think the trend will continue? Explain.
d) Use each model to predict when the glaciers on Mount Kilimanjaro will disappear.
e) How similar are the predictions? Which prediction, if any, do you think is most accurate? Explain.

Reflect

- Why does it make sense to fit more than one regression model to the sets of data in this *Inquire*?
- In Part A, as you change models, how much do the predictions for both 2015 and 2050 change? Explain.
- Even if a regression model is an excellent fit to data, why might the actual future values be significantly different from those predicted by the model?

Study Guide

Rate of Change

- Average rate of change = $\dfrac{\text{Change in dependent variable}}{\text{Change in independent variable}}$

  $= \dfrac{\text{Change in temperature}}{\text{Change in time}}$

  $= \dfrac{5°C - 20°C}{3\,h - 1\,h} = \dfrac{-15°C}{2\,h} = -7.5°C/h$

Temperature Change over Time

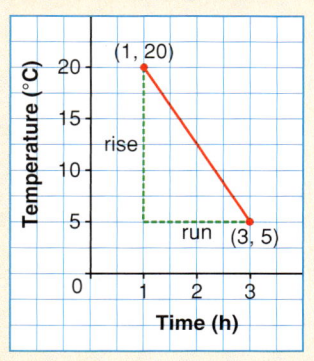

- If the graph is ... then the rate of change
  - a horizontal line — is zero
  - a straight line — is constant
  - not a straight line — varies (changes)

Linear, Quadratic, and Exponential Models

Linear Graphs $y = mx + b$		Quadratic Graphs $y = ax^2 + bx + c$		Exponential Graphs $y = ab^x$	
$m > 0$ 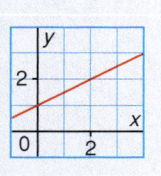	$m < 0$	$a > 0$ 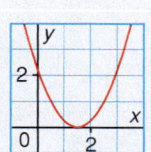	$a < 0$	$a > 0$ and $b > 1$ 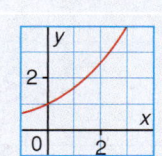	$a > 0$ and $0 < b < 1$
The first differences are constant.		The second differences are constant.		The decay/growth factors are constant.	

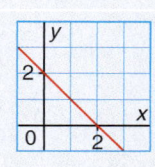

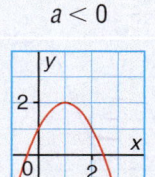

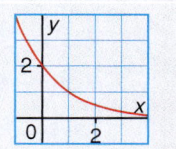

Regression

We can use the regression feature on a graphing calculator to fit a line or curve to data.

- Plot the data on a scatter plot.
- Choose a regression equation to model the data.

  Linear model: $y = ax + b$

  Quadratic model: $y = ax^2 + bx + c$

  Exponential model: $y = ab^x$

- Graph the regression equation on the scatter plot to see how well it fits the data.
- Use the regression equation to make predictions.

Chapter Review

5.1

1. Describe the relationship between population and time in each graph.

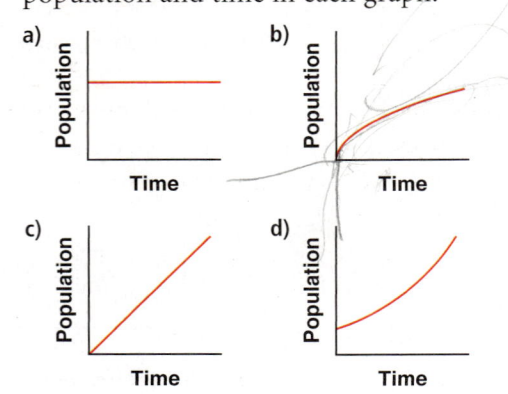

   a) [Graph: Population vs Time, horizontal line]
   b) [Graph: Population vs Time, curve increasing then leveling]
   c) [Graph: Population vs Time, straight line increasing]
   d) [Graph: Population vs Time, curve increasing more steeply]

2. This graph shows the annual energy consumption of a new 22 cu. ft. refrigerator for each year.

   Energy Consumption of a 22 cu. ft. Refrigerator
   [Graph: Energy consumption (kWh/year) vs Year, decreasing from ~1000 in 1990 to ~500 in 2002]

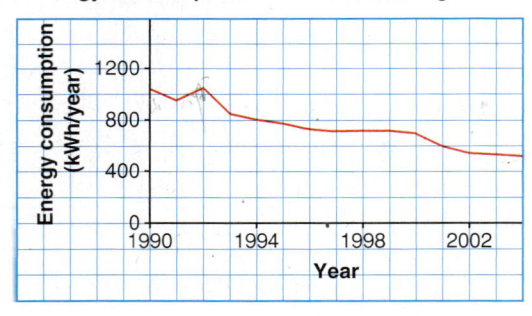

   a) Describe the trends in the graph.
   b) In which year were refrigerators the most inefficient? How do you know?
   c) Predict the annual energy consumption of a new refrigerator in 2010. Explain your prediction.

5.2

3. Refer to the graphs in question 1.
   a) Which graph illustrates each rate of change?
      i) Zero   ii) Constant   iii) Varying
   b) Identify the units of the rate of change for each graph.

4. Describe the rate of change indicated in each graph.

   a) **Loudness of Television Program**
   [Graph: Loudness (sone) vs Time (s)]

   b) **Volume of Sand in a Beach**
   [Graph: Volume of sand gained since October 2006 (cm^3) vs Month, Jan 07 – Jul 07]

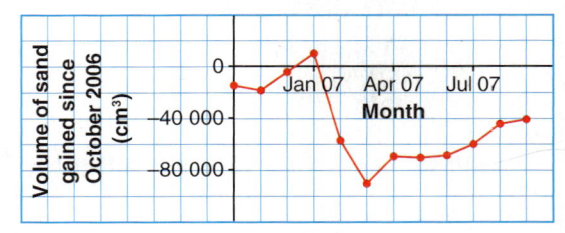

5. This graph shows the growth in Zoltan's hourly wages.

   Zoltan's Hourly Wage
   [Graph: Hourly wage ($) vs Years since started job, linear increase]

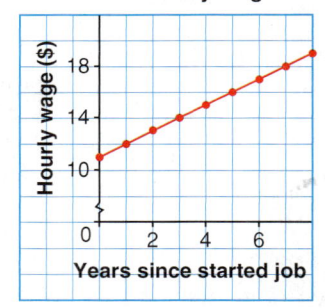

   a) Determine the rate of change in Zoltan's wages with respect to time.
   b) What does the rate of change represent in this situation?
   c) Does it matter which two points you use to calculate the rate of change? Explain.

5.3

6. Which of these tables of values models a linear relation? How do you know?

a)
s	0	10	20	30	40
t	62.4	58.6	54.8	51.0	47.2

b)
x	0	2	4	6	8
y	1	2	6	24	120

7. Kimiko and Atsuko have a catering business. The amount they charge, C dollars, depends on the event and the number of people, n, being served.
Lunch: $C = 100 + 15n$
Appetizers: $C = 100 + 10n$

a) In each formula, which number represents the rate of change? Explain.
b) What do the rates of change represent?
c) What other information does each formula give? Explain.

8. As China becomes more industrialized, its population consumes more plant oils.

Year	1991	1993	1997	2000	2004
Per person consumption (g/day)	23	25	29	33	35

a) Create a scatter plot of the data, and describe any trends.
b) Determine the equation of the line of best fit. Graph the line on the scatter plot.
c) Predict China's consumption of plant oils in 2010.
d) What factors could cause this consumption pattern to change? Explain.

5.4

9. Which of these equations models a quadratic relation? How do you know?
a) $y = 4x^2 + x - 1$
b) $y = 2x - 1$
c) $y = 3(2)^x$
d) $y = 3x^2$

10. The energy, E joules, of an object with mass m kilograms moving at a speed of v metres per second can be modelled by the formula $E = \frac{1}{2}mv^2$.

a) Use a graph to investigate the relationship between E and m when $v = 6$ m/s.
b) Describe the relationship between E and m. Justify your answer.
c) Use a graph to investigate the relationship between E and v when $m = 6$ kg.
d) Describe the relationship between E and v. Justify your answer.

11. A company designs custom prints for promotional beach balls. This table shows the surface areas of different balls.

Diameter (cm)	15	30	45	60	75
Surface area (cm²)	707	2827	6362	11 310	17 671

a) Create a scatter plot of the data, and describe any trends.
b) Determine the equation of the parabola of best fit. Graph the parabola on the scatter plot.
c) Predict the surface area of a ball with diameter 122 cm.

Chapter Review **333**

12. Use this graph.

Percent of Uranium Isotopes Remaining Since Their Creation

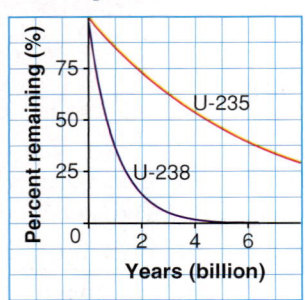

a) Compare the initial quantities of each radioactive isotope. What do you notice? Explain why this makes sense.
b) Which isotope has the greater decay factor? Explain.

13. Nyarai deposits $2500 in an account that earns a constant percent interest compounded annually. This table shows the amount in the account after 4 years.

Year	Amount
0	$2500.00
1	$2650.00
2	$2809.00
3	$2977.54
4	$3156.19

a) Create a scatter plot of the data, and describe any trends.
b) Determine the equation of the exponential regression curve. Graph the curve on the scatter plot.
c) Predict the amount in Nyarai's account after 10 years.

14. Identify which type of relation each equation represents.
a) $y = 5x + 2$
b) $y = 5(2)^x$
c) $y = 5x^2$
d) $y = x^2 + x + 6$
e) $y = 1.8x$
f) $y = -2x - 7$

15. The average house price in a popular neighbourhood has been increasing.

Year	2001	2002	2003	2004	2005
Price (thousands of dollars)	364	464	547	594	644

a) Draw a scatter plot of the data.
b) What type of relationship seems to exist between the average house price and the year? Justify your answer.
c) Use the model you chose in part a to generate a regression equation for the data, where y represents the average price of a house x years after 2000.
d) Estimate the average house price in 1995, 2000, and 2010. What factors may affect the reliability of these estimates?

16. Natural gas production in Norway has been increasing with the development of North Sea projects.

Year	2000	2001	2002	2003	2004
Production (million tonnes)	43	48	57	64	76

a) Draw a scatter plot of the data.
b) Perform linear, quadratic, and exponential regressions on the data.
c) Which model best represents the data? Justify your answer.
d) Predict the natural gas production in Norway in 2010 and 2020. What factors may affect the accuracy of your predictions?

Practice Test

Multiple Choice: Choose the correct answer for questions 1 and 2. Justify each choice.

1. Which of these graphs best represents the relationship between a person's age and height?

   A. B. 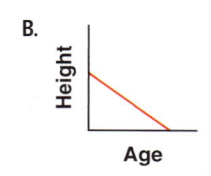 C. 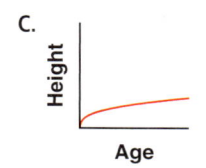 D.

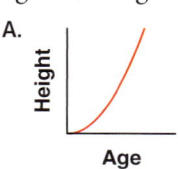

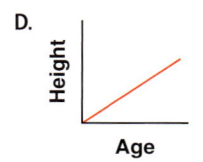

2. Which variable in $d = \frac{1}{2}at^2$ should be set constant to obtain a linear relation?
   A. d B. T C. a D. t

3. **Knowledge and Understanding** This table shows Raluca's height at different ages. What was the average rate of change in her height?

Age (months)	Height (cm)
3	59.8
9	70.1

4. **Thinking** Jason deposited money in a savings account. He let the money earn interest for 6 years, and then made regular withdrawals. Explain how the graph would change under each scenario. Include a sketch in your explanation.
   i) The account earned a lesser interest rate.
   ii) Jason made greater regular withdrawals.
   iii) Jason deposited a greater principal.

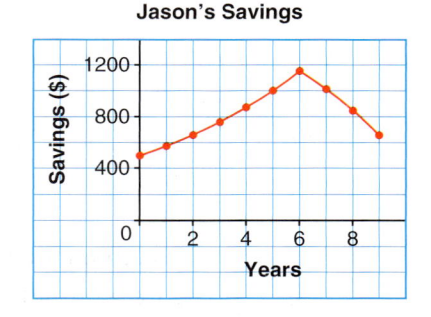

5. **Application** The United States imports aluminum cans from Canada for recycling.
   a) Select a regression model to represent the trend in the data. Justify your selection.
   b) Use your regression model from part a. Predict the mass of cans imported in 2007.

Year	1995	1997	1999	2001	2003	2005
Mass of cans imported (thousand tonnes)	27.2	34.4	41.5	43.3	47.1	55.4

6. **Communication** This graph compares the mean weight change over time for three popular diets.
   a) Describe the trends in each graph.
   b) Suppose you planned to start a diet. Which diet would you choose? Explain.

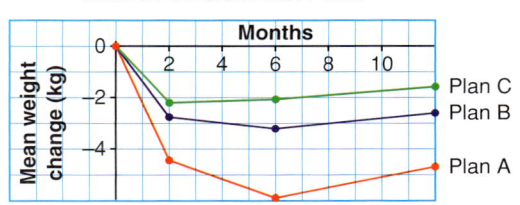

Chapter Problem: Modelling the Price of Diamonds

Diamonds are graded by the five C's: carats, clarity, cut, colour, and certification. A carat is a measure of the size of the diamond. So, for diamonds with similar clarity, cut, colour, and certification, the cost will depend on the size.

Size (carats)	Cost ($)
0.23	610
0.25	639
0.28	679
0.30	821
0.32	899
0.39	1056
0.42	1136
0.46	1104
0.74	2879
1.01	4066
1.04	5618

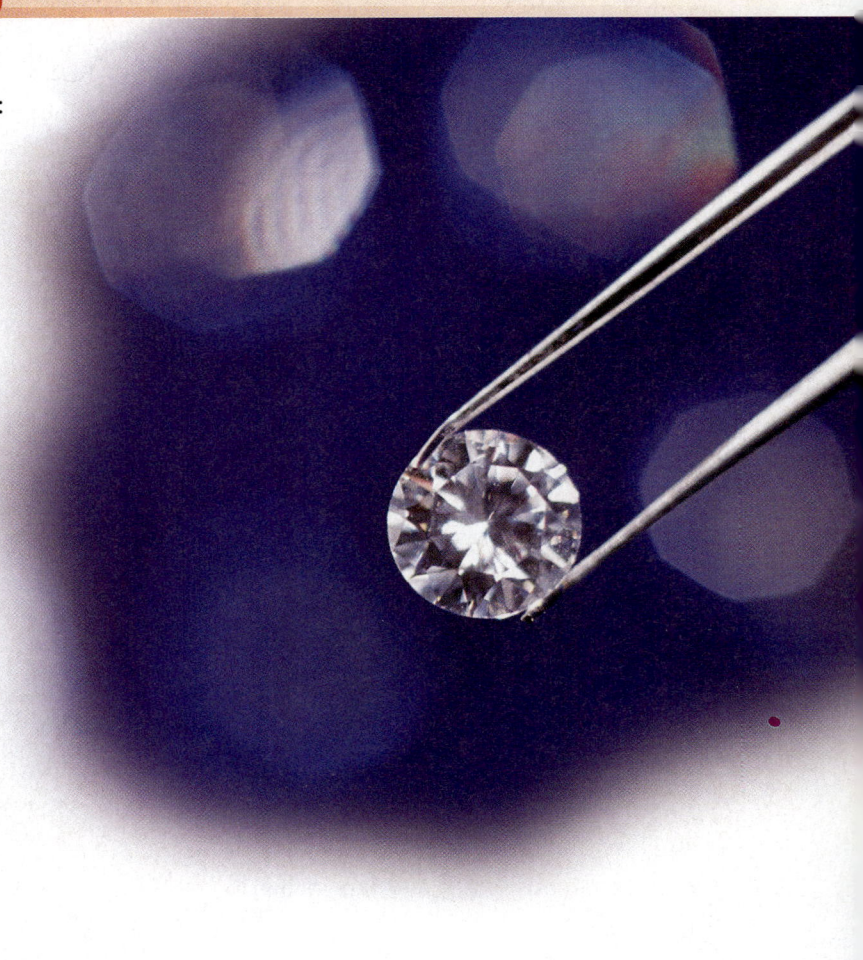

1. Bella thinks a linear relation is best to predict the price of a diamond. Why might she think this?

2. Max thinks a quadratic relation is best to predict the price of a diamond. Why might he think this?

3. Carmen thinks an exponential relation is best to predict the price of a diamond. Why might she think this?

4. Which do you think is best for modelling the relation between the size in carats and the cost of a diamond: linear, quadratic, exponential relation? Justify your answer.

Decide what tools you can use to represent data.

6 Algebraic Models

What You'll Learn

How to rearrange formulas, evaluate powers, and solve exponential equations

And Why

These skills can help you solve real-world problems in fields such as business, construction, design, medicine, science, and technology.

Key Words

- square root
- formula
- inverse operations
- base
- exponent
- power
- cube root
- rational exponent
- exponential equation

CHAPTER 6

Activate Prior Knowledge

Square Roots

Prior Knowledge for 6.1

The **square** of a number is the number multiplied by itself.
Finding the **square root** is the **inverse operation** of squaring.
For example, since $5^2 = 5 \times 5 = 25$
and $(-5)^2 = (-5) \times (-5) = 25$,
both 5 and -5 are square roots of 25.
We write $\sqrt{25} = 5$, and $-\sqrt{25} = -5$.

> $\sqrt{25}$ is the positive square root of 25, while $-\sqrt{25}$ is the negative square root of 25. So, $\sqrt{25} = 5$ and $-\sqrt{25} = -5$.

Example

Materials
- scientific calculator

Using a TI-30X IIS scientific calculator, press:
3 [2nd] [x²] 5 [)] [ENTER]
If you are using a different calculator, refer to the user's manual.

Evaluate. Round to the nearest hundredth where necessary.
a) $\sqrt{36}$
b) $-\sqrt{100}$
c) $\sqrt{123}$
d) $3\sqrt{5}$

Solution

a) $\sqrt{36} = 6$ since $6^2 = 36$
b) $-\sqrt{100} = -10$ since $(-10)^2 = 100$
c) Use the square root key on a calculator: $\sqrt{123} \doteq 11.09$
d) Use the square root key on a calculator: $3\sqrt{5} \doteq 6.71$

CHECK ✓

1. Evaluate. Round to the nearest hundredth where necessary.
   a) $\sqrt{49}$
   b) $-\sqrt{64}$
   c) $\sqrt{10}$
   d) $-\sqrt{81}$
   e) $2\sqrt{7}$
   f) $-\sqrt{9}$
   g) $3\sqrt{16}$
   h) $\sqrt{\dfrac{8}{\pi}}$

   For which parts did you use a calculator? Explain.

2. Integers whose square roots are also integers are called *perfect squares*.
   a) Explain why 81 is a perfect square, but 82 is not.
   b) Write the first 12 perfect square integers and their square roots.

3. The formula $T = 2\pi\sqrt{\dfrac{L}{9.8}}$ gives the time, T seconds, for one complete swing of a pendulum with length L metres. A clock pendulum is 22 cm long. Determine, to the nearest tenth of a second, the time it takes to complete one swing.

338 CHAPTER 6: Algebraic Models

Solving Linear Equations

Prior Knowledge for 6.2

Use a balance model to solve linear equations in one variable.
Perform the same operation on both sides of the equation
until the variable is isolated on one side.

Example

Materials
- scientific calculator

Solve.
a) $2x + 1 = 7$ b) $7x + 3 = -2x + 9$

In a linear equation, all variables are raised to the first power ($x = x^1$).

Solution

a) $2x + 1 = 7$
   Isolate $2x$ first, then solve for x.
   $2x + 1 - 1 = 7 - 1$ Subtract 1 from each side.
   $2x = 6$
   $\dfrac{2x}{2} = \dfrac{6}{2}$ Divide each side by 2.
   $x = 3$

To check, substitute $x = 3$ in $2x + 1 = 7$
L.S.	R.S.
$2(3) + 1$	7
$= 6 + 1$	
$= 7$	

L.S. = R.S., so the solution is correct.

b) $7x + 3 = -2x + 9$
   Collect the variable terms on the left side, and the numbers on the right side.
   $7x + 3 + 2x = -2x + 9 + 2x$ Add $2x$ to each side.
   $9x + 3 = 9$
   $9x + 3 - 3 = 9 - 3$ Subtract 3 from each side.
   $9x = 6$
   $\dfrac{9x}{9} = \dfrac{6}{9}$ Divide each side by 9.
   $x = \dfrac{2}{3}$

CHECK ✓

1. Solve.
   a) $x - 12 = -5$ b) $-3x = -54$ c) $5x - 3 = 12$ d) $-3x + 4 = 25$

2. Solve and check.
   a) $13x + 8 = 6x + 22$ b) $3x - 11 = -2x + 9$ c) $-2x + 8 = -7x - 2$
   Why should you always check your solution in the original equation?

3. The equation $T = 10d + 20$ gives the temperature, T degrees Celsius, at a depth of d kilometres below the surface of the Earth. Determine the depth of a mine shaft in which the temperature is 40°C. How do you know that your answer is correct? Explain.

Activate Prior Knowledge

Evaluating Powers with Integer Exponents

Prior Knowledge for 6.3

Positive integer exponent
$a^n = \underbrace{a \times a \times a \times \cdots \times a}_{n \text{ factors}}$

Zero exponent
$a^0 = 1, a \neq 0$

Negative integer exponent
$a^{-n} = \dfrac{1}{a^n}, a \neq 0$

a^{-n} is the reciprocal of a^n.

Example

Materials
- scientific calculator

Evaluate.

a) 3^4 b) 2^{-5} c) $(-5)^0$ d) $\left(\dfrac{4}{5}\right)^{-2}$ e) 0.4^{-3}

Solution

a) $3^4 = (3)(3)(3)(3) = 81$

b) $2^{-5} = \dfrac{1}{2^5} = \dfrac{1}{32}$

c) $(-5)^0 = 1$

d) $\left(\dfrac{4}{5}\right)^{-2} = \left(\dfrac{5}{4}\right)^2$ The reciprocal of $\dfrac{4}{5}$ is $\dfrac{5}{4}$.

$= \dfrac{5}{4} \times \dfrac{5}{4}$ Raise $\dfrac{5}{4}$ to the exponent 2.

$= \dfrac{25}{16}$

The product of a non-zero number and its reciprocal is 1. The reciprocal of $\dfrac{4}{5}$ is $\dfrac{5}{4}$ since $\dfrac{5}{4} \times \dfrac{4}{5} = 1$.

e) Use the exponent key on a calculator to obtain $0.4^{-3} = 15.625$

Press: 0.4 ^ (−) 3 ENTER

CHECK ✓

1. Evaluate without using a calculator.

a) 2^3 b) 4^3 c) $(-5)^2$ d) 3^{-2}

e) 8^0 f) $\left(\dfrac{1}{2}\right)^3$ g) $(-7)^{-1}$ h) $\left(\dfrac{3}{5}\right)^{-2}$

2. Evaluate with a calculator. Round to the nearest hundredth.

a) 0.95^7 b) 1.6^{-3} c) $200(1.04)^5$

d) $500(0.95)^{-3}$ e) $\left(\dfrac{2}{3}\right)^6$ f) $\left(\dfrac{5}{6}\right)^{-3}$

3. Explain the difference between the expressions in each pair and determine their values.

a) 3^2 and 2^3 b) 4^3 and $(-4)^3$ c) 5^2 and 5^{-2}

340   CHAPTER 6: Algebraic Models

Learning with Others

At college, in an apprenticeship program, and in the workplace, people learn new skills and solve problems by working with others.

These strategies will help you succeed in team environments.

- Make sure everyone understands the situation in the task.
- Be creative with models and technology tools.
- Contribute ideas and information. Express your ideas clearly.
- Be open to, and listen actively to, the ideas of others.
- Analyse ideas and ask questions.
- Do your fair share of the work, and help your partners.
- Make sure everyone can explain the solution.

The goal is for each person to learn, so everyone needs to cooperate as a team to help each other understand the math thinking and communicate.

1. What is another strategy you would suggest for learning with others?

2. When working on this chapter, choose an *Investigate* and apply the strategies for learning with others.

3. Then, explain whether the strategies for learning with others were useful.
    - Did the strategies help you understand the math? Did they help others understand the math? Include examples.
    - Would you use some of the strategies in the next chapter? Explain your thinking.
    - What suggestions would you give someone else for learning as a team?

4. Imagine being an apprentice, an employee, or a college student. How do you think you might use strategies for learning with others in this role?

6.1 Using Formulas to Solve Problems

Forensic scientists and anthropologists use formulas to predict the height of a person from the lengths of their bones. They can use the radius bone or the femur bone.

Investigate: Estimating Height from the Lengths of Bones

Materials
- metre stick
- scientific calculator

Work with a partner.

These formulas give the height, h, of an adult in terms of the lengths of the radius bone, r, and femur bone, f.

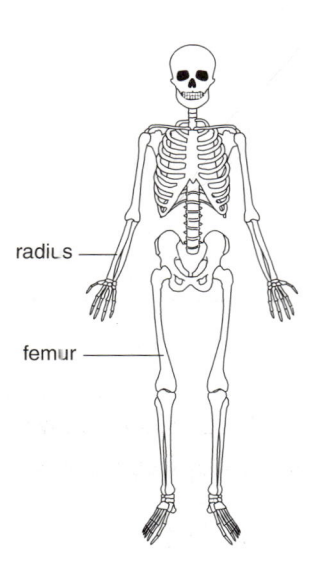

Male
$h = 3.65r + 80.41$
$h = 2.24f + 69.09$

Female
$h = 3.88r + 73.50$
$h = 2.32f + 61.41$

All measurements are in centimetres.
- Predict the height of a female whose femur has length 40.6 cm.
- Predict the height of a male whose radius has length 28.1 cm.
- Have your partner measure the length of your radius and femur bones. Use each measure and the appropriate formula to estimate your height.
- Which formula gave the more accurate prediction of your height? Explain.

342 CHAPTER 6: Algebraic Models

> **Reflect**
> - How do you think the formulas were obtained?
> - Why is there a different set of formulas for males and females?
> - What might account for the difference between your actual height and the heights predicted by the formulas?

Connect the Ideas

Formulas

A **formula** is a mathematical equation that relates two or more variables representing real-world quantities. Rules and procedures in many occupations are expressed as formulas.

Example 1 — Substituting into a Formula

Materials
- scientific calculator

Pediatric nurses use *Young's formula*, $C = \dfrac{Ag}{g + 12}$, to calculate a child's dose of medicine, C milligrams, when the adult dose, A milligrams, and the child's age, g years, are known. Suppose the adult dose of a certain medication is 600 mg. Determine the corresponding dose for a 3-year-old child.

Solution

Substitute $A = 600$ and $g = 3$ in the formula $C = \dfrac{Ag}{g + 12}$.

$$C = \dfrac{(600)(3)}{3 + 12}$$

$$= \dfrac{1800}{15}$$

$$= 120$$

The child's dose is 120 mg.

Example 2 — Using a Formula to Solve a Problem

Materials
- scientific calculator

A landscaper wants to estimate the cost of fertilizing a triangular lawn with side lengths 150 m, 200 m, and 300 m. One bag of fertilizer costs $19.98 and covers an area of 900 m².
She uses *Heron's formula* to determine the area of the lawn:
The area A of a triangle with side lengths a, b, and c is given by
$A = \sqrt{s(s - a)(s - b)(s - c)}$, where $s = \dfrac{a + b + c}{2}$.
Estimate the cost of fertilizing the lawn.

6.1 Using Formulas to Solve Problems   **343**

Solution

Plan the solution.

To find the:	We need to know the:
…cost of the fertilizer	…number of bags needed
…number of bags needed	…area of the lawn
…area of the lawn	…formula for the area

Plan your solution by working backward from what you are trying to find to what you are given. Write the solution by working forward from what you are given to what you are trying to find.

The formula for the area of the lawn is:
$A = \sqrt{s(s-a)(s-b)(s-c)}$, where $s = \dfrac{a+b+c}{2}$

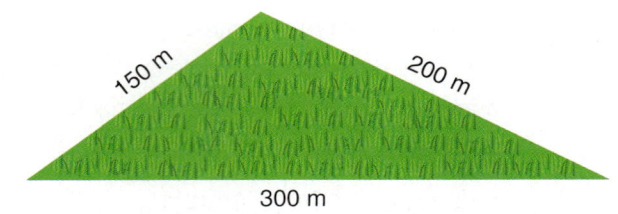

To calculate s, substitute: $a = 150$, $b = 200$, and $c = 300$
$s = \dfrac{150 + 200 + 300}{2}$, or 325

Calculate A. Substitute: $s = 325$, $a = 150$, $b = 200$, and $c = 300$
$A = \sqrt{325(325-150)(325-200)(325-300)}$
$A = \sqrt{325(175)(125)(25)}$
$\doteq 13\,331.71$

The area of the field is approximately 13 331.71 m².
Each bag of fertilizer covers an area of 900 m².
The number of bags needed to cover 13 331.71 m² is: $\dfrac{13\,331.71}{900} \doteq 14.8$
So, about 15 bags of fertilizer are needed.
The cost of the 15 bags of fertilizer is: 15($19.98) = $299.70
It costs $299.70 to fertilize the lawn.

Example 3

Choosing Formulas and Converting Measures

Materials
- scientific calculator

A landscaper uses a bucket with radius 18 cm and height 18 cm to pour soil into a rectangular planter that measures 1 m by 40 cm by 20 cm.

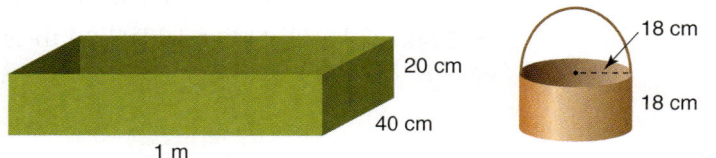

How many buckets of soil are needed to fill the planter?

344 CHAPTER 6: Algebraic Models

> Problems in landscaping, construction, and design often involve the use of geometric formulas. The measurements substituted into these formulas must be in the same units.

Solution

Convert all measurements to the same units.
1 m = 100 cm
Find the volume of soil each object can hold.

Planter

Use the formula for the volume of a rectangular prism: $V = lwh$ where V is the volume, l is the length, w is the width, and h is the height
Substitute: $l = 100$, $w = 40$, and $h = 20$

$V = (100)(40)(20)$
$ = 80\,000$

The planter holds 80 000 cm³ of soil.

Bucket

Use the formula for the volume of a cylinder: $V = \pi r^2 h$ where V is the volume, r is the radius, and h is the height
Substitute: $r = 18$ and $h = 18$

$V = \pi(18)^2(18)$
$ \doteq 18\,321.77$

The bucket holds about 18 321.77 cm³ of soil.
So, the number of buckets of soil needed is: $\dfrac{80\,000}{18\,321.77} \doteq 4.4$
About 4 buckets of soil are needed.

Planning and organizing your solution

Organization is an important part of solving multi-step problems. The answers to these questions may be helpful in planning your solution.
- What formulas or relationships can be used?
- What numerical information is given?
- What numerical information do you need to find or estimate?
- What units of measurement are used?
   Do you need to convert from one set of units to another?

These problem-solving strategies may also be helpful.
- Work backward to determine what information you need.
- Work forward from the formulas and information you are given.
- Make a checklist of variables and their values.

Practice

A

For help with question 1, see Example 1.

1. The area, A, of a rectangle with length l and width w is $A = lw$. Find the area of a rectangle with each length and width.
   a) $l = 10$ m, $w = 4$ m
   b) $l = 6$ cm, $w = 8$ cm
   c) $l = 9.5$ m, $w = 4.2$ m
   d) $l = 8.4$ cm, $w = 7.2$ cm

2. The density, D, of an object with mass M and volume V is $D = \frac{M}{V}$. Determine the density of an object with each mass and volume.
   a) $M = 200$ g, $V = 10$ cm^3
   b) $M = 45$ g, $V = 7$ cm^3
   c) $M = 7.8$ kg, $V = 2.6$ L
   d) $M = 10$ kg, $V = 5.4$ L

3. The formula $S = 0.6T + 331.5$ gives the approximate speed of sound in air, S metres per second, when the air temperature is T degrees Celsius. Determine the speed of sound at each air temperature.
   a) 30°C
   b) −15°C
   c) 10°C
   d) −25°C

In Canada, temperatures are given in degrees Celsius, but in the United States, they are given in degrees Fahrenheit.

4. We can use the formula $C = \frac{5(F - 32)}{9}$ to convert degrees Fahrenheit, F, to degrees Celsius, C. Determine the Celsius equivalent of each Fahrenheit temperature.
   a) 77°F
   b) 212°F
   c) 50°F
   d) −4°F

5. The approximate pressure, P kilopascals, exerted on the floor by the heel of a shoe is given by the formula $P = \frac{100\,m}{h^2}$, where m kilograms is the wearer's mass and h centimetres is the width of the heel. Determine the pressure exerted by the heel of each person's shoe.

	Person's mass (kg)	Shoe heel width (cm)
a)	80	6
b)	60	1.5
c)	55	3
d)	75	4.5

B

6. A doughnut and an inner tube are examples of a *torus*. The volume of a torus is given by the formula $V = 2\pi^2 a^2 b$. A dog chew toy is a torus with $a = 1$ cm and $b = 5$ cm. Determine the volume of rubber in the toy.

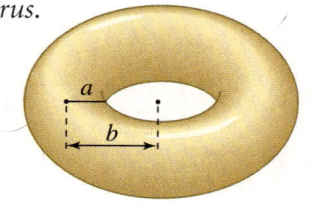

346 CHAPTER 6: Algebraic Models

7. In a round-robin tournament, each team plays every other team once. The formula $G = \frac{n(n-1)}{2}$ gives the number of games G that must be scheduled for n teams.
   a) How many games must be scheduled for 6 teams?
   b) Will the number of games double if the number of teams doubles? Justify your answer.

■ For help with question 8, see Example 2.

8. Vinh makes and sells T-shirts. The cost, C dollars, to produce n T-shirts is given by $C = 300 + 7n$. The revenue, R dollars, earned when n T-shirts are sold is given by $R = n\left(15 - \frac{n}{200}\right)$.
   a) Determine the cost of making 200 T-shirts.
   b) Profit is the difference between revenue and cost. Determine the profit from making and selling 1000 shirts.

9. A fuel storage tank consists of a cylinder with radius 1.25 m and length 7.20 m, with hemispheres of radius 1.25 m at each end.

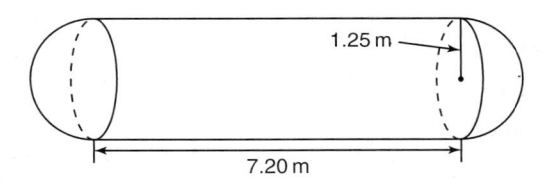

   a) Determine the surface area of the tank. Use the formula $SA = 4\pi r^2 + 2\pi rl$, where SA is the surface area of the tank, r is its radius, and l is the length of the cylinder.
   b) Determine the cost to cover the tank with 2 coats of paint. One can of paint costs $34.99 and covers an area of 29 m².

10. Body surface area is used to calculate drug dosages for cancer chemotherapy. The formula $B = \sqrt{\frac{w \times h}{3600}}$ gives the body surface area, B square metres, of an individual with height h centimetres and mass w kilograms.
   a) Determine the body surface area of a child 102 cm tall with a mass of 21 kg.
   b) The recommended child's dosage of a chemotherapy drug is 20 mg/m². How much medicine should the child in part a receive?

6.1 Using Formulas to Solve Problems **347**

■ For help with question 11, see Example 3.

11. The bottle on an office water dispenser is a cylinder with radius 13.5 cm and height 49.1 cm. The paper cones from which people drink are 9.5 cm high with radius 3.5 cm. How many full cones of water can be dispensed?

12. A paving contractor has been hired to lay 6 cm of compacted asphalt on a road 12-m wide and 3.5-km long.
Each cubic metre of compacted asphalt has mass 2.5 t.
How many tonnes of asphalt should the contractor order?

13. Assessment Focus A hard rubber ball with radius 2.0 cm sells for $1.25.
   a) Calculate the volume of the ball.
   b) Suppose the radius is doubled. Does the volume double? Explain.
   c) What would you charge for a ball with double the radius?
      Justify your answer.

14. Samuel owns a pool maintenance company. One of his jobs is to chlorinate pool water. A single chlorine treatment requires 45 g of powdered chlorinator per 10 000 L of water. The chlorinator is sold in a 11.4-kg bucket that costs $54.99.
   a) One of Samuel's clients has a swimming pool 18 m long and 10 m wide with an average depth of 2.5 m. How many litres of water does the pool hold? Explain.
   b) How many grams of powdered chlorinator are required for a single treatment?
   c) What is the cost of a single treatment? Explain.

348 CHAPTER 6: Algebraic Models

15. In the forestry industry, it is important to estimate the volume of wood in a log. One formula that is used is $V = \frac{1}{2}L(B + b)$, where V cubic metres is the volume of wood, L metres is the length of the log, and B and b are the areas of the ends in square metres.

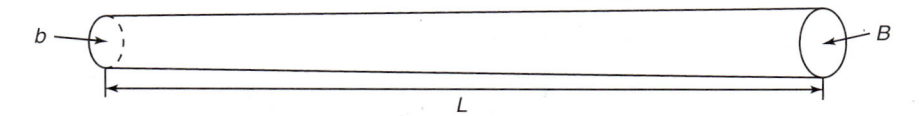

Estimate the volume of wood in a log with length 3.7 m and end diameters 30 cm and 40 cm.

16. Literacy in Math Create a matrix or checklist for the quantities in question 15. Write the given numerical values. Explain how you found the other values.

C

17. *Example 1* introduced *Young's formula* for calculating a child's medicine dose, C milligrams: $C = \frac{Ag}{g + 12}$, where A represents the adult dose in milligrams and g represents the child's age, in years.
  a) For a 6-year-old child, what fraction of the adult's dose is the child's dose? Explain how this fraction changes for older children.
  b) For a given age, is the relationship between a child's dose and an adult's dose linear? Justify your answer.

18. *Euler's formula* relates the number of vertices (V), faces (F), and edges (E) of a polyhedron. Determine the value of $V + F - E$ for each polyhedron.
  a) square pyramid   b) cube   c) octahedron

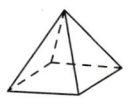

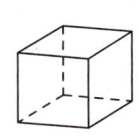

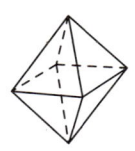

What do you think Euler's formula is? Explain.

In Your Own Words

Explain what is meant by this statement.
"The thinking and organizing you do to solve a multi-step problem is often backward from the presentation of the final solution."
Use an example to illustrate your explanation.

6.2 Rearranging Formulas

Travel agents make sure that their clients know what weather to expect at their destination. The formula $C = \frac{5(F-32)}{9}$ converts a temperature in degrees Fahrenheit, F, to degrees Celsius, C.

Investigate — Inverse Operations

We can convert from Celsius to Fahrenheit by rearranging the formula $C = \frac{5(F-32)}{9}$ to isolate F.

One way to do this is to use inverse operations.
Inverse operations "undo" or reverse each other.

Work with a partner.
For each arrow diagram:

- Which operations will "undo" the sequence of operations in the top row of the diagram?
- Copy and complete the diagram.

Changing a flat tire

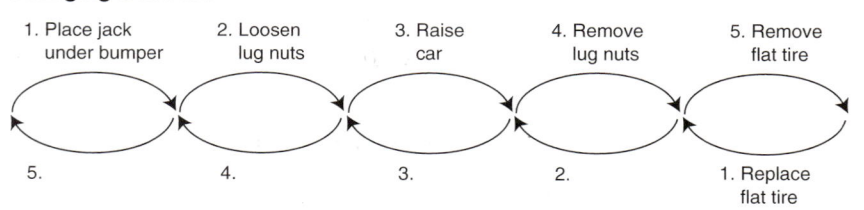

1. Place jack under bumper → 2. Loosen lug nuts → 3. Raise car → 4. Remove lug nuts → 5. Remove flat tire

5. → 4. → 3. → 2. → 1. Replace flat tire

Converting between degrees Fahrenheit and degrees Celsius

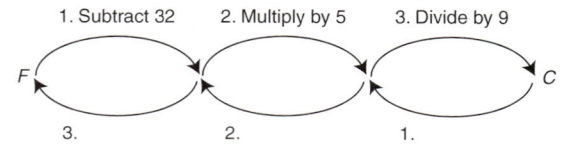

1. Subtract 32 → 2. Multiply by 5 → 3. Divide by 9

F → → → → C

3. → 2. → 1.

350 CHAPTER 6: Algebraic Models

> **Reflect**
>
> - How are the steps and operations in the top row of each arrow diagram related to the steps and operations in the bottom row of the diagram? Why are they related this way?
> - List three different mathematical operations and their inverses.

Connect the Ideas

Rearranging formulas

Formulas usually express one variable in terms of one or more variables. We can use our knowledge of equations and inverse operations to rewrite the formula in terms of a different variable.

Example 1 — Isolating a Variable

Rearrange each formula to isolate the indicated variable.

a) The amount, A dollars, of an investment is given by the formula $A = P + I$, where P dollars is the principal and I dollars is the interest earned. Isolate P.

b) The volume, V cubic metres, of a rectangular prism with length l metres, width w metres, and height h metres, is given by the formula $V = lwh$. Isolate h.

c) Ohm's Law, $I = \dfrac{V}{R}$, relates the current, I amperes, running along an electrical circuit to the voltage, V volts, and the resistance, R ohms. Isolate V.

Solution

Use an arrow diagram to determine the inverse operations needed.

a) $A = P + I$

To isolate P, subtract I from each side.

$A - I = P + I - I$
$A - I = P$

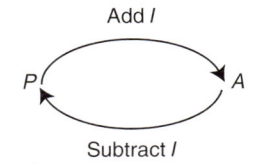

b) $V = lwh$

To isolate h, divide each side by lw.

$\dfrac{V}{lw} = \dfrac{lwh}{lw}$

$\dfrac{V}{lw} = h$

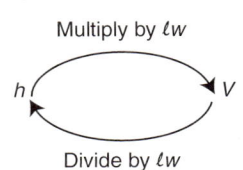

6.2 Rearranging Formulas   **351**

c) $I = \dfrac{V}{R}$

To isolate V, multiply each side by R.

$I \times R = \dfrac{V}{R} \times R$

$IR = V$

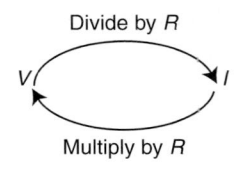

Example 2

Solving Problems by Rearranging a Formula

Convert 30°C to degrees Fahrenheit. Use the formula $C = \dfrac{5(F-32)}{9}$.

Solution

Rearrange the formula to isolate F. Use an arrow diagram to determine the inverse operations required.

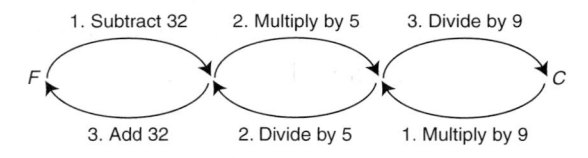

1. Subtract 32   2. Multiply by 5   3. Divide by 9

$F \rightarrow \cdots \rightarrow C$

3. Add 32   2. Divide by 5   1. Multiply by 9

Method 1	**Method 2**
Isolate F, then substitute.	Substitute, then solve for F.
$C = \dfrac{5(F-32)}{9}$	$C = \dfrac{5(F-32)}{9}$
Multiply each side by 9.	Substitute: $C = 30$
$C \times 9 = \dfrac{5(F-32)}{9} \times 9$	$30 = \dfrac{5(F-32)}{9}$
$9C = 5(F-32)$	Multiply each side by 9.
Divide each side by 5.	$30 \times 9 = \dfrac{5(F-32)}{9} \times 9$
$\dfrac{9C}{5} = \dfrac{5(F-32)}{5}$	$270 = 5(F-32)$
$\dfrac{9C}{5} = F - 32$	Divide each side by 5.
Add 32 to each side.	$\dfrac{270}{5} = \dfrac{5(F-32)}{5}$
$\dfrac{9C}{5} + 32 = F - 32 + 32$	$54 = F - 32$
$\dfrac{9C}{5} + 32 = F$	Add 32 to each side.
Substitute: $C = 30$	$54 + 32 = F - 32 + 32$
$F = \dfrac{9(30)}{5} + 32$	$86 = F$
$F = 86$	
30°C is equivalent to 86°F.	

352   CHAPTER 6: Algebraic Models

Choosing a strategy

The "isolate, then substitute" and the "substitute, then solve" strategies produce the same result. Sometimes, one strategy is more efficient than the other.
- Isolate the variable first if you will have to calculate it several times.
- Substitute first if the numbers are simple or rearranging the formula is difficult.

Example 3

Materials
- scientific calculator

Solving Problems Involving Powers

a) The area, A, of a circle with radius r is $A = \pi r^2$.
   Use the formula $A = \pi r^2$ to determine the radius of a circular oil spill that covers an area of 5 km².

b) The volume, V, of a sphere with radius r is $V = \frac{4}{3}\pi r^3$.
   Use the formula $V = \frac{4}{3}\pi r^3$ to determine the radius of a Nerf ball with volume 1 m³.

Solution

Powers and roots are inverse operations.
To "undo" squaring, take the square root.
To "undo" cubing, take the cube root.

a) Draw an arrow diagram.

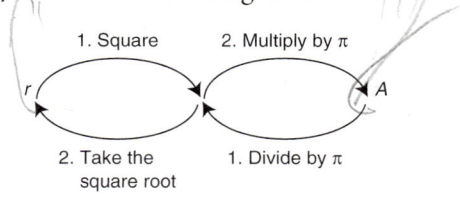

1. Square   2. Multiply by π
r → → A
2. Take the   1. Divide by π
square root

In general, the inverse of the nth power is the nth root: $\sqrt[n]{}$

For example, $\sqrt[3]{64} = 4$ since $4^3 = 64$.

In real-world situations, variables usually represent positive quantities, so take the positive root.

Substitute $A = 5$ in the formula $A = \pi r^2$.

$5 = \pi r^2$ Divide each side by π.

$\dfrac{5}{\pi} = r^2$ Take the square root of each side.

$\sqrt{\dfrac{5}{\pi}} = r$ Evaluate the left side.

$1.26 \doteq r$

The radius of the oil spill is about 1.26 km.

b) Draw an arrow diagram.

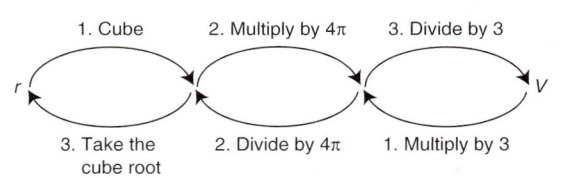

1. Cube   2. Multiply by 4π 3. Divide by 3
r → → → V
3. Take the   2. Divide by 4π 1. Multiply by 3
cube root

To evaluate $\sqrt{\dfrac{5}{\pi}}$ press:
[2nd] [x²] 5 [÷] π [)] [ENTER]

6.2 Rearranging Formulas   **353**

Substitute $V = 1$ in the formula $V = \frac{4}{3}\pi r^3$.

$1 = \frac{4}{3}\pi r^3$ Multiply each side by 3.

$3 = 4\pi r^3$ Divide each side by 4π.

$\frac{3}{4\pi} = r^3$ Take the cube root of each side.

$\sqrt[3]{\frac{3}{4\pi}} = r$ Evaluate the left side.

$0.62 \doteq r$

The radius of the Nerf ball is about 0.62 m.

To evaluate $\sqrt[3]{\frac{3}{4\pi}}$ press:

3 [2nd] [^] [(] 3 [÷]
[(] 4 π [)] [)] [ENTER]

Practice

For questions 1 to 4, use an arrow diagram to determine the inverse operations needed.

For help with question 1, see Example 1.

1. The accounting formula $A = L + E$ relates assets A, liability L, and owners' equity E.
   a) Isolate L.   b) Isolate E.

2. The profit, P, earned by a business is given by the equation $P = R - C$, where R is the revenue and C is the cost.
   a) Isolate R.   b) Isolate C.

3. The area, A, of a parallelogram is given by the equation $A = bh$, where b is the length of the base and h is the height.
   a) Isolate b.   b) Isolate h.

4. The density, D, of an object is given by the equation $D = \frac{M}{V}$, where M is the object's mass and V is the object's volume.
   a) Isolate M.   b) Isolate V.

For help with question 5, see Example 2.

5. A company uses the formula $a + s = 90$ to determine when an employee can retire with a full pension. In the formula, a is the employee's age and s is the number of years of service.
   a) Solve for s when $a = 58$.
   b) Solve for a when $s = 27$.

6. The formula $E = Rt$ gives the money earned, E dollars, for working at a rate of R dollars per hour for t hours. Jennie earns $12 per hour. How many hours does she have to work to earn each amount?
   a) $420   b) $126   c) $504

354 CHAPTER 6: Algebraic Models

B   **7.** Use the formula $E = Rt$ from question 6.

     a) Drew works 35 h and earns $542.50. What is his hourly rate of pay?

     b) Did you substitute and solve or isolate and substitute? Explain.

8. The formula $S = 0.6T + 331.5$ gives the speed of sound in air, S metres per second, at an air temperature of T degrees Celsius.

     a) Draw an arrow diagram to show the steps needed to isolate T in the formula.

     b) Isolate T.

     c) Determine the air temperature for each speed of sound.

         i) 343.5 m/s   **ii)** 336 m/s   **iii)** 328.5 m/s

9. A shoe store uses the formula $s = 3f - 21$ to model the relationship between a woman's shoe size, s, and her foot length, f, in inches. Nalini wears a size $7\frac{1}{2}$ shoe. Estimate her foot length to the nearest tenth of an inch.

10. The formula $H = nl + b$ gives the height, H, of n stacked containers, where each container has lip height l and base height b. Zoë is stacking flower pots with an 8-cm lip height and 50-cm base height at a garden centre. For safety reasons, the maximum allowable height of the stack is 1.3 m. How many pots can Zoë put in one stack? Justify your answer.

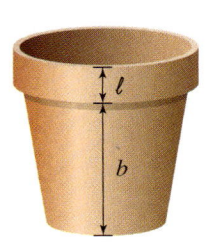

11. Office placement agencies use the formula $s = \frac{w - 10e}{m}$ to determine keyboarding speed. In the formula, s is the keyboarding speed in words per minute, w is the number of words typed, e is the number of errors made, and m is the number of minutes of typing.

     a) Mark types 450 words in 5 min and makes 12 errors. What is his keyboarding speed?

     b) If Rana makes no errors, how many words would she have to type in 5 min to have the same keyboarding speed as Mark?

12. Airplane pilots use the formula $s = \frac{d}{t}$ to estimate flight times. In the formula, s is the average speed, d is the distance travelled, and t is the flight time.

     a) Estimate the flight time from Ottawa to Thunder Bay, a distance of 1100 km. Assume that the airplane flies at an average speed of 350 km/h.

     b) Describe the operations you used to isolate t.

13. In house construction, the safe load, m kilograms, that can be supported by a beam with length l metres, thickness t centimetres, and height h centimetres is given by the formula $m = \frac{4th^2}{l}$.
   a) Determine t when $m = 500$ kg, $l = 4$ m, and $h = 10$ cm.
   b) Determine l when $m = 250$ kg, $t = 10$ cm and $h = 5$ cm.
   c) How are the steps used to solve for a variable in the denominator of a fraction similar to the steps used to solve for a variable in the numerator? How are they different?

For help with question 14, see Example 3.

14. The equation $V = \frac{1}{6}\pi d^3$ gives the volume, V, of a sphere in terms of its diameter, d. Use the formula to determine the diameter of a ball with volume 1000 cm³.

1 kW = 1000 W

15. The formula $P = \frac{r^2 s^3}{2}$ gives the approximate power, P watts, generated by a wind turbine with radius r metres in a wind of speed s metres per second. The Exhibition Place Wind Turbine in Toronto has radius about 24 m. Determine the wind speed when the turbine generates 500 kW of power.

356 CHAPTER 6: Algebraic Models

16. **Assessment Focus** The volume of a cylinder is given by the formula $V = \pi r^2 h$, where V is the volume, r is the radius, and h is the height.
    a) Rearrange the formula to isolate r.
       Explain your choice of inverse operations.
    b) Determine the radius of a cylindrical fuel tank that is 16 m high and holds 200 m³ of fuel.
    c) Determine the height of a cylindrical mailing tube with volume 2350 cm³ and radius 5 cm. Justify your choice of strategy.

17. **Literacy in Math** Use a graphic organizer to summarize the pairs of inverse operations that can be used to rearrange a formula. Explain the reason for your choice of organizer.

18. A police officer uses the formula $S = 15.9\sqrt{df}$ to estimate the speed of a vehicle when it crashed. In the formula, S kilometres per hour is the speed of the vehicle, d metres is the length of the skid marks left on the road, and f is the coefficient of friction, a measure of the traction between the road surface and the vehicle's tires.
    a) The skid marks left on a dry road measure 40 m.
       What was the speed of the vehicle if $f \doteq 0.85$ for a dry road?
    b) A car travelling at 30 km/h skids and crashes in an icy parking lot.
       Estimate the length of the skid marks at the crash site if $f \doteq 0.35$ for an icy road.

19. In Chapter 1, you used the Cosine Law $a^2 = b^2 + c^2 - 2bc \cos A$ to solve oblique triangles.
    a) Rearrange the formula to isolate cos A.
    b) Determine the measure of the greatest angle in a triangle with side lengths 5 m, 6 m, and 7 m.
    c) Why did we rearrange the formula for cos A instead of ∠A?

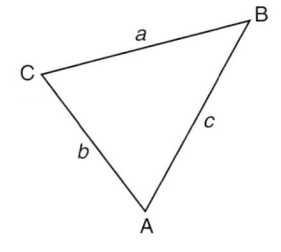

> **In Your Own Words**
>
> Choose a reversible routine from daily life such as setting the table or getting dressed. Explain why reversing the routine means undoing each step in the opposite order. Explain how this idea is used to rearrange a formula. Include an example in your explanation.

6.2 Rearranging Formulas **357**

6.3 Laws of Exponents

Many formulas in science, business, and industry involve integer exponents. For example, the formula $V = 0.05hc^2$ is used in the forestry industry to estimate the volume of wood in a tree. In the formula, V is the volume of wood in the tree, h is the height of the tree, and c is the circumference of the trunk.

Investigate — Simplifying Products and Quotients of Powers

Materials
- TI-89 calculator (optional)

Part A: Expanding Products and Quotients of Powers

- Copy and complete each table.
- Describe the relationship between the exponents in the original expression and the exponent in the expression as a single power.

Multiplying powers

Original expression	Powers in expanded form	Expression as a single power
$b^2 \times b^4$		
$b^5 \times b^{-3}$	$(\cancel{b} \times \cancel{b} \times \cancel{b} \times b \times b)\left(\dfrac{1}{\cancel{b} \times \cancel{b} \times \cancel{b}}\right)$	b^2
$b^{-8} \times b^5$		
$b^4 \times b^2 \times b^{-1}$		

Power of a power

Original expression	Powers in expanded form	Expression as a single power
$(b^2)^3$		
$(b^4)^3$		
$(b^{-3})^5$	$b^{-3} \times b^{-3} \times b^{-3} \times b^{-3} \times b^{-3}$	b^{-15}
$(b^4)^{-1}$		

358 CHAPTER 6: Algebraic Models

Dividing powers

Original expression	Powers in expanded form	Expression as a single power
$\dfrac{b^7}{b^2}$		
$b^5 \div b^2$		
$\dfrac{b^2}{b^3}$	$\dfrac{\overset{1}{\cancel{b}} \times \overset{1}{\cancel{b}}}{\underset{1}{\cancel{b}} \times \underset{1}{\cancel{b}} \times b} = \dfrac{1}{b}$	b^{-1}
$b^3 \div b^7$		

Part B: Using a CAS

The expressions in the tables in Part A were entered in a computer algebra system (CAS). These results were obtained.

Multiplying powers

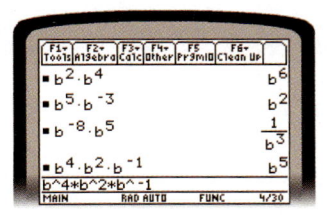

Power of a power

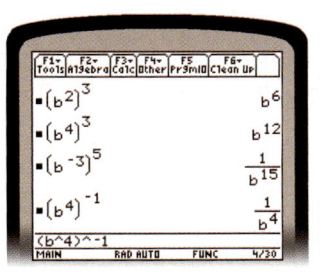

Dividing powers

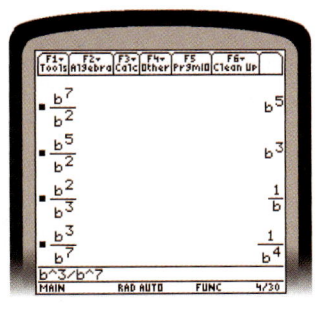

- Compare your answers in Part A with those from the CAS. Explain any differences in the answers.
- Complete these exponent laws.

  Multiplying powers: $\quad a^m \times a^n = a^?$

  Power of a power: $\quad (a^m)^n = a^?$

  Dividing powers: $\quad a^m \div a^n = a^?$

Reflect

- How does the CAS display powers with negative exponents? Why do you think it displays them that way?
- Suppose you forget the exponent laws or are not sure how to apply them. What strategies can you use to help?

6.3 Laws of Exponents

Connect the Ideas

Definitions of integer exponents

a^{-n} is the reciprocal of a^n.

The definition of a power depends on whether the exponent is a positive integer, zero, or a negative integer.

- Positive integer exponent $\quad a^n = \underbrace{a \times a \times a \times \cdots \times a}_{n \text{ factors}}$

- Zero exponent $\quad a^0 = 1, a \neq 0$

- Negative integer exponent $\quad a^{-n} = \frac{1}{a^n}, a \neq 0$

Laws of exponents

The definitions of integer exponents lead to general rules for working with exponents.

The exponent laws apply to numerical and variable bases. When the base is a variable, we assume that it is not 0.

Laws of exponents
- Multiplication law $\quad a^m \times a^n = a^{m+n}$
- Division law $\quad a^m \div a^n = a^{m-n}, a \neq 0$
- Power of a power law $\quad (a^m)^n = a^{mn}$

m and n are any integer.

The laws can be used to evaluate numerical expressions and to simplify algebraic expressions.

Example 1

Applying the Laws of Exponents

Simplify. Evaluate where possible.

a) $5^4 \times 5^{-2}$

b) $\dfrac{(-6)^2}{(-6)^{-1}}$

c) $(m^5)^{-3}$

d) $\dfrac{a^2 a^{-5}}{(a^{-2})^3}$ $\quad$ $a^2 a^{-5}$ means $a^2 \times a^{-5}$.

Solution

a) $5^4 \times 5^{-2} = 5^{4 + (-2)}$
$= 5^2$
$= 25$

b) $\dfrac{(-6)^2}{(-6)^{-1}} = (-6)^{2 - (-1)}$
$= (-6)^{2+1}$
$= (-6)^3$
$= -216$

By convention, a simplified algebraic expression contains only positive exponents. So, we write m^{-15} as $\dfrac{1}{m^{15}}$.

c) $(m^5)^{-3} = m^{5 \times (-3)}$
$= m^{-15}$
$= \dfrac{1}{m^{15}}$

d) $\dfrac{a^2 a^{-5}}{(a^{-2})^3} = \dfrac{a^{2 + (-5)}}{a^{-2 \times 3}}$
$= \dfrac{a^{-3}}{a^{-6}}$
$= a^{-3 - (-6)}$
$= a^3$

Example 2 Using Exponents in an Application

Materials
- scientific calculator

The number of hybrid vehicles sold in the United States, S, can be modelled by the formula $S = 199\,148(2.39)^n$, where n is the number of years since 2005.

a) Evaluate $S = 199\,148(2.39)^n$ when $n = 0$. What does the answer represent?
b) Estimate the number of hybrid vehicles sold in 2004.
c) Predict the number of hybrid vehicles that will be sold in 2007.

> This article is an excerpt of a *CBS News* article from May 4, 2006.
>
> **Hybrid Vehicle Sales More than Double**
> Registrations in the United States for new hybrid vehicles rose to 199 148 in 2005, a 139% increase from the year before...

> The model assumes that hybrid car sales will continue to grow at the rate of increase given in the article.

Solution

a) $S = 199\,148(2.39)^0$
   $= 199\,148$

   $(2.39)^0 = 1$

   This represents the number of vehicles sold in 2005.

b) Substitute $n = -1$ in $S = 199\,148(2.39)^n$.
   $S = 199\,148(2.39)^{-1}$
   $\doteq 83\,326$

   2004 is 1 year before 2005.

   About 83 000 hybrids were sold in 2004.

c) Substitute $n = 2$ in $S = 199\,148(2.39)^n$.
   $S = 199\,148(2.39)^2$
   $\doteq 1\,137\,553$

   2007 is 2 years after 2005.

   If the rate of growth given in the article continues, more than 1 million hybrids will be sold in 2007.

> A zero exponent represents an initial value. Positive exponents represent going forward in time. Negative exponents represent going back in time.

6.3 Laws of Exponents   **361**

Example 3 — Simplifying Expressions

Evaluate each expression for $a = 1$, $b = -2$, and $c = 3$.

a) $(a^{-2}b)(a^3b^4)$

b) $\dfrac{a^{-4}b^5c^2}{ab^3c}$

c) $(a^5b^2)^3$

d) $(2a^2b)^5$

Solution

Simplify first, then evaluate.

a) $(a^{-2}b)(a^3b^4) = a^{-2+3}b^{1+4}$
$= a^1 b^5$
$= (1^1)(-2)^5$
$= -32$

b) $\dfrac{a^{-4}b^5c^2}{ab^3c} = a^{-4-1}b^{5-3}c^{2-1}$
$= a^{-5}b^2c^1$
$= \dfrac{b^2c}{a^5}$ ($c = c^1$)
$= \dfrac{(-2)^2(3)}{(1)^5}$
$= 12$

c) $(a^5b^2)^3 = a^{5\times 3}b^{2\times 3}$
$= a^{15}b^6$
$= (1)^{15}(-2)^6$
$= 64$

d) $(2a^2b)^5 = 2^{1\times 5}a^{2\times 5}b^{1\times 5}$
$= 2^5 a^{10} b^5$
$= 32(1)^{10}(-2)^5$
$= -1024$

Practice

A

■ For help with questions 1 to 3, see Example 1.

A simplified expression contains only positive exponents.

1. Simplify, but do not evaluate.

a) $2^3 \times 2^4$

b) $3^1 \times 3^{-4}$

c) $(1.05)^{-3} \times (1.05)^4$

d) $c^5 c^4$

e) $\left(\dfrac{1}{2}\right)^3 \times \left(\dfrac{1}{2}\right)^5$

f) $a^4\, a^{-2} a$

2. Simplify, but do not evaluate.

a) $4^5 \div 4^2$

b) $\dfrac{5^3}{5^7}$

c) $(1.02)^{13} \div (1.02)^{10}$

d) $\dfrac{d^5}{d}$

e) $(-3)^7 \div (-3)^{-4}$

f) $\dfrac{h^{30}}{h^{20}}$

3. Simplify, but do not evaluate.

a) $(5^3)^2$

b) $(3^{-2})^5$

c) $[(-2)^{-4}]^3$

d) $(m^5)^4$

e) $(r^{-10})^{-2}$

f) $(a^3)^3$

4. Evaluate without a calculator.
   a) 10^4
   b) 9^0
   c) 3^{-2}
   d) 2^{-3}
   e) $\left(\dfrac{2}{3}\right)^2$
   f) $\left(\dfrac{1}{5}\right)^{-2}$

5. Evaluate.
   a) 3^9
   b) 4^{-2}
   c) $(-4)^{-2}$
   d) -2^4
   e) 0.5^{-2}
   f) $\left(\dfrac{2}{5}\right)^3$
   g) 1.05^{27}
   h) $(-1)^{55}$

   Which expressions could you evaluate without a calculator? Explain.

6. Simplify each expression.
   Which exponent laws did you use?
   a) $d^5 d^{-2}$
   b) $(x^{-5})^2$
   c) $\dfrac{c^{11}}{c^{-3}}$
   d) $\left(\dfrac{1}{z^3}\right)^{-6}$
   e) $n^4 n^{-13} n^7$
   f) $w^{-8}(w^3)^2$
   g) $\dfrac{s^5 s^4}{s^{-3}}$
   h) $\dfrac{(t^4)^{-5}}{t^6}$

■ For help with questions 7 and 8, see Example 2.

7. Evaluate $N = 400(2)^n$ for each value of n.
   a) $n = 3$
   b) $n = 0$
   c) $n = -3$

8. Computer power has been doubling approximately every 2 years as more and smaller transistors have been integrated to build better computer chips. The number of transistors, T, in a chip has increased according to $T = 4500\,(1.4)^n$, where n is the number of years since 1974. Determine the number of transistors in a computer chip in each year.
   a) 1974
   b) 1972
   c) 2002

For help with questions 9 to 11, see Example 3.

9. Evaluate for $x = 2$ and $y = -3$ without a calculator.
   a) x^{-4}
   b) 5^y
   c) x^y
   d) y^x

10. a) Substitute $x = 2$ in the expression $\dfrac{x^5 x^4}{(x^2)^3}$. Evaluate without simplifying.
   b) Simplify $\dfrac{x^5 x^4}{(x^2)^3}$, and then evaluate at $x = 2$.
   c) Compare the methods in parts a and b. Describe the advantages of each method.

11. Use the CAS calculator screen below.

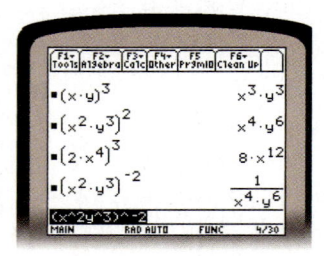

   a) How are the exponents of the original expressions related to the exponents of the simplified expressions?
   b) Explain the relationship by writing the original expressions in expanded form and simplifying.
   c) Complete the law that generalizes the pattern: $(a \times b)^n = a^? b^?$
   d) Simplify.
      i) $(2f)^4$
      ii) $(a^3 b)^4$
      iii) $(s^{-3} v^4)^5$
      iv) $(5h)^{-2}$

12. Evaluate for $x = 2$, $y = -3$, and $z = 5$.
   a) $x^2 y^4 x^3 y^{-2}$
   b) $\dfrac{x^3 y^3 z}{x y^4 z^{-2}}$
   c) $\dfrac{(5x)^2 (2y)^3}{10xy^2}$
   d) $(xyz)^4 x^{-5} y^7 z^{-5}$

13. Assessment Focus
   a) Evaluate $\dfrac{a^2 b^5 c^5}{ab^{-3} c^4}$ for $a = 6$, $b = 2$, and $c = -10$. Explain your method.
   b) Terry rewrote $(5r)^3$ as $5r^3$ and $5r^{-2}$ as $\dfrac{1}{5r^2}$ on a test. Explain the mistakes Terry made. What strategies might Terry use to help him avoid making these mistakes in the future?

364 CHAPTER 6: Algebraic Models

14. The formula $V = \pi r^2 h$ gives the volume, V, of a cylinder with radius r and height h.
    a) Determine the volume of a cylindrical gift tube with radius $2x$ and height $7x$.
    b) Calculate the volume of the gift tubes when $x = 5$ cm and $x = 12$ cm.

15. **Literacy in Math** An excerpt of a *CBS News* article from May 4, 2006 is shown at the right.
    a) Explain the phrase "has grown exponentially."
    b) What quantities can you calculate from the given information?
    c) Explain why the numbers in the second sentence are reasonably consistent with each other.
    d) In *Example 2*, the 2004 sales estimate was 83 326 hybrids. Is this inconsistent with the estimate given in the article? Explain.

    > **Hybrid Vehicle Sales More than Double**
    > Hybrids accounted for 1.2 percent of the 16.99 million vehicles sold in the United States last year. In 2004, the 83 153 hybrids sold were 0.5 percent of the 16.91 million vehicles sold. The U.S. hybrid market has grown exponentially since 2000, when 7781 were sold.

C

16. Refer to *Example 2* and question 15.
    a) Show that hybrid sales did not increase by 139% each year between 2000 and 2004.
    b) Estimate the actual average growth rate between 2000 and 2004. Explain your method.

17. The formula $A = P(1 + i)^n$ gives the amount, A dollars, of a compound interest investment. In the formula, P dollars is the principal invested, i is the annual interest rate as a decimal, and n is the number of years.
    a) Rearrange the formula to isolate P.
    b) Rewrite the formula in part a using a negative exponent.
    c) Evaluate the formulas in parts a and b for P when $A = \$1000$, $i = 6\%$, and $n = 5$ years. Which formula did you find easier to evaluate? Explain.

In Your Own Words

What are some mistakes you have made when working with exponents? Why do you think you made these mistakes? How might you avoid making these mistakes in the future? Include examples in your explanation.

6.3 Laws of Exponents

6.4 Patterns in Exponents

Coffee, tea, cola, and chocolate each contain caffeine. The formula $P = 100(0.87)^n$ models the percent, P, of caffeine left in your body n hours after you drink a caffeinated beverage. After half an hour, the percent of caffeine remaining in your body is given by the equation $P = 100(0.87)^{\frac{1}{2}}$.

Inquire: Exploring Rational Exponents

Materials
- TI-83 or TI-84 graphing calculator

An exponent that can be written as a fraction of integers is a **rational exponent**.

Work with a partner.

Part A: Exploring the Meaning of $a^{\frac{1}{n}}$

1. The expressions in the table use the exponents 2, −2, and $\frac{1}{2}$.

$1^2 = 1$	$1^{-2} = 1$	$1^{\frac{1}{2}} = 1$
$2^2 = 4$	$2^{-2} = \frac{1}{4}$	$4^{\frac{1}{2}} = 2$
$3^2 = 9$	$3^{-2} = \frac{1}{9}$	$9^{\frac{1}{2}} = 3$
$4^2 = 16$	$4^{-2} = \frac{1}{16}$	$16^{\frac{1}{2}} = 4$

   a) Determine the next 3 rows in the table. Explain your reasoning.

   b) Compare the numbers in the first and second columns. Describe any relationships you see. What does it mean to raise a number to the exponent 2? To the exponent −2?

   c) Think of any number, a. What can you say about the value of $a^{\frac{1}{2}}$?

366 CHAPTER 6: Algebraic Models

d) Compare the numbers in the first and third columns. Notice that the exponent $\frac{1}{2}$ appears to "undo," or reverse, the exponent 2.

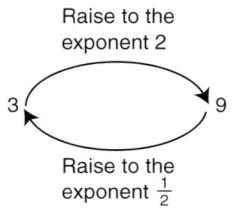

What do you think it means to raise a number to the exponent $\frac{1}{2}$? Confirm your answer by trying other examples on your calculator.

2. The patterns in the table use the exponents 3, −3, and $\frac{1}{3}$.

$1^3 = 1$	$1^{-3} = 1$	$1^{\frac{1}{3}} = 1$
$2^3 = 8$	$2^{-3} = \frac{1}{8}$	$8^{\frac{1}{3}} = 2$
$3^3 = 27$	$3^{-3} = \frac{1}{27}$	$27^{\frac{1}{3}} = 3$
$4^3 = 64$	$4^{-3} = \frac{1}{64}$	$64^{\frac{1}{3}} = 4$

a) Complete the next three lines in each pattern. Explain your reasoning.

b) Compare the numbers in the first and second columns. Describe any relationships you see. What does it mean to raise a number to the exponent 3? To the exponent −3?

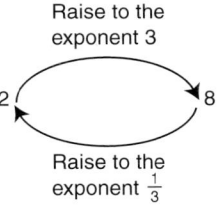

c) Compare the numbers in the first and third columns. Notice that the exponent $\frac{1}{3}$ appears to "undo" or reverse the exponent 3. What do you think it means to raise a number to the exponent $\frac{1}{3}$? Confirm your answer by trying other examples on your calculator.

3. a) You have explored the meaning of $a^{\frac{1}{2}}$ and $a^{\frac{1}{3}}$. What do you think $a^{\frac{1}{4}}$ and $a^{\frac{1}{5}}$ mean? Use a calculator to test your predictions.

b) How would you define $a^{\frac{1}{n}}$? Explain your reasoning.

6.4 Patterns in Exponents   **367**

a	$a^{\frac{1}{2}}$	$a^{\frac{3}{2}}$	$a^{\frac{5}{2}}$
1	1		
4	2		
9	3		
16	4		

Brackets are needed around the exponent so that the calculator evaluates $4^{\frac{3}{2}}$, not $4^3 \div 2$.

Part B: Exploring the Meaning of $a^{\frac{m}{n}}$

4. a) Copy the table.

b)
> Use your graphing calculator to complete the third column of the table. For example, to determine $4^{\frac{3}{2}}$, press:
> 4 [^] [(] 3 [÷] 2 [)] [ENTER]
>
> 4^(3/2)
>   8

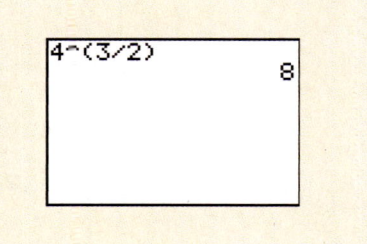

c) Compare the numbers in corresponding rows of the second and third columns of the table.
How do the values of $a^{\frac{3}{2}}$ appear to be related to the values of $a^{\frac{1}{2}}$? Explain.

d) We can think of the exponent $\frac{3}{2}$ as the product $\frac{1}{2} \times 3$.

Explain why this allows us to rewrite $4^{\frac{3}{2}}$ as $(4^{\frac{1}{2}})^3$.
Evaluate each expression to show that they produce the same result.
How does this explain the relationship in part c?

e) How do you think the values of $a^{\frac{5}{2}}$ will be related to the values of $a^{\frac{1}{2}}$? Explain your reasoning. Use a calculator to complete the fourth column of the table. Were you correct? Explain.

a	$a^{\frac{1}{3}}$	$a^{\frac{2}{3}}$	$a^{\frac{5}{3}}$
1	1		
8	2		
27	3		
64	4		

5. Copy the table.

a) What do you think $a^{\frac{2}{3}}$ and $a^{\frac{5}{3}}$ mean? Explain your reasoning.

b) How do you think the values of $a^{\frac{2}{3}}$ and $a^{\frac{5}{3}}$ will be related to the value of $a^{\frac{1}{3}}$? Justify your answer.

c) Use a calculator to complete the table.
Were you correct in part b? Explain.

6. Use the results of questions 4 and 5.
How do you think $a^{\frac{m}{n}}$ is defined?
Explain your reasoning.

368   CHAPTER 6: Algebraic Models

Practice

A

1. Explain the meaning of the exponent in each expression.
  a) 8^3
  b) 8^{-3}
  c) $8^{\frac{1}{3}}$
  d) $8^{\frac{2}{3}}$

2. Evaluate each expression without using a calculator.
  a) $9^{\frac{1}{2}}$
  b) $49^{\frac{1}{2}}$
  c) $64^{\frac{1}{2}}$
  d) $27^{\frac{1}{3}}$
  e) $(-8)^{\frac{1}{3}}$
  f) $1000^{\frac{1}{3}}$

  How do you know your answers are correct?

3. a) Evaluate.
  i) $25^{\frac{1}{2}}$
  ii) $25^{\frac{2}{2}}$
  iii) $25^{\frac{3}{2}}$
  iv) $25^{\frac{4}{2}}$
  v) $25^{\frac{5}{2}}$

  b) What pattern do you notice in the answers? Explain.
  c) Write, then evaluate the next 3 terms in the pattern. Justify your answers.

4. a) Explain why $100^{\frac{1}{2}} = 10$.
  b) How will the values of $100^{\frac{3}{2}}$, $100^{\frac{5}{2}}$, and $100^{\frac{7}{2}}$ be related to the value of $100^{\frac{1}{2}}$?
  c) Use a calculator to determine the value of $100^{\frac{3}{2}}$, $100^{\frac{5}{2}}$, and $100^{\frac{7}{2}}$.
     Were you correct in part b? Explain.

5. Rewrite using radicals and evaluate without a calculator.
  a) $32^{\frac{1}{5}}$
  b) $81^{\frac{1}{4}}$
  c) $16^{\frac{3}{2}}$
  d) $9^{\frac{5}{2}}$
  e) $100^{\frac{3}{2}}$
  f) $16^{\frac{3}{4}}$
  g) $8^{\frac{4}{3}}$
  h) $27^{\frac{3}{3}}$

  How do you know that your answers are correct?

6.4 Patterns in Exponents

6. Use the table of values and graph of $y = 2^x$ shown here.

   a) Explain why the value of $2^{\frac{1}{2}}$ must be between 1 and 2.

   b) Use the graph to estimate the value of $2^{\frac{1}{2}}$ to the nearest tenth.

   c) Which two whole numbers is $2^{\frac{3}{2}}$ between? Repeat for $2^{\frac{5}{2}}$ and $2^{\frac{7}{2}}$.

   d) Estimate the value of $2^{\frac{3}{2}}$, $2^{\frac{5}{2}}$, and $2^{\frac{7}{2}}$ to the nearest tenth.

x	$y = 2^x$
0	1
1	2
2	4
3	8
4	16

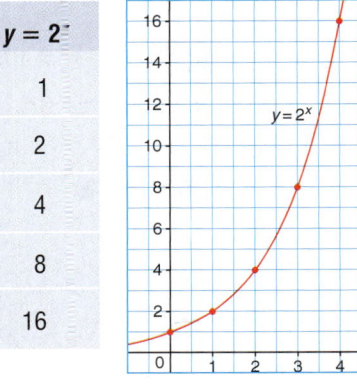

7. The equation $P = 100(0.87)^x$ models the percent, P, of caffeine left in your body x hours after you consume it. Determine the value of P after each time.

   a) $\frac{1}{2}$ h

   b) $\frac{3}{2}$ h

   c) 40 min

   How do you know your answers are reasonable?

Reflect

- In the power, $x^{\frac{m}{n}}$, what does the numerator represent? What does the denominator represent? Explain.

- What steps do you take to evaluate the power $x^{\frac{m}{n}}$? Use examples in your explanation.

Mid-Chapter Review

6.1

1. The area, A, of a diamond shape with diagonal lengths d and D is $A = \frac{1}{2}dD$. Find the area of a diamond with each of these diagonal lengths.
  a) $d = 4$ m, $D = 3$ m
  b) $d = 47$ cm, $D = 68$ cm

2. During aerobic exercise, the maximum desirable heart rate, h beats per minute, is given by the formula $h = 198 - 0.9a$, where a is the person's age in years. Determine your maximum desirable heart rate.

3. Zan is planning to waterproof a rectangular driveway that is 12 m long and 5.5 m wide.
  a) What is the area of the driveway?
  b) One can of waterproofing sealer costs $15.99 and covers an area of 30 m². How much will it cost Zan to waterproof the driveway?

4. The area, A, of a triangle with side lengths a, b, and c is given by the formula $A = \sqrt{s(s-a)(s-b)(s-c)}$, where $s = \frac{a+b+c}{2}$. The sides of a triangular plot of land measure 500 m, 750 m, and 1050 m. Land is priced at $5400 per hectare (1 ha = 10 000 m²). Determine the value of the plot.

6.2

5. a) Describe the steps to rearrange the equation $y = 3x + 5$ to isolate x. Use an arrow diagram to determine the inverse operations needed.
  b) Isolate x.

6. A car accelerates away from a stop sign. The formula $d = \frac{1}{2}at^2$ gives the distance, d metres, that the car travels in t seconds at an acceleration of a metres per second squared.
  a) Find d when $a = 2$ m/s² and $t = 15$ s.
  b) Find a when $d = 100$ m and $t = 5$ s.
  c) Find t when $a = 0.01$ m/s² and $d = 20.48$ m.

7. The formula $I = Prt$ gives the simple interest, I dollars, earned on a principal of P dollars invested at an annual interest rate of r percent for t years.
An investment of $1000 earns $131.25 interest in 2.5 years. What annual rate of interest was paid?

6.3

8. Evaluate without a calculator.
  a) $(-3)^2$ b) -3^2 c) -3^{-2} d) $(-3)^{-2}$

9. Simplify.
  a) $p^4 \times p^{-2}$ b) $p^3 \div p^8$ c) $(p^{-2})^5$

10. Simplify and evaluate for $x = -3$, $y = 4$, and $z = 5$.
  a) $x^7 y^{-2} x^3$ b) $\frac{x^5 y^2 z^3}{x^{-1} y^0 z}$ c) $(2x^3)^2$

11. a) Write the next three terms in the pattern. Describe the pattern.
$$4^0 \quad 4^{\frac{1}{2}} \quad 4^1 \quad 4^{\frac{3}{2}} \quad 4^2$$
  b) Evaluate each power in the pattern as a whole number or a fraction. Describe the pattern in the answers.

12. Evaluate without a calculator.
  a) $16^{\frac{1}{2}}$ b) $64^{\frac{1}{3}}$ c) $25^{\frac{3}{2}}$

Mid-Chapter Review **371**

6.5 Rational Exponents

Many formulas in biology involve rational exponents. The formula $v = 0.783 \left(\dfrac{s^{10}}{h^7} \right)^{\frac{1}{6}}$ approximately relates an animal's speed, v metres per second, to its stride length, s metres, and its hip height, h metres.

Investigate

Calculating the Speed of a Dinosaur

Materials
- scientific calculator

Paleontologists use measurements from fossilized dinosaur tracks to estimate the speed at which the dinosaur travelled.
- The stride length, s, of a dinosaur is the distance between successive footprints of the same foot.
- The hip height, h, of a dinosaur is about 4 times the foot length, f.

Work with a partner.

Use the measurements on the diagram.
■ Estimate the speed of the dinosaur.
  Use the formula $v = 0.783 \left(\dfrac{s^{10}}{h^7} \right)^{\frac{1}{6}}$.

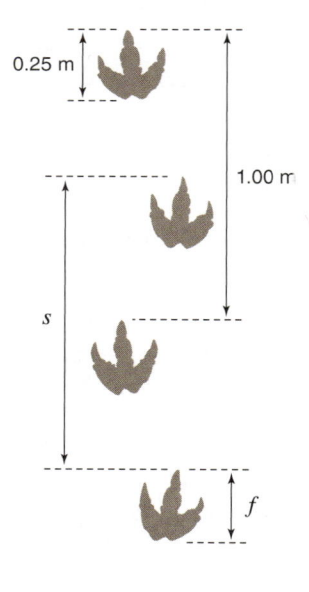

Reflect

■ Describe your strategy. Explain how well your strategy worked.
■ Compare your strategy with another pair's strategy. How are they the same? How are they different?

CHAPTER 6: Algebraic Models

Connect the Ideas

Definition of $a^{\frac{1}{n}}$

You explored the meaning of rational exponents in Lesson 6.4. Mathematicians chose these meanings by extending the exponent laws to rational exponents.

Extending the exponent law $(a^m)^n = a^{mn}$ to include rational exponents gives:

$$(6^{\frac{1}{2}})^2 = 6^{\frac{1}{2} \times 2} \qquad (-6^{\frac{1}{2}})^2 = [(-1)^1 \times 6^{\frac{1}{2}}]^2$$
$$= 6^1 \qquad\qquad\qquad = (-1)^{1 \times 2} \times 6^{\frac{1}{2} \times 2}$$
$$= 6 \qquad\qquad\qquad\quad = (-1)^2 \times 6^1$$
$$\qquad\qquad\qquad\qquad\qquad = 6$$

$\sqrt{n}$ is the positive root of n, so $\sqrt{36} = 6$.

But: $(\sqrt{6})^2 = 6$ and $(-\sqrt{6})^2 = 6$

So, mathematicians defined: $6^{\frac{1}{2}} = \sqrt{6}$ and $-6^{\frac{1}{2}} = -\sqrt{6}$.

They also defined $6^{\frac{1}{3}} = \sqrt[3]{6}$, $6^{\frac{1}{4}} = \sqrt[4]{6}$, and so on.

> **Definition of $a^{\frac{1}{n}}$**
>
> $a^{\frac{1}{n}}$ is the nth root of a. That is, $a^{\frac{1}{n}} = \sqrt[n]{a}$.
>
> n is a natural number.
>
> $a \geq 0$ if n is even.

A natural number is any number in the set 1, 2, 3, ...

Definition of $a^{\frac{m}{n}}$

The expression $a^{\frac{m}{n}}$ can be interpreted in two ways.

- $a^{\frac{m}{n}} = (a^{\frac{1}{n}})^m = (\sqrt[n]{a})^m$

  Take the nth root of a, then raise the result to the exponent m.

  For example, $8^{\frac{2}{3}} = (\sqrt[3]{8})^2 = (2)^2 = 4$

- $a^{\frac{m}{n}} = (a^m)^{\frac{1}{n}} = \sqrt[n]{a^m}$

  Raise a to the exponent m, then take the nth root.

  For example, $8^{\frac{2}{3}} = \sqrt[3]{8^2} = \sqrt[3]{64} = 4$

> **Definition of $a^{\frac{m}{n}}$**
>
> $a^{\frac{m}{n}} = (\sqrt[n]{a})^m = \sqrt[n]{a^m}$
>
> m is an integer.
>
> n is a natural number.
>
> $a \geq 0$ if n is even.

6.5 Rational Exponents

Example 1

Evaluating Powers with Rational Exponents

Evaluate without a calculator.

a) $49^{\frac{1}{2}}$ b) $(-64)^{\frac{1}{3}}$ c) $32^{\frac{4}{5}}$ d) $0.04^{\frac{3}{2}}$

Solution

Rewrite each expression in radical form.

a) $49^{\frac{1}{2}} = \sqrt{49}$
$= 7$ since $7^2 = 49$

b) $(-64)^{\frac{1}{3}} = \sqrt[3]{-64}$
$= -4$ since $(-4)^3 = -64$

c) $32^{\frac{4}{5}} = (32^{\frac{1}{5}})^4$
$= (\sqrt[5]{32})^4$
$= 2^4$
$= 16$

d) $0.04^{\frac{3}{2}} = (0.04^{\frac{1}{2}})^3$
$= (\sqrt{0.04})^3$
$= 0.2^3$
$= 0.008$

Using rational exponents to solve equations

Rational exponents are useful for solving equations involving powers. For example, take both sides of the equation $x^3 = 125$ to the power $\frac{1}{3}$ to find the solution $x = 5$.

Example 2

Solving for the Base in a Power

Solve for x. Assume x is positive.

a) $x^4 = 16$ b) $x^{\frac{3}{2}} = 27$

Solution

Use inverse operations to "undo" the exponents.

a) $x^4 = 16$ Raise both sides to the exponent $\frac{1}{4}$.

$(x^4)^{\frac{1}{4}} = 16^{\frac{1}{4}}$

$x = \sqrt[4]{16}$
$= 2$

> By the power of a power rule,
> $(x^4)^{\frac{1}{4}} = x^{4 \times \frac{1}{4}} = x^1 = x.$

To check, substitute $x = 2$ in $x^4 = 16$.

L.S.	R.S.
$(2)^4 = 16$	16

L.S. = R.S., so the solution is correct.

b) $x^{\frac{3}{2}} = 27$ Raise both sides to the exponent $\frac{2}{3}$.

$(x^{\frac{3}{2}})^{\frac{2}{3}} = 27^{\frac{2}{3}}$

$x = (\sqrt[3]{27})^2$
$= 3^2$
$= 9$

> By the power of a power rule,
> $(x^{\frac{3}{2}})^{\frac{2}{3}} = x^{\frac{3}{2} \times \frac{2}{3}} = x^1 = x.$

374 CHAPTER 6: Algebraic Models

Example 3

Materials
- scientific calculator

Solving a Financial Problem

Under annual compounding, a principal of $700 grows to $900 in 5 years. Determine the annual interest rate.

Solution

Use the formula for compound interest: $A = P(1 + i)^n$. Substitute: $A = 900$, $P = 700$, and $n = 5$ to obtain
$$900 = 700(1 + i)^5$$
Draw an arrow diagram to determine how to isolate i.

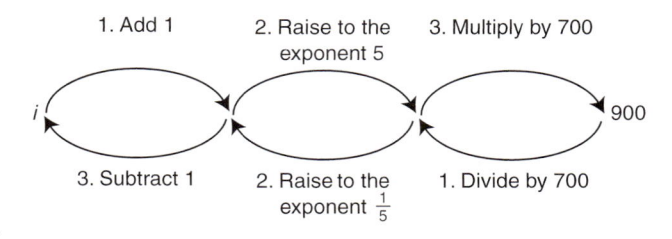

1. Add 1   2. Raise to the exponent 5   3. Multiply by 700

i → → → 900

3. Subtract 1   2. Raise to the exponent $\frac{1}{5}$ 1. Divide by 700

$900 = 700(1 + i)^5$ Divide each side by 700.

$\dfrac{900}{700} = (1 + i)^5$ Raise each side to the exponent $\frac{1}{5}$.

$\left(\dfrac{900}{700}\right)^{\frac{1}{5}} = (1 + i)$ Evaluate the left side.

$1.0515 \doteq 1 + i$ Subtract 1 from each side.

$0.0515 \doteq i$

The interest rate is approximately 5.15%.

6.5 Rational Exponents **375**

Practice

A

For help with questions 1, 2, and 6, see Example 1.

1. Determine each value without using a calculator.
   a) $36^{\frac{1}{2}}$
   b) $81^{\frac{1}{2}}$
   c) $144^{\frac{1}{2}}$
   d) $0.25^{\frac{1}{2}}$

2. Determine each value without using a calculator.
   a) $8^{\frac{1}{3}}$
   b) $64^{\frac{1}{3}}$
   c) $0.027^{\frac{1}{3}}$
   d) $(-125)^{\frac{1}{3}}$

3. Rewrite each expression using rational exponents.
   a) $\sqrt{64}$
   b) $\sqrt{1.21}$
   c) $\sqrt[3]{216}$
   d) $\sqrt[3]{-343}$

4. Determine the value of each expression in question 3.

5. Determine the value of each expression.
   a) $\sqrt[4]{16}$
   b) $\sqrt[4]{0.0256}$
   c) $\sqrt[5]{-243}$
   d) $\sqrt[6]{64}$

B

6. Write each expression in radical form, and then evaluate without a calculator.
   a) $243^{\frac{1}{5}}$
   b) $9^{\frac{3}{2}}$
   c) $8^{\frac{5}{3}}$
   d) $81^{\frac{3}{4}}$
   e) $0.0625^{\frac{1}{4}}$
   f) $(-32)^{\frac{3}{5}}$
   g) $0.01^{\frac{3}{2}}$
   h) $(-27)^{\frac{4}{3}}$

7. The expression $a^{\frac{m}{n}}$ can be interpreted as $(\sqrt[n]{a})^m$ or $\sqrt[n]{a^m}$.
   a) Evaluate $16^{\frac{3}{2}}$ as $(\sqrt{16})^3$.
   b) Evaluate $16^{\frac{3}{2}}$ as $\sqrt{16^3}$.
   c) Which form did you find easier to evaluate? Explain.

8. Scientists use the formula $D = 0.099M^{\frac{9}{10}}$ to give the drinking rate, D litres per day, for a mammal with mass M kilograms.
   a) Rewrite the equation using radicals.
   b) Determine the drinking rate of each mammal.
      i) a 35-kg dog
      ii) a 520-kg moose
      iii) a 28-g mouse

9. John and Maria are comparing their solutions to the equation $x^3 = 8$. Whose solution is correct? Justify your answer.

John's solution
$x^3 = 8$
To undo the exponent 3, raise each side to the exponent −3.
$(x^3)^{-3} = 8^{-3}$
$x = \frac{1}{8^3}$
$x = \frac{1}{512}$

Maria's solution
$x^3 = 8$
To undo the exponent 3, raise each side to the exponent $\frac{1}{3}$.
$(x^3)^{\frac{1}{3}} = 8^{\frac{1}{3}}$
$x = \sqrt[3]{8}$
$x = 2$

10. Solve for x. Assume x is positive. How do you know that your answers are correct?
    a) $x^{\frac{1}{2}} = 7$
    b) $x^2 = 9$
    c) $x^3 = 64$
    d) $x^{\frac{3}{2}} = 8$
    e) $x^3 = \frac{27}{64}$
    f) $x^{\frac{4}{3}} = 625$

11. Determine the annual interest rate needed to double the value of a $500 investment in 7 years. Assume that the interest is compounded annually.

12. Honeybees came to North America in the early 1600s with English settlers. In one region, the area, A hectares, inhabited by honeybees increased according to the formula $A = 0.5(9)^{\frac{t}{2}}$, where t is the number of years since introduction.
    a) Determine the area inhabited by honeybees after 1 year.
    b) Determine the area inhabited by honeybees after 3 years.

13. The equation $V = \frac{1}{3}\pi r^3$ gives the volume, V, of a cone whose height and base radius, r, are equal. Determine the radius of the cone if its volume is 1000 cm³.

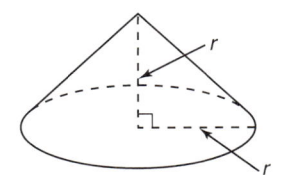

6.5 Rational Exponents

14. **Assessment Focus** The brain mass and body mass of mammals are approximately related by the formula $b = 0.011 m^{\frac{2}{3}}$. In the formula, b is the brain mass in kilograms and m is the body mass in kilograms.
   a) Determine the brain mass of a 512-kg giraffe.
   b) Determine the brain mass of a 420-g chinchilla.
   c) The brain mass of a cat is about 0.025 kg. Determine its body mass. Explain your strategy.

15. **Literacy in Math** Use a Frayer Model or a graphic organizer of your choice. Explain what a rational exponent is and how to simplify an expression involving a rational exponent. Use examples in your explanation.

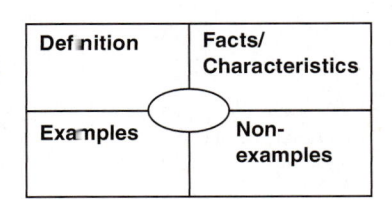

Definition	Facts/Characteristics
Examples	Non-examples

16. The formula $P = \frac{r^2 s^3}{2}$ gives the approximate power, P watts, generated by a wind turbine with radius r metres when the wind speed is s metres per second.
   a) Rearrange the formula to isolate r. Give your answer in rational exponent form and radical form.
   b) Repeat part a for s.

17. The speed, s metres per second, at which a liquid flows from a small hole in a container is given by the formula $s \doteq (19.6 h)^{\frac{1}{2}}$, where h metres is the height of the liquid above the hole.
   a) Determine the speed at which the liquid flows when the liquid is 1.0 m above the hole.
   b) What height corresponds to a flow speed of 2 m/s? Round your answer to the nearest centimetre.

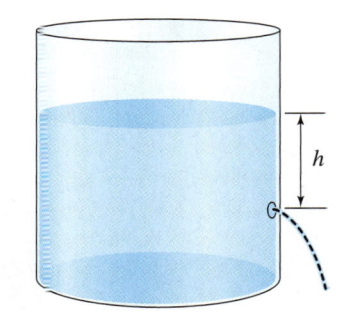

In Your Own Words

Explain what a rational exponent represents.
Describe how rational exponents can help you solve equations.
Explain how to decide which rational exponent to use in solving the equation. Include examples in your explanation.

378 CHAPTER 6: Algebraic Models

Power Dominoes

Materials

- 15 power domino tiles

Play with a partner.

■ Shuffle the power domino tiles. Spread them out face down.

■ Each player takes seven tiles. Turn the remaining tile face up.

■ Players take turns matching an end of one of their tiles to an end of a tile on the table. Tile ends match if they simplify to the same expression. For example, $a^2 a^{-4}$ matches $\frac{a^3}{a^5}$ since both expressions simplify to a^{-2}.

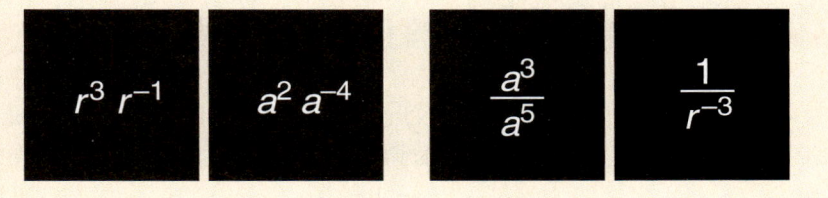

| $r^3 \, r^{-1}$ | $a^2 \, a^{-4}$ | $\dfrac{a^3}{a^5}$ | $\dfrac{1}{r^{-3}}$ |

■ If a player cannot make a match or makes an incorrect match, play passes to the other player.

■ The player who uses all seven tiles first wins.

Reflect

■ What is a pair of expressions that does not simplify to the same expression? Tell how you know.

6.6 Exponential Equations

Salmonella is a bacterium that causes food poisoning. Under favourable conditions, it takes 1 salmonella bacterium about 20 min to divide into 2 new salmonella.

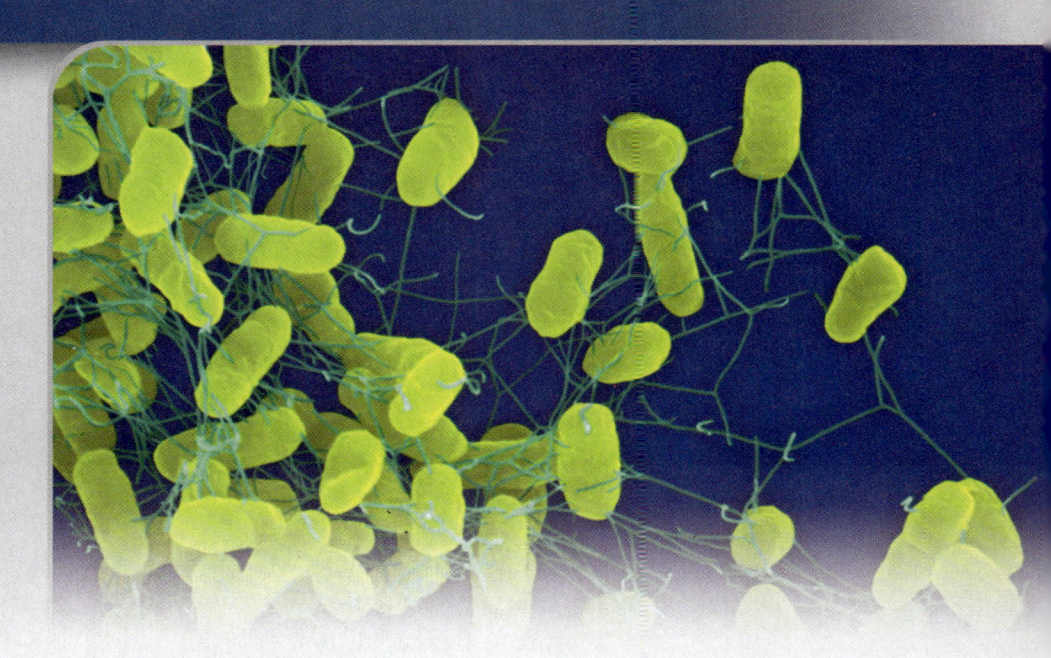

Investigate

Solving an Exponential Equation

Materials
- grid paper or graphing calculator
- scientific calculator

> In an exponential equation, the unknown is an exponent.

A lab technician starts with 1 salmonella bacterium. She uses the equation $P = 2^{3t}$ to model the population, P, of salmonella after t hours. To determine when there will be 1000 salmonella, she solves the **exponential equation** $1000 = 2^{3t}$.

Work with a partner.

Three students are discussing how to solve the equation $2^{3t} = 1000$.
- Jawad suggests substituting different values for t in 2^{3t} until the correct value is obtained.
- Lily suggests using a graph of $P = 2^{3t}$.
- Max suggests isolating t by raising each side of the equation to the exponent $\frac{1}{3}$.

Will each of these strategies work?
Solve for t using each strategy that works.
Explain why the other strategy or strategies will not work.

380 CHAPTER 6: Algebraic Models

Reflect

Compare the exponential equation $2^{3t} = 1000$ to the equations in *Example 2* of Lesson 6.5.
- How are the equations the same? How are they different?
- Is it possible to use the same strategy to solve both types of equations? Justify your answer.

Connect the Ideas

An exponential equation is an equation that contains a variable in the exponent. Some examples of exponential equations are:

$2^x = 32$ $\qquad\qquad 9^{x+1} = 27^x \qquad\qquad (0.8)^x = 0.18$

Without technology

Some exponential equations can be solved by writing both sides of the equation as powers of the same base. This allows us to use the following property.

> **Equality of powers with a common base**
> If $a^m = a^n$, then $m = n$ $(a > 0, a \neq 1)$

For example, since 4^x and 4^3 are both powers of 4, the solution to $4^x = 4^3$ is $x = 3$.

Example 1

Finding a Common Base

Solve.
a) $5^x = 5^6$
b) $2^x = 32$
c) $7^{3x-4} = 49$
d) $3^{5x+8} = 27^x$
e) $2^{2(x-5)} = 4^{3x-1}$

Solution

a) $5^x = 5^6$ Equate the exponents.
$\quad x = 6$

b) $2^x = 32$ Write 32 as a power of 2.
$\quad 2^x = 2^5$ Equate the exponents.
$\quad x = 5$

6.6 Exponential Equations   **381**

> To check, substitute $x = 2$ in 7^{3x-4}.
>
> | L.S. | R.S. |
> |---|---|
> | 7^{3x-4} | 49 |
> | $= 7^{3(2)-4}$ | |
> | $= 7^2$ | |
> | $= 49$ | |
>
> L.S. = R.S., so the solution is correct.

> When we take a power of a power, we multiply the exponents. So $(3^3)^x = 3^{3x}$.

c) $7^{3x-4} = 49$ Write 49 as a power of 7.
$7^{3x-4} = 7^2$ Equate the exponents.
$3x - 4 = 2$ Solve for x.
$3x = 6$
$x = 2$

d) $3^{5x+8} = 27^x$ Write 27 as a power of 3.
$3^{5x+8} = (3^3)^x$ Simplify the right side.
$3^{5x+8} = 3^{3x}$ Equate the exponents.
$5x + 8 = 3x$ Solve for x.
$8 = -2x$
$-4 = x$

e) $2^{2(x-5)} = 4^{3x-1}$ Write 4 as a power of 2.
$2^{2(x-5)} = (2^2)^{3x-1}$ Simplify each side.
$2^{2x-10} = 2^{6x-2}$ Equate the exponents.
$2x - 10 = 6x - 2$ Solve for x.
$-4x = 8$
$x = -2$

With technology

Most exponential equations cannot be easily expressed as powers of the same base. We use technology to solve these equations.

Example 2 — Using Systematic Trial

Materials
- scientific calculator

Use systematic trial to solve $3^x = 7$ to 2 decimal places.

Solution

7 is between $3^1 = 3$ and $3^2 = 9$, but closer to 9.
So, the solution to $3^x = 7$ is between 1 and 2, but closer to 2.

Try $x = 1.6$: $3^{1.6} \doteq 5.80$ (too small)
Try $x = 1.7$: $3^{1.7} \doteq 6.47$ (still too small)
Try $x = 1.8$: $3^{1.8} \doteq 7.22$ (too large)
Try $x = 1.78$: $3^{1.78} \doteq 7.07$ (still too large)
Try $x = 1.77$: $3^{1.77} \doteq 6.99$ (close enough)
So, $x \doteq 1.77$.

Example 3

Materials
- TI-83 or TI-84 graphing calculator

Using a Graph

Use a graph to solve $3^x = 7$ to 2 decimal places.

Solution

Graph $y = 3^x$ and $y = 7$ on the same screen, and determine the x-coordinate of the point of intersection.

Enter the equations.
Press [Y=].
At $Y_1 =$, press 3 [^] [X,T,Θ,n].
At $Y_2 =$, press 7.

Set the viewing window.
Press [WINDOW].
Change the settings to those shown at the right.

Graph the equations.
Press [GRAPH].
The graph of $y = 3^x$ is the curve. The graph of $y = 7$ is the horizontal line.

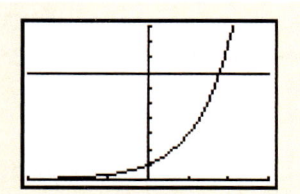

Determine the x-coordinate of the point of intersection.
- Use the INTERSECT feature in the CALC menu.
  Press [2nd] [TRACE] 5.
- At each prompt, press [ENTER].
- The x-coordinate of the point of intersection is $x \doteq 1.7712437$.

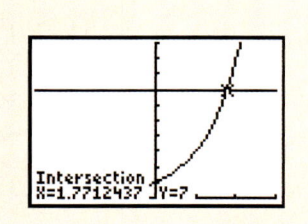

To the nearest hundredth, the solution is $x \doteq 1.77$.

Exact versus approximate solutions

Compare the answers in *Example 1* to the answer in *Examples 2* and *3*. The solutions to $2^x = 32$, $7^{3x-4} = 49$, and $3^{5x+8} = 27^x$ are exact since 32, 49, and 27 are powers of 2, 7, and 3 respectively. We can only approximate the solution of $3^x = 7$ since 7 is not a power of 3.

6.6 Exponential Equations **383**

Practice

A

For help with questions 1 and 5, see Example 1.

1. Solve each equation.
   a) $4^x = 4^3$
   b) $7^x = 7^2$
   c) $2^x = 2^7$
   d) $5^{2x} = 5^3$

2. Solve each equation.
   a) $x - 8 = 7$
   b) $4x + 1 = 9$
   c) $11 - 2x = 5 + x$
   d) $2(x - 6) = 3x$

3. Solve each equation.
   a) $3^{x+3} = 3^8$
   b) $10^{x-3} = 10^{-2}$
   c) $2^{3x} = 2^{8-x}$
   d) $6^{3x-7} = 6^{x+2}$

4. Express each number as a power.
   a) 36 as a power of 6
   b) 16 as a power of 2
   c) 125 as a power of 5
   d) 1000 as a power of 10

5. Express the right side of the equation as a power of 3, then solve the equation.
   a) $3^x = 9$
   b) $3^x = \dfrac{1}{9}$
   c) $3^{2x} = 81$
   d) $3^{x+5} = 27$

   How do you know that your answers are correct?

B

6. a) Solve $2^x = 16$. Explain your strategy.
   b) Can you use the same strategy to solve $2^x = 25$? Explain.

7. Solve each equation algebraically.
   a) $5^x = 125$
   b) $4^{2x} = 64$
   c) $2^{x+1} = 8$
   d) $6^{x-1} = 36$
   e) $7^{2x-1} = 49$
   f) $10^{1-2x} = 100$
   g) $3^{x+1} = \dfrac{1}{9}$
   h) $2^{3x+6} = 1$

8. Choose two equations from question 7. Explain the steps in the solution. Check your solution by substituting for x.

For help with question 9, see Example 2.

9. a) Use the base of the power on the left side of each equation. Between which two integer powers of the base does the solution lie? Justify your answers.
   i) $2^x = 30$
   ii) $5^x = 100$
   iii) $3^x = 75$
   iv) $2^x = \dfrac{1}{5}$

   b) Use systematic trial to solve each equation in part a. Round to 2 decimal places.

384 CHAPTER 6: Algebraic Models

10. Solve each equation algebraically.
    a) $9^{x+1} = 27^x$
    b) $4^{3x} = 32^{x-1}$
    c) $3^{2(x+2)} = 27^{x+2}$
    d) $7^{3x-5} = 49^{-x}$
    e) $100^{2x-3} = 1000^{3x+1}$
    f) $5^{2(x-5)} = 125^{x-1}$

11. Choose two equations from question 10. Explain your choice of strategy.

For help with question 12, see Example 3.

12. Use the graph of $y = 4^x$ to solve the equation $4^x = 12$. Describe the steps you used to find the solution. Verify your answer numerically.

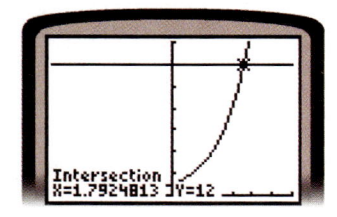

13. Determine the approximate solutions using graphing technology.
    Round to 2 decimal places.
    a) $3^x = 14$
    b) $7^x = 100$
    c) $10^x = 50$
    d) $2^x - 36 = 0$
    e) $3^{2x} = 300$
    f) $(1.06)^x = 2$
    Include a sketch of the graphing calculator screen in your solution.

14. Consider the equation $5^{2x-1} = 45$.
    a) Will the solution be exact or approximate? Justify your answer.
    b) Solve the equation. Explain your choice of strategy.

15. A strain of bacteria doubles every hour. A lab technician starts with 100 bacteria. He uses the equation $B = 100(2)^t$ to model the number of bacteria, B, after t hours.
    a) Write an exponential equation that can be used to determine when there are 6400 bacteria in the culture.
    b) Solve the equation. Explain your choice of strategy.

16. **Assessment Focus**
    a) Solve each equation algebraically: i) $2^{2x+3} = 8$ ii) $9^x = 27^{-x+2}$
    b) Explain how you solved the equation $2^{2x+3} = 8$.
       How do you know that your answer is correct?
    c) Solve $9^x = 27^{-x+2}$ graphically.
       Include the graphing calculator screen in your solution.
       How is the solution related to the graphs of $y = 9^x$ and $y = 27^{-x+2}$?
       Explain.

17. **Literacy in Math** List all of the strategies you have used in this chapter to solve equations. Include an example for each strategy. Use a graphic organizer to present your work.

6.6 Exponential Equations   **385**

18. The planning department of a township is responsible for estimating the number of utility hook-ups needed in new subdivisions. The current capacity is about 729 hook-ups per year. The actual demand, D, for hook-ups is given by $D = 27(3^n)$, where n is the number of years since 2007.
   a) Use algebra to determine when the demand reaches the capacity.
   b) Use graphing technology to verify your answer.

19. A cross-country skier forgets a mug of coffee and a muffin in a snowbank. Their temperatures, in degrees Celsius, after t minutes can be modelled by the formulas:
$T_{coffee} = 81 \times 3^{-2t}$ and $T_{muffin} = 27 \times 3^{-t}$
   a) Use algebra to determine when the coffee cools to the same temperature as the muffin. What is the common temperature?
   b) Verify your answer by using graphing technology.

In Your Own Words

You have learned several strategies for solving exponential equations. How do you decide which strategy to use in a given situation? Include examples in your explanation.

6.7 Applications of Exponential Equations

Carbon-14 (C-14) is a radioactive element that is absorbed from the atmosphere by plants and animals while they are alive. When a plant or animal dies, the C-14 in the organism's remains decays exponentially over time.

Investigate

Modelling Carbon-14 Decay

Materials
- container with a lid
- paper plate
- 100 pennies
- 100 nickels
- grid paper or graphing calculator

C-14 atoms decay to nitrogen atoms. We use pennies to represent C-14 atoms and nickels to represent nitrogen atoms.

Each penny that lands heads up represents a C-14 atom that has decayed into a nitrogen atom.

The symbol for microgram is µg. 1000 µg = 1 mg

Approximately every 5730 years, the amount of C-14 in the remains of an organism is reduced by a factor of one-half. Scientists can estimate when the organism was alive by comparing the amount of C-14 in the remains to the amount of C-14 in a living organism.

Work in a group of 4.

Trial number	Number of C-14 atoms
0	100

- Place 100 pennies in the container. Copy this table.
- Shake the container and empty it onto the paper plate. Replace each penny that lands heads up with a nickel. Then, record the trial number and the number of pennies left on the plate.
- Pour the coins on the plate into the container. Repeat the previous step for 4 more trials.
- Work individually. Graph the data in the table.
- Explain why each trial represents a time span of approximately 5730 years.
- Suppose a bone that originally contained 100 µg of C-14 now contains 10 µg of C-14. Estimate the age of the bone. Justify your answer.

Reflect

- What fraction of the pennies would you expect to remain after each trial? How closely did your results match this expected result? Explain.
- Did you use the table or graph to estimate the age of the bone? Explain the reason for your choice.
- C-14 dating is only used for objects less than 50 000 years old. Use your table or graph to explain why.

Connect the Ideas

Solving $y = ab^x$ for x

Exponential relations can be modelled by $y = ab^x$.
  a is the initial value
  b is the growth/decay factor
   • $b > 1$ models growth
   • $0 < b < 1$ models decay

Real-world applications of exponential growth or decay may require solving the equation $y = ab^x$ for x.

Example 1

Using a Table of Values

Materials
- TI-83 or TI-84 graphing calculator

The variable t is often used to represent time.

The population of Ontario was 9.3 million in 1985 and has been growing at an annual rate of 1.5%. This situation can be modelled by the equation $P = 9.4(1.0125)^t$, where P million represents the population t years after 1985. In which year did Ontario's population first exceed 10 million?

Solution

Use the equation $P = 9.4(1.0125)^t$.
Substitute: $P = 10$
$10 = 9.4(1.0125)^t$
Use a table of values to solve for t.

Use the variable X on the calculator to represent t in the equation.

Enter the right side of the equation in the equation editor. Press [Y=].
Move the cursor to $Y_1 =$.
Press 9.4 [(] 1.0125 [)] [^] [X,T,Θ,n].

388 Chapter 6: Algebraic Models

t = 0 corresponds to the year 1985.

Set up the TABLE editor.
Press [2nd] [WINDOW].
Set TblStart = 0 so that the table starts at X = 0.
Set ΔTbl = 1 so that X increases in steps of 1.

Generate the table.
Press [2nd] [GRAPH]. Scroll through the table until Y_1 is greater than 10. This occurs at X = 5.

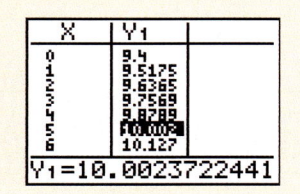

The population first exceeds 10 million approximately five years after 1985; that is, in the year 1990.

Example 2

Materials
- TI-83 or TI-84 graphing calculator

Using a Graph

Redo *Example 1*.
Use a graph to solve for *t*.

Solution

We wish to solve $10 = 9.4(1.0125)^t$ for *t*.
Graph the left and right sides of the equation on the same screen, and determine the X-coordinate of the point of intersection.

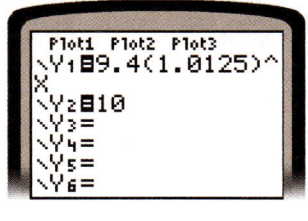

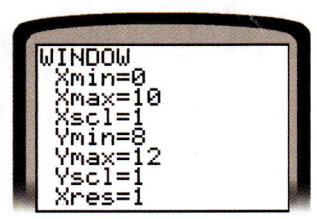

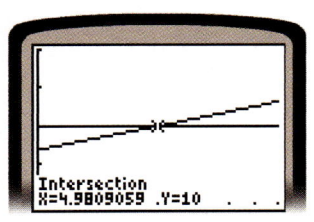

At the point of intersection, X ≐ 5. The population exceeds 10 million five years after 1985; that is, in the year 1990.

6.7 Applications of Exponential Equations   **389**

Exponential equations involving doubling time and half-life

In Chapter 5, you learned that quantities that grow or decay exponentially increase or decrease at a constant percent rate. These quantities have a constant doubling time or half-life. When the doubling time, d, or half-life, h, is known, the relationship between the initial amount, A_0, and the amount A after time t can be modelled by these equations.

Exponential growth
$$A = A_0(2)^{\frac{t}{d}}$$

Exponential decay
$$A = A_0(0.5)^{\frac{t}{h}}$$

Example 3

Solving an Application Involving Half-Life

Materials
- TI-83 or TI-84 graphing calculator

Caffeine has a half-life of approximately 5 h. Suppose you drink a cup of coffee that contains 200 mg of caffeine. How long will it take until there is less than 10 mg of caffeine left in your bloodstream? Give your answer to 1 decimal place.

Solution

Use the equation $A = A_0(0.5)^{\frac{t}{h}}$.

Substitute: $A = 10$, $A_0 = 200$, and $h = 5$

$$10 = 200(0.5)^{\frac{t}{5}}$$

Create a table of values to solve for t.
Scroll down until Y_1 is less than 10.

Every 5 h, the amount of caffeine in your bloodstream is reduced by a factor of one-half.

You could also use a graph to solve for t.

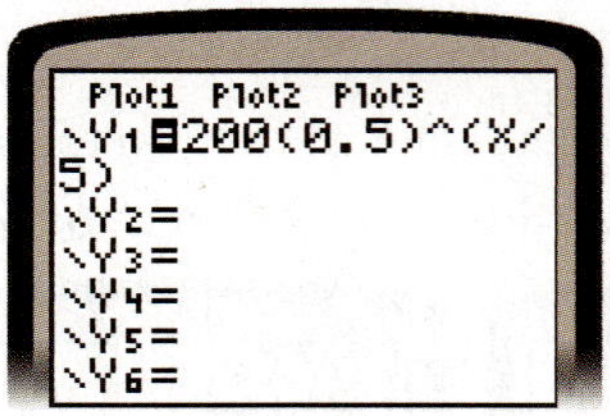

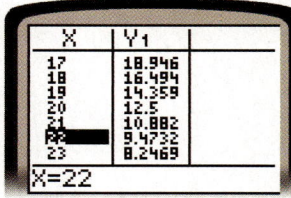

Y_1 is greater than 10 at $X = 21$ and less than 10 at $X = 22$.
To find the solution to the nearest tenth, go back to the table setup and change it so that X starts at 21 and increases by 0.1.

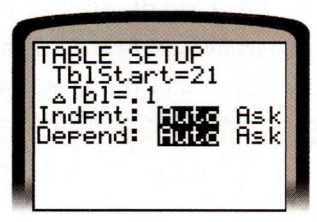

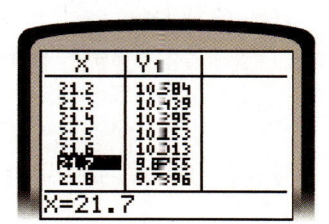

To the nearest tenth, $X \doteq 21.7$.
After 21.7 h, there is less than 10 mg of caffeine left in your bloodstream.

Practice

A

For help with questions 1 and 2, see Example 1.

1. A new car decreases in value exponentially after it is purchased. The value, V dollars, of a certain car t years after it was purchased is given by $V = 20\,000(0.84)^t$. Write an exponential equation that can be used to determine when the value of the car is equal to each amount.
   a) $10 000
   b) $15 000
   c) $7500
   d) $17 500

2. Use the table to determine in which year the value of the car will first be less than each amount in question 1.

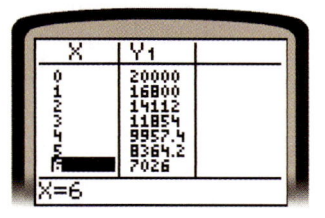

X	Y1
0	20000
1	16800
2	14112
3	11854
4	9957.4
5	8364.2
6	7026

   X=6

3. The population, P million, of Alberta between 1987 and 2005 can be modelled by the equation $P = 2.4(1.017)^t$, where t is the number of years since 1987. Write an exponential equation that can be used to estimate when the population equalled each number of people.
   a) 2.5 million
   b) 2.7 million
   c) 3.0 million

4. Use the tables to estimate in which year the population first exceeded each number of people in question 3.

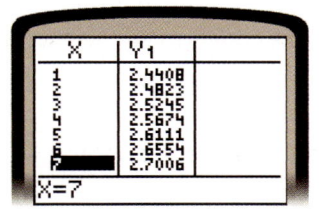

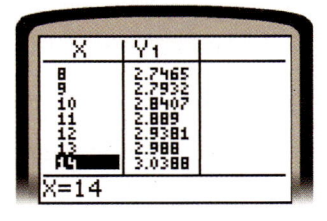

X	Y1
1	2.4408
2	2.4823
3	2.5245
4	2.5674
5	2.6111
6	2.6554
7	2.7006

   X=7

X	Y1
8	2.7465
9	2.7932
10	2.8407
11	2.889
12	2.9381
13	2.988
14	3.0388

   X=14

5. Match each equation with its table of values.
   a) $y = 4^x$
   b) $y = 3(4)^x$
   c) $y = 3(2)^x$

   i)
X	Y1
0	3
1	6
2	12
3	24
4	48
5	96
6	192

   X=0

   ii)
X	Y1
0	1
1	4
2	16
3	64
4	256
5	1024
6	4096

   X=0

   iii)
X	Y1
0	3
1	12
2	48
3	192
4	768
5	3072
6	12288

   X=0

6.7 Applications of Exponential Equations

6. Use the table of values.
   a) What is the value of y for each value of x?
      i) $x = 2$ ii) $x = 5$
   b) What is the value of x for each value of y?
      i) $y = 2$ ii) $y = 0.5$

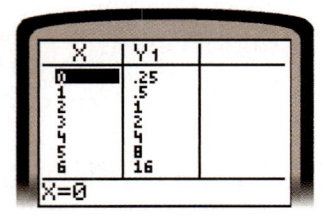

X	Y1
0	.25
1	.5
2	1
3	2
4	4
5	8
6	16

X=0

B   7. The table shows the growth of a culture of bacteria over time under laboratory conditions. The variable X represents the time in hours and the variable Y_1 represents the number of bacteria.
   a) How many bacteria were present initially? How do you know?
   b) How long does it take for the population to double? Justify your answer.

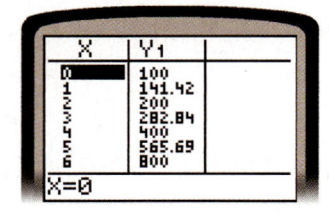

X	Y1
0	100
1	141.42
2	200
3	282.84
4	400
5	565.69
6	800

X=0

8. A principal of $500 is invested at 8% per year, compounded annually. After n years, the amount of the investment, A dollars, is given by $A = 500(1.08)^n$. Write an exponential equation that can be used to determine how long it takes for the investment to:
   a) grow to $600   b) double in value   c) triple in value

■ For help with questions 9 and 10, see Example 2.

9. Blue jeans fade with repeated washing. The equation $P = 100(0.98)^n$ models the percent, P, of colour left after n washings.
   a) Write an exponential equation that can be used to determine the number of washings until 50% of the colour remains.
   b) Use the graph to solve the equation. Justify your answer.

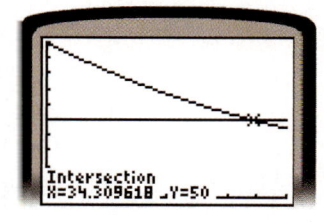

Intersection
X=34.309618 Y=50

392   Chapter 6: Algebraic Models

10. Lupine is a wildflower that attracts honeybees and butterflies. The equation $N = 100(1.4)^t$ models the number, N, of wild lupine seeds in a wildflower seed bank t years after collection began.
   a) Write an exponential equation that can be used to estimate when there will be 2000 seeds in the bank.
   b) Use graphing technology to solve the equation in part a. Include a sketch of the graphing calculator screen in your solution.

■ For help with question 11, see Example 3.

11. Suppose you invest $500 at 6% per year, compounded annually. The value, A dollars, of your investment after n years is given by $A = 500(1.06)^n$.
   a) Use graphing technology to graph $A = 500(1.06)^n$.
   b) Estimate the number of years that it will take for your investment to grow to each amount.
      i) $600   ii) $1000   iii) $1200

Use the formula $A = A_0(0.5)^{\frac{t}{h}}$ when you know the half-life.

12. The mortality rate from heart attack can be modelled by the relation $M = 88.9(0.9418)^t$, where M is the number of deaths per 100 000 people and t is the number of years since 1998.
   a) Has the mortality rate increased or decreased since 1998? Justify your answer.
   b) When will the mortality rate be one-half the rate in 1998?
   c) When will the mortality rate decrease to 22.2 deaths per 100 000? Justify your answers.

13. **Assessment Focus** Dye is injected to test pancreas function. The mass, R grams, of dye remaining in a healthy pancreas after t minutes is given by $R = I(0.96)^t$, where I grams is the mass of dye initially injected.
   a) Suppose 0.50 g of dye is injected into a healthy pancreas. How long will it take until 0.35 g of dye remain? Justify your answer.
   b) Describe the steps used to solve part a.
   c) Find the half-life of the dye in a healthy pancreas.
   d) How would the half-life change for a patient with an overactive pancreas? Explain.

6.7 Applications of Exponential Equations   **393**

14. Airplane cabins are pressurized because air pressure decreases as the height above sea level increases. The equation $P = 100(0.917)^h$ models the air pressure, P kilopascals, at a height of h kilometres above sea level.
   a) Determine the air pressure at a height of 10 km
   b) At what altitude is the air pressure 50% of its value at sea level?
   c) For which part did you have to solve an exponential equation? Explain.

15. Tritium, a radioactive gas that builds up in CANDU nuclear reactors, is collected, stored in pressurized gas cylinders, and sold to research laboratories. Tritium decays into helium over time. Its half-life is about 12.3 years.
   a) Write an equation that gives the mass of tritium remaining in a cylinder that originally contained 500 g of tritium.
   b) Estimate the time it takes until less than 5 g of tritium is present.

16. An archaeologist uses radiocarbon dating to determine the age of a Viking ship. Suppose that a sample that originally contained 100 mg of Carbon-14 now contains 85 mg of Carbon-14. What is the age of the ship to the nearest hundred years?

> Use the formula $A = A_0(2)^{\frac{t}{d}}$ when you know the doubling time.

17. A colony of bacteria doubles in size every 20 min. How long will it take for a colony of 20 bacteria to grow to a population of 10 000?

18. Compare the formula for doubling time to the formula for half-life. How are they the same? How are they different? Explain.

19. The formula $P = 29.6(1.0124)^t$ models Canada's population, where P is the population in millions and t is the number of years since 1995.
   a) Determine the doubling time for Canada's population.
   b) Use the result of part a and the formula $A = A_0(2)^{\frac{t}{d}}$ to model the growth of Canada's population in terms of its doubling time.
   c) Use both models to determine when Canada's population will first reach 40 million. Why should the answers be the same?

20. **Literacy in Math** The words *exponent, exponential growth, exponential decay, exponential relation, exponential regression,* and *exponential equation* are used when working with real-world situations involving growth and decay. Define each term and give an example of it.

C **21.** The exponential relations $P = 100(0.87)^t$ and $P = 100(0.5)^{\frac{t}{5}}$ can be used to model the percent of caffeine in your bloodstream t hours after you drink a beverage containing caffeine.
  a) What do the numbers in the relation $P = 100(0.87)^t$ represent?
  b) What do the numbers in the relation $P = 100(0.5)^{\frac{t}{5}}$ represent?
  c) Explain why the expression $(0.5)^{\frac{t}{5}}$ can be rewritten as $[(0.5)^{\frac{1}{5}}]^t$. Evaluate $(0.5)^{\frac{1}{5}}$.
  d) Explain how the result of part c shows that the relations $P = 100(0.5)^{\frac{t}{5}}$ and $P = 100(0.87)^t$ are equivalent.

22. When a nuclear reactor is shut down, the core contains many radioactive isotopes which continue to decay. These exponential relations model the activity, A becquerels, for two isotopes after t days.

Iodine-131: $A_I = (3 \times 10^{18}) \times (0.5)^{\frac{t}{8}}$,   Xenon-133: $A_{Xe} = (6 \times 10^{18}) \times (0.5)^{\frac{t}{5}}$

  a) Which isotope has the greater initial activity? Which decays more quickly? Explain.
  b) After how many days will the two isotopes have the same activity?
  c) Predict what will happen to the activities of the two isotopes after the time you found in part b.

In Your Own Words

Explain the difference between an exponential relation and an exponential equation. Explain how an exponential equation can be used to solve a situation that can be modelled by an exponential relation. Include an example in your explanation.

6.8 Occupations Using Mathematical Modelling

Many careers require you to be able to create and use graphical and algebraic models of real-world situations.

Inquire: Researching Careers Involving Mathematical Modelling

Materials
- computer with Internet access

Mathematical modelling is used in a variety of careers. For example:
- The manager of a timber company may use a mathematical model to determine the best age at which to harvest a stand of trees.
- A landscaper may use mathematical formulas to estimate the materials and labour needed to complete a project.
- A workplace safety officer may use a mathematical model to determine the effects of exposure to airborne asbestos.
- A sales manager may use graphical models to analyse trends and patterns in sales data.

Work in a small group.

Part A: Creating a List of Careers

Create a list of careers that use tables, graphs, or formulas to model relationships.

- Brainstorm to start the list.
  Briefly describe how mathematical modelling could be used in each career.
- Continue the list by going through each lesson in Chapters 5 and 6. List the careers referred to in the lessons.
  Briefly describe how mathematical modelling could be used in each career.

CHAPTER 6: Algebraic Models

- Each member of the group should select two careers from the list to research further.

Part B: Researching Careers

- Investigate the careers you chose in Part A.
  You may choose to:
  - Read about the career on the Internet or in printed material.
  - Contact and interview someone who works in the career.
- If you use the Internet, you may type phrases like these into a search engine such as Google or Yahoo to help you start your research.

  Search words
  ☐ nursing formulas
  ☐ media planner math
  ☐ police math modelling

- Summarize your research. Include details that answer questions such as:
  - How is mathematical modelling used in this career?
    What is each model supposed to simulate, describe, or predict?
  - What technology is used to graph and analyse data or to evaluate formulas?
  - Is data collected to develop and test models?
  - What measurement system (metric and/or imperial) is used in this career? How important are accurate measurements and calculations?
  - What is a typical wage or salary for someone in this career? Are employees paid on an hourly or a salary basis?

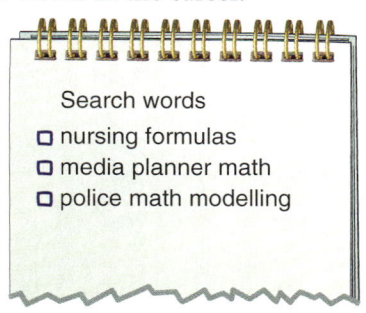

Part C: Researching Educational Requirements

- Research the educational requirements for the careers you investigated in Part B.

  You may choose to:
  - Go to the Web sites of community colleges or other post-secondary institutions.
  - Use course calendars of post-secondary institutions or other information available through your school's guidance department.
- Record your findings.
- Be prepared to present your research from Parts A and B to the class.

Reflect

- What other areas of mathematics would likely be applied in a career that uses modelling?
- What types of written or graphical communication are used in these careers?
- What opportunities for advancement, leadership, or self-employment exist in these careers?

Study Guide

Rearranging Formulas

Rearrange the formula $V = \frac{4}{3}\pi r^3$ to isolate r.

$V = \frac{4}{3}\pi r^3$ Multiply each side by 3.

$3V = 4\pi r^3$ Divide each side by 4π.

$\frac{3V}{4\pi} = r^3$ Take the cube root of each side.

$\sqrt[3]{\frac{3V}{4\pi}} = r$

Use inverse operations and a balance strategy.

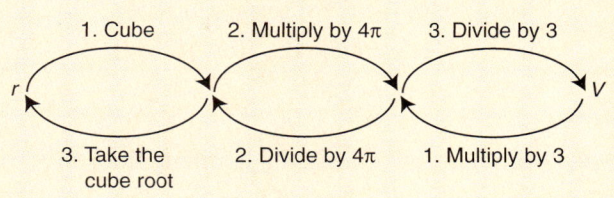

1. Cube 2. Multiply by 4π 3. Divide by 3

$r \rightleftharpoons \rightleftharpoons \rightleftharpoons V$

3. Take the cube root 2. Divide by 4π 1. Multiply by 3

Exponents

Definitions		Examples
• Positive integer exponents	$a^n = \underbrace{a \times a \times a \times \cdots \times a}_{n \text{ factors}}$	$4^3 = 4 \times 4 \times 4 = 64$
• Zero exponents	$a^0 = 1, a \neq 0$	$4^0 = 1$
• Negative integer exponents	$a^{-n} = \frac{1}{a^n}, a \neq 0$	$4^{-3} = \frac{1}{4^3} = \frac{1}{64}$
• Rational exponents	$a^{\frac{1}{n}} = \sqrt[n]{a}, a > 0$ if n is even $a^{\frac{m}{n}} = (\sqrt[n]{a})^m = \sqrt[n]{a^m}, a > 0$ if n is even	$64^{\frac{1}{3}} = \sqrt[3]{64} = 4$ $64^{\frac{2}{3}} = (\sqrt[3]{64})^2 = 4^2 = 16$

Laws of Exponents		Examples
• Multiplication law	$a^m \times a^n = a^{m+n}$	$x^5 \times x^{-7} = x^{-2}$
• Division law	$a^m \div a^n = a^{m-n}, a \neq 0$	$\frac{x^4 y^4}{(x^2)^{-3} y^3} = \frac{x^4 y^4}{x^{-6} y^3} = x^{4-(-6)} y^{4-3} = x^{10} y$
• Power of a power law	$(a^m)^n = a^{mn}$	$(y^4)^2 = y^8$

Solving Exponential Equations

Common base

Set the exponents equal to each other and solve the resulting equation.

$2^{x+3} = 32$
$2^{x+3} = 2^5$
$x + 3 = 5$
$x = 2$

Different bases

Solve for x: $10(0.8)^x = 5$
Plot $Y_1 = 10(0.8)^x$ and $Y_2 = 5$. Determine the X-coordinate of the point of intersection.
$x \doteq 3.1$

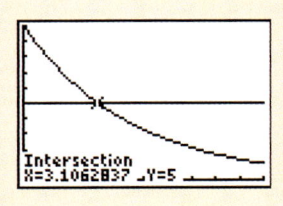

Intersection
X=3.106283 Y=5

Study Guide **399**

Chapter Review

6.1

1. Suzie works in a boutique. Each month, her earnings, E dollars, are given by the formula $E = 2500 + 0.10s$, where s dollars are her sales for the month. One month, Suzie's sales were $8000. How much did she earn that month?

2. Bonnie wants to install a laminate floor in a room that measures 5.0 m by 3.0 m. The flooring comes in bundles. Each bundle costs $47.76 and covers an area of 2.23 m². Determine the cost of the flooring.

3. Houd plans to replace the light bulbs in his house with energy-saver light bulbs. The formula $C = 0.006h + 0.45$ gives the cost, C dollars, of using a regular bulb for h hours. The cost for an energy-saver bulb is $C = 0.004h + 1.05$. How much money does Houd save with the energy saver bulb after 1000 h of use?

4. The formula $E = Pt$ gives the energy, E watt hours, used when an electrical appliance of power P watts is used for t hours. Determine the energy used in each situation.
  a) A 17-W fluorescent bulb is left on for 1 h.
  b) A 1400-W blow dryer is used for 10 min.
  c) How long would the fluorescent bulb need to be left on so that it consumes as much energy as using the blow dryer for 10 min?

6.2

5. The circumference, C, of a circle with diameter d is $C = \pi d$.
  a) Solve the formula for d.
  b) Calculate the diameter of a circle with circumference 8 m.

6. The power in an electrical circuit is given by the formula $P = I^2R$, where P is the power in watts, I is the current in amperes, and R is the resistance in ohms. Determine R when $P = 1000$ W and $I = 12$ A. Explain your method.

7. The aspect ratio of a hang glider describes its performance during flight. The formula $R = \frac{s^2}{A}$, gives the aspect ratio, R, for a hang glider with wingspan s and wing area A.

  a) Rearrange the formula to isolate s. Use an arrow diagram to show the inverse operations you can use.
  b) Jake wants to design a hang glider with an aspect ratio of 2.7 and a wing area of 30 square feet. What will be the wingspan of the glider?

8. The formula $A = P(1 + rt)$ gives the amount, A dollars, of a simple interest investment. In the formula, P dollars is the principal invested, r is the annual interest rate expressed as a decimal, and t is the number of years.
   a) Solve the formula for P.
   b) The amount of a simple interest investment is $2000 after 3.5 years. The principal earned 5% interest per year. Calculate the principal invested.

9. Evaluate without a calculator.
   a) -5^2 b) 5^{-2} c) $(-5)^2$
   Explain how you know whether the sign of the answer is positive or negative.

10. Simplify, then evaluate.
   a) $10^2 \times 10^3$
   b) $\dfrac{7^6}{7^{-2}}$
   c) $(1.12^2)^3$
   d) $\dfrac{2^3 \times 2^2}{2^5}$
   For which parts did you use a calculator? Explain.

11. Simplify using the exponent rules.
   a) $k^4 l^3 k^{-3} l^{-2}$
   b) $(a^5)^{-2}(a^3)^4$
   c) $\dfrac{xy^{-2}}{x^3 y^2}$
   d) $\left(\dfrac{s}{n}\right)^4 \left(\dfrac{n}{s}\right)^{-3}$

12. Evaluate when $c = 5$ and $d = -3$.
   a) c^d
   b) d^c
   c) $\dfrac{cd^3}{c^3 d}$
   d) $(2cd^2)^3$

13. Evaluate.
   a) $16^{\frac{1}{2}}$ b) $(-27)^{\frac{1}{3}}$ c) $32^{\frac{1}{5}}$
   d) $8^{\frac{5}{3}}$ e) $625^{\frac{3}{4}}$ f) $0.81^{\frac{3}{2}}$

14. The formula $m = 3.5(0.5)^{\frac{t}{8}}$ gives the mass, m milligrams, of radioactive Iodine in a sample t days after the initial measurement. Determine the mass of radioactive Iodine after each number of days.
   a) 2 days   b) 4 days
   c) 16 days

15. Solve for x. Assume x is positive. Show your work.
   a) $x^2 = 100$
   b) $x^3 = 64$
   c) $x^{\frac{1}{2}} = 9$
   d) $x^{\frac{3}{2}} = 8$

16. The formula $A = P(1 + i)^n$ gives the amount, A dollars, of a compound interest investment. In the formula, P dollars is the principal invested, i is the annual interest rate expressed as a decimal, and n is the number of years. Shelley invests $500 with interest compounded annually. The amount of the investment after 6 years is $651. Determine the annual interest rate.

17. The formula $B = 0.4089M^{\frac{3}{4}}$ gives the bird inhalation rate, B cubic metres of air per day, for a bird with mass M kilograms.
   a) Rewrite the formula using radicals.
   b) Calculate the inhalation rate for each bird.
      i) a 4.5-kg bald eagle
      ii) a 8.0-kg Canada goose
   c) Determine the mass of a bird whose inhalation rate is twice that of the bald eagle.
   d) Is the mass in part c twice that of the bald eagle? Explain.

Chapter Review **401**

18. Rewrite each set of numbers as powers of the same base.
   a) 4, 16, 32
   b) $\frac{1}{5}$, 25, 125
   c) 9, 81, 243

19. Solve algebraically.
   a) $4^{2x} = 4^6$
   b) $5^x = 625$
   c) $3^{2x+1} = 9$
   d) $10^{x+1} = 10^{2x-3}$
   e) $4^{3x-2} = 32^{x+1}$
   f) $25^{x+1} = 125^{x-2}$

20. Choose 2 equations from question 19.
   a) Explain how you solved the equations.
   b) Use graphing technology to verify your answers to the equations in part a.

21. Determine the value of n to the nearest tenth. Explain your strategy.
   a) $10^n = 125$
   b) $3^n = 6$
   c) $5^n = 0.01$
   d) $250(1.03)^n = 400$
   e) $1000(0.85)^n = 300$

22. Choose 2 equations from question 21. Verify your answers numerically.

23. Gillian takes a 300-mg tablet of pain medication. The amount, A milligrams, of medication remaining in her body after t hours is given by the formula $A = 300(0.8)^t$.
   a) Write an exponential equation that can be used to determine when 1 mg of medication remains in Gillian's body.
   b) Solve the equation in part a.

24. An archaeologist estimates the age of an artifact by measuring the amount of Carbon-14 it contains. The artifact originally contained 200 µg of Carbon-14. Now it contains 25 µg of Carbon-14. The half-life of Carbon-14 is about 5730 years. What is the age of the artifact to the nearest hundred years?

25. The resale value of a used vehicle is given by the formula $V = C(1 - r)^n$, where V dollars is the resale value, C dollars is the original price, r is the rate of depreciation as a decimal, and n is the age of the vehicle.
   a) A certain vehicle depreciates at a rate of 20% per year. The original price of the vehicle was $36 000 and its resale value after n years is $7549.74. Write an exponential equation that can be used to determine the age of the vehicle.
   b) Solve the equation in part a.
   c) How do you know your answer in part b is correct?

26. Name two occupations where mathematical modelling is applied.

Practice Test

Multiple Choice: Choose the correct answer for questions 1 and 2. Justify each choice.

1. Which formula is equivalent to $E = \dfrac{mv^2}{2}$?

   A. $m = \dfrac{2v^2}{E}$ B. $v = \sqrt{\dfrac{2E}{m}}$ C. $v = \sqrt{\dfrac{m}{2E}}$ D. $m = \dfrac{E}{2v^2}$

2. Which is the solution to $3^{2x+3} = \dfrac{1}{9}$?

   A. $x = 0.5$ B. $x = 2$ C. $x = -0.5$ D. $x = -2.5$

Show your work for questions 3 to 6.

3. **Knowledge and Understanding**
   a) Evaluate without a calculator.
      i) $(-2)^{-3}$ ii) $\left(\dfrac{25}{9}\right)^{\frac{1}{2}}$ iii) $81^{\frac{3}{4}}$
   b) Evaluate $\dfrac{2a^4 b^{-3}}{a^2 b}$ when $a = -5$ and $b = 2$.

4. **Communication** Sandy wants to estimate the cost of installing 2 layers of fibreglass insulation in an attic. The floor plan of the attic is shown. The insulation comes in batts that measure 23 inches by 47 inches. There are 10 batts in a bag, and a bag costs $48.27. Make a plan for solving this problem, but *do not solve it*. Provide any formulas needed.

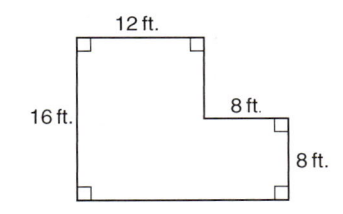

5. **Application** The population, P, of the city of Hazelton has grown according to the mathematical model $P = 32\,000(1.09)^t$, where t is the number of years since 1990.
   a) What do the numbers 32 000 and 1.09 represent?
   b) If this trend continues, in what year will the population reach 100 000? Justify your answer.

6. **Thinking** A biologist uses the formula $E = km^{\frac{3}{4}}$ to model the relationship between the mass of a bird's egg, E grams, and the mass of the bird, m grams. The number k is a constant that is close to 0.25 for healthy eggs.
   a) A ruby-throated hummingbird has a mass of 3.4 g. What is the mass of its egg?
   b) An ostrich egg has a mass of 1.4 kg. What is the mass of the mother bird?
   c) A 7.3-kg whooping crane lays an egg whose mass is 208 g. Does this bird appear to be healthy? Explain.

Chapter Problem: A Butterfly Conservatory

A glass butterfly conservatory is a square-based pyramid. The height of the conservatory is one-half the side length of the base. The volume is 900 m³.

1. Demetra showed how the formula $V = \frac{1}{6}c^3$ gives the volume of the conservatory. Explain her thinking. Include your own diagram.

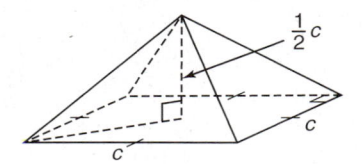

$$V = \frac{1}{3}b^2h$$
$$= \frac{1}{3}c^2 \times \frac{1}{2}c$$
$$= \frac{1}{3} \times \frac{1}{2}c^2c$$
$$= \frac{1}{6}c^3$$

2. a) Describe how to use a formula to determine the height and the side lengths of the base of the butterfly conservatory. Calculate these dimensions to the nearest tenth of a metre.
   b) Substitute the dimensions into the formula to check. Are your results reasonable? Explain.

3. Describe a different way to determine the side lengths of the base and the height.

4. Suppose the height of the conservatory is doubled, but the base remains the same. Predict what would happen to the volume. Justify your prediction. Check your prediction. Explain how to extend your reasoning to make a generalization about the relationship between the height and the volume of a square-based pyramid.

404 CHAPTER 6: Algebraic Models

7 Annuities and Mortgages

What You'll Learn

To use technology to solve problems involving annuities and mortgages and to gather and interpret information about annuities and mortgages

And Why

Annuities are used to save and pay for expenses such as a car, a home, education, and retirement. Understanding the mathematics of annuities and mortgages will help you manage your money more effectively so that you can achieve your financial goals.

Key Words

- simple interest
- compound interest
- annuity
- ordinary simple annuity
- amount of an annuity
- present value of an annuity
- mortgage
- amortization period
- amortization table

CHAPTER 7

Activate Prior Knowledge

Simple and Compound Interest

Prior Knowledge for 7.1

Simple interest is money earned on a starting principal.
- $I = Prt$, where I dollars is the interest earned, P dollars is the principal, r is the annual interest rate, and t is the time in years.

Compound interest is money earned on both the principal and previous interest.
- In compound interest problems, $A = P(1 + i)^n$, where A dollars is the amount, P dollars is the principal, i is the interest rate per compounding period as a decimal, and n is the number of compounding periods.

Example

Materials
- scientific calculator

The quarterly interest rate is $\frac{1}{4}$ of the annual interest rate.

Tezer invests $3000 for 3 years in a bond that earns 6% per year compounded quarterly. How much interest does the bond earn?

Solution

Substitute $P = 3000$, $i = \frac{0.06}{4} = 0.015$, and $n = 3 \times 4 = 12$ in the formula:
$A = P(1 + i)^n$

$A = 3000(1 + 0.015)^{12} \doteq 3586.85$

Press: 3000 × (1 + 0.015) ^ 12 ENTER

The amount after 3 years is $3586.85.
To determine the interest earned, subtract the principal from the amount.
$3586.85 - 3000 = 586.85$
The bond earns $586.85 in interest.

CHECK ✓

1. Vishnu borrows $4500 for 3 months at an annual rate of 7.5%. What amount will Vishnu repay at the end of the 3 months?

2. Nancy invests $10 000 at 5% per year compounded semi-annually. Determine the amount after 3 years.

   Semi-annually means 2 times a year.

3. Refer to question 2. Will the amount double in each case?
   a) The principal invested is twice as great, $20 000.
   b) The interest rate is twice as great, 10%.
   c) The term is twice as long, 6 years.
   Justify your answers.

Present Value

Prior Knowledge for 7.2

The principal that must be invested today to obtain a given amount in the future is the **present value** of the amount.
To calculate the present value in a situation involving compound interest, rearrange the formula $A = P(1 + i)^n$ to solve for P.

Example

Materials
- scientific calculator

What principal should Yvette invest today at 4.6% per year compounded semi-annually to have $6500 three years from now?

Solution

The semi-annual interest rate is $\frac{1}{2}$ of the annual interest rate,
so $i = \frac{0.046}{2} = 0.023$.
Interest is compounded 2 times a year for 3 years,
so $n = 3 \times 2 = 6$.
Isolate P, then substitute $A = 6500$, $i = 0.023$, and $n = 6$ in the formula:

$$A = P(1 + i)^n$$
$$\frac{A}{(1 + i)^n} = P$$
$$P = A(1 + i)^{-n}$$
$$P = 6500(1 + 0.023)^{-6}$$
$$\doteq 5671.00$$

Yvette should invest $5671.00.

You could also substitute, then solve for P.

Since $\frac{1}{x} = x^{-1}$, $\frac{1}{(1 + i)^n} = (1 + i)^{-n}$

Press: 6500 × (1 + 0.023) ^ (-) 6 ENTER

CHECK ✓

1. Jeremy plans to go on a cruise 4 years from now. He will need $7500 at that time. What principal should Jeremy invest now at 8.4% per year compounded monthly to obtain the required amount?

2. Rami invested money at 3.6% per year compounded semi-annually. He received $12 387.21 at the end of a 6-year term. How much interest did Rami earn?

3. Suppose you are solving a problem involving compound interest. How do you know whether to determine the amount or present value of a given sum of money? Explain.

Activate Prior Knowledge

Evaluations

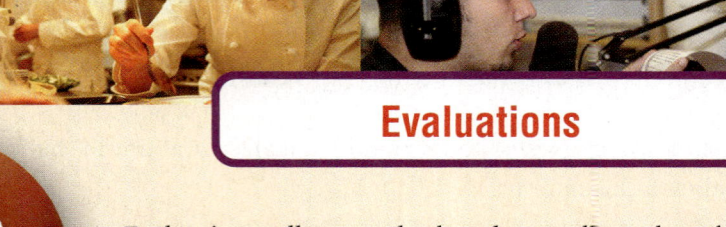

Evaluations tell you and others how well you have learned and performed at school or in the workplace. They can be used to determine whether a person has the skills needed for an opportunity, such as acceptance to a college or apprenticeship program, or to assess how well a person is doing in a course or on the job.

The Workplace

In the workplace, there are many evaluation methods. For example, some employers use performance appraisals to evaluate employees' progress. A supervisor keeps a detailed record of how the employee performs tasks at work and writes an appraisal or report. Then, the supervisor and employee meet to discuss the report. Learn more about workplace evaluation.

1. Choose several jobs such as a factory worker, police officer, hairdresser in a large salon, or bank clerk. Interview people in these fields and use the Internet to answer these questions about the jobs.

    - How are the employees evaluated?
    - How often are they evaluated?
    - What are the criteria for evaluation?
    - What are the rewards for a good evaluation? What happens to those who get negative comments on their work?

College

Most college math courses have quizzes, tests, and projects. Quizzes and tests are often multiple choice. Although you may need to complete several steps to figure out your answer, in most cases, there are no part marks. Making a study sheet can help you prepare for these evaluations.

2. Create a study sheet for this chapter.
    - Make it fit on one side of a sheet of paper.
    - You might use the computer to create symbols and diagrams.
    - Use graphic organizers such as a Venn diagram or Frayer model.
    - Include notes and examples so that your study guide can help you. Refer to the instructions for *Collecting Important Ideas* in Chapter 1.
    - Don't just copy from the text!

7.1 The Amount of an Annuity

Hiroshi plans to buy a motorcycle in 5 years. He saves for the down payment by making regular deposits into his investment account.

Investigate — Determining the Accumulated Value of Regular Deposits

Materials
- scientific calculator

Work with a partner.

Suppose Hiroshi deposits $1000 at the end of each year for 5 years. His account earns at 6% per year compounded annually. How much has Hiroshi saved at the end of 5 years?

■ Determine the balance in the account at the end of each year. Organize your calculations in a table like the one below.

Year	Starting balance	Interest earned (6%)	Deposit	Ending balance
1	$0.00	$0.00	$1000.00	$1000.00
2	$1000.00	$60.00	$1000.00	$2060.00

The interest is compounded before the deposit is made.

Reflect

■ Why does the interest earned increase each year?
■ What is an advantage and a disadvantage of using a table to determine Hiroshi's savings after 5 years?

Connect the Ideas

Ordinary simple annuities

An **annuity** is a series of equal payments made at regular intervals. In an **ordinary simple annuity**, payments are made at the end of each compounding period. The **amount of an annuity** is the sum of the regular deposits plus interest.

Example 1

Materials
- scientific calculator

This annuity is an ordinary simple annuity because a deposit is made at the end of each quarter and the interest is compounded quarterly.

Interest earned =
Starting balance
× 0.025

Ending balance =
Starting balance
+ Interest earned
+ Deposit

Using a Table

Suppose $450 is deposited at the end of each quarter for 1.5 years in an investment account that earns 10% per year compounded quarterly.
a) What is the amount of the annuity?
b) How much interest does the annuity earn?

Solution

The annual interest rate is 10%, so the quarterly rate is:
$$\frac{10\%}{4} = 2.5\%$$
The number of quarters in 1.5 years is:
$1.5 \times 4 = 6$
Use a table to organize the calculations.

Quarter	Starting balance	Interest earned (2.5%)	Deposit	Ending balance
1	$0.00	$0.00	$450.00	$450.00
2	$450.00	$11.25	$450.00	$911.25
3	$911.25	$22.78	$450.00	$1384.03
4	$1384.03	$34.60	$450.00	$1868.63
5	$1868.63	$46.72	$450.00	$2365.35
6	$2365.35	$59.13	$450.00	$2874.48
	Total	$174.48	$2700.00	

a) The amount of the annuity is $2874.48.
b) The interest earned is $174.48.

It is time-consuming to create a table to determine the amount of an annuity. A simpler alternative is to use a formula.

410 CHAPTER 7: Annuities and Mortgages

Amount of an ordinary simple annuity
$$A = \frac{R[(1 + i)^n - 1]}{i}, \text{ where:}$$
- A is the amount in dollars
- R is the regular payment in dollars
- i is the interest rate per compounding period as a decimal
- n is the number of compounding periods

The amount formula can only be used when:
- The payment interval is the same as the compounding period.
- A payment is made at the end of each compounding period.
- The first payment is made at the end of the first compounding period.

Example 2

Materials
- scientific calculator

Using the Amount Formula

In the annuity in *Example 1*, $450 is deposited at the end of each quarter for 1.5 years at 10% per year compounded quarterly.
a) What is the amount of the annuity?
b) How much interest does the annuity earn?

Solution

a) The regular payment is $450, so $R = 450$.
$i = \frac{0.10}{4} = 0.025$; $n = 1.5 \times 4 = 6$

Substitute $R = 450$, $i = 0.025$, and $n = 6$ into the amount formula.

$$A = \frac{R[(1 + i)^n - 1]}{i}$$

$$A = \frac{450[(1 + 0.025)^6 - 1]}{0.025}$$

Press: 450 ((1 + 0.025) ^ 6 − 1) ÷ 0.025 ENTER

$\doteq 2874.48$

The amount is $2874.48.

b) The amount, $2874.48, is the total of the deposits, plus interest.
6 deposits of $450: $6 \times \$450 = \$2700$
So, the interest earned is: $2874.48 − $2700 = $174.48

7.1 The Amount of an Annuity   **411**

Annuity calculations can also be performed on a TI-83 or TI-84 graphing calculator, we can use a financial application called the TVM (Time Value of Money) Solver.

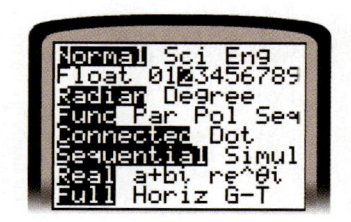

- Set the calculator to 2 decimal places.
  Press: MODE ▼ ▶ ▶ ▶ ENTER
- Open the TVM Solver by pressing: APPS 1 1
  The variables represent the following quantities.

N	Total number of payments
I%	Annual interest rate as a percent
PV	Principal or present value
PMT	Regular payment
FV	Amount or future value
P/Y	Number of payments per year
C/Y	Number of compounding periods per year
PMT:	Indicates whether payments are made at the beginning or end of the payment period

- The calculator displays either positive or negative values for PV, PMT, and FV. Negative values indicate that money is *paid out*, while positive values mean that money is *received*.
- In annuity calculations, only one of the amount (FV) or present value (PV) is used. Enter 0 for the variable not used.

412 CHAPTER 7: Annuities and Mortgages

Example 3

Materials
- TI-83 or TI-84 graphing calculator

Using the TVM Solver

In the annuity in *Example 1*, $450 is deposited at the end of each quarter for 1.5 years at 10% per year compounded quarterly.
a) What is the amount of the annuity?
b) How much interest does the annuity earn?

Solution

a)

Enter the known values.
Set PMT to END because payments are made at the end of each period in an ordinary simple annuity.

```
N=6.00
I%=10.00
PV=0.00
PMT=-450.00
FV=0.00
P/Y=4.00
C/Y=4.00
PMT:END BEGIN
```

To enter −450, press the negative key [(−)], not the subtract key [−].

Solve for the amount.
Move the cursor to FV.
Press: [ALPHA] [ENTER]

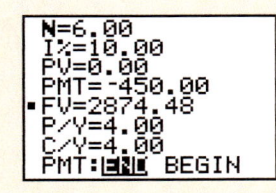

```
N=6.00
I%=10.00
PV=0.00
PMT=-450.00
•FV=2874.48
P/Y=4.00
C/Y=4.00
PMT:END BEGIN
```

The amount after 1.5 years is $2874.48.

b)

ΣInt(1,6) means the sum of the interest for payments 1 to 6.

Press [2nd] [MODE] to exit the TVM Solver. Use the ΣInt command to determine the total interest earned.
Press [APPS] 1 [ALPHA] [MATH] to show ΣInt(.
Press: 1 [,] 6 [)] [ENTER]

```
ΣInt(1,6)
          174.48
```

The interest earned is $174.48.

Annuities and regular savings

Annuities are often used to save money for expenses such as a car, a down payment on a house, or a vacation. They are also used to save for education and retirement. Relatively small, regular deposits of money can accumulate to large sums of money over time.

7.1 The Amount of an Annuity   **413**

Example 4

Comparing Retirement Plans

Materials
- TI-83 or TI-84 graphing calculator

Tom and Beth are twins. They save for retirement as follows.
- Starting at age 25, Tom deposits $1000 at the end of each year for 40 years.
- Starting at age 45, Beth deposits $2000 at the end of each year for 20 years.

Suppose each annuity earns 8% per year compounded annually. Who will have the greater amount at retirement?

Solution

Determine the amount of each annuity.

Tom's retirement fund
- Enter: N = 40, I% = 8, PV = 0, PMT = −1000, FV = 0, P/Y = 1, and C/Y = 1
- Move the cursor to FV. Press: ALPHA ENTER

Beth's retirement fund
- Enter: N = 20, I% = 8, PV = 0, PMT = −2000, FV = 0, P/Y = 1, and C/Y = 1
- Move the cursor to FV. Press: ALPHA ENTER

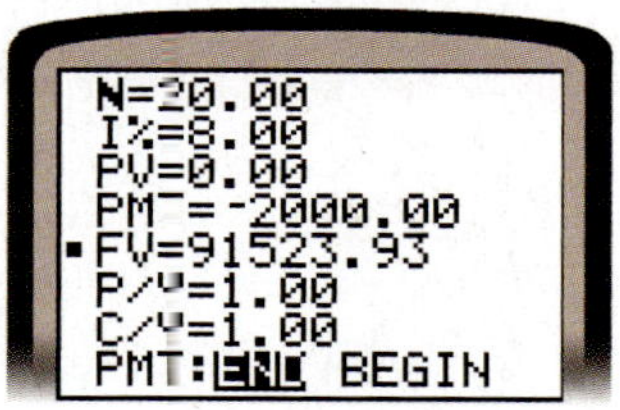

```
N=40.00
I%=8.00
PV=0.00
PMT=-1000.00
■FV=259056.52
P/Y=1.00
C/Y=1.00
PMT:END BEGIN
```

```
N=20.00
I%=8.00
PV=0.00
PMT=-2000.00
■FV=91523.93
P/Y=1.00
C/Y=1.00
PMT:END BEGIN
```

We could have used the annuity formula instead of the TVM Solver.

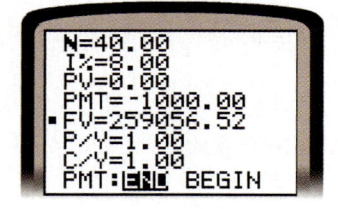

Tom will have $259 056.52.   Beth will have $91 523.93.
Tom will have the greater amount saved at retirement.

Example 4 illustrates the power of time on the value of money and the advantage of starting to save early. This advantage is also illustrated in these graphs.

Both Tom and Beth deposit a total of $40 000. But, by starting earlier, Tom earns an additional $167 532.59 in interest.

Tom's Retirement Fund

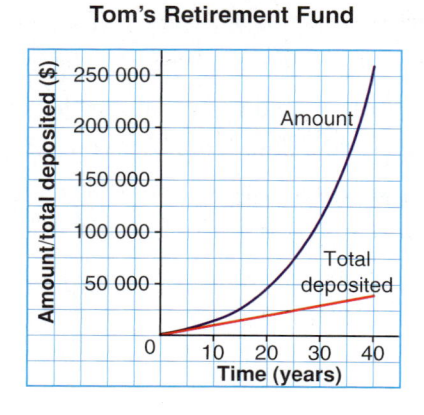

Beth's Retirement Fund

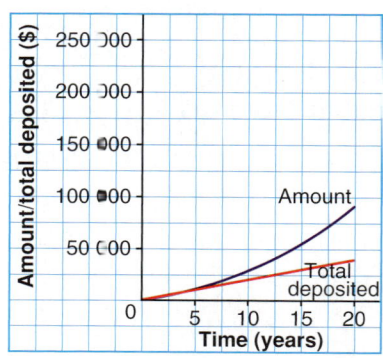

414 CHAPTER 7: Annuities and Mortgages

Practice

For help with question 1, see Example 1.

1. Complete each table. What is the amount of each annuity?

a) $1000 deposited at the end of each year at 8% per year compounded annually

Year	Starting balance	Interest earned	Deposit	Ending balance
1	$0.00	$0.00	$1000.00	$1000.00
2	$1000.00	$80.00	$1000.00	$2080.00
3			$1000.00	
4			$1000.00	

b) $100 deposited at the end of each month at 6% per year compounded monthly

Month	Starting balance	Interest earned	Deposit	Ending balance
1	$0.00	$0.00	$100.00	$100.00
2	$100.00	$0.50	$100.00	$200.50
3			$100.00	
4			$100.00	

For help with questions 2 and 3, see Example 2.

2. Determine i, the interest rate per compounding period as a decimal, and n, the number of compounding periods for each annuity.

	Time of payment	Length of annuity	Interest rate per year	Frequency of compounding
a)	end of each year	7 years	3%	annually
b)	end of every 6 months	12 years	9%	semi-annually
c)	end of each quarter	8 years	2.4%	quarterly
d)	end of each month	5 years	18%	monthly

3. Use the formula: $A = \dfrac{R[(1+i)^n - 1]}{i}$

Calculate A for each set of values.

a) $R = \$200$, $i = 0.05$, $n = 3$
b) $R = \$1000$, $i = 0.08$, $n = 7$
c) $R = \$700$, $i = 0.02$, $n = 12$

■ For help with questions 4 and 5, see Example 3.

4. For each annuity in question 2, what values would you enter into the TVM Solver for N, I%, P/Y, and C/Y?

5. Bill entered these values into the TVM Solver to determine the amount of an annuity.
   a) What is the regular payment?
   b) How often are the regular payments made?
   c) How many payments are made?
   d) What is the annual interest rate?
   e) How often is the interest compounded?
   f) What is the amount of the annuity?

   ```
   N=6.00
   I%=8.00
   PV=0.00
   PMT=-300.00
   ■FV=1830.27
   P/Y=12.00
   C/Y=12.00
   PMT:END BEGIN
   ```

B 6. Determine the amount of each ordinary simple annuity.
   a) $3000 deposited every year for 10 years at 7% per year compounded annually
   b) $650 deposited every 6 months for 8 years at 9% per year compounded semi-annually
   c) $1450 deposited every quarter for 9 years at 6.25% per year compounded quarterly
   d) $375 deposited every month for 6 years at 5.9% per year compounded monthly

7. Determine the interest earned by each annuity in question 6.

8. Use a different method to verify your answers to questions 6 and 7. Which method do you prefer? Explain.

9. Shen Wei wants to save $10 000 for his first year of college. He deposits $300 at the end of each month in an account that earns 5.6% per year compounded monthly. Will Shen Wei have enough money saved at the end of 2.5 years? Justify your answer.

416   CHAPTER 7: Annuities and Mortgages

10. Geneva's parents saved for her college education by depositing $1200 at the end of each year in a *Registered Education Savings Plan* (RESP) that earns 6% per year compounded annually.
   a) What is the amount of the RESP at the end of 18 years?
   b) How much interest is earned?
   c) How much extra interest would have been earned at an interest rate of 7% per year compounded annually?

11. Verena is saving for a new computer. She deposits $100 at the end of each month into an account that earns 4% per year compounded monthly.
   a) Determine the amount in the account after 3 years.
   b) Does the amount in part a double with each of these changes?
      i) The deposits are twice as great, $200.
      ii) The interest rate is twice as great, 8%.
      iii) The time period is twice as long, 6 years.
      Justify your answers.
   c) Which scenario in part b produced the greatest amount? Explain.

■ For help with question 12, see Example 4.

12. Jackson and Abina save money for retirement.

   ```
   Investment Plan
   Name: Jackson
   Monthly investment: $40
   Start: Now
   Time period: 30 years
   Annual interest rate: 6%
   Compounding period: monthly
   ```

   ```
   Investment Plan
   Name: Abina
   Monthly investment: $80
   Start: 15 years from now
   Time period: 15 years
   Annual interest rate: 6%
   Compounding period: monthly
   ```

   a) Compare the amount of each annuity with the total investment.
   b) Determine the interest earned by each annuity.
   c) Use the results of parts a and b to explain why financial planners recommend saving for retirement from an early age.

13. **Assessment Focus** Consider these two annuities.
   Annuity 1: $100 deposited at the end of each month for 5 years at 4% per year compounded monthly
   Annuity 2: $300 deposited at the end of each quarter for 5 years at 4% per year compounded quarterly
   a) Calculate the total deposit and the amount of each annuity.
   b) Why are the amounts different even though the total deposit is the same?

7.1 The Amount of an Annuity

14. Consider an annuity of $1000 deposited at the end of each year for 5 years at 3.5% per year compounded annually.
   a) Predict which of the following changes to the annuity would produce the greatest amount.
      i) Doubling the regular deposit
      ii) Doubling the interest rate
      iii) Doubling the time period
      iv) Doubling the frequency of the deposit and the compounding
   b) Calculate the total deposit and the amount of each annuity. Compare your results. Was your prediction correct?

15. **Literacy in Math** Use a graphic organizer to explain how to determine the amount of an annuity. Include an example in your explanation.

16. Suppose you deposit $250 at the end of every 6 months in an investment account that earns 8% per year compounded semi-annually.
   a) Make a graph to illustrate the growth in the amount over a 30-year period.
   b) Gareth says the growth appears to be exponential. Is he correct? Justify your answer.

17. Kishore and Giselle save for their retirement in an investment account that earns 10% per year compounded annually.
   • Kishore starts saving at age 20. He invests $2000 at the end of each year for 10 years. Then he leaves the money to earn interest for the next 35 years.
   • Giselle starts saving at age 35. She invests $2000 at the end of each year for 30 years.
   a) Who do you think will have the greater amount at retirement? Explain.
   b) Calculate the total investment and the amount of each annuity. Compare your results.
   c) Are you surprised by the results? Explain.

In Your Own Words

Many young people delay saving for retirement because they think they can make up the difference by investing more money later. Explain the benefits of saving early. Use examples to illustrate your explanation.

418 CHAPTER 7: Annuities and Mortgages

7.2 The Present Value of an Annuity

Lottery winners are often given the choice of receiving their winnings over time in an annuity or as an immediate cash payment.

Investigate

Calculating the Cash Payment for Winning a Lottery

Materials
- TI-83 or TI-84 graphing calculator

Bi-weekly means every 2 weeks.

Work with a partner.

Top prize winners of the PayDay lottery in British Columbia can receive their winnings as an annuity of $2000 every 2 weeks for 20 years.

What cash payment received today is equivalent to receiving $2000 every 2 weeks for 20 years? Assume money can be invested at an annual interest rate of 5.6% compounded bi-weekly.

Reflect

- Compare your strategies with another group.
- The actual cash payment for the PayDay lottery is $625 000. Compare this value to the one you calculated. What might account for the difference?

Connect the Ideas

Present value of an annuity

The **present value of an annuity** is the principal that must be invested today to provide the regular payments of an annuity.

> **Present value of an ordinary simple annuity**
> $$PV = \frac{R[1 - (1 + i)^{-n}]}{i}, \text{ where:}$$
> - PV is the present value in dollars
> - R is the regular payment in dollars
> - i is the interest rate per compounding period, as a decimal
> - n is the number of compounding periods

The present value formula can only be used when:
- The payment interval is the same as the compounding period.
- A payment is made at the end of each compounding period.
- The first payment is made at the end of the first compounding period.

Example 1

Materials
- scientific calculator
- TI-83 or TI-84 graphing calculator

Providing for an Annuity

Victor wants to withdraw $700 at the end of each month for 8 months, starting 1 month from now. His bank account earns 5.4% per year compounded monthly. How much must Victor deposit in his bank account today to pay for the withdrawals?

Solution

The principal that Victor must deposit today is the present value of an annuity of $700 per month for 8 months at 5.4% per year compounded monthly.

Method 1: Use the present value formula

Substitute $R = 700$, $i = \frac{0.054}{12} = 0.0045$, and $n = 8$ into the present value formula.

$$PV = \frac{R[1 - (1 + i)^{-n}]}{i}$$

$$PV = \frac{700[1 - (1 + 0.0045)^{-8}]}{0.0045}$$

$$\doteq 5488.28$$

Victor must deposit $5488.28.

Press: 700 (1 − (1 + 0.0045) ^ (−) 8) ÷ 0.0045 ENTER

420 CHAPTER 7: Annuities and Mortgages

Method 2: Use the TVM Solver

- Enter: N = 8, I% = 5.4,
  PV = 0, PMT = −700, FV = 0,
  P/Y = 12, and C/Y = 12
- Move the cursor to PV.
  Press: [ALPHA] [ENTER]

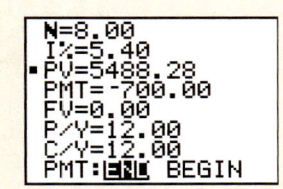

```
N=8.00
I%=5.40
•PV=5488.28
PMT=-700.00
FV=0.00
P/Y=12.00
C/Y=12.00
PMT:END BEGIN
```

Victor must deposit $5488.28.

Example 2

Materials
- scientific calculator
- TI-83 or TI-84 graphing calculator

Payments every 3 months are quarterly payments.

Calculating the Amount Needed at Retirement

Azadeh plans to retire at age 60. She would like to have enough money saved in her retirement account so she can withdraw $7500 every 3 months for 25 years, starting 3 months after she retires. How much must Azadeh deposit at retirement at 9% per year compounded quarterly to provide for the annuity?

Solution

The principal that Azadeh must deposit at retirement is the present value of the annuity payments.

Method 1: Use the present value formula

Substitute $R = 7500$, $i = \frac{0.09}{4} = 0.0225$, and $n = 25 \times 4 = 100$ into the present value formula.

$$PV = \frac{R[1 - (1 + i)^{-n}]}{i}$$

$$PV = \frac{7500[1 - (1 + 0.0225)^{-100}]}{0.0225}$$

Press: 7500 [(] 1 [−] [(] 1 [+] 0.0225 [)] [∧] [(−)] 100 [)] [÷] 0.0225 [ENTER]

$\doteq 297\,313.05$

Azadeh needs $297 313.05 at retirement to pay for the annuity.

Method 2: Use the TVM Solver

- Enter: N = 25 × 4, I% = 9,
  PV = 0, PMT = −7500, FV = 0,
  P/Y = 4, and C/Y = 4
- Move the cursor to PV.
  Press: [ALPHA] [ENTER]

```
N=100.00
I%=9.00
•PV=297313.05
PMT=-7500.00
FV=0.00
P/Y=4.00
C/Y=4.00
PMT:END BEGIN
```

Azadeh needs $297 313.05 at retirement to pay for the annuity.

Repaying loans

Most loans are repaid by making equal monthly payments over a fixed period of time. These payments form an annuity whose present value is the principal borrowed. When all of the payments are made, both the principal borrowed and the interest due will have been paid.

Example 3

Materials
- TI-83 or TI-84 graphing calculator

Calculating the Principal Borrowed for a Loan

Seema plans to buy a used car. She can afford monthly car loan payments of $300. The car dealer offers Seema a loan at 6.9% per year compounded monthly, for 3 years. The first payment will be made 1 month from the date she buys the car.

a) How much can Seema afford to borrow?
b) How much interest will Seema pay on the loan?

Solution

a) Use the TVM Solver to determine the present value of a loan with $300 monthly payments.

We could have used the formula for the present value of an annuity instead of the TVM Solver.

- Enter: N = 3 × 12, I% = 6.9, PV = 0, PMT = −300, FV = 0, P/Y = 12 and C/Y = 12
- Move the cursor to PV.
  Press: [ALPHA] [ENTER]

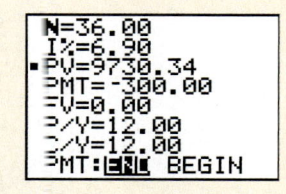

```
N=36.00
I%=6.90
■PV=9730.34
PMT=-300.00
FV=0.00
P/Y=12.00
C/Y=12.00
PMT:END BEGIN
```

Seema can afford to borrow $9730.34.

b) Seema pays a total of 36 × $300 = $10 800.
   The original loan is $9730.34.
   So, the interest paid is $10 800 − $9730.34 = $1069.66.
   This result can be verified with the TVM Solver.

The 2¢ difference in answers is due to rounding.

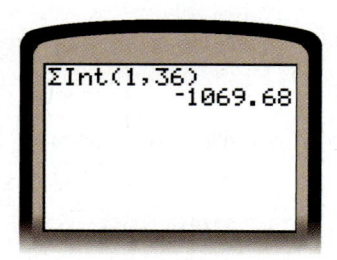

```
ΣInt(1,36)
        -1069.68
```

Press [2nd] [MODE] to exit the TVM Solver.
To determine the interest earned, press:
[APPS] 1 [ALPHA] [MATH] 1 [,] 36 [)] [ENTER]

422 CHAPTER 7: Annuities and Mortgages

Practice

A

For help with questions 1 to 5, see Examples 1 and 2.

1. Evaluate each expression. Write each answer to 2 decimal places.

   a) $\dfrac{45[1 - (1 + 0.02)^{-24}]}{0.02}$

   b) $\dfrac{575[1 - (1 + 0.003)^{-48}]}{0.003}$

   c) $\dfrac{2000[1 - (1 + 0.0065)^{-14}]}{0.0065}$

   d) $\dfrac{95[1 - (1 + 0.12)^{-8}]}{0.12}$

2. Use the formula: $PV = \dfrac{R[1 - (1 + i)^{-n}]}{i}$
   Calculate the value of PV for each set of values.
   a) $R = \$200$, $i = 0.05$, $n = 3$
   b) $R = \$1000$, $i = 0.08$, $n = 7$
   c) $R = \$750$, $i = 0.02$, $n = 12$

3. Maeve wants to set up an annuity to help with her college expenses. She uses the TVM Solver to explore a possible plan.
   a) What regular withdrawal does Maeve plan to make?
   b) How often will she make these withdrawals?
   c) What is the total number of withdrawals Maeve will make?
   d) How much will Maeve have to deposit to provide for the withdrawals?

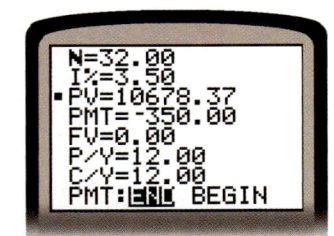

   ```
   N=32.00
   I%=3.50
   •PV=10678.37
   PMT=-350.00
   FV=0.00
   P/Y=12.00
   C/Y=12.00
   PMT:END  BEGIN
   ```

4. Use the present value formula to determine the present value of each annuity.

	Payment	Interest rate	Frequency of compounding	Length of annuity
a)	$300	12%	monthly	2 years
b)	$4500	4%	annually	6 years
c)	$900	9%	semi-annually	4 years
d)	$800	8%	quarterly	5 years

5. Use the TVM Solver to verify your answers to question 4.

7.2 The Present Value of an Annuity

B 6. Determine the present value of each ordinary simple annuity.
   a) Payments of $75 for 10 years at 9.6% per year compounded annually
   b) Payments of $240 for 15 years at 7.25% per year compounded semi-annually
   c) Payments of $8500 for 25 years at 6.3% per year compounded annually
   d) Payments of $50 for 4.5 years at 4.8% per year compounded quarterly

7. Determine the interest earned by each annuity in question 6.

8. A contest offers a prize of $1000 every month for 1 year. The first payment will be made 1 month from now. If money can be invested at 8% per year compounded monthly, what cash payment received immediately is equivalent to the annuity?

9. Tam is setting up an income fund for her retirement. She wishes to receive $1500 every month for the next 20 years, starting 1 month from now. The income fund pays 6.25% per year compounded monthly. How much must Tam deposit now to pay for the annuity?

10. **Assessment Focus** Isabel receives a disability settlement. She must choose one of these payment plans.
    - A single cash payment of $80 000 to be received immediately
    - Monthly disability payments of $1200 for 10 years

    Assume that money can be invested at 4.8% per year compounded monthly. Which settlement do you think Isabel should accept? Justify your answer.

11. Terence's parents want to set up an annuity to help him with his college expenses. The annuity will allow Terence to withdraw $300 every month for 4 years. The first withdrawal will be 1 month from now. The annuity earns 3.5% per year compounded monthly.
    a) What principal should Terence's parents invest now to pay for the annuity?
    b) In which of these scenarios will Terence's parents deposit the least principal?
       i) Terence's withdrawals are twice as great, $600.
       ii) The interest rate is twice as great, 7%.
       iii) The time period is twice as long, 8 years.
       Justify your answers.

For help with question 12, see Example 3.

12. Jeongsoo borrows money to buy a computer. She will repay the loan by making monthly payments of $112.78 per month for the next 2 years at an interest rate of 7.75% per year compounded monthly.
   a) How much did Jeongsoo borrow?
   b) How much interest does Jeongsoo pay?

13. Angela's annuity pays $600 per month for 5 years at 9% per year compounded monthly. Becky's annuity pays $300 per month for 10 years at 9% per year compounded monthly.
The total of the regular payments is the same for each annuity.
Do both annuities have the same present value? Justify your answer.

14. **Literacy in Math** Create Frayer models for the amount and present value of an annuity.

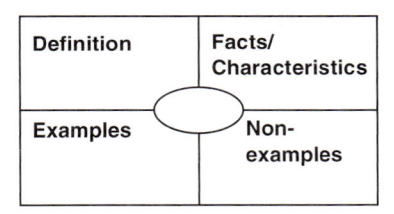

Definition	Facts/Characteristics
Examples	Non-examples

15. Piers wins a talent contest. His prize is an annuity that pays $1000 at the end of each month for 2 years, and then $500 at the end of each month for the next 3 years. How much must the contest organizers deposit in a bank account today to provide the annuity? Assume that money can be invested at 8% per year compounded monthly.

16. Chloe borrowed money from the bank to renovate her home. She will repay the loan by making 24 monthly payments of $64.17 at 12.5% per year compounded monthly.
   a) How much did Chloe borrow?
   b) How much would it cost Chloe to pay off the loan after the 12th payment?
   c) How much interest does Chloe save by paying off the loan early?

In Your Own Words

How are problems involving the present value of an annuity similar to problems involving the amount of an annuity? How are they different? Include examples in your explanation.

7.3 The Regular Payment of an Annuity

Many financial experts caution that young people today cannot count on government or employer pension plans to provide a comfortable retirement. They recommend that young people plan for their retirement by saving early and regularly.

Investigate

Million Dollar Retirement

Materials
- TI-83 or TI-84 graphing calculator

Work with a partner or in a small group.

Some financial experts suggest that a comfortable retirement requires savings of $1 000 000.

■ What monthly payment would you have to make at ages 20, 30, 40, 50, or 60 to accumulate a $1 000 000 retirement fund at age 65? Assume that the fund earns 9% per year compounded monthly.

You may want to organize your work in a table like this.

Age	Years until retirement	Number of monthly payments	Monthly payment
20			

■ Suppose you have $1 000 000 saved in a retirement fund. What regular withdrawal can you make from the fund at the end of each year for 25 years if the fund earns 8% per year compounded annually?

426   CHAPTER 7: Annuities and Mortgages

Reflect

- How does the monthly payment to accumulate $1 000 000 change as the years to retirement decrease?
- At what age do you think it becomes unrealistic to expect to save $1 000 000 for retirement? Explain.
- What would you consider a "comfortable retirement"? What annual income do you think you will need when you retire to have a comfortable retirement?

Connect the Ideas

In Lessons 7.1 and 7.2, we used two different formulas to solve problems involving annuities.

- We used the amount formula to determine the accumulated value of the regular payments at the *end* of an annuity.
- We used the present value formula to determine the money needed at the *beginning* of an annuity to provide regular annuity payments.

When we know the amount or the present value, we can solve for the regular payment. To do this, we rearrange the appropriate formula to isolate R.

Amount formula

$$A = \frac{R[(1 + i)^n - 1]}{i}$$

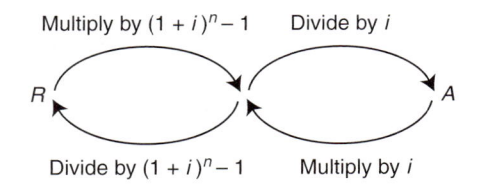

Multiply by $(1 + i)^n - 1$ Divide by i

$R \longrightarrow \longrightarrow A$

Divide by $(1 + i)^n - 1$ Multiply by i

Present value formula

$$PV = \frac{R[1 - (1 + i)^{-n}]}{i}$$

Multiply by $1 - (1 + i)^{-n}$ Divide by i

$R \longrightarrow \longrightarrow PV$

Divide by $1 - (1 + i)^{-n}$ Multiply by i

Example 1 — Determining Payments Given the Amount

Materials
- scientific calculator

Payments every 6 months are semi-annual payments.

Brianne wants to save $6000 for a trip she plans to take in 5 years. What regular deposit should she make at the end of every 6 months in an account that earns 6% per year compounded semi-annually?

Solution

The $6000 represents the money to be accumulated by the regular deposits. So, the $6000 is the amount of the annuity.

We could also isolate R, then substitute.

Substitute, then solve for R.

Substitute $A = 6000$, $i = \frac{0.06}{2} = 0.03$, and $n = 5 \times 2 = 10$ into the amount formula.

$$A = \frac{R[(1+i)^n - 1]}{i}$$

$$6000 = \frac{R[(1+0.03)^{10} - 1]}{0.03}$$ Multiply each side by 0.03.

$6000 \times 0.03 = R(1.03^{10} - 1)$ Divide each side by $1.03^{10} - 1$.

$$\frac{6000 \times 0.03}{1.03^{10} - 1} = R$$

Press: (6000 × 0.03) ÷
(1.03 ^ 10 − 1) ENTER

$R \doteq 523.38$

Brianne will have to make semi-annual deposits of $523.38.

Example 2 — Determining Payments Given the Present Value

Materials
- TI-83 or TI-84 graphing calculator

Donald borrows $1200 from an electronics store to buy a computer. He will repay the loan in equal monthly payments over 3 years, starting 1 month from now. He is charged interest at 12.5% per year compounded monthly. How much is Donald's monthly payment?

Solution

The equal monthly payments Donald makes form an annuity whose present value is $1200.

Use the TVM Solver to determine his monthly payment.

- Enter: $N = 3 \times 12$, $I\% = 12.5$, $PV = 1200$, $PMT = 0$, $FV = 0$, $P/Y = 12$, and $C/Y = 12$
- Move the cursor to PMT.
  Press: ALPHA ENTER

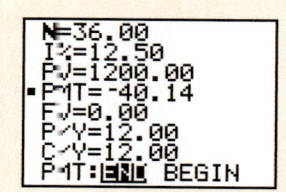

```
N=36.00
I%=12.50
PV=1200.00
•PMT=-40.14
FV=0.00
P/Y=12.00
C/Y=12.00
PMT:END BEGIN
```

Donald's monthly payment is $40.14.

428 CHAPTER 7: Annuities and Mortgages

Example 3

Materials
- TI-83 or TI-84 graphing calculator

Choosing between Two Loan Options

Sheri borrows $9500 to buy a car. She can repay her loan in 2 ways. The interest is compounded monthly.
- **Option A:** 36 monthly payments at 6.9% per year
- **Option B:** 60 monthly payments at 8.9% per year

a) What is Sheri's monthly payment under each option?
b) How much interest does Sheri pay under each option?
c) Give a reason why Sheri might choose each option.

Solution

a) ■ Enter the known values into the TVM Solver.
   ■ Solve for PMT.

Option A

N=36.00
I%=6.90
PV=9500.00
■PMT=-292.90
FV=0.00
P/Y=12.00
C/Y=12.00
PMT:**END** BEGIN

Sheri's monthly payment is $292.90.

Option B

N=60.00
I%=8.90
PV=9500.00
■PMT=-196.74
FV=0.00
P/Y=12.00
C/Y=12.00
PMT:**END** BEGIN

Sheri's monthly payment is $196.74.

> We could have used the formula for the present value of an annuity instead of the TVM Solver.

b) The interest paid is the difference between the total amount paid and the principal borrowed.

Option A
36 payments of $292.90: 36 × $292.90 = $10 544.40
The principal borrowed is $9500.
So, the total interest paid is: $10 544.40 − $9500.00 = $1044.40

Option B
60 payments of $196.74: 60 × $196.74 = $11 804.40
The principal borrowed is $9500.
The total interest paid is: $11 804.40 − $9500.00 = $2304.40

c) Sheri might choose Option A because she will pay less total interest. She might choose Option B because the monthly payments are smaller.

Example 3 illustrates a relationship that is true in general. Extending the time taken to repay a loan reduces the regular payment, but increases the total interest paid.

Practice

A

For help with questions 1 to 5, see Examples 1 and 2.

1. Rearrange each formula to isolate R.

   a) $A = \dfrac{R[(1+i)^n - 1]}{i}$

   b) $PV = \dfrac{R[1 - (1+i)^{-n}]}{i}$

2. Evaluate each expression. Write each answer to 2 decimal places.

   a) $\dfrac{7800 \times 0.03}{1.03^{14} - 1}$

   b) $\dfrac{35\,500 \times 0.025}{1.025^{72} - 1}$

3. Imran plans to finance a new home entertainment system. He uses the TVM Solver to determine his monthly payment.
   a) How much will Imran borrow?
   b) What interest rate will he be charged?
   c) What is the monthly payment?
   d) How many payments will Imran make?
   e) How many years will it take Imran to repay the loan?

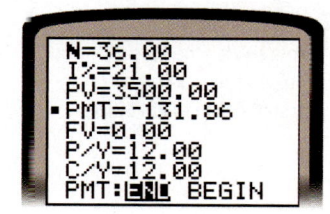

   ```
   N=36.00
   I%=21.00
   PV=3500.00
   ■PMT=-131.86
   FV=0.00
   P/Y=12.00
   C/Y=12.00
   PMT:END BEGIN
   ```

4. Determine the regular payment of each annuity. Each payment is made at the end of the compounding period.

	Amount	Present value	Interest rate	Frequency of compounding	Length of annuity
a)	$4500	–	8%	semi-annually	6 years
b)	$25 000	–	6%	annually	12 years
c)	–	$4000	8%	quarterly	5 years
d)	–	$3500	24%	monthly	2 years

5. Use a different method to verify your answers to question 4.

B

6. Determine whether each situation involves the amount or present value of an annuity. Explain your reasoning.
   a) Steven plans to repay his student loan of $15 000 by making equal annual payments.
   b) Winnie saves $5000 by making equal weekly payments at her bank.
   c) Sergio plans to retire a millionaire by making equal monthly deposits into his retirement savings plan.
   d) Veronika plans to make equal quarterly withdrawals from her $300 000 retirement income fund.

7. Carolyn gets a small business loan for $75 000 to start her hair salon. She will repay the loan with equal monthly payments over 5 years at 8.4% per year compounded monthly.
   a) What is Carolyn's monthly loan payment?
   b) What is the total amount Carolyn repays?
   c) How much of the amount repaid is interest?

8. Shahrzad starts a savings program to have $23 000 in 10 years. She makes deposits at the end of each quarter in an investment account that earns 6.2% per year compounded quarterly.
   a) Determine Shahrzad's quarterly deposit.
   b) Does Shahrzad's deposit double under each change? Justify your answers.
      i) The amount is twice as great, $46 000.
      ii) The interest rate is twice as great, 12.4%.
      iii) The time period is twice as long, 20 years.

9. Boza will need $35 000 in 5 years to start his own business. He plans to save the money by making semi-annual deposits in an account earning 7.8% per year compounded semi-annually.
   a) What semi-annual deposit must Boza make?
   b) How much interest does Boza earn?

■ For help with question 10, see Example 3.

10. Chandra finances a car loan of $18 000 at 9.9% per year compounded monthly. She can repay the loan in 36 months or 48 months. The first payment will be made 1 month after the car is purchased.
   a) What is Chandra's monthly payment for each loan?
   b) How much interest does Chandra save by repaying the loan in 36 months instead of 48 months?

11. David and Ulani each arrange a 3-year car loan for $20 000.
   • David is charged interest at 9.3% per year compounded monthly
   • Ulani is charged interest at 12.5% per year compounded monthly
   a) Determine the monthly payment for each loan.
   b) How much extra interest does Ulani pay? Explain.

12. Create a problem involving the regular payment of an annuity whose solution is given by the TVM Solver screen.

```
N=24.00
I%=7.50
PV=4000.00
■PMT=-180.00
FV=0.00
P/Y=12.00
C/Y=12.00
PMT:END BEGIN
```

13. **Assessment Focus** Lincoln wants to have $10 000 in 6 years by making equal regular deposits into a bank account. He can:
    - Make a deposit at the end of each month in an account that earns 7.8% per year compounded monthly
    - Make a deposit at the end of each quarter in an account that earns 8.0% per year compounded quarterly

    Which option should Lincoln choose? Justify your answer.

14. Megan and Nancy each want to save $250 000 for their retirement in 40 years.
    a) Nancy begins her regular deposits immediately. How much must she deposit at the end of each year at 12% per year compounded annually to achieve her goal?
    b) Megan decides to wait 10 years before she starts her regular deposits. What annual deposit does she need to make?
    c) Compare Megan and Nancy's total deposits. How much less does Nancy deposit by starting early?

15. **Literacy in Math** Use a concept map to summarize what you have learned about annuities. Add to the concept map as you work through the chapter.

16. Chukwuma deposits $1500 in a retirement savings plan at the end of every 6 months for 20 years. The money earns 11% per year compounded semi-annually. After 20 years, Chukwuma converts the retirement savings plan into an income fund that earns 7% per year compounded monthly. He plans to make equal withdrawals at the end of every month for 15 years. What regular withdrawals can Chukwuma make?

In Your Own Words

Suppose you are asked to determine the regular payment of an annuity. How do you know which formula to use, or whether to enter a value for FV or PV in the TVM Solver? Use examples in your explanation.

7.4 Using a Spreadsheet to Investigate Annuities

Spreadsheets are an important tool in business and personal finance. We can use a spreadsheet to change the features of an annuity and analyse the effect of the change.

Inquire: Analysing Annuities with a Spreadsheet

Materials
- Microsoft Excel
- Amount.xls
- Loan.xls

Work with a partner.

Part A: Analysing the Amount of an Annuity

Kiran deposits $500 every 6 months into an investment account that earns 7% per year compounded semi-annually. What is the amount in the account after 3 years?

- If you are using the file *Amount.xls*, open it and begin at question 2.
- If you are not using the file *Amount.xls*, start at question 1.

1. **Creating an investment template**

   a) • Open a new spreadsheet document.
      • Copy the headings, values, and formulas shown below.

	A	B	C	D	E
1	Amount of an Annuity				
2	Regular payments		500		
3	Annual interest rate		0.07		
4	Compounding periods per year		2		
5	Number of years		3		
6	Interest rate per period		=C3/C4		
7	Number of periods		=C4*C5		
8					
9	Period	Starting balance	Interest	Deposit	Ending balance
10	1	0	0	=$C$2	=B10+C10+D10
11	=A10+1	=E10	=ROUND($C$6*B11,2)	=$C$2	=B11+C11+D11

   • Format cells C2 and B10 to E11 as currency.
   • Format cells C3 and C6 as percents to 2 decimal places.

7.4 Using a Spreadsheet to Investigate Annuities **433**

b) Refer to the formulas in the spreadsheet in part a.
- Explain the formulas in cells C6 and C7.
- Why is the interest 0 in the first period?
- The formula =$C$2 appears in cells D10 and D11. The $ sign indicates that cell address C2 should not change when we copy the formula to other cells in column D. Explain why this makes sense.
- Explain the formulas for the ending and starting balances.

c)
- Select cells A11 to E11.
- **Fill Down** to copy the formulas in row 11 through row 15. Your spreadsheet should look like this.

	A	B	C	D	E
1	Amount of an Annuity				
2	Regular payments		$500.00		
3	Annual interest rate		7.00%		
4	Compounding periods per year		2		
5	Number of years		3		
6	Interest rate per period		3.50%		
7	Number of periods		6		
8					
9	Period	Starting balance	Interest	Deposit	Ending balance
10	1	$0.00	$0.00	$500.00	$500.00
11	2	$500.00	$17.50	$500.00	$1017.50
12	3	$1017.50	$35.61	$500.00	$1553.11
13	4	$1553.11	$54.36	$500.00	$2107.47
14	5	$2107.47	$73.76	$500.00	$2681.23
15	6	$2681.23	$93.84	$500.00	$3275.07

The amount of the annuity is $3275.07.

> Record the amount under each change in questions 2 to 5. You will compare these amounts in question 6.

2. Changing the payment

a) Suppose the regular payment is doubled from $500 to $1000. Predict how the amount will change. Explain your reasoning.

> Be as specific as you can in your prediction.

b) Check your prediction. Were you correct? Explain.

c) Repeat parts a and b when the regular payment is halved from $500 to $250.

Change the regular deposit back to $500.00.

3. Changing the interest rate

a) Suppose the annual interest rate is doubled from 7% to 14%. Predict how the amount will change. Explain your reasoning.

b) Check your prediction. Were you correct? Explain.

c) Repeat parts a and b when the annual interest rate is halved from 7% to 3.5%.

Change the interest rate back to 7%.

4. Changing the term

> Use the value in cell C7 to determine the number of periods to display.

a) Suppose the term is doubled from 3 years to 6 years. Predict how the amount will change. Explain your reasoning.
b) Check your prediction. Were you correct? Explain.
c) Repeat parts a and b when the term is halved from 3 years to 1.5 years.

Change the term back to 3 years.

5. Changing the payment frequency and the compounding

> Use the value in cell C7 to determine the number of periods to display.

a) Suppose the payment frequency and compounding are doubled from semi-annually to quarterly. Predict how the amount will change. Explain your reasoning.
b) Check your prediction. Were you correct? Explain.
c) Repeat parts a and b when the payment frequency and the compounding is halved from semi-annually to annually.
d) Compare the amount under each payment and compounding frequency. Which produces the greatest amount? Why does this result make sense?

Change the number of compounding periods back to 2.

6. Comparing the effect of each change

Refer to your answers in questions 2 to 5.
Which of the following changes produced the greatest amount?
- Doubling the payments
- Doubling the interest rate
- Doubling the term
- Doubling the payment frequency and compounding

7.4 Using a Spreadsheet to Investigate Annuities **435**

Part B: Repaying a Loan

Devon plans to borrow $500 for 12 months at 18% per year compounded monthly. He will repay the loan with equal monthly payments.

Devon uses an online loan calculator to determine his monthly payment.

```
Calculate:  ● Payment
            ○ Loan Amount

Monthly payment:  $45.84
Loan amount:      $500
Term in months:   12
Interest rate:    18.000%

Loan amount:       $500
Total of payments: $550.08
Total interest paid: $50.08
```

We can use a spreadsheet to analyse how Devon's loan is repaid over the 12 months.

- If you are using the file *Loan.xls*, open it and begin at question 8.
- If you are not using the file *Loan.xls*, start at question 7.

7. **Creating a loan repayment template**

    a) • Open a new spreadsheet document.
       • Copy the headings, values, and formulas shown below.

	A	B	C	D	E
1	Loan Repayment Schedule				
2	Principal borrowed		500		
3	Annual interest rate		0.18		
4	Compounding periods per year		12		
5	Interest rate per period		=C3/C4		
6	Number of payments		12		
7	Monthly payment		=PMT(C5, C6, -C2)		
8					
9	Payment	Payment	Interest	Principal	Outstanding
10	number		paid	paid	balance
11	0				=C2
12	1	=ROUND($C$7,2)	=ROUND(E11*$C$5,2)	=B12-C12	=E11-D12

In cell C7, enter -C2 to produce a positive number for the payment.

   • Format cells C2, C7, E11, and B12 to E12 as currency.
   • Format cells C3 and C5 as percents to 2 decimal places.

   b) Refer to the formulas in the spreadsheet in part a.
      • The **PMT** function in cell C7 calculates the regular payment of the loan. What do the numbers in the brackets represent?
      • Explain the remaining formulas in the table.

436 CHAPTER 7: Annuities and Mortgages

c) • Select cells A12 to E12.
• **Fill Down** to copy the formulas in row 12 through row 23. Your spreadsheet should look like this.

	A	B	C	D	E
1	Loan Repayment Schedule				
2	Principal borrowed		$500.00		
3	Annual interest rate		18.00%		
4	Compounding periods per year		12		
5	Interest rate per period		1.50%		
6	Number of payments		12		
7	Monthly payment		$45.84		
8					
9	Payment	Payment	Interest	Principal	Outstanding
10	number		paid	paid	balance
11	0				$500.00
12	1	$45.84	$7.50	$38.34	$461.66
13	2	$45.84	$6.92	$38.92	$422.74
14	3	$45.84	$6.34	$39.50	$383.24
15	4	$45.84	$5.75	$40.09	$343.15
16	5	$45.84	$5.15	$40.69	$302.46
17	6	$45.84	$4.54	$41.30	$261.16
18	7	$45.84	$3.92	$41.92	$219.24
19	8	$45.84	$3.29	$42.55	$176.69
20	9	$45.84	$2.65	$43.19	$133.50
21	10	$45.84	$2.00	$43.84	$89.66
22	11	$45.84	$1.34	$44.50	$45.16
23	12	$45.84	$0.68	$45.16	$0.00

8. **Explaining how the interest paid and the outstanding balance are calculated**

In a loan, the principal borrowed plus interest must be repaid. So, part of each loan payment is interest, and the rest reduces the principal.

Payment 1
- When payment 1 is made, a month's interest is owed on the outstanding balance, $500.
  From cell C5, the monthly rate is 1.50%.
  So, the monthly interest charge is: $500 \times 0.015 = \$7.50$
- The monthly payment is $45.84. Since $7.50 is interest, the part that repays principal is: $\$45.84 - \$7.50 = \$38.34$
- The outstanding balance at the end of the first month is: $\$500 - \$38.34 = \$461.66$

Repeat these calculations for two other payments in the table. Explain your work.

9. **Analysing the repayment schedule**
   a) Why do you think that the interest is paid before the principal is reduced?

b) As the outstanding balance decreases, the interest paid decreases and the principal repaid increases. Explain why this happens.
c) Has the principal of the loan been reduced by one-half after 6 of the 12 payments? Explain.
d) We can calculate the total interest paid over the life of the loan by adding the values in cells C12 to C23.
- In cell C24, type: =SUM(
- Select cells C12 to C23.
  The formula should now read: =SUM(C12:C23
- Complete the formula by typing:)

How does this value compare with the value Devon obtained on the online calculator?
e) How does the total interest paid compare to the original principal of the loan?

10. Making changes to the loan
a) Predict how each of the following affect the regular payment and total interest paid over the life of the loan.
- Changing the principal borrowed to $1000
- Changing the interest rate to 9%
- Changing the term of the loan to 6 months
- Changing the term of the loan to 24 months

b) Check each prediction. Remember to change the spreadsheet back to its original form after each change.
c) Suppose Devon doubles his monthly payment. Will the time taken to repay the loan decrease by one-half? Justify your answer.
d) Summarize your results. What features of the loan could Devon change to accomplish each of these goals?
- Decrease the monthly payment
- Reduce the total interest paid
- Reduce the time taken to repay the loan

Can Devon accomplish all three goals at the same time? Explain.

Reflect

- What is an advantage and disadvantage of using a spreadsheet?
- Which do you prefer to use: formulas, TVM Solver, or a spreadsheet? Explain.

7.5 Saving for Education and Retirement

Jim is a financial planner. He encourages his clients to save regularly to obtain the money needed for future expenses such as their retirement or their children's education. Most people find it easier to save by setting aside a portion of each paycheque rather than coming up with large sums of money to invest.

Inquire: Researching Savings Plans for Education and Retirement

Materials
- computer with Internet access
- print materials about registered plans or a financial planner

Work in a small group.

Post-secondary education and retirement typically involve large sums of money. The Canadian government offers *Registered Education Savings Plans* (RESPs) and *Registered Retirement Savings Plans* (RRSPs) to encourage Canadians to save for these goals.

1. **Planning the research**
   - What sources will you use to find information about RESPs and RRSPs?
   - If you use the Internet, what other search words might you use?
   - How will you record your research? Which graphic organizers may be helpful?
   - How will you share the work among group members?

Search words
- ☐ Canada Revenue Agency RESPs RRSPs
- ☐ Human Resources Canada RESPs
- ☐ RESPs and RRSPs explained

2. Gathering information about RESPs

Use your own words to answer these questions.
- Research background information
  - What is an RESP?
  - What are the benefits of saving money in an RESP?
  - Where can you open an RESP?
  - What is a qualifying educational program?
  - What tax rules apply to an RESP?
- Research contribution rules
  - Who can contribute to an RESP?
  - What is the maximum lifetime contribution allowed?
  - How much money does the government contribute to an RESP?
- Research withdrawal rules
  - How is money withdrawn from an RESP when the student starts post-secondary education?
  - What happens to the money in the RESP if the student decides not to pursue post-secondary education?
- Record any other information you discovered in your research that you think is important. Why do you think it is important?

3. Gathering information about RRSPs and RIFs

Use your own words to answer these questions.
- Research background information
  - What is an RRSP?
  - What are the benefits of saving money in an RRSP?
  - Where can you open an RRSP?
  - What tax rules apply to an RRSP?

- Research contribution rules
  - Who can contribute to an RRSP?
  - How often and how much can a person contribute to an RRSP?
  - For how long can a person contribute to an RRSP?
- Research withdrawal rules
  - What happens when a person needs some of the money saved in an RRSP for an emergency?
  - What do you do with the money in an RRSP when you retire?
  - What is the difference between an RRSP and a *Retirement Income Fund* (RIF)?
- Record any other information you discovered in your research that you think is important. Why do you think it is important?

4. **Using online financial calculators**

   Many financial institutions offer online RESP, RRSP, and RIF calculators.
   - Find an RESP calculator.
     - What information does the calculator require?
       Try some sample calculations.
       Record or print the calculator screen.
   - Find an RRSP calculator.
     - Think of the largest contribution you can afford to make every month from age 18 to age 65. What will be the amount of these contributions when you retire? Use realistic interest rates.
     - Investigate how the amount at retirement changes if you:
       – Double your contributions
       – Wait 10 years until you start contributing
       Record or print the calculator screen.

Reflect

- Why is it important to start contributing early to any savings plan?
- What was the most important fact you learned about RESPs and RRSPs? Why is it important?
- How easy was it to find and use an online calculator? Explain.

Mid-Chapter Review

7.1

1. Babette spends $225 a year on lottery tickets. After 15 years, her total winnings are $1200. Suppose Babette had invested the money she spends on lottery tickets in an account that earns 6% per year compounded annually. How much would Babette have accumulated after 15 years?

2. Harvey deposits $2500 at the end of each year into an RRSP that earns 9.6% per year compounded annually.
   a) Determine the amount in the RRSP at the end of each number of years.
      i) 10 years   ii) 20 years   iii) 40 years
   b) Determine the interest earned after each time period in part a.
   c) Use your answers in parts a and b to explain the advantages of saving early.

7.2

3. Allison wins a lottery. She can receive $25 000 at the end of every 6 months for 20 years or an equivalent cash payment immediately. Determine the value of the cash payment if money can be invested at 8.5% per year compounded semi-annually.

4. Create an example to show how the present value of an annuity changes in each situation.
   a) The regular payment is doubled.
   b) The interest rate is doubled.
   c) The number of years is doubled.
   d) The compounding period is doubled.

7.3

5. Pilar needs $2500 three years from now. How much should she deposit at the end of each quarter at 4% per year compounded quarterly to obtain the required amount?

6. Florine borrowed $25 000 at 9.6% per year compounded monthly to buy a new houseboat. She can repay the money by making equal monthly payments for 7 years or 10 years.
   a) Determine the monthly payment for each time period.
   b) How much would Florine save in interest by choosing the 7-year loan?
   c) Why might Florine choose the 10-year loan even though the interest costs are greater?

7.4

7. Malcolm plans to invest $500 at the end of every 6 months in a savings account that earns 5% per year compounded semi-annually.
   a) Use a spreadsheet to determine the amount in the account after 2 years.
   b) How much more would Malcolm have at the end of the 2 years under each change?
      i) The monthly deposits are $600.
      ii) The interest rate is 6% per year.

8. Elyse borrows $8000 at 12% per year compounded monthly. She will repay the loan by making monthly payments of $177.96 for the next 5 years.
   a) Use a spreadsheet to create a payment schedule for Elyse's loan.
   b) How much does Elyse have left to repay after 3 years?

7.5

9. What do you think are the two main benefits of using an RESP?

442   CHAPTER 7: Annuities and Mortgages

7.6 What Is a Mortgage?

Kyle and Tea want to arrange a **mortgage** to buy a house. They talk to the mortgage specialist at their bank and research mortgages on the Internet so they can make an informed decision about the mortgage best suited to their personal and financial goals.

Inquire

Researching Mortgages

Materials
- computer with Internet access

Work in small groups.

1. **Planning the research**
   - What sources can you use to find information about Canadian mortgages?
   - What search words might you use to research mortgage vocabulary? What search words might you use to learn about the various types of mortgages available?
   - How will you record your research? Which graphic organizers may be helpful?
   - How will you share the work among group members?

Search words
- ☐ Canada Mortgage and Housing Corporation
- ☐ Ontario Real Estate Association
- ☐ Mortgages explained
- ☐ Mortgage glossary

2. **Gathering information about mortgages**
   - Research general information about mortgages
     - What is a mortgage?
     - Where can a mortgage be obtained?
     - What financial requirements must be met to qualify for a mortgage?
     - How is a mortgage generally repaid?
   - Research down payments
     - What is the minimum down payment required for a mortgage?
     - What are the advantages and disadvantages of a large down payment instead of a small down payment?
     - What are the advantages and disadvantages of a large or small down payment?
   - Research features of a mortgage
     - What are the current interest rates on mortgages?
     - How often is the interest compounded?
     - What is the difference between the *amortization period* and *term* of a mortgage? What amortization periods and terms are commonly available?
     - How often can mortgage payments be made?
     - Some financial institutions allow accelerated payments. What does this mean? What is the benefit of repaying a mortgage with accelerated payments?

3. **Comparing different types of mortgages**
   Explain the difference between each type of mortgage. Why might a homeowner choose one type of mortgage over the other?
   - Conventional mortgage or high-ratio mortgage
   - Open or closed mortgage
   - Fixed-rate or variable-rate mortgage
   - Short-term or long-term mortgage

4. **Completing the research**
   Record any other terms or information you came across in your research that you think are important.

Practice

You may need to do additional research to answer these questions.

A

1. Kyle and Tea purchase a house for a selling price of $155 000. They plan to make a down payment of 25% and arrange a mortgage for the rest.
   a) How much is their down payment?
   b) How much will they borrow for the mortgage?

B

2. Suppose Kyle and Tea cannot afford a 25% down payment.
   a) What additional costs will Kyle and Tea pay by making the lesser down payment? Explain.
   b) How would a greater initial down payment end up saving money over the life of the mortgage?

3. Kyle and Tea may have to pay a number of other costs when they purchase their house. These costs are usually given as a percent of the selling price.
   Land-transfer tax: 0.75%   *Mortgage loan insurance premium*: 2.75%
   Building inspection: 0.25%   *Legal fees*: 1%
   Determine the total of these costs for a house with selling price $155 000.

4. Kyle and Tea's bank offers them a 3% cash back incentive on their mortgage. They can use this money towards their down payment.
   a) How much cash back would they receive on a $155 000 mortgage?
   b) Why do you think banks offer a cash back option to customers?

5. Kyle and Tea obtain a mortgage for $139 500 with an amortization period of 25 years and a 5-year term.
   a) What does this mean in everyday language?
   b) Use an online mortgage calculator to determine Kyle and Tea's monthly payment. Use current interest rates.
   c) What other payment frequencies are available to repay a mortgage?

Reflect

- Are mortgages annuities? Explain.
- Which sources of information did you find most helpful? Why were they helpful?
- Which of the costs associated with buying a house surprised you the most? Explain.

7.7 Amortizing a Mortgage

Geri is a mortgage specialist at a bank. When a customer arranges a mortgage, she provides a payment schedule that gives a detailed breakdown of how the mortgage will be repaid.

Investigate

Analysing the Repayment of a Mortgage

Materials
- scientific calculator

Work with a partner.

The Babiaks will repay a $100 000 mortgage loan, plus interest, over 15 years by making equal monthly payments. The mortgage calculator on their bank's Web site displays a table and graph that show how the mortgage is repaid year by year over the 15 years.

Year	Total of payments	Principal paid	Interest paid	Ending principal balance
				$100 000.00
1	$9457.44	$4612.42	$4845.02	$95 387.58
2	$9457.44	$4845.92	$4611.52	$90 541.66
3	$9457.44	$5091.26	$4366.18	$85 450.40
4	$9457.44	$5348.99	$4108.45	$80 101.41
5	$9457.44	$5619.80	$3837.64	$74 481.61
6	$9457.44	$5904.29	$3553.15	$68 577.32
7	$9457.44	$6203.20	$3254.24	$62 374.12
8	$9457.44	$6517.23	$2940.21	$55 856.89
9	$9457.44	$6847.17	$2610.27	$49 009.72
10	$9457.44	$7193.81	$2263.63	$41 815.91
11	$9457.44	$7557.98	$1899.46	$34 257.93
12	$9457.44	$7940.61	$1516.83	$26 317.32
13	$9457.44	$8342.61	$1114.83	$17 974.71
14	$9457.44	$8764.94	$692.50	$9209.77
15	$9458.52	$9209.77	$248.75	$0.00

446 CHAPTER 7: Annuities and Mortgages

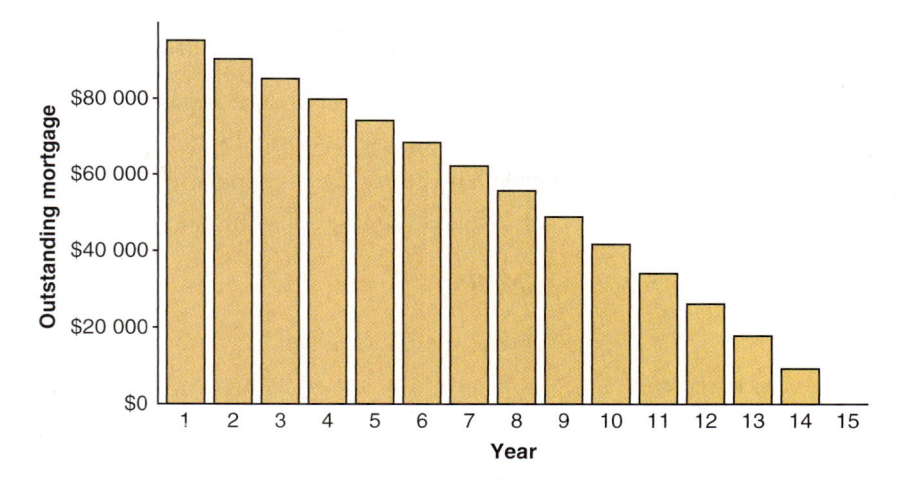

- How do the total paid, interest, principal, and balance change over the life of the mortgage?
- What is the total amount paid and the total interest paid over the lifetime of the mortgage? How do these values compare with the principal originally borrowed?

Reflect

- What patterns do you see in the table or the graph?
- Why do you think these patterns occur?

Connect the Ideas

Mortgage interest rates

Under Canadian law, interest on mortgages can be compounded at most semi-annually. However, mortgage payments are often made monthly. These monthly payments form an annuity whose present value is the principal originally borrowed.

Since the payment period and compounding period are different, we cannot calculate the monthly payment on a mortgage by using the formula for the present value of an ordinary simple annuity. We use the TVM Solver instead. To represent monthly payments and semi-annual compounding, we set P/Y = 12 and C/Y = 2.

7.7 Amortizing a Mortgage **447**

Example 1

Determining the Monthly Mortgage Payment

Materials
- TI-83 or TI-84 graphing calculator

The Cafirmas take out a mortgage of $210 000 at 5% per year compounded semi-annually for 25 years.

a) What is their monthly payment?
b) What is the total interest paid over the 25 years?

Solution

a) Use the TVM Solver.
   - Enter the known values.
   - Solve for PMT.

   The Cafirmas' monthly payment is $1221.37.

N is the total number of payments.

```
N=300.00
I%=5.00
PV=210000.00
•PMT=-1221.37
FV=0.00
P/Y=12.00
C/Y=2.00
PMT:END BEGIN
```

b) 300 payments of $1221.37: 300 × $1221.37 = $366 411.00
   The principal borrowed is $210 000.
   So, the total interest paid over the 25 years is
   $366 411.00 − $210 000 = $156 411.00
   This result can be verified with a graphing calculator.

The 25¢ difference is due to rounding.

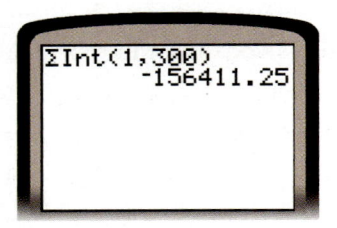

```
ΣInt(1,300)
      -156411.25
```

Amortizing a mortgage

A mortgage is *amortized* when both the principal and interest are paid off with a series of equal, regular payments. For example, the mortgage in *Example 1* was amortized by making monthly payments of $1221.37 over an **amortization period** of 25 years.

To simplify the math, we assumed that the interest rate is fixed for the entire amortization period.

In reality, mortgage interest rates are fixed for a shorter length of time called the *term* of the mortgage. The term normally ranges from 6 months to 10 years. At the end of the term, the mortgage must be paid off or renewed at the current rate of interest.

Amortization table

We can use an **amortization table** to analyse how a mortgage is repaid. The amortization table gives a detailed breakdown of the interest and principal paid by each payment and the loan balance after the payment.

Example 2

Reading and Interpreting an Amortization Table

Here is a partial amortization table for the Cafirmas' mortgage.

Payment number	Monthly payment	Interest paid	Principal paid	Outstanding balance
0				$210 000.00
1	$1221.37	$866.02	$355.35	$209 644.65
2	$1221.37	$864.56	$356.81	$209 287.84
3	$1221.37	$863.09	$358.28	$208 929.56
4	$1221.37	$861.61	$359.76	$208 569.80
5	$1221.37	$860.12	$361.25	$208 208.55
6	$1221.37	$858.63	$362.74	$207 845.81
⋮	⋮	⋮	⋮	⋮
295	$1221.37	$29.79	$1191.58	$6032.26
296	$1221.37	$24.88	$1196.49	$4835.77
297	$1221.37	$19.94	$1201.43	$3634.34
298	$1221.37	$14.99	$1206.38	$2427.96
299	$1221.37	$10.01	$1211.36	$1216.60
300	$1221.62	$5.02	$1216.60	$0.00
Total	$366 411.25	$156 411.25	$210 000.00	

a) How much interest and principal is paid in the 5th payment?
   How much do the Cafirmas still owe after this payment?
b) What is the outstanding balance after 6 months?
c) Compare the interest and principal paid in the first 6 months of the mortgage with the interest and principal paid in the last 6 months of the mortgage. What do you notice?
d) Why is the monthly payment increased for the 300th payment?
e) What percent of the total amount paid is interest?

Solution

a) The interest paid in the 5th payment is $860.12.
   The principal paid is $361.25.
   The Cafirmas still owe $208 208.55 after this payment.
b) The outstanding balance after 6 months is $207 845.81.
c) In the first 6 months, the payments mostly cover interest, while in the last 6 months, the payments mostly cover principal.

d) The mortgage is paid off in the 300th payment, so the outstanding balance should be $0.00. The outstanding principal after the 299th payment is $1216.60 and the interest charge for this payment is $5.02. So, the 300th payment increases to $1216.60 + $5.02 = $1221.62.

e) The total paid over the life of the mortgage is $366 411.25.
The total interest paid is $156 411.25.
The interest as a percent of the total amount paid is:
$$\frac{\$156\,411.25}{\$366\,411.25} \times 100\% \doteq 43\%$$

Example 2 illustrates some key points about the amortization of a mortgage.
- Although the monthly payments are equal, the split between interest and principal changes with each payment.
- With each payment, the outstanding balance on the mortgage decreases. So, the part of each payment that covers interest decreases.
- As the portion of each payment that covers interest decreases, the part that repays principal increases.

Practice

A

For help with questions 1 to 4, see Example 1.

1. Nadia uses the TVM Solver to estimate the monthly payment for her mortgage.
a) How much does Nadia plan to borrow?
b) What interest rate is Nadia charged?
c) What is Nadia's monthly payment?
d) What is the total number of payments Nadia will make?
e) Why is P/Y = 12 and C/Y = 2?

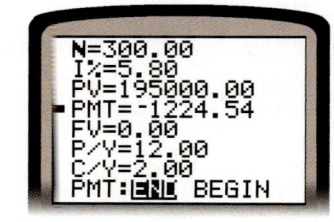

```
N=300.00
I%=5.80
PV=195000.00
PMT=-1224.54
FV=0.00
P/Y=12.00
C/Y=2.00
PMT:END BEGIN
```

2. For each TVM Solver screen shown:
   i) What was the principal borrowed?
   ii) How many payments will it take to repay the mortgage?
   iii) What is the total of the monthly payments over the life of the mortgage?
   iv) What is the total interest paid?

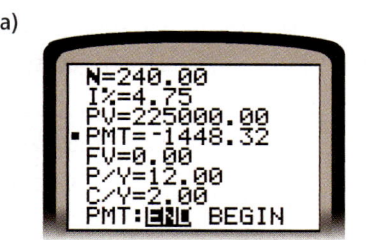

a)
```
N=240.00
I%=4.75
PV=225000.00
•PMT=-1448.32
FV=0.00
P/Y=12.00
C/Y=2.00
PMT:END BEGIN
```

b)
```
N=180.00
I%=7.50
PV=80000.00
•PMT=-736.41
FV=0.00
P/Y=12.00
C/Y=2.00
PMT:END BEGIN
```

3. Determine the monthly payment for each mortgage. The interest is compounded semi-annually.

	Principal borrowed	Interest rate	Length of mortgage
a)	$65 000	4%	15 years
b)	$150 000	5%	25 years
c)	$190 000	7.5%	20 years
d)	$289 000	6.25%	30 years

4. Determine the total interest paid over the life of each mortgage in question 3.

■ For help with question 5, see Example 2.

5. This amortization table shows the first 3 payments on the Parks' mortgage.

Payment number	Monthly payment	Interest paid	Principal paid	Outstanding balance
0				$125 000.00
1	$799.76	$617.33	$182.43	$124 817.57
2	$799.76	$616.43	$183.33	$124 634.24
3	$799.76	$615.52	$184.24	$124 450.00
Total	$2399.28	$1849.28	$550.00	

a) How much money did the Parks borrow?
b) What is their monthly payment?
c) How much of the 1st payment is interest?
d) How much of the 2nd payment is principal?
e) What is the outstanding balance after the 3rd payment?
f) Compare the total interest and total principal paid in the first 3 payments. What do you notice?

B This amortization table shows the first 3 payments and last 3 payments on a mortgage. Use the table to answer questions 6 and 7.

Payment number	Monthly payment	Interest paid	Principal paid	Outstanding balance
0				$120 000.00
1	$1104.62	$738.54	$366.08	$119 633.92
2	$1104.62	$736.29	$368.33	$119 265.59
3	$1104.62	$734.02	$370.60	$118 894.99
:	:	:	:	:
178	$1104.62	$20.14	$1084.48	$2187.86
179	$1104.62	$13.47	$1091.15	$1096.71
180	$1103.46	$6.75	$1096.71	$0.00
Total	$198 830.44	$78 830.44	$120 000.00	

■ For help with question 6, see Example 3.

6. a) What is the principal borrowed and the monthly payment?
   b) What is the amortization period? Justify your answer.
   c) How much of the 1st payment is interest and how much repays principal?
   d) What is the total interest paid over the life of the mortgage? How does this compare to the principal originally borrowed?
   e) What is the outstanding balance after the first 3 payments?

7. a) How does the interest paid in the first 3 months compare with the principal paid? Explain.
   b) How does the interest paid in the last 3 months compare with the principal paid? Explain.
   c) What percent of the first 3 payments pays interest and repays principal? Show your calculations.

8. The Smiths would like to buy a new cottage. They have negotiated a selling price of $175 000. They will make a down payment of 15% and arrange a mortgage at 4.5% per year compounded semi-annually over 15 years to finance the rest.
   a) Determine the down payment and the principal of the mortgage loan.
   b) What is the Smiths' monthly payment?
   c) What is the total amount the Smiths pay over the life of the mortgage? How does this compare to the principal originally borrowed? Explain.

9. Literacy in Math Create a flowchart that shows the steps used to calculate a monthly mortgage payment on the TVM Solver. Identify the steps in which the values entered are always the same.

10. **Assessment Focus** Joseph has arranged a mortgage to purchase his first home. The mortgage will be repaid over 25 years at 6% per year compounded semi-annually. The amortization table shows Joseph's first 12 payments.

Payment number	Monthly payment	Interest paid	Principal paid	Outstanding balance
0				$185 000.00
1	$1183.64	$913.65	$269.99	$184 730.01
2	$1183.64	$912.31	$271.33	$184 458.68
3	$1183.64	$910.97	$272.67	$184 186.01
4	$1183.64	$909.63	$274.01	$183 912.00
5	$1183.64	$908.27	$275.37	$183 636.63
6	$1183.64	$906.91	$276.73	$183 359.90
7	$1183.64	$905.55	$278.09	$183 081.81
8	$1183.64	$904.17	$279.47	$182 802.34
9	$1183.64	$902.79	$280.85	$182 521.49
10	$1183.64	$901.40	$282.24	$182 239.25
11	$1183.64	$900.01	$283.63	$181 955.62
12	$1183.64	$898.61	$285.03	$181 670.59

a) Use the TVM Solver to verify the monthly mortgage payment. Record the screen. Justify the values you enter for the variables.
b) How much of the 6th payment is interest? How much repays principal?
c) What is the outstanding balance after 8 payments?
d) What percent of the mortgage has been repaid at the end of the first 12 payments? Explain.

11. A mortgage for $225 000 at 5.25% per year compounded semi-annually will be repaid with equal monthly payments.
a) Deanna thinks that the monthly payment for a 15-year amortization period should be double the monthly payment for a 30-year amortization period since the amortization period is one-half as long. Do you agree? Explain your reasoning.
b) Calculate the monthly payment for each amortization period. Were you correct? Explain.
c) These amortization tables show the first 6 payments for each mortgage.

15-year amortization

Payment number	Interest paid	Principal paid	Outstanding balance
0			$225 000.00
1	$973.78	$828.27	$224 171.73
2	$970.19	$831.86	$223 339.87
3	$966.59	$835.46	$222 504.41
4	$962.98	$839.07	$221 665.34
5	$959.35	$842.70	$220 822.64
6	$955.70	$846.35	$219 976.29

30-year amortization

Payment number	Interest paid	Principal paid	Outstanding balance
0			$225 000.00
1	$973.78	$260.81	$224 739.19
2	$972.65	$261.94	$224 477.25
3	$971.52	$263.07	$224 214.18
4	$970.38	$264.21	$223 949.97
5	$969.23	$265.36	$223 684.61
6	$968.09	$266.50	$223 418.11

Explain why the 15-year loan is paid off in half the time of a 30-year loan even though the monthly payment for the 15-year loan is only $567.46 more than the monthly payment for the 30-year loan.

7.7 Amortizing a Mortgage

12. a) The TVM Solver can be used to create an amortization table. Read the user manual or research on the Internet to learn how to do this.

b) Use the TVM Solver to create an amortization table for the first 6 payments of a $200 000 mortgage at 5.3% per year compounded semi-annually for 25 years.

13. In the United States, mortgage payments can be compounded monthly. Consider a mortgage of $145 000 at 6.5% per year with an amortization period of 25 years.

a) These amortization tables show the first 6 payments for each mortgage.

US mortgage

Payment number	Monthly payment	Interest paid	Principal paid	Outstanding balance
0				$145 000.00
1	$979.05	$785.42	$193.63	$144 806.37
2	$979.05	$784.37	$194.68	$144 611.69
3	$979.05	$783.31	$195.74	$144 415.95
4	$979.05	$782.25	$196.80	$144 219.15
5	$979.05	$781.19	$197.86	$144 021.29
6	$979.05	$780.12	$198.93	$143 822.36

Canadian mortgage

Payment number	Monthly payment	Interest paid	Principal paid	Outstanding balance
0				$145 000.00
1	$971.24	$774.99	$196.25	$144 803.75
2	$971.24	$773.94	$197.30	$144 606.45
3	$971.24	$772.88	$198.36	$144 408.09
4	$971.24	$771.82	$199.42	$144 208.67
5	$971.24	$770.76	$200.48	$144 008.19
6	$971.24	$769.69	$201.55	$143 806.64

Compare the total interest paid in the first 6 months under each compounding period. Why do you think that mortgage interest rates are compounded semi-annually in Canada?

b) Use the TVM Solver to verify the monthly payment for each payment.

c) How much interest is saved over the life of the mortgage with semi-annual compounding instead of monthly compounding?

In Your Own Words

What information about a mortgage can you learn from an amortization table?
Why is this information important?

Mortgage Tic-Tac-Toe

This game is for 2 players. One player is X and the other player is O.

- Draw a 3 by 3 tic-tac-toe board.
- Players take turns rolling 3 dice.
- Both players have 15 s to use the table to secretly choose the principal, interest rate, and amortization period for a mortgage.
  The goal is to obtain the lower monthly payment.

Number rolled	Amortization period	Annual interest rate	Principal borrowed
1 or 6	20 years	5%	$150 000
2 or 5	25 years	6%	$175 000
3 or 4	30 years	7%	$200 000

The interest is compounded semi-annually.

Materials
- 3 dice
- TI-83 or TI-84 graphing calculator
- timer

Players then have 1 min to determine their monthly payments. For example, suppose 3, 1, and 2 are rolled.

Player X's choices

- Interest rate of 7%
- Principal of $150 000
- Amortization period of 25 years

Player O's choices

- Principal of $200 000
- Amortization period of 20 years
- Interest rate of 6%

```
N=300.00
I%=7.00
PV=150000.00
■PMT=-1050.62
FV=0.00
P/Y=12.00
C/Y=2.00
PMT:END BEGIN
```

```
N=240.00
I%=6.00
PV=200000.00
■PMT=-1424.38
FV=0.00
P/Y=12.00
C/Y=2.00
PMT:END BEGIN
```

If both players have the same monthly payment, they roll again.

- Player X has the lower monthly payment and used the TVM Solver correctly, so she writes an X on the tic-tac-toe board.
- The first player to mark 3 Xs or 3 Os in a vertical, horizontal, or diagonal line wins.

Reflect

- Explain how you decided which number to use for the principal borrowed, interest rate, and amortization period.

GAME: Mortgage Tic-Tac-Toe **455**

7.8 Using Technology to Generate an Amortization Table

An amortization table can be generated quickly and efficiently with a spreadsheet. Spreadsheets also allow us to change the features of a mortgage and see the immediate effect of the change in the amortization table.

Inquire: Creating an Amortization Table

Materials
- computer with Internet access or a TI-83 or TI-84 graphing calculator
- *Microsoft Excel*
- *Amortization.xls*

Work with a partner.

The Lees are finalizing the purchase of their home. They arrange a mortgage of $175 000 at 6.25% per year compounded semi-annually to be repaid monthly over 25 years.

1. **Determining the monthly payment**

   Use an online mortgage calculator to determine the monthly payment. If you have access to the TVM Solver, you may wish to use it instead.

Mortgage amount:	$ 175 000
Amortization period:	25 years
Interest rate:	6.25 %
Calculate	
Monthly mortgage payment:	$1145.80

   The Lees' monthly payment is $1145.80.

2. **Creating an amortization table**
   - If you are using the file *Amortization.xls*, open it and begin at part b.
   - If you are not using the file *Amortization.xls*, start at part a.

456   CHAPTER 7: Annuities and Mortgages

a) • Open a new spreadsheet document.
   • Copy the headings, values, and formulas shown.

	A	B	C	D	E
1	Amortization Table				
2	Principal		175000		
3	Annual interest rate		0.0625		
4	Equivalent monthly rate		=(1+C3/2)^(1/6)-1		
5	Amortization period in years		25		
6	Number of payments		=C5*12		
7	Monthly payment		=PMT(C4, C6, -C2)		
8					
9	Payment	Monthly	Interest	Principal	Outstanding
10	number	payment	paid	paid	balance
11	0				=C2
12	=A11+1	=ROUND($C$7,2)	=ROUND($C$4*E11,2)	=B12-C12	=E11-D12

   • Format cells C2, C7, E11, and B12 to E12 as currency.
   • Format cell C3 as a percent to 2 decimal places and cell C4 as a percent to 7 decimal places.

*To view the formulas in the spreadsheet, hold on **Ctrl** and press `.*

b) Refer to the formulas in the spreadsheet in part a.
   • The formula in cell C4 converts the annual interest rate compounded semi-annually into an equivalent monthly rate. This monthly rate is used to calculate the values in the *Interest paid* column in the amortization table.
   • The **PMT** function in cell C7 calculates the regular payment of the loan. What do the numbers in the brackets represent?
   • Explain the remaining formulas in the table.

c) • Select cells A12 to E12.
   • **Fill Down** to copy the formulas in row 12 through row 311. The first 10 rows of your spreadsheet should look like this.

	A	B	C	D	E
1	Amortization Table				
2	Principal		$175 000.00		
3	Annual interest rate		6.25%		
4	Equivalent monthly rate		0.5141784%		
5	Amortization period in years		25		
6	Number of payments		300		
7	Monthly payment		$1145.80		
8					
9	Payment	Monthly	Interest	Principal	Outstanding
10	number	payment	paid	paid	balance
11	0				$175 000.00
12	1	$1145.80	$899.81	$245.99	$174 754.01
13	2	$1145.80	$898.55	$247.25	$174 506.76
14	3	$1145.80	$897.28	$248.52	$174 258.24
15	4	$1145.80	$896.00	$249.80	$174 008.44
16	5	$1145.80	$894.71	$251.09	$173 757.35
17	6	$1145.80	$893.42	$252.38	$173 504.97
18	7	$1145.80	$892.13	$253.67	$173 251.30
19	8	$1145.80	$890.82	$254.98	$172 996.32
20	9	$1145.80	$889.51	$256.29	$172 740.03
21	10	$1145.80	$888.19	$257.61	$172 482.42

3. Explaining how the interest paid and the outstanding balance are calculated

Part of each monthly payment is interest and the rest is principal.

Payment 1
- When payment 1 is made, a month's interest is owed on the outstanding balance, $175 000.
  From cell C4, the monthly rate, 0.5141784%, corresponds to an annual rate of 6.25% per year compounded semi-annually.
  So, the monthly interest charge is:
  $175\ 000 \times 0.005141784 \doteq \$899.81$
- The monthly payment is $1145.80. Since $899.81 is interest, the part that repays principal is: $1145.80 - \$899.81 = \$245.99$
- The outstanding balance at the end of the first month is:
  $175\ 000 - \$245.99 = \$174\ 754.01$

Repeat these calculations for two other payments in the table.

4. **Adjusting the last payment**
   The mortgage is paid off in the 300th payment, so the outstanding balance should be $0.00. The outstanding balance after the 299th payment is $1138.53 and the interest charge for this payment is $5.85. So, the 300th payment should be:
   $1138.53 + \$5.85 = \$1144.38$
   - Enter $1144.38 as the 300th payment to obtain an outstanding balance of $0.00 after the 300th payment.

5. **Calculating the total repaid and the total interest paid**
   a) We can calculate the total paid over the life of the mortgage by adding the values in cells B12 to B311.
      - In cell B312, type: **=SUM(**
      - Select cells B12 to B311.
        The formula should now read: **=SUM(B12:B311**
      - Complete the formula by typing: **)**
   b) Repeat part a to determine the total interest paid and the total principal paid over the life of the mortgage.

6. **Interpreting the amortization table**
   a) How much of the 4th payment is interest? the 8th payment?
   b) What is the outstanding balance after half of the payments have been made? Is the outstanding balance also reduced by one-half? Explain.
   c) What cash payment will pay off the mortgage after the 200th payment?
   d) What is the total interest paid over the life of the mortgage? How does this compare with the principal originally borrowed?

Practice

A

1. Use an online Canadian mortgage calculator to determine the monthly payment on each mortgage.

	Principal borrowed	Interest rate	Amortization period
a)	$100 000	5.25%	15 years
b)	$156 000	6.75%	25 years
c)	$230 000	8.5%	20 years

B

2. Anita and Kenny are arranging a mortgage for their new home. They will be taking out a $145 000 mortgage at 5% per year compounded semi-annually for 25 years. They will repay the mortgage with monthly payments.
   a) Determine the monthly payment.
   b) What is the equivalent monthly interest rate?
   c) Create an amortization table for the mortgage.
   d) What is the total interest paid over the life of the mortgage?

3. Refer to the mortgage in question 2. Choose any two payments. Explain how the interest paid, principal paid, and outstanding balance for these payments are calculated.

7.8 Using Technology to Generate an Amortization Table   **459**

4. Claude purchases a home for $225 000 and applies a $25 000 down payment. He arranges a mortgage for the outstanding balance with monthly payments for 25 years at 6.5% per year compounded semi-annually.
   a) Determine the monthly payment.
   b) Create an amortization schedule for Claude's first six monthly payments.
   c) How much principal is paid down in the first six months?
   d) What is the total interest paid in the first six months?

5. The Ugars arrange a mortgage for a new condominium for $245 000. They decide on a mortgage for 5.75% per year compounded semi-annually to be repaid with monthly payments over 30 years.
   a) Determine the regular monthly payment.
   b) Create an amortization schedule showing the first 6 payments.
   c) What is the total interest paid during the 6 months?

6. Paul and Kaori arrange a mortgage for $168 000 to be paid back monthly over 20 years at 4.5% per year compounded semi-annually.
   a) Determine their regular monthly payment.
   b) Create an amortization schedule for the first 2 years of the mortgage.
   c) Use the amortization table in part b to answer these questions.
       i) Determine the total interest paid and principal repaid after 2 years.
       ii) What is the outstanding balance after 2 years?
       iii) Determine the percent of the original mortgage that has been repaid after 2 years. How long will it take to repay the mortgage completely if this rate of payment remains constant? Explain why, in reality, it does not take this long.

Reflect

- Why is it necessary to change the semi-annual interest rate into an equivalent monthly rate before completing the mortgage amortization table?
- How can you check that the values in the amortization table are correct?
- Suppose you have to determine the total interest paid on a mortgage. Do you find it easier to use a spreadsheet or the TVM Solver? Explain.

7.9 Reducing the Interest Costs of a Mortgage

The total interest paid over the lifetime of a mortgage is a considerable sum of money, often in the hundreds of thousands of dollars. There are a number of strategies a homeowner can use to reduce the interest costs of a mortgage.

Inquire | Analysing Interest-Saving Strategies

Materials
- TI-83 or TI-84 graphing calculator

Work with a partner or in a small group.

■ Use the TVM Solver to determine the monthly payment.
  Use the ΣInt command to determine the total interest paid.
■ Discuss the questions below each table as a group and record your answers.

1. Changing the amortization period

Most homeowners choose an amortization period of 25 years, but amortization periods of 15, 20, and 30 years are also allowed.

a) Copy and complete this table.

Use a mortgage of $100 000 at 5% per year compounded semi-annually.

Amortization period	Number of payments (N)	Monthly payment (PMT)	Total interest paid	Interest saved
25 years				
15 years				
20 years				
30 years				—

b) How does the monthly payment change as the amortization period increases? Explain.
c) How does the total interest paid change as the amortization period increases? Explain.
d) Why might a homeowner choose a shorter amortization period? Why might a homeowner choose a longer amortization period?
e) Compare the difference in the monthly payments with the difference in the interest saved for different pairs of amortization periods. Does the interest saved justify paying more each month? Explain.

2. **Changing the interest rate**

Mortgage interest rates are largely determined by economic conditions. They change frequently over time.

a) Copy and complete this table.

Use a mortgage of $100 000 amortized over 25 years.

The interest rate is an annual rate compounded semi-annually.

Interest rate	Year	Monthly payment (PMT)	Total interest paid	Interest saved
5%	1951			–
6%	2007			
10%	1969			
14%	1990			
21.5%	1982			

b) How do the monthly payment and the total interest paid change as the interest rate increases? Explain.
c) Does an increase of 4% in the interest rate result in a 4% increase in the total interest paid? Explain.

3. Changing the payment frequency

Many mortgages are repaid with monthly payments, but more frequent payments are also allowed. This online mortgage calculator screen shows other commonly used payment periods.

> We will examine accelerated payments in question 4.

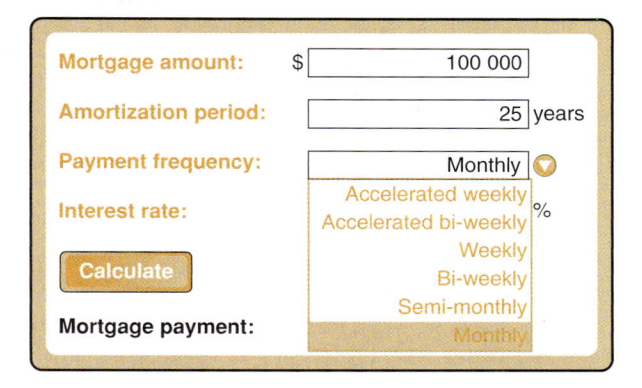

Mortgage amount:	$ 100 000
Amortization period:	25 years
Payment frequency:	Monthly ▼
	Accelerated weekly
Interest rate:	Accelerated bi-weekly %
	Weekly
Calculate	Bi-weekly
	Semi-monthly
Mortgage payment:	Monthly

- semi-monthly (twice a month; 24 payments a year)
- bi-weekly (every 2 weeks; 26 payments a year)
- weekly (every week; 52 payments a year)

a) This online calculator screen shows the bi-weekly payment on a mortgage of $100 000 amortized over 25 years at 5% per year compounded semi-annually.

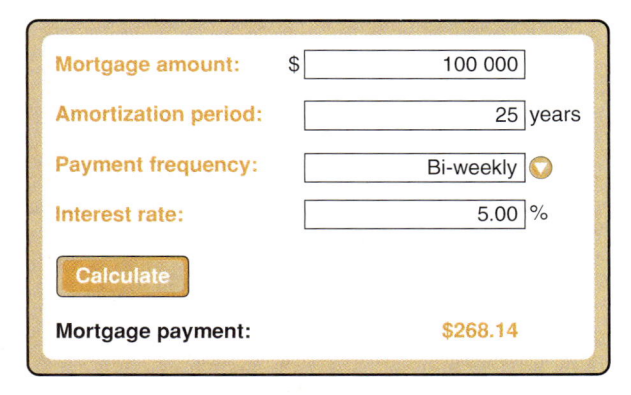

Mortgage amount:	$ 100 000
Amortization period:	25 years
Payment frequency:	Bi-weekly ▼
Interest rate:	5.00 %
Calculate	
Mortgage payment:	$268.14

- Use the TVM Solver to verify the payment.
- Explain the values you entered for the variables.

b) Copy and complete this table.

> Use the mortgage from part a.

Payment frequency	Payments per year (P/Y)	Number of payments (N)	Payment (PMT)	Total interest	Interest saved
Monthly					
Semi-monthly					
Bi-weekly					
Weekly					—

7.9 Reducing the Interest Costs of a Mortgage **463**

c) Is the interest saved significant when payments are made more often? Explain.

d) Why might a homeowner choose to make semi-monthly, bi-weekly, or weekly payments instead of monthly payments?

4. Making accelerated payments

Most financial institutions allow "accelerated" weekly and bi-weekly payments.

With this option, the weekly payment is one-quarter of the monthly payment, while the bi-weekly payment is one-half of the monthly payment.

a) The online calculator screens below show the monthly payment and accelerated bi-weekly payment for a mortgage of $100 000 amortized over 25 years at 5% per year compounded semi-annually.

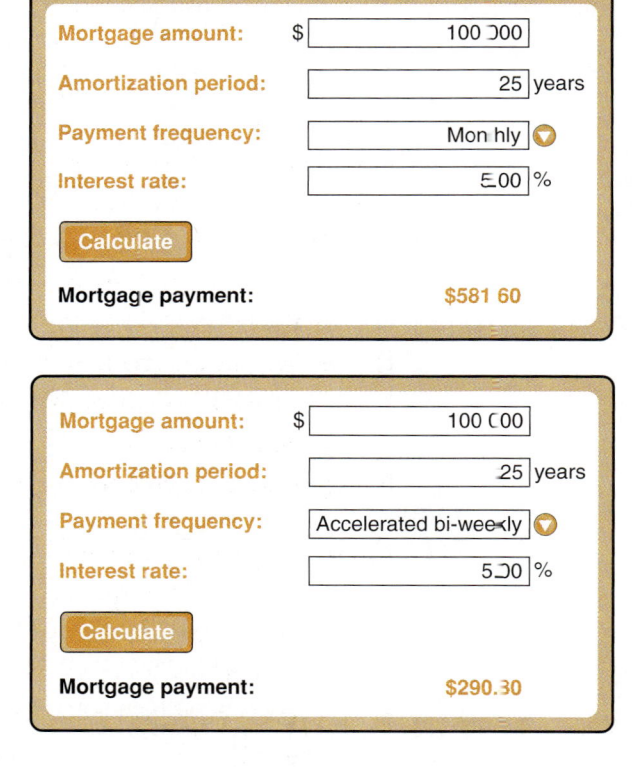

Mortgage amount:	$ 100 000
Amortization period:	25 years
Payment frequency:	Monthly
Interest rate:	5.00 %
Calculate	
Mortgage payment:	$581.60

Mortgage amount:	$ 100 000
Amortization period:	25 years
Payment frequency:	Accelerated bi-weekly
Interest rate:	5.00 %
Calculate	
Mortgage payment:	$290.30

Verify the bi-weekly payment.

b) Copy this table.

Complete the row for monthly payments and the second and third columns of the accelerated bi-weekly and accelerated weekly payments.

Payment frequency	Payments per year (P/Y)	Payment (PMT)	Number of payments (N)	Total interest	Interest saved
Monthly					–
Accelerated bi-weekly					
Accelerated weekly					

c) Compare the accelerated weekly and accelerated bi-weekly payments with the regular weekly and bi-weekly payments in question 3a. Why will the mortgage be paid off more quickly with accelerated payments?

d) Use the TVM Solver to determine the number of payments it takes to pay off the mortgage with accelerated bi-weekly payments.
- Open the TVM Solver.
- Enter the known values for I%, PV, PMT, FV, P/Y, and C/Y.
- Move the cursor to N, and press [ALPHA] [ENTER].
  Round this result to the nearest whole number and record it in the table.

e) Use the ΣInt command to determine the total interest paid with accelerated bi-weekly payments. Use the value of N you determined in part c.

f) Repeat parts c and d for the accelerated weekly payments.

5. Comparing regular and accelerated payments

Compare the tables in questions 3 and 4.

a) Why are the interest savings much greater with accelerated payments than with regular payments?

b) How many payments are saved by making accelerated payments?
How much time does this represent in years and months?

c) In Canada, the most popular payment frequency is the accelerated bi-weekly option. Why do you think this is the most popular option?

Practice

A

1. Calculate each regular payment for a mortgage of $130 000 amortized over 20 years at 8.5% per year compounded semi-annually.
   a) Monthly payment
   b) Accelerated bi-weekly payment
   c) Accelerated weekly payment

2. What effect do each of the following have on the regular payment and the total interest paid on a mortgage? Explain.
   a) Increasing the amortization period
   b) Making more frequent payments
   c) Making accelerated payments

B

3. The Thompsons borrow $179 000 for their new home. They plan to repay the mortgage by making monthly payments for 25 years at 6% per year compounded semi-annually. Calculate the Thompsons' monthly payment and the total interest they will pay over the life of the mortgage.

4. Refer to the mortgage in question 3. Calculate the Thompsons' new regular payment and the total interest saved under each scenario.
   a) They arrange a 20-year mortgage instead of a 25-year mortgage.
   b) They receive an interest rate of 5.75% by applying for their mortgage over the Internet.
   c) They make weekly payments instead of monthly payments.
   d) They make accelerated bi-weekly payments instead of monthly payments.

5. Compare your answers to question 4.
   Which change resulted in the greatest interest saved? Explain.

Reflect

- What are some strategies a homeowner can use to reduce the total interest paid on a mortgage? Why will these strategies reduce the interest costs?
- How do age, family circumstances, income, and lifestyle factors affect the strategies used to reduce the interest costs of a mortgage? Explain.

Study Guide

Ordinary Simple Annuities

- An annuity is a series of equal, regular payments. In an ordinary simple annuity, payments are made at the end of each compounding period. The interest is compounded just before the payment is made.
- The amount of an annuity is the sum of the regular payments plus the interest.
- The present value of an annuity is the money that must be deposited today to provide regular payments in the future.

Amount of an ordinary simple annuity	**Present value of an ordinary simple annuity**
$A = \dfrac{R[(1+i)^n - 1]}{i}$, where	$PV = \dfrac{R[1 - (1+i)^{-n}]}{i}$, where
• A is the amount	• PV is the present value
• R is the regular payment	• R is the regular payment
• i is the interest rate per compounding period as a decimal	• i is the interest rate per compounding period as a decimal
• n is the number of compounding periods	• n is the number of compounding periods

- The interest earned on an annuity is the difference between the total of the regular payments and the amount or present value of the annuity.

Mortgages

- A mortgage is a loan that is used to buy property.
- In Canada, the interest rate on mortgages can be compounded at most semi-annually. However, mortgage payments are usually made monthly or bi-weekly.
- Since the payment period is different from the compounding period, mortgages are not ordinary simple annuities.

In the TVM Solver:
- Solve for FV to determine the amount.
- Solve for PV to determine the present value.
- Solve for PMT to determine the regular payment.
- Use the ΣInt command to determine the interest earned.

Chapter Review

7.1

1. Determine the amount of each ordinary simple annuity.
   a) Deposits of $2550 for 7 years at 9.7% per year compounded annually
   b) Deposits of $1380 for 4 years at 10% per year compounded semi-annually
   c) Deposits of $750 for 5 years at 12.6% per year compounded monthly

2. Calculate the interest earned on each annuity in question 1.

3. Yvonne and Teresa each make regular deposits into an annuity.
   - Yvonne deposits $150 at the end of each month at 8% per year compounded monthly
   - Teresa deposits $450 at the end of each quarter at 8% per year compounded quarterly
   a) Who do you think will have the greater amount at the end of 4 years? Explain your reasoning.
   b) Verify your prediction by calculating each amount.

4. Carlos and Renata each invest money at the end of each year in an RRSP.
   - Carlos invests $4500 for 30 years at 7.5% per year compounded annually.
   - Renata invests $9000 for 15 years at 7.5% per year compounded annually.
   a) Determine the amount in each RRSP.
   b) Does the amount remain the same when the regular deposit is doubled and the time period is halved? Explain.

7.2

5. Determine the principal that must be deposited today to provide for each ordinary simple annuity.
   a) Payments of $3500 for 7 years at 6.5% per year compounded annually
   b) Payments of $3575 for 12 years at 9% per year compounded semi-annually

6. Calculate the interest earned on each annuity in question 5.

7. Shawn buys a new computer. He will make monthly payments of $72 for the next 2 years, starting 1 month from now. He is charged 15% per year compounded monthly.
   a) How much did Shawn borrow to purchase the computer?
   b) How much interest will Shawn pay?

8. Consider these 3 annuities.
   Annuity A: $50 per quarter for 4 years at 6% per year compounded quarterly
   Annuity B: $100 per quarter for 4 years at 6% per year compounded quarterly
   Annuity C: $50 per quarter for 8 years at 6% per year compounded quarterly
   a) Determine the present value of each annuity.
   b) Which of the following has the greater effect on the present value of an annuity?
      i) Doubling the payments
      ii) Doubling the time period
      Justify your answer.

CHAPTER 7: Annuities and Mortgages

9. Adrian wants to have $15 000 in 3 years to start a mechanic shop. He plans to save the money by making regular deposits into an annuity that earns 11.7% per year compounded semi-annually. What semi-annual deposits does Adrian have to make?

10. Tarak uses the TVM Solver to compare two annuities.

```
N=24.00            N=12.00
I%=7.50            I%=7.80
PV=0.00            PV=0.00
PMT=-2478.12       PMT=-4969.91
FV=74250.00        FV=74250.00
P/Y=4.00           P/Y=2.00
C/Y=4.00           C/Y=2.00
PMT:END BEGIN      PMT:END BEGIN
```

a) Describe each annuity.
b) Which annuity do you think Tarak should choose? Justify your answer.

11. Anisha obtains a small business loan for $6500 to start her roofing business. She can repay the loan in 24 months or 36 months. She is charged interest at 6.2% per year compounded monthly.
a) How much more will Anisha pay each month if she repays the loan in 24 months instead of 36 months?
b) How much interest will Anisha save if she repays the loan in 24 months instead of 36 months?
c) Why does Anisha pay less interest with a 24-month loan even though her monthly payments are greater than with the 36-month loan?

12. Suppose you make regular quarterly deposits to amount to $5000 over 12 years at 9.4% per year compounded quarterly.

a) Use a spreadsheet to determine the quarterly deposit required.
b) How much more could you save if the interest rate is 10% per year compounded quarterly?
c) Suppose you could increase your regular payments by $100. How much more quickly could you reach your goal?

13. Suppose you borrow $8000 at 7.5% per year compounded monthly for 6 years. You repay the loan by making monthly payments of $138.32.
a) Use a spreadsheet to create a loan repayment schedule.
b) What is the total interest paid on the loan? How does this compare to the principal originally borrowed?

14. Use a Venn diagram to illustrate the similarities and differences between RESPs and RRSPs.

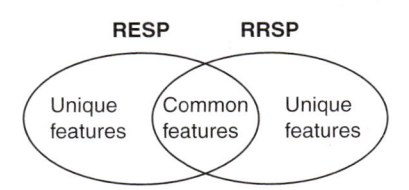

15. Wenfeng's friend has just purchased a house. She tells Wenfeng that she has arranged "a closed mortgage with bi-weekly payments at 6% per year compounded semi-annually with an amortization period of 30 years and a 5-year term." Wenfeng does not understand the terminology his friend is using. Explain the details of the mortgage in everyday language.

Chapter Review **469**

16. For each Canadian mortgage, determine the monthly payment and total interest paid.

	Principal borrowed	Interest rate	Length of mortgage
a)	$97 000	4.5%	25 years
b)	$145 000	3.25%	20 years
c)	$207 000	10%	15 years
d)	$299 000	5.5%	30 years

17. The amortization table shows the first 6 monthly payments of a mortgage.

Payment numbers	Monthly payment	Interest paid	Principal paid	Outstanding balance
0				$150 000.00
1	$959.71	$740.79	$218.92	$149 781.08
2	$959.71	$739.71	$220.00	$149 561.08
3	$959.71	$738.63	$221.08	$149 340.00
4	$959.71	$737.53	$222.18	$149 117.82
5	$959.71	$736.44	$223.27	$148 894.55
6	$959.71	$735.33	$224.38	$148 670.17

a) What is the principal borrowed?
b) What is the monthly payment?
c) How much of the 4th payment is interest?
d) What is the outstanding balance after the 3rd payment?
e) What is the total interest paid in the first 6 payments?

18. Part of an amortization table is shown below. Determine the missing values in the table. Justify your answers.

Payment numbers	Monthly payment	Interest paid	Principal paid	Outstanding balance
0				$180 000.00
1	$1,077.84	$786.36	$291.48	$179 708.52
2	$1,077.84	$785.09	$292.75	$179 415.77
3	$1,077.84	$783.81	$294.03	$179 121.74
4		$782.52	$295.32	$178 826.42
5	$1,077.84	$781.23		$178 529.81
6	$1,077.84		$297.90	

19. A mortgage of $190 000 is amortized over 15 years at 5.25% per year compounded semi-annually.
a) Use the TVM Solver or an online calculator to determine the monthly payment.
b) Use a spreadsheet to create an amortization table for the first 12 payments.
c) Determine the total amount and interest paid over the life of the mortgage.

20. How many payments will you make in 1 year if you repay your mortgage with:
a) Semi-monthly payments
b) Bi-weekly payments
c) Weekly payments
d) Accelerated bi-weekly payments

21. The Mahers were pre-approved for a mortgage of $250 000 amortized over 30 years at 7% per year compounded semi-annually. They will make monthly payments. If the Mahers apply for a mortgage over the Internet, they can choose to make one of these changes.
Option I: A 25-year amortization period
Option II: A 6.5% interest rate
Option III: Semi-monthly payments
Option IV: Accelerated bi-weekly payments
a) Determine the total interest under the original terms of the mortgage.
b) Predict which option would save the Mahers the most interest over the life of their mortgage. Explain your reasoning.
c) Verify your prediction in part b by calculating the interest saved under each option. Was your prediction correct? Explain.

470 CHAPTER 7: Annuities and Mortgages

Practice Test

Multiple Choice: Choose the correct answer for questions 1 and 2. Justify each choice.

1. Karyan deposits $675 at the end of each year into an account that earns 5.4% per year compounded annually. Which is the amount after 3 years?
   A. $2025.00   B. $2136.32   C. $1791.01   D. $784.35

2. Which is the monthly payment on a mortgage of $230 000 amortized over 30 years at 5.75% per year compounded semi-annually?
   A. $1320.93   B. $1332.34   C. $1342.22   D. $1277.78

Show your work for questions 3 to 6.

3. **Knowledge and Understanding** Determine the present value of quarterly payments of $250 for 2.5 years at 2.4% per year compounded quarterly.

4. **Application** The Zaidis arrange a mortgage amortized over 30 years at 9% per year compounded semi-annually. Here is part of an amortization table for the mortgage.

Payment number	Monthly payment	Interest paid	Principal paid	Outstanding balance
0				$80 000.00
1	$634.27	$589.05		$79 954.78
2				$79 909.23

   a) How much did the Zaidis borrow?
   b) How much of the 1st payment is principal? How much is interest?
   c) Complete the row for payment 2. Explain why the interest payments decrease each month while the principal payments increase each month.
   d) How much interest will the Zaidis pay over the life of the mortgage?
   e) Suggest two strategies that the Zaidis can use to reduce the interest costs on their mortgage. Why will these strategies reduce the interest costs?

5. **Thinking** Mila will need $10 000 when she goes to college 5 years from now. She has 2 options for saving the money.
   Option A: A regular deposit at the end of each month into an account that earns 7% per year compounded monthly
   Option B: A regular deposit at the end of each year into an account that earns 7.25% per year compounded annually
   Which option should Mila choose? Make a recommendation, then justify it.

6. **Communication** Explain the advantages of saving as early as possible for large expenses. Include examples to support your explanation.

Chapter Problem

Planning Ahead

Materials
- TI-83 or TI-84 graphing calculator

Jamie and Sam have been saving money each month for the past 5 years. They now have $10 000 saved.

1. How much did Jamie and Sam set aside each month?
   Assume an average interest rate of 4.8% per year compounded monthly.

2. Jamie and Sam plan to use the $10 000 as a down payment for a house. One house they are interested in buying has a selling price of $179 900. What monthly mortgage payment will Jamie and Sam make if they arrange a mortgage at 5.25% per year compounded annually for 20 years?

3. Jamie and Sam have different opinions about whether they should buy a house now.
   - Sam suggests that they save the difference between the monthly payment in part b and their current monthly rental payment of $750 in an annuity that earns 3.5% per year compounded monthly for 5 years. He says that at the end of that time, they will have larger down payment and can afford a nicer house.
   - Jamie suggests that they buy the house now since house values are expected to increase by an average of 3.5% per year for the next 5 years.
   What advice would you give Jamie and Sam?
   How would you convince them to follow your advice?
   Include calculations in your answer.

472 CHAPTER 7: Annuities and Mortgages

8 Budgets

What You'll Learn

To determine the costs related to owning or renting your own house or apartment and to design a monthly budget that reflects these costs

And Why

To make the most of the money you earn, it is important to be aware of all of the costs associated with living on your own.

Key Words

- fixed cost
- variable cost
- utilities
- tenant
- landlord
- budget
- income
- expenses
- savings
- balance

CHAPTER 8

Activate Prior Knowledge

Fixed and Variable Costs

Prior Knowledge for 8.1

A **fixed cost** is the same amount charged at regular intervals. Monthly car payments and insurance payments are fixed costs associated with owning a vehicle.

A **variable cost** changes and may not follow a regular schedule. Gasoline and maintenance costs are variable costs associated with owning a vehicle.

Example

Philippa predicts the costs associated with owning and operating her vehicle. She lists them as fixed costs or variable costs. Calculate the total cost of owning the vehicle this year, based on her predictions and estimates.

Fixed cost	Variable cost
Insurance: $87/month	Gasoline: $1.15/L, 18 500 km driven/year, 9.4 L/100 km fuel efficiency
Car loan payment: $385/month	Oil change: $34.95 every three months
Licence and ownership: $75/year	Repairs: $750/year

Solution

Annual fixed cost	Annual variable cost
Insurance: $87/month × 12 months = $1044	Gasoline: Gasoline used: 18 500 km ÷ 100 km = 185 185 × 9.4 L = 1739 L Cost of gasoline: $1.15/L × 1739 L = $1999.85
Car loan payment: $385/month × 12 months = $4620	Oil change: $34.95 ÷ 3 months × 12 months = $139.80
Licence and ownership: $75	Repairs: $750
Total fixed cost: $1044 + $4620 + $75 = $5739	Total variable cost: $1999.85 + $139.80 − $750 = $2889.65

Total cost: $5739.00 + $2889.65 = $8628.65

For Philippa's predictions, the total cost of owning and operating the vehicle this year is $8628.65.

CHECK ✓

1. Greg has driven 21 540 km this year. The average cost of gas was $1.21/L. His car has an average fuel efficiency rating of 10.7 L/100 km. Determine the total fuel cost for one year.

2. Determine the total annual cost of owning and operating a vehicle with these costs.

Fixed cost	Variable cost
Insurance: $95/month	Gasoline: $1.24/L, 19 700 km driven/year, 9.2 L/100 km fuel efficiency
Car loan payment: $415/month	Oil change: $39.95 every three months
Licence and ownership: $75/year	Repairs: $550/year

3. Jazmine is thinking of buying a dog. To estimate what it will cost, she prepares this list of expenses to research.
   a) Classify each expense as fixed or variable.
   b) Which will be one-time costs?

   Dog Expenses
   - Buying the dog
   - Dog licence
   - Obedience training course
   - Food
   - Annual vet visit and shots
   - Vet visits for illness
   - Grooming products or services
   - Toys and other supplies

4. Pedro has always been a good athlete. He decides to begin training and competing in triathlons. To estimate what it will cost, he prepares this list of expenses. Classify each expense as fixed or variable.

   Triathlon Expenses
   - bicycle
   - bike maintenance and parts
   - biking and running clothes
   - swim goggles
   - membership in triathlon club
   - helmet
   - sunglasses
   - bathing suits
   - fitness training
   - bicycle shoes
   - running shoes
   - wet suit
   - race entrance fees

5. Use the *Example* on page 474. How likely do you think each of Philippa's predicted costs is? Explain your thinking.

Transitions

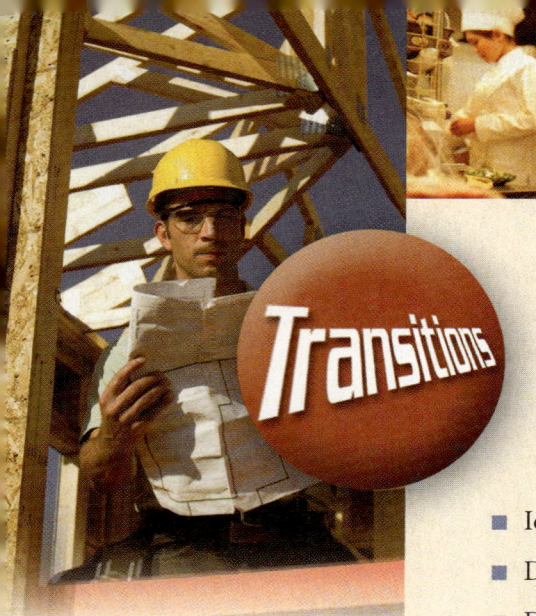

Your Financial Future

Whether you are at school or working, you will have financial goals. You may want to save for items such as a computer, a car, a vacation, education, retirement, or a down payment on a home. These steps can help you save.

- Identify your savings goals.
- Decide when you hope to reach each of your goals.
- Design a savings plan to meet your goals.
  - Set up a separate bank account for your savings.
  - Pay yourself first. Whenever you receive money, put some of it into your savings account. For example, you may decide to save 10% of any money you receive.
  - Make a monthly savings amount an expense that you include in your monthly budget.
  - Money in a savings account will earn interest. You can also decide to invest the money you save in some other way that will help it grow, such as by buying stocks or real estate.

Search words
- personal banking
- personal financial services
- accounts and services
- payment services
- investments
- financial planning

1. Choose a Canadian bank. You will need access to the Internet to search the bank's Web site, print materials about the savings alternatives offered at the bank, or a representative from the bank. Research the services the bank offers to help individuals save. For example, you can arrange to have a certain amount of money automatically transferred from your chequing account to your savings account each month, or arrange for pre-authorized RSP contributions.
2. Apply strategies for saving as you design and adjust budgets in this chapter.
3. How can you use a budget to help you save to meet your financial goals? Explain.
4. Which of the strategies described above do you think you may use to save in the future?

• Complete questions 3 and 4 after you complete Chapter 8.

476 Transitions: Your Financial Future

8.1 Choosing a Home

Depending on your housing needs, your income, and where you live, you may have many housing options: a duplex, detached house, townhouse, condominium, apartment, mobile home, or room.

Inquire: Choosing Properties to Rent or Buy

Materials
- access to the Internet or print materials such as newspapers or magazines with real estate ads
- maps of your town or city
- TI-83 or TI-84 graphing calculator

Work in small groups.

- Describe what you would look for if you were choosing a home in your community for each person or family described below. Include details such as location, number of bedrooms and bathrooms, parking, and special features.
  - A single working adult with a vehicle
  - A college student with no vehicle
  - A single parent with a young child and a teenager; the parent has a vehicle but the teenager uses public transit
  - A couple with three children, aged 5, 9, and 12; the family has two vehicles

- Find one property to rent and one property to purchase in your community that you think would be suitable for each person or family described above.

Include at least one of each of these types of dwellings among the properties you choose.
- Room
- Apartment
- Condominium
- Townhouse
- Detached house

Search words
- ☐ Ontario Real Estate Association
- ☐ MLS Canada
- ☐ Ontario rental properties

[Note: try replacing Ontario with the name of your community]

> **Utilities** are services such as heat, water, and electricity.

> Save the housing cost data you find for use in Lesson 8.6.

> See Chapter 7 for information on using the TVM Solver.

■ Record the monthly rental cost of each rental property you picked and answer these questions:
  - Are utilities and parking included in the cost?
  - Does the person renting have to pay first and last month's rent as part of the initial agreement?
  - Will the person renting have to sign a lease agreeing to rent for a certain period of time?
■ For each property you picked to purchase, determine the monthly mortgage payment. Assume a 10% down payment, a 7.95% annual interest rate, and a term of 15 years. Use an online mortgage calculator or the TVM Solver on a TI-83 or TI-84 graphing calculator to perform the calculations.
■ Gather information about other expenses involved in moving to a new home, such as moving costs, legal fees, mortgage insurance, home inspection, condo fees, and so on. Decide whether buyers, renters, or both would have to pay these costs.
■ Choose one of the properties you researched. Write a persuasive letter to the person or family for whom you chose the property. Describe why you think the home is suitable and outline the costs involved in moving to it.

Reflect

■ Will renting or owning require a greater initial investment? Justify your answer.
■ You only considered the initial costs of buying a home. What additional expenses might an owner have each month?
■ How could you use this research in your own life?

478 CHAPTER 8: Budgets

8.2 The Costs of Owning or Renting a Home

Your mortgage or rent payment is just one of the costs of housing. You must also think about insurance, utilities, appliances, furnishings, maintenance, and repairs. These expenses may contribute significantly to your housing cost.

Investigate

Comparing Renting and Owning a Home

Work in small groups.
Read the case studies on page 480.
Use the information to answer these questions.

- What are the fixed costs for each household?
  What are the variable costs?

- Determine how much the household will spend on housing in a year.

- Does renting or owning appear to involve greater costs?
  Which will likely leave you more money for other things, such as vacations or savings?

- If you want to move, will it be easier if you rent or own?
  What factors may affect this? Justify your answer.

- In each case study, who do you think would have to pay for repairs if the stove or another appliance were to break? Who would pay if the roof needed repairs?

Fixed costs are the same every time.
Variable costs vary depending on factors such as level of use of a service.

CASE STUDY 1

Aziza works full time. Her husband Hassan is a student with a part-time job.

- They rent a 2-bedroom loft apartment.
- They pay $975 a month in rent.
- The rent includes heat, water, electricity, parking, laundry, and cable TV.
- They have a phone plan that includes unlimited calling within Canada and the US and 400 min of overseas calling for $59.95 a month.
- They do not have a lease, and pay rent month to month.

CASE STUDY 2

The Martins are a retired couple with no children living at home. They own a 2-bedroom condo.

- Their mortgage payments are $981.72 a month.
- They pay condo fees of $450 per month. These fees include heat, electricity, water, and maintenance.
- They pay $387.92 to insure their condo unit each year.
- Their property taxes are $1976.47 a year.
- They have a cable/phone/Internet bundle that costs $99 a month.

CASE STUDY 3

Tom and Hyo-Jin are a working couple with a 5-year-old child. They own a 3-bedroom detached house.

- Their mortgage payments are $1465.33 a month.
- They pay $336.52 twice a year for home insurance.
- Their property taxes are $2358.60 a year.
- They pay for water, sewer, and electricity every two months. The bills for the last year were: $210.23, $199.51, $186.88, $188.76, $213.75, $193.69
- Their home is heated by natural gas. Their gas bills for the last year were: $145.82, $103.44, $83.12, $78.71, $31.69, $22.02, $25.07, $23.66, $25.03, $48.26, $93.51, $120.96
- They rent a water heater for $52.95 every 3 months.
- They have a phone plan that includes unlimited calling within Canada and the US and costs $49.95 a month.

480 CHAPTER 8: Budgets

Reflect

- Were the costs for renting or owning higher than, lower than, or about what you expected? Explain.
- How are the housing needs of the people in each case study different?
- Do you think you will rent or buy your first home? Explain.

Connect the Ideas

Costs of renting
- A **tenant** uses a property owned by another person, the **landlord**, and pays rent to the landlord for the use of the property.
- Depending on the rental agreement, utilities and services such as heat, water, electricity, and parking may be included in the rent. Otherwise, the tenant will pay for them.
- The tenant usually pays for her or his phone, cable, and Internet service.
- A tenant may buy tenant insurance to protect her or his belongings.

Costs of owning
- Most Canadians who buy property use a cash down payment and a mortgage loan. This loan must be repaid.
- Owners pay property taxes.
- Most owners buy home insurance to protect their home and its contents.
- Condo owners usually pay a monthly fee to cover operating, maintenance, administrative, and improvement costs.
- Owners pay for all utilities, maintenance, and services they use.

Example 1

Calculating Annual Expenses

Determine how much each expense will cost for one year.
a) Monthly rent of $665.
b) Bi-weekly mortgage payments of $856.21.
c) Semi-annual home insurance payments of $546.75.

Solution

a) There are 12 months in a year.
   $12 \times \$665 = \$7980$
   The annual cost is $7980.

b) Bi-weekly means every two weeks.
There are 26 two-week periods in a year.
26 × $856.21 = $22 261.46
The annual cost is $22 261.46.

c) Semi-annual means twice a year.
2 × $546.75 = $1093.50
The annual cost is $1093.50.

In Ontario, the laws that cover landlord-tenant relations are part of the Residential Tenancies Act. Here are some of the details from the law.

Tenants' rights

- Your rental home must be safe and in good repair.
- You must have access to services such as heat, hot and cold water, electricity, and fuel.
- You have a right to privacy, though the landlord can enter the home for repairs, to show it to prospective tenants, or in an emergency.
- Your landlord can raise the rent once every 12 months and must give you 90 days written notice of the increase.

Tenants' responsibilities

If a tenant does not uphold her or his responsibilities, the landlord can give a notice of termination stating the problem and asking the tenant to fix it or move out.

- You must pay your rent on time.
- You must keep the home reasonably clean.
- You are responsible for repairing any damage you cause to the unit.
- If you sign a lease agreeing to rent the property for a specified period of time, you must honour the lease and give 60 days notice before the end of the lease if you do not intend to renew.
- If you have a month-to-month rental, you must give 60 days written notice before moving out.

Example 2

Knowing Your Rights

Katrina is two months behind on her rent.
Her landlord threatens to cut off the electricity and heat.
Explain whether or not the landlord can legally carry out her threat.

Solution

The landlord cannot carry out her threat.
By law, the tenant has a right to services such as electricity and heat.
The landlord can give Katrina notice to either pay the money owed or move out.

Example 3 — Comparing Rental Properties

Adisa will be moving to Sarnia to attend college.

Option 1
- 1-bedroom apartment within walking distance from the college
- Rent: $625 a month, including heat, electricity, water, and parking
- Must sign a 1-year lease

Option 2
- College residence
- Share a room and washroom with another student
- Room includes: beds, desks, shelves, a small fridge, a microwave, cable service (but no TV), and local telephone service
- Shared kitchen available
- Cost: $6800 for September to April, including utilities and an $1800 meal card

a) What would be the annual rent for the apartment? Is it more or less expensive than living in residence?

b) What additional expenses might Adisa have in setting up the apartment that he would not have for the residence room?

c) What, if any, benefits might there be to living in the apartment? To living in residence?

d) Which option would you recommend? Justify your answer.

Solution

a) $12 \times \$625 = \$7500$. The annual rent would be $7500.
The apartment is more expensive than living in residence.

b) Adisa would have to purchase furnishings if he can't bring them from home. He may also want to get telephone and cable services.

c) In the apartment, Adisa would have more privacy. He could have visitors more easily, and would have a place to stay in the summer if he gets a job in Sarnia. Living in residence, he might find it easier to meet new people. He would need less money since he would not have to provide furnishings, or do grocery shopping or cooking.

d) To save money, I would recommend staying in residence the first year. Adisa may meet people with whom he could find shared off-campus housing the next year.

Practice

A

■ For help with question 1, see Example 1.

Use the *Course Study Guide* at the end of the book to recall the meaning of the common payment periods.

1. Determine how much each expense will cost for one year.
   a) Monthly rent of $1250
   b) Bi-weekly mortgage payments of $975.56
   c) Semi-annual home insurance payments of $328.14
   d) Monthly cable bill of $32.95

2. Classify each of the following as either fixed or variable expenses.
   a) Monthly rent
   b) Monthly charges for natural gas
   c) Bi-weekly mortgage payments
   d) Bi-monthly charges for water usage

3. Determine the total monthly housing cost.
   a) Rent of $650 plus $88 for utilities
   b) Rent of $980 plus $179 for utilities
   c) Rent of $320 plus a one-third share of $150 for utilities
   d) Rent of $235 plus a one-quarter share of $200 for utilities

■ For help with questions 4 and 5, see Example 2.

4. Ashley broke a window in the apartment she rents. Explain whether she or her landlord is responsible for replacing it.

5. Paul is a landlord. He includes the cost of utilities in the rent he charges. Because utility costs are higher than Paul expected, he asks the tenants to pay an extra $30 per month. Explain whether or not he can legally do this.

484 CHAPTER 8: Budgets

Use the following information to answer questions 6 to 8 and part a of question 9. Rob is a single parent who owns a 2-bedroom house.

Rob's Annual Housing Expenses	
Expense	**Cost**
Bi-weekly mortgage payments	$607.79
Quarterly home insurance payments	$68.75
Annual property taxes	$1738.50
Water, sewer, and electricity bills (received bi-monthly)	$180.45, $164.32, $156.74, $149.76, $167.23, $185.18
Natural gas bills (received monthly)	$123.02, $99.47, $64.82, $70.51, $24.16, $19.79, $23.09, $21.58, $22.14, $36.44, $78.58, $112.56
Monthly phone and TV bill	$62.95

B

6. Classify Rob's expenses as fixed and variable.

7. How much does Rob spend each year on mortgage payments?

8. What are Rob's total annual housing expenses?

9. Some natural gas companies allow their customers to pay in twelve equal payments based on the previous year's bills. Adjustments are made later for actual usage levels.
  a) Calculate Rob's monthly gas bill if he uses an equal payment plan.
  b) Use the data in Case Study 3, page 480. Calculate Tom and Hyo-Jin's monthly gas bill if they use an equal payment plan.
  c) Why might this plan appeal to some homeowners?

8.2 The Costs of Owning or Renting a Home   **485**

10. Use the data in Case Study 3, page 480. Tom and Fyo-Jin have a bachelor apartment in the basement of their home. To help with household expenses, they decide to rent it. They will include utilities in the monthly rent.

   a) They calculate their annual costs for heat, electricity, water, and sewage. They decide to charge $350 plus one-quarter of these utility costs. What will the rent be?

   b) They research other bachelor apartments in their neighbourhood. They find one that costs $600 per month and another that costs $570. Both are all inclusive. Is the rent they plan to charge reasonable? Justify your answer.

■ For help with questions 11 and 12, see Example 3.

11. Samayah is moving to Ottawa to attend college. She finds two apartments within walking distance from the campus. Both require her to sign a 1-year lease.

Option 1   A 1-bedroom apartment in a low-rise.
The rent is $635 per month including heat, water, and parking.
Samayah will have to pay for electricity.
No smoking and no pets are allowed.

Option 2   A 1-bedroom apartment that is the top floor of a house.
The rent is $650 a month including heat, electricity, water, and parking.
No smoking is allowed, but pets are allowed.

   a) What would be the annual rent for each apartment?
   b) What additional expense will Samayah have in the first apartment?
   c) Which option would you recommend to Samayah? Justify your answer.

12. **Assessment Focus** Tyrell has just graduated from college and is moving to Hamilton to start a full-time job. He has found two apartments he likes.

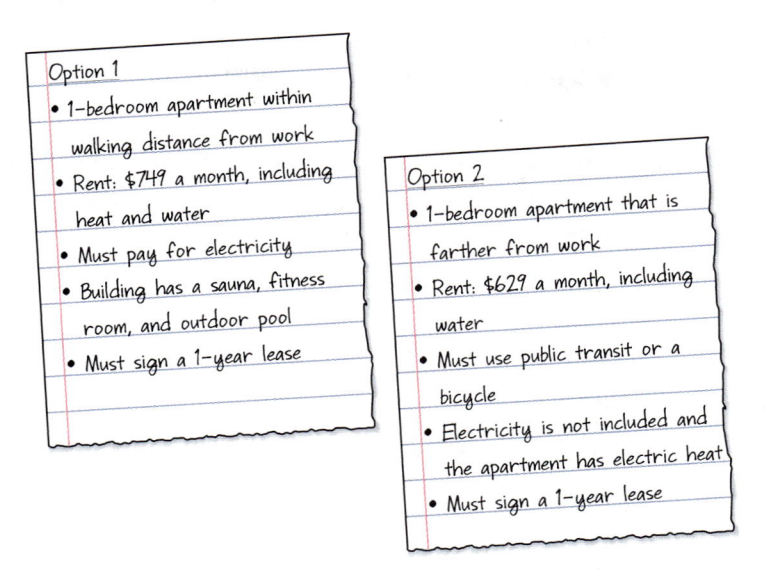

Option 1
- 1-bedroom apartment within walking distance from work
- Rent: $749 a month, including heat and water
- Must pay for electricity
- Building has a sauna, fitness room, and outdoor pool
- Must sign a 1-year lease

Option 2
- 1-bedroom apartment that is farther from work
- Rent: $629 a month, including water
- Must use public transit or a bicycle
- Electricity is not included and the apartment has electric heat
- Must sign a 1-year lease

a) What would be the annual rent for each apartment?
b) What additional costs will Tyrell have with each apartment? How do you think they will compare?
c) Which option would you recommend to Tyrell? Justify your answer.

13. **Literacy in Math** Imagine you were looking for a rental property in your community. List five factors that would affect your choice. Rank the factors from most to least important. Explain how you decided on the order.

14. Create a case study similar to those in *Investigate*. Write and answer two questions about owning or renting a home related to your case study.

In Your Own Words

Decide whether you think this statement is true. Justify your answer.
Usually tenants have lower housing costs than homeowners, so most people should rent their home.

8.2 The Costs of Owning or Renting a Home   **487**

8.3 Estimating Living Costs

The costs of running a household vary greatly depending on the number, ages, and activities of people living in the home. Where you live and the things that are important to you may affect your living costs as well.

Inquire: Researching Living Costs Using E-STAT

Materials
- computer with Internet access
- E-STAT user name and password

If you are working from home, you will need to get a user name and password from your teacher.

You will need a computer with access to the Internet and E-STAT.

Go to the Statistics Canada Web site.
- Click **English**.
- Select **Learning resources** from the menu on the left.
- Click on **E-STAT** in the golden box on the right.
- Click on **Accept and enter**.

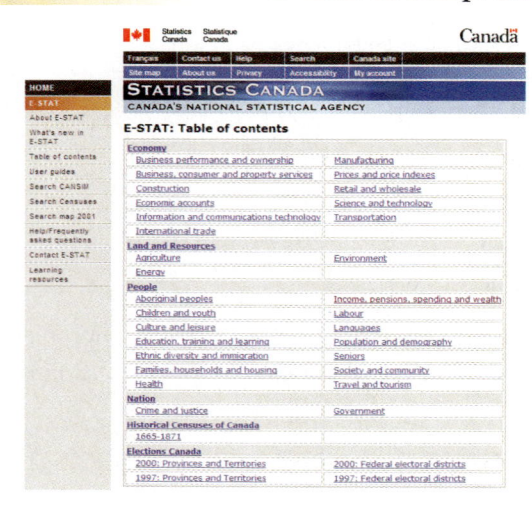

The E-STAT table of contents will be displayed.
- In the *People* section, click on **Income, pensions, spending and wealth**.
- From the list of CANSIM data, click on **Household spending and savings**.
- Click on table **203-0001**.

This table contains data about the amount of money households throughout Canada spend on a variety of expenses in a year.

488 CHAPTER 8: Budgets

- Under *Geography*, select **Ontario**.
  Under *Household expenditure summary*, you will be selecting 14 items, but there are too many choices to see all at once. Click on **View checklist** to display all the items more conveniently.
  Scroll down and select all 14 items in the *Total current consumption* category.

  1. On which four of these items do you think a typical Ontario household spends the most? Justify your predictions.

- Scroll up to the top of the page and click **Return to pick list**.
  Under *Statistics*, select **Median expenditure per household reporting (dollars)**. This measure gives the median amount a household spent on an item for all households reporting spending money on the item.
  Set both reference periods to the most recent year available.
  Click on **Retrieve as a Table**.

- Select the output format.
  In the *SCREEN OUTPUT formats* section, under *Graph (maximum 13 series)*: select **Pie chart (last observation)**.
  Click on **Retrieve now** for the following display:

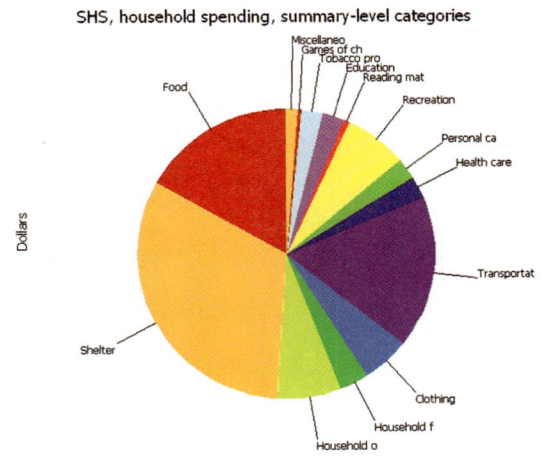

> The *median* is the middle number when a set of numbers are in numerical order, or the mean of the two middle numbers.

8.3 Estimating Living Costs

2. a) What were the six greatest expenditures for Ontario households? How do you know?
  b) How accurate were your predictions in question 1?
  c) Estimate the fraction of household consumption represented by each item in part a.

■ To view the data as a table, click **Back** on the tool bar.
  In the SCREEN OUTPUT formats section, under HTML Table: select **Time as columns**.
  Click on **Retrieve now**.

> You will use the data again in Lesson 8.4, Practice question 13.

3. Use the data in the table to check your estimates in part c of question 2.

■ Print the table for use in question 4 and in Lesson 8.4.
■ Click on the **Back** button twice to return to the *Subset selection* page. Retrieve data about a different province.

4. Repeat questions 1 to 3 using the new data.
  Compare your answers for the two provinces.

The data you have been using are the median expenditures for households that participated in the survey and reported spending money on the items you selected.

■ Click on the back button twice to return to the *Subset selection* page.
  Under *Geography*, select **Ontario**.
  Under *Statistics*, select **Average expenditure (dollars)**, then hold down the **Ctrl** key and select **Median expenditure per household reporting (dollars)**.
  Display the data in a table.

> The *average* gives the mean amount per household for all households in the survey.

5. a) How do the average and median data compare?
  b) Which categories appear to show the greatest differences? Why do you think this is?
  c) Which data do you think provide a better estimate of typical household expenditures? Justify your answer.

490 CHAPTER 8: Budgets

6. Describe how, if at all, you think each of these factors might affect household expenditures on food, shelter, clothing, or transportation. Justify your answers.
   a) The number of people in a household
   b) The number of teenagers in a household
   c) A family member with special dietary requirements
   d) A family that needs a wheelchair-accessible home and vehicle

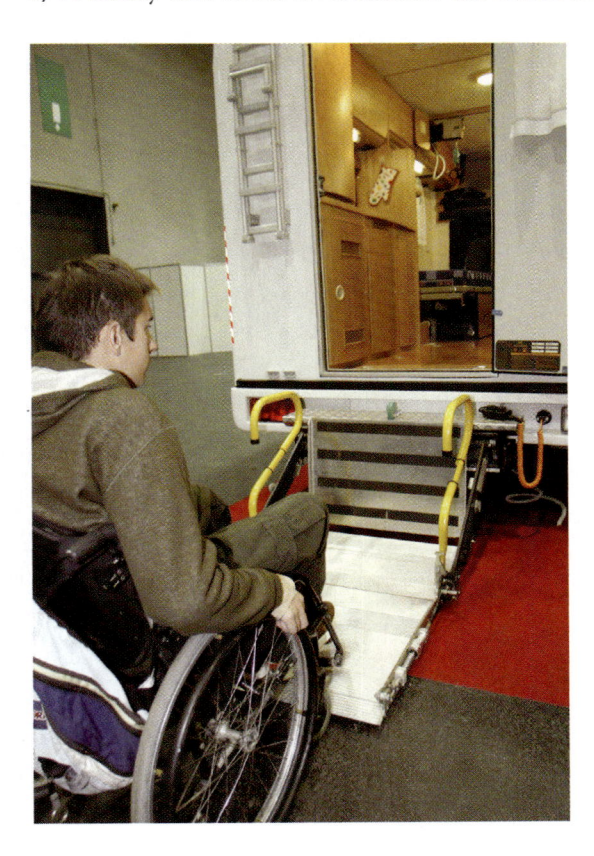

7. How might the neighbourhood, size of town or city, or province you live in affect your living costs? Give examples to support your ideas from E-STAT data or personal observations.

Reflect

- Did you prefer to work with the household expense data in table form or in graph form? Give reasons for your choice.
- How might personal requirements and lifestyle choices affect the costs of running a household? Justify your answer.

Mid-Chapter Review

8.1

1. a) List two housing expenses that both renters and owners may have to pay.
   b) List three expenses that only owners would pay.

2. Mark and Jen live in your community with their two children. They rent a three-bedroom condominium.
   a) List the housing expenses that Mark and Jen would incur.
   b) Estimate each monthly expense that you listed in part a for your community. Explain how you estimated.

8.2

3. Determine how much each housing expense will cost for one year.
   a) Weekly mortgage payments of $346.78
   b) Monthly phone bill of $29.95
   c) Monthly rent of $765
   d) Bi-annual insurance bill of $427.60
   e) Quarterly water heater rental of $68.95

4. Karim's landlord phones to tell Karim that he will be raising the rent by 2.7% next month. Explain whether or not this is legal.

5. Ivor estimated his monthly housing costs. Calculate the total cost for one year.

Description	Frequency	Cost ($)
Rent	monthly	950.00
Contents insurance	bi-monthly	35.00
Cable/Internet	monthly	95.00
Electricity	monthly	45.00
Home furnishings	semi-annually	750.00

6. Melissa and Farideh will be moving to Owen Sound to attend the same college and plan to share an apartment.
   - They find a 2-bedroom apartment within walking distance from the college. The rent is $750 a month, including utilities.
   - In the same building, they notice a 3-bedroom apartment for rent for $795 a month, including utilities.
   - Both apartments require a 1-year lease.
   a) Calculate each girl's annual housing cost if they rent the 2-bedroom apartment and share costs equally.
   b) Farideh suggests they rent the 3-bedroom apartment and find another roommate. If they share expenses, what amount would each girl save in a year?
   c) Suppose they rent the 3-bedroom apartment, but the third roommate leaves after 6 months and they cannot find another roommate. Will the girls still have saved money? Justify your answer.

8.3

7. Go to the Statistics Canada Web site. Follow the steps from Lesson 8.3 to access table **203-0001** and make category selections, but under *Geography*, choose two cities.
   a) What are the 5 greatest expenses in each city?
   b) How do the amounts spent on these items compare?
   c) Which city appears to be more expensive? Justify your answer.

492 CHAPTER 8: Budgets

8.4 Designing Monthly Budgets

A **budget** is a detailed plan that compares what you earn with what you spend for a set period of time. Budgets can help you see where your money is being spent and can highlight areas where you might be able to reduce spending.

Investigate | Creating a Budget

Materials
- scientific calculator
- *Microsoft Word*
- BudgetTemplate.doc

Work with a partner or in a small group.

Mikayla rents a small apartment. She is starting to think about buying her first home.

- Her net annual income is $29 550.
- Her monthly rent is $650, which includes heat and water.
- Her electric bill averages about $50 a month.
- She leases a car. The lease includes maintenance. Her monthly payments for the car and insurance are $350.
- She spends an average of about $120 a month on gasoline.
- Her weekly expenses for groceries and household supplies average $120.
- She spends up to $200 a month on eating out and entertainment.
- Her gym membership costs $30 each month.
- Clothing, home furnishings, vacations, her pet dog, and other miscellaneous costs total about $4000 each year.

Net income is the income you have after income tax and other **payroll deductions** (Canada Pension Plan, Employment Insurance) have been subtracted.

- Open file *BudgetTemplate.doc* or create your own budget template. Create a budget to organize Mikayla's financial information.
- Compare Mikayla's total monthly income with her total monthly expenses.
- List some new costs Mikayla will need to budget for if she owns a home.
- About how large a monthly mortgage payment do you think Mikayla could afford? Explain your reasoning.
- Mikayla has no money saved for a down payment. In which spending areas do you think she could cut back to build her savings? How much do you think she could save over the next year? Justify your responses.
- If Mikayla's net annual income increased by $5000, how would you distribute the extra income each month? Justify your response.

Budget Item	Monthly Amount ($)
INCOME	
Salary	
Total Income	
EXPENSES	
Housing	
Subtotal	
Transportation	
Subtotal	
Food	
Subtotal	
Other	
Subtotal	
Savings	
Total Expenses	
Total Income − Total Expenses	

Reflect

- How did you decide how to organize Mikayla's financial information to create a budget?
- Explain how a monthly budget could help you manage your money.

Connect the Ideas

Income and expenses

Income is the money you earn. **Expenses** are the money you spend. A budget is an organized list that compares income and expenses. Budgets are used as a tool in financial planning.
- To make sure you don't spend more than you earn
- To help you understand exactly where your money is going
- To help you focus your spending on the things that are most important to you
- To help you save to meet a financial goal

494 CHAPTER 8: Budgets

Personal budgets are usually planned monthly. Monthly amounts can then be multiplied by 12 to determine annual amounts.

Budget categories

Income and expenses are recorded by type of income or expense. Money set aside for a future use is called **savings** and is shown as an expense.

A broad category, such as *Housing*, may be broken down into more specific categories, such as *Mortgage payment*, *Utilities*, and *Insurance*. Or, several types of expenses may be grouped under one category, such as *Entertainment*.

The categories chosen depend on how a person or family earns and spends money. Every source of income and every regular expense should fit in a category.

Example 1

Interpreting a Budget

This circle graph represents Jamila's expenses for one month.
a) List the expenses from greatest to least.
b) The total of Jamila's monthly expenses was $3000. How much did she spend on transportation?

Jamila's Monthly Expenses
- Miscellaneous 15%
- Housing and utilities 30%
- Savings 10%
- Recreation and education 7%
- Transportation 13%
- Health and personal care 4%
- Food and clothing 21%

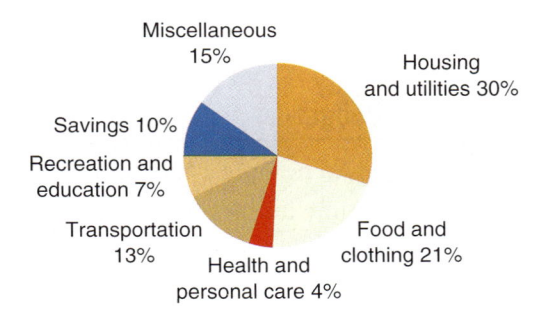

Solution

a) Housing and utilities, Food and clothing, Miscellaneous, Transportation, Savings, Recreation and education, Health and personal care

b) Write 13% as a decimal: 0.13
$0.13 \times \$3000 = \$390$
Jamila spent $390 on transportation.

8.4 Designing Monthly Budgets **495**

Balancing a budget

The **balance** is the difference when total expenses are subtracted from total income.
If the balance is negative, you are spending more than you earn.
You need to adjust some expenses so that the balance is zero.
This is called **balancing** the budget.
If the balance is positive, you can save the money for unexpected expenses or spend some of it.

Predicted and actual budgets

Expenses and income are often estimated, or predicted. The actual amounts earned and spent are then tracked and compared with the predicted amounts.

Example 2

Balancing a Budget

Materials
- scientific calculator

Bi-weekly means every two weeks.

Raul is studying to become a cabinetmaker and furniture technician. He worked full time in the summer to pay for his tuition, books, and supplies. However, he was not able to save enough for his living expenses during the school year.
He receives a $3500 scholarship each school year.
He also earns $500 bi-weekly at a part-time job.
He has payroll deductions of $74 bi-weekly.
His other expenses are: rent and utilities at $400 each month, transportation $80 each month, food $75 each week, entertainment $25 each week, clothing $110 each month, and $100 bi-weekly for miscellaneous items.

a) Use the data provided to design a monthly budget for the school year.
b) Is Raul earning enough to cover his expenses? If not, suggest how he could balance his budget. Explain your reasoning.

496 CHAPTER 8: Budgets

There are 52 weeks in a year and 26 two-week periods.

Solution

a) Group income items together and expense items together. Determine a monthly amount for each item. Round amounts to the nearest dollar.

To convert a weekly expense to a monthly expense, multiply by 52, then divide by 12.

To convert a bi-weekly amount to a monthly amount, multiply by 26, then divide by 12.

Income	Monthly amount ($)
Scholarship ($3500 ÷ 12)	292
Part-time job ($500 × 26 ÷ 12)	1083
Total income	1375
Expenses	
Payroll deductions ($74 × 26 ÷ 12)	160
Rent and utilities	400
Transportation	80
Food ($75 × 52 ÷ 12)	325
Entertainment ($25 × 52 ÷ 12)	108
Clothing	110
Miscellaneous ($100 × 26 ÷ 12)	217
Total expenses	1400
Balance (Total income − total expenses)	−25

b) Raul is spending $25 more than he earns each month.
He is also not saving money for unexpected expenses.
He should reduce his expenses by $25 to break even. He should try to reduce even more to have money for unexpected expenses. The easiest categories to cut back on are entertainment, clothing, and miscellaneous costs.

Fixed and variable costs

Sometimes budgets show fixed and variable costs separately to make it easier to see which categories of expenses can be adjusted, if necessary.

Example 3

Designing a Budget to Meet a Savings Goal

Materials
- scientific calculator

Anika needs to save $7500 over the next 12 months to start her own small business. Her monthly paycheque after payroll deductions is $3000. She earns an average of $50 a month from investments. She prepares a list of her expenses.

```
Anika's Expenses
Housing
 • $850 for rent each month, which includes heat
   and water
 • an average of $75 each month for electricity
 • $105 a month for a phone/cable/Internet bundle
Transportation
 • monthly car payment of $420
 • annual vehicle licence fee of $75
 • $420 for car insurance twice a year
 • $150 a month for gas
 • $30 every three months for oil changes
 • $450 per year for maintenance and repairs
Other
 • $250 for groceries
 • $200 for clothes and personal care
 • $100 a week for restaurant meals
 • $20 automatically deducted from her bank account
   for charitable donation
```

> The amounts for groceries, clothes, personal care, and charitable donation are monthly expenses.

a) Convert all amounts to monthly amounts.
b) Which costs are fixed?
c) Are there expenses that you think Anika has forgotten to include? Explain. Estimate the monthly amount for any missing expenses.
d) Create a monthly budget that shows fixed and variable expenses in each budget category. Group items where appropriate. Include any estimates from part c.
e) What percent of Anika's costs are fixed? Why is this important?
f) Can Anika meet her savings goal with her current income and expenses? If not, in which categories might she be able to cut back on costs?
g) How did the budget help you answer parts e and f? Explain.

498 CHAPTER 8: Budgets

Solution

Round all monthly amounts to the nearest dollar.

First convert weekly or bi-weekly costs to annual costs, then covert to monthly costs.

a) Vehicle licence fee:
   $75 \div 12$ months $\doteq$ $6/month
   Oil changes:
   $30 \div 3$ months = $10/month
   Vehicle maintenance and repairs:
   $450 \div 12$ months $\doteq$ $38/month
   Restaurant meals:
   $100/week $\times$ 52 weeks $\div$ 12 months $\doteq$ $433/month
   Car insurance:
   $420 \div 6$ months = $70/month
   Savings:
   $7500 \div 12$ months = $625/month
   All other amounts given are for one month.

b) Rent, phone/cable/Internet, car payment, vehicle licence fee, car insurance, charitable donations, and her monthly savings amount are fixed costs.

c) Anika has not included any costs for entertainment. Even if she spends very little on this, she would likely spend at least $30 a month.

d) Organize the given data by grouping expenses in categories. Display fixed and variable costs separately. See page 500.

e) From the table on page 500, Anika's total expenses are $3282, of which $2096 are fixed.
   To express as a percent, divide the fixed costs by the total costs and multiply the result by 100%.
   $2096 \div $3282 $\times$ 100% $\doteq$ 64%
   About 64% of Anika's expenses are fixed. This is important because it is harder to cut back on fixed costs than variable.

f) No, Anika needs to spend at least $232 less each month. She will need to reduce some of her variable expenses, such as clothing and restaurant meals. She could also consider cutting her cable service.

g) The budget shows the total expenses and total fixed expenses for the month. This information is needed to calculate the percent in part e. The budget also shows how much Anika is overspending. You need to know this to determine how much she has to reduce spending.

8.4 Designing Monthly Budgets   **499**

	Monthly amount ($)		
INCOME	Fixed	Variable	Total
Salary	3000		3000
Investments		50	50
Total income	3000	50	**3050**
EXPENSES			
Housing			
Rent	850		850
Utilities		75	75
Phone/cable/Internet	105		105
Subtotal	955	75	**1030**
Transportation			
Car payment	420		420
Fuel cost		150	150
Insurance	70		70
Other	6	48	54
Subtotal	496	198	**694**
Food			
Groceries		250	250
Restaurant meals		433	433
Subtotal	0	683	**683**
Other			
Clothing/personal care		200	200
Entertainment		30	30
Charitable donations	20		20
Subtotal	20	230	**250**
Savings	625		**625**
Total expenses	2096	1186	**3282**
Total income − Total expenses			−232

Oil changes and other car maintenance expenses are grouped.

Practice

A

For help with question 1, see Example 1.

1. The circle graph represents Luca's expenses for one month.
   a) Which three categories represent his greatest monthly expenses?
   b) Luca spent a total of $1600 for the month. What amount did he spend on books and supplies?

Luca's Monthly Expenses
- Personal expenses 8%
- Health and dental 1%
- Travel 6%
- Tuition 35%
- Meals 15%
- Residence 25%
- Books and supplies 10%

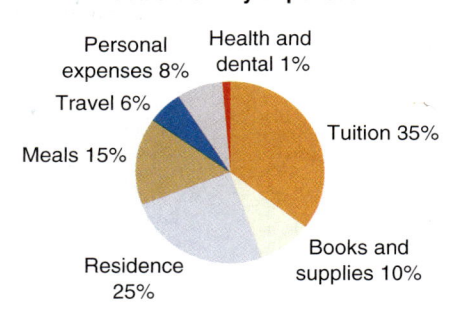

2. Classify each item as either income or an expense.
   a) rent
   b) loan payment
   c) wages from part-time job
   d) water bill
   e) child care
   f) scholarship

3. Ingrid works as a legal assistant. She has a net yearly income of $32 500.
   a) Calculate Ingrid's monthly income.
   b) Ingrid has monthly expenses of $2400. Is Ingrid in a position to save money or is she in debt at the end of the month? Explain how you know.

For help with questions 4 and 5, see Example 2.

4. Classify each item as income or an expense. Then convert the item to a monthly amount.
   a) groceries, $140/week
   b) net salary, $56 000 a year
   c) online auction sales, about $275 bi-monthly
   d) property tax, $800 quarterly

B

As you work on the questions, round all monthly amounts to the nearest dollar.

5. Yvonne is a student at a community college. She has a part-time job with take-home pay of $260 each week. She has also received a student loan of $5000 for the year. Her parents pay her tuition.
   a) Design a monthly budget using the data provided. Show your calculations.
   b) Is Yvonne earning enough to cover her expenses? If not, suggest how she could balance her budget.

Yvonne's Expenses
- Rent and utilities: $550/month
- Food: $85/week
- Bus pass: $65/month
- Cell phone: $25/month
- Books and supplies: $1200/year
- Miscellaneous: $250/2 weeks

8.4 Designing Monthly Budgets **501**

Use the following information to answer questions 6–8.

Over the coming year, Jim and Iona want to save $5000 for a vacation. Their total net income is $65 000/year. They earn about $75/month from investments. They currently have the following expenses:

Jim and Iona's Expenses
- bi-weekly mortgage payment and property tax of $675
- $650/year for home insurance and $1050/year for car insurance
- utility costs that average $230/month
- phone/cable costs of $75/month
- vehicle lease of $410/month, which includes maintenance
- gasoline costs of $175/month
- retirement savings plan contributions of $225 bi-weekly
- grocery costs of $160/week
- clothing costs of $3000/year
- entertainment costs of $120/week
- charitable donations of $1000/year
- miscellaneous costs of $150 bi-weekly

■ For help with questions 6 and 7, see Example 3.

6. Convert all of the income and expense items into monthly amounts. Which are fixed and which are variable?

7. a) Design a monthly budget for Jim and Iona that shows fixed and variable costs.
   b) Can Jim and Iona meet their savings goal with their current income and expenses? If not, in which categories might they be able to cut back on costs?

8. Jim has to cut back on his work hours to look after an ill parent. As a result, Jim and Iona's net income is reduced to $58 000/year. Can they still meet their savings goal? If not, suggest how they could balance their budget.

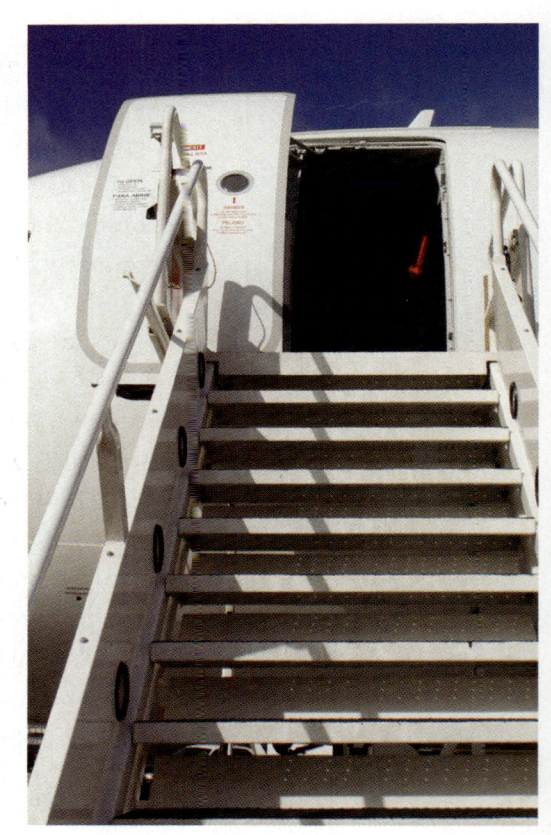

502 CHAPTER 8: Budgets

9. **Assessment Focus** Halima works part time as a waitress. She earns a take home salary of $425 a week including tips. Her expenses are $650 a month for rent and utilities, $95 a week for groceries, $80 a month for a bus pass, and $150 bi-weekly for miscellaneous expenses. She is also taking a course at a community college and must set aside $100 a month for her education expenses.
   a) Design a monthly budget using the information provided. Show your calculations.
   b) Is Halima earning enough to cover her expenses? Explain.
   c) Halima would like to take a vacation. She needs to save $1000 over the next 4 months. Will this be possible with her current budget? If not, how could she adjust her budget to save this money?

10. Nihal is an apprentice arborist. From January until the end of August, he will be earning $350 per week in net pay from a part-time job. Starting in September, he will spend 12 weeks in school. He will not work during this period. He will return to work at the beginning of December.
    a) By the end of August, Nihal needs to save $900 to pay for his tuition and equipment. Design a monthly budget for the first 8 months of the year to help him meet this goal. How much money will Nihal save each month?
    b) During the 3 months that Nihal is at school, he will receive a total of $2310 in Employment Insurance. Design a monthly budget for this period. How much of his savings will Nihal need to use to meet his monthly expenses during this period?
    c) Nihal also wants to save $1500 by the end of December for a trip to British Columbia. Can he do it? Explain.

    Nihal's Monthly Expenses

Expense	Cost ($)
Rent	450
Utilities	120
Food	320
Clothing	30
Bus pass	60
Entertainment	40
Phone and Internet	60
Contact lenses	25

11. a) Think of your own finances or imagine the finances of a "typical" Grade 12 student in your community. Create a monthly budget that includes all income and expenses.
    b) Are there any items that you could change to save more money on a monthly basis? Is this realistic?

8.4 Designing Monthly Budgets **503**

12. Literacy in Math Write a short paragraph on the features of a "good" budget plan. Possible topics to mention may include: how to set goals, track expenses, decide on areas to cut back, and monitor progress.

> Use the E-STAT data you researched in Lesson 8.3.

13. In Lesson 8.3, you researched the median annual expenses of Ontario households across 14 categories. Two categories were not included in the data: personal insurance payments and pension contributions of about $325/month and gifts of money and contributions of about $60/month.
   a) Which categories would you combine if you were creating a budget? Explain your thinking, then determine the combined annual expenses for these categories.
   b) Calculate the median monthly expenses for all the categories.
   c) Suppose a household had expenses equal to these provincial medians. What would the total monthly expenses be?

14. Clive works part time from the beginning of September to the end of April, while he is attending college, and then full time through the summer months. He lives in a college residence during the school year and at home during the summer. What difficulties might he encounter when trying to create a monthly budget? Suggest at least one way he could solve the problem.

C

> What tools can you use to help you draw the circle graph?

15. a) Choose one of the monthly budgets you have designed. Determine the percent of the total monthly expenses that each individual monthly expense represents.
   b) Create a circle graph that shows the breakdown of the monthly expenses.

16. Think about how your financial situation will change once you graduate from high school. Create a monthly budget that incorporates your vision of your financial situation one year from now.

In Your Own Words

What factors make it difficult for you to follow a budget at this point in your own life? How important do you think it is for high school students to create a monthly budget?

8.5 Creating a Budget Using a Spreadsheet

A budget can help you save for a financial goal, whether that goal is paying for a vacation, a wedding, or college tuition. Spreadsheet software is an effective tool for creating the budget because it allows you to quickly see the effects of changes you make.

Inquire: Designing a Budget Using Technology

Materials
- *Microsoft Excel*
- Budget.xls (optional)
- access to the Internet (for Part C only)

Work with a partner.

Part A: Creating a Monthly Budget

Nuri is in his third year of employment as a firefighter. Nuri has been tracking his spending for several weeks so that he can create a monthly budget.

Income: Annual net income of $36 850
Investment income of $500 every six months

Expenses: $750 per month for rent, utilities, and household services
$425 per month for transportation
$110 each week for food and household supplies
$50 each week for entertainment
$150 bi-weekly for miscellaneous items

1. Open the file *Budget.xls*. If you do not have the file, start a new spreadsheet file and enter the text and formulas shown here. Format the cells in column B as shown.

	A	B
1	Nuri's Monthly Budget	
2	Description of Income	Monthly Amount ($)
3		
4		
5		
6		
7	Total	0
8		
9	Description of Expense	Monthly Amount ($)
10		
11		
12		
13		
14		
15		
16		
17		
18		
19		
20	Total	0
21		
22	Balance	0

Save a copy of Nuri's budget.

2. Enter each income and expense in the spreadsheet, changing income and expenses to monthly amounts as needed. Use the calculation capabilities of the spreadsheet software to complete this task. For example, to calculate Nuri's net monthly income, enter the formula: = **36850/12** in the appropriate cell in column B.

3. What is the balance? Does this mean Nuri is spending too much or has money left over? How do you know?

If the balance is shown in brackets, it is negative.

Part B: Adjusting a Budget

Reviewing your budget is an important step in keeping an accurate record of your financial situation. If there are changes in your earnings, expenses, or long-term plans and goals, it will be necessary to make adjustments to your budget.

4. Nuri receives a raise. His net salary increases to $37 955. Open the budget worksheet that you completed in Part A. Change Nuri's budget worksheet to reflect this change.
   a) What effect does this change have on the balance?
   b) If the change leads to monthly debt, suggest what Nuri could do to avoid debt. Test your ideas by adjusting the expenses in the spreadsheet.

5. Repeat question 4 for each of these changes. Begin with the spreadsheet you saved in Part A each time.
   a) Nuri gets engaged and wants to save for a wedding and honeymoon trip. He wants to save $250/week for the next several months.
   b) Nuri would like to save $12 000 over the next year towards a down payment on a cottage.
   c) Nuri's sister plans to enter a paramedic program at college. Nuri wants to save $2000 in the next 2 months to lend her for tuition.
   d) Nuri is hurt at work and needs 2 months to recover. During this time he earns 80% of his usual net salary.

6. Do any of the changes in question 5 carry other costs that Nuri should budget for? Might any increase his income? Explain with details.

Part C: Budgeting for a Goal of Your Choice

Choose a goal from this list.

- Making a down payment for a townhouse, condo, or duplex in your community
- Paying for a college education in the program of your choice
- Paying for a two-week vacation of your choice
- Buying a small boat or making a down payment on a larger boat

Research the costs involved with this goal. Decide an approximate length of time Nuri will save for the goal. Revise Nuri's original budget to reflect this change.

Practice

A

1. Describe what each formula in cells B7, B20, and B22 of Nuri's budget spreadsheet is calculating.

2. List four lifestyle changes that would require a person to revise her or his monthly budget.

Use the following information to answer questions 3 and 4.

Rebecca is a second-term apprentice electrician. She has 3 more months of work before she starts a 2-month period of study at school. During this time, she needs to save $525 to pay for school courses, books, and supplies.

Her net weekly income is $481.12. Her expenses are:

$480 per month for rent and utilities
$75 per week for food and household supplies
$40 per week for entertainment
$15 per week for transportation
$200 bi-weekly for miscellaneous items

B

3. Create a monthly budget for Rebecca using the spreadsheet template.
   a) How much money will she have to set aside each month for schooling?
   b) How much extra money can she save during the 3-month period?

4. Rebecca has just started her 2 months of class time. She will receive employment insurance (EI) benefits of 55% of her net income. She will have the same expenses, except she will not need to save for schooling.
   a) Revise the budget that you created in question 3 to reflect these new conditions.
   b) Can Rebecca pay her monthly expenses with the money she earns from EI? If not, she will use her extra savings from part b of question 3. How much will she have left after two months?

5. Oliver is just starting his last year of high school.
   He has take-home pay of $275 bi-weekly from a part-time job.
   He lives at home and pays $30 per week for room and board.
   His other expenses are:
   • $30 per month for a cell phone
   • $25 per week for entertainment
   • $70 bi-weekly for miscellaneous items
   a) Create a monthly budget for Oliver using the spreadsheet template. How much money can he save each month?
   b) Oliver is applying to a chef training program at college. If he is accepted, over the next 10 months he needs to save $2000 to pay for the first term's tuition and supplies. Can he save this with his current expenses? If not, suggest how he could change his spending to meet this goal.

Reflect

- Do you prefer to create a budget on paper or use technology? Explain your choice.
- Choose one way you adjusted a budget in this lesson. Describe how you made the decision about how to adjust the budget. Explain the effect of that decision in your budget.

8.5 Creating a Budget Using a Spreadsheet **509**

GAME

Budget Shuffle Challenge

Play in two teams of 1 to 3 players.

The template shows Elishea's monthly budget. In this game, you will be reworking this budget to meet Elishea's savings requirements and new financial situations.

- Shuffle the *Budget Change Cards* and place them in a pile.
  One team chooses a card from the pile.
  The other team rolls the die and multiplies the result by 100. The result is the amount of savings Elishea wants at the end of the month.

- Each team reworks the budget to reflect both of these new requirements.

- When both teams are finished, check the accuracy and reasonableness of each other's work.

- If mistakes are found, the team that made the budget must correct them.

- Each team that creates an accurate and reasonable budget without having to correct it scores 10 points. A team that creates an accurate budget after one chance to correct it scores 5 points. If a team is unable to produce an accurate budget, it scores no points.

- Take turns drawing *Budget Change Cards* and rolling the die until each team has created three new budgets. Each turn begin with Elishea's original budget.

- Each team adds the total score for all three budgets. The winner is the team with the greater score.

Materials
- *Budget Shuffle Challenge* template with original data and space for three reworked budgets
- *Budget Change Cards*
- 1 die
- scientific calculator

A *Budget Change Card* explains a change that has occurred in Elishea's circumstances or lifestyle.

Reflect

- Which items in Elishea's budget could you not change? Justify your answer.
- Which items in Elishea's budget were the most flexible? Explain whether you think this is true for most household budgets.

8.6 Making Decisions about Buying or Renting

Deciding whether to rent or buy a home depends on many factors: availability, cost, your housing needs, income, savings, and expenses. You also should consider whether changes may occur in your lifestyle that will affect your future needs and finances.

Inquire: Deciding Whether to Rent or Buy

Work in small groups.

> In Part B, you will need the housing cost data you researched in Lesson 8.1.

Part A: Factors Influencing Housing Decisions

- Brainstorm a list of at least five factors that would influence a decision whether to rent or buy property.
- Identify the expenses from the list below that only someone purchasing property would pay, that only a renter would pay, and that both purchasers and renters may pay. Which are one time expenses and which are ongoing?

down payment	insurance	legal fees
home inspection	moving costs	appliances
mortgage payments	furnishings	utilities
land transfer tax	property taxes	landscaping
repairs and maintenance	painting	renovations
yard equipment such as lawnmower		

8.6 Making Decisions about Buying or Renting   **511**

- Which appears to be more expensive: buying or renting a home? Justify your answer.
- Once a buyer pays off the mortgage, he or she owns the property. How is this different from renting? Why does this make owning property appealing?

Part B: Making a Recommendation about Renting or Buying

Consider this case study.

CASE STUDY

- Jane and Greg Dillon have two children, aged 6 and 10.
- The family rents a townhouse in Calgary and pays $1350 monthly rent, plus utilities.
- With their current income and expenses, the Dillons are able to save $1230 each month.
- They have $45 000 in savings.
- Greg has accepted a two-year contract as a process operator at a refinery in Ontario, with an option to renew for 2 additional years. His net monthly salary will be $4285, which is $585 greater than his current monthly salary. His employer will pay the family's moving expenses.
- Jane is a teacher. She hopes to find a permanent full-time position within a year. Until then, she thinks that she will be able to supply teach two days a week on average. This will represent a drop of $315 each week from her current net salary.

The Dillons are deciding whether to rent or buy a home in Ontario.

512 CHAPTER 8: Budgets

Use the data you researched in Lesson 8.1.

- Assume Jane is able to supply teach two days a week. By how much will the net monthly family income increase or decrease because of the move?
- What one-time expenses might the Dillons have if they buy a home? How much of their savings should they set aside for these costs?
- What additional monthly expenses might the Dillons have if they buy a home? Estimate the cost of these expenses.
- How much do you think the Dillons can afford to spend on housing each month after they move? Justify your answer.
- If they were to buy a home, how much of their savings would you recommend Jane and Greg use as a down payment? Justify your answer.

Suppose the Dillons are moving to your community.

- Which of the properties you researched should the Dillons buy or rent? Justify your choice by listing the major advantages and any important disadvantages of what you are recommending. Present your recommendation in a form that you think would convince the Dillon family to consider your recommendation.

Practice

A

A home inspection is optional and is sometimes provided by the seller.

1. The Khamvosa family is buying a house in August. Their initial costs of purchasing the home are shown. Determine the total initial cost.

   - Down payment: $25 000
   - Legal fees and disbursements: $1200
   - Land transfer tax: $1500
   - Property tax for Aug. to Dec.: $900
   - Home inspection: $350
   - Moving costs: $800
   - Six months of home insurance: $300
   - Washer and dryer: $1800

2. The Huether family is moving to a rental townhouse. Their initial costs of renting the home are shown. Determine the total initial cost.

   - First and last month's rent: $2500
   - Moving costs: $800
   - Six months of tenant's insurance: $175

In questions 3, 5, and 7, determine whether renting or buying is the better option. Justify your answer by describing the factors that led you to your recommendation.

B 3. Mario and Julia have a net annual combined income of $75 000. They have no children. The total of their monthly expenses, not including any expenses related to owning or renting, is $2000. They have savings of $35 000.

Option 1: Buying a 1-bedroom condo
Down payment of $20 000, monthly mortgage payments of $1700, and monthly condo fees of $300

Option 2: Renting a 1-bedroom condo
$1050 per month, plus utilities

4. Refer to question 3. Suppose that Mario and Julia will have a baby in a few months. When Julia takes maternity leave, their combined net income will drop to $51 500.
   a) How will having a baby affect their housing needs?
   b) Will Mario and Julia still be able to afford these housing options? Explain.
   c) What recommendation would you make to them about housing? Justify your answer.

5. The Walters family includes 2 children aged 12 and 18. Both parents are working full time and have a bi-weekly net income of $2750. The total of their monthly expenses, not including expenses related to owning or renting, is $3150. They have savings of $40 000.

Option 1: Buying a 3-bedroom detached house
Down payment of $30 000, bi-weekly mortgage payments of $1050

Option 2: Renting a 3-bedroom detached house
$1270 per month, including utilities

6. Refer to question 5. Suppose that the Walters' older child will be leaving home to attend college in a few months. The Walters will give her an allowance of $450 per month to help pay for schooling and living expenses.
   a) How will this affect the money they can spend on housing?
   b) Would you still make the same recommendation about renting or buying? Justify your answer.

7. Narges is starting a 4-year nursing program offered cooperatively by a college and university. Her parents have offered to help her buy a house that she could share with other students. She has also found an apartment

she could rent. Which option would you recommend? Justify your recommendation. What factors other than financial considerations may affect her decision?

Option 1: Buying a 3-bedroom bungalow

Narges' parents would pay a $35 000 down payment and the initial moving costs.

Narges would be responsible for monthly mortgage payments of $1945.81.

The house has a basement apartment that is rented at $750/month, all inclusive.

Narges plans to rent 2 bedrooms in the main floor to other nursing students for $400/month each, including utilities.

Option 2: Renting a 1-bedroom apartment in a high-rise

$929 per month, including utilities

8. A credit rating tells the lender how likely a person is to be able to pay back a loan. It is based on the person's financial history and affects how much a person might be loaned and the interest rate charged. Explain how your credit rating might affect your choice of whether to rent or own a home.

9. Explain how owning a home could be considered a type of long-term investment. If a person decides to rent, what other options does he or she have for creating a long-term savings plan?

Reflect

"Over the long term, buying a home is always a better decision than renting one." Debate this statement. What factors would lead a person to support this opinion?

8.7 Occupations Involving Finance

Financial mathematics plays an important role in a variety of occupations. Many major life decisions also require an understanding of personal finance or the advice of financial professionals.

Inquire: Researching Occupations and College Programs

Materials
- computer with Internet access
- print materials about college programs
- job advertisements from newspapers

Financial mathematics is used in a variety of occupations. For example:

- A mortgage broker finds the mortgage that is best for her client.

- An auto mechanic creates a detailed budget when he applies for a loan to set up his own repair shop.

- A certified financial planner analyses his client's assets, debts, and financial goals and recommends how the client can reach her goals.

- A sales manager sets and monitors sales targets for the sales representatives she supervises.

Work in a small group.

Part A: Creating a List of Occupations

List occupations that involve finance. Include a few words that you think are important about each occupation.

- Use your general knowledge about occupations.
- Think of people you know who use finance at work.
- Use search words to research on the Internet. Specify Canadian or Ontario sites.
- Look at job advertisements on the Internet and in newspapers. Decide whether finance is required. Note the education and experience expected.

Occupations Involving Finance
Bookkeeper
Real estate agent/assistant
Small business owner
Loan officer
Credit counsellor

Part B: Researching Occupations

■ Each group member should select two occupations from the list developed in Part A to research further. You may choose to:
- Read about the occupation on the Internet or in printed material.
- Contact and interview someone who works in the occupation.

■ Summarize your research. Include details such as:
- What are the main duties involved in this occupation?
- How is financial mathematics used in this occupation?
- What is a typical wage or salary for someone in this occupation?
- What are the job opportunities for this occupation?
- Are there part-time or summer jobs for students related to this occupation?

8.7 Occupations Involving Finance **517**

Part C: Researching College and Apprenticeship Programs

> Many apprenticeship programs involve a combination of work terms and study terms at college.

- Research college and apprenticeship programs that prepare for work in the occupations you researched.

- Use information for colleges in Ontario or in a region where you would like to study. You might:
  - Read program and course descriptions on college Web sites, in libraries, at colleges, or in your guidance department.
  - Talk with people who work at colleges or with college students.

- Summarize your research. Include details about these topics:
  - College name and location
  - Prerequisites for programs and for courses
  - Course descriptions
  - Work experience during the program
  - Length of time for programs, diplomas, or certificates
  - Further education in the field after the program

Part D: Preparing Your Report

- Share your research with the other group members.

- Together, combine your research about occupations and about education for these occupations. You might prepare a visual display, a magazine article, or an oral presentation.

Reflect

- Why is knowledge of financial mathematics important even if you do not take a college course or work at an occupation related to finance?
- How might you use the knowledge of personal finance that you have gained in this chapter after you graduate from high school?
- What advice would you give someone who wanted an occupation in finance? How would you convince them to consider your advice?

Study Guide

Fixed and Variable Costs
- Fixed costs are the same amount every time.
- Variable costs vary depending on factors such as level of use of a service.

Renting and Owning Accommodation
- A tenant uses a property owned by another person, the landlord, and pays rent to the landlord for the use of the property.
- The landlord is usually responsible for paying for property taxes, property insurance, and maintenance. Depending on the rental agreement, the landlord may also pay for utilities.
- A home owner has purchased a property to live in. This usually involves making a down payment and getting a mortgage that is paid back to the bank over many years.
- A homeowner is responsible for paying for repairs, utilities, taxes, insurance, and any other costs involved in maintaining the property.

Income and Expenses
- Income is the amount of money you earn.
- Net income is the income you have after income tax and other payroll deductions, such as Canada Pension Plan and Employment Insurance, have been subtracted.
- Expenses are the money you spend.

Budgets
- A budget is an organized list that records income and expenses. It can be a valuable tool to help you manage your savings.
- Income and expenses are organized and recorded in categories.
- A broad category, such as *Housing*, may be broken down into more specific items, such as *Mortgage payment*, *Utilities*, and *Insurance*. Or, different types of expenses may be grouped under one category, such as *Entertainment*.
- The categories depend on how a person or family earns money and spends it. There should be a category for each source of income and regular expense.

Chapter Review

8.1

1. What would you look for if you were choosing a home in your community for each person or family described below? Include the location, number of bedrooms and bathrooms, and other features.
   a) A single working adult who uses a bike and public transit
   b) A single parent who drives a car and has one young child
   c) A couple with two children, aged 1 and 5; the family has a van

2. Spence and Lia live in your community with their twin two-year-old daughters. They own a 3-bedroom townhouse.
   a) List the housing expenses Spence and Lia would have.
   b) Estimate how much each expense you listed in part a might cost per month. Explain how you estimated.

8.2

3. Determine the amount of each expense in one year.
   a) Monthly rent of $875
   b) Bi-weekly mortgage payments of $623.94
   c) Semi-annual home insurance payments of $268.55
   d) Monthly phone bill of $37.95

4. Determine the total monthly housing cost.
   a) Rent of $715 plus $65 for utilities
   b) Rent of $938 plus $102 for utilities
   c) Rent of $280 plus a one-third share of $210 for utilities
   d) Rent of $335 plus a one-quarter share of $180 for utilities

5. Toby listed his monthly housing costs. Calculate his total housing cost for one year.

Description	Frequency	Estimated cost ($)
Rent	monthly	875.00
Contents insurance	bi-monthly	80.00
Cable/Internet	monthly	79.00
Utilities	bi-weekly	90.00
Home furnishings	semi-annually	600.00

6. In the apartment above Rashid's, a water pipe breaks, which affects Rashid's apartment. Does the landlord have the right to enter Rashid's apartment to assess the damage?

7. Tori is moving to Kitchener to attend college. She finds two housing options. Which option would you recommend? Justify your answer.
   a) A room in a furnished apartment shared with two other students. The rent is $450 per month including utilities. Tori would have to sign a 1-year lease.
   b) A room in a furnished 2-bedroom suite in residence with shared kitchenette and bathroom. The cost is $5450 for September to April and includes utilities, cable, Internet, and local phone service.

520   CHAPTER 8: Budgets

8. Go to the Statistics Canada Web site. Follow the steps from Lesson 8.3, but select table **203-0003** instead. Retrieve data about Ontario and one other province. Compare the median amounts spent on **Rented living quarters**, **Owned living quarters**, and **Water, fuel, and electricity for principal accommodation** in the most recent year available. In which province did people spend more on housing? Justify your answer.

9. a) In which province from question 8 would you expect people to spend more on food? Justify your prediction.

b) Follow the steps from Lesson 8.3, but select table **203-0002** instead. Check your prediction in part a by comparing the median amounts spent on **Food purchased from stores** and **Food purchased from restaurants**. Were you correct?

10. This circle graph shows the monthly expenses for the Aboulnaga family.
a) Which category represents the greatest expense? The least?
b) The Aboulnaga family spends a total of $5000 each month. What amount do they spend on day care each month?

The Aboulnagas' Monthly Expenses

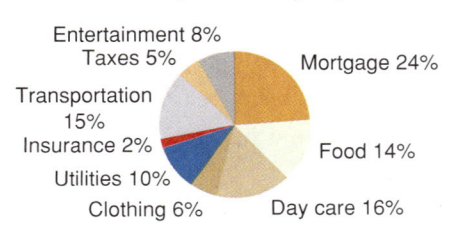

Entertainment 8%
Taxes 5%
Transportation 15%
Insurance 2%
Utilities 10%
Clothing 6%
Mortgage 24%
Food 14%
Day care 16%

11. Classify each item as income or an expense. Convert the item to a monthly amount.
a) Allowance: $30/week
b) Tuition: $2000 every 6 months
c) Gasoline: $45 every 2 weeks
d) Canada Savings Bond interest: $815 a year

12. Luigi lives in a townhouse with his wife Olga and baby Olivia.
- Luigi's take-home pay: $2400 per month
- Rent: $975 per month, including utilities
- Phone plan: An average of $50 per month
- Car payment: $325 per month
- Gasoline: An average of $50 per week
- Food: $120 per week
- Car and contents insurance: $30 per week
- Entertainment: $170 per 2 weeks
- Miscellaneous: $200 per month

a) Design a monthly budget using the data provided. Show your calculations.
b) Is the family earning enough to cover its expenses? If not, suggest how Luigi and Olga could balance their budget.

Use the case study presented below to answer questions 13 to 15.

Phillip and Teresa have two school-aged children. Both parents work full time. Their bi-weekly take-home pay is $1270 and $1500. These are their regular expenses:

Description	Expense
Mortgage payment	$1380 monthly
Food and restaurants	$435 bi-weekly
Utilities and services	$550 monthly
Transportation	$700 monthly
Property tax	$750 quarterly
Vacation expenses	$1925 annually
Entertainment	$90 weekly
Insurance and RSPs	$400 monthly
Miscellaneous	$500 monthly

13. a) Design a monthly budget using the data provided. Show your calculations.
b) Is the family earning enough to cover its expenses? If not, suggest how Phillip and Teresa could balance their budget.

14. Teresa cuts back her work from 5 days a week to 4. Her salary decreases by 20%.
a) Create a new budget that reflects this change. If the budget is not balanced, suggest changes the family could make.
b) The family would like to begin saving $300 each month for an RESP fund for their children. Suggest changes that could be made to their budget to make this possible.

15. Verify your results from questions 13 and 14 by using spreadsheet software to create a budget.

16. Hamid and Farah Feiz have 3 children. Hamid works full time and has a bi-weekly net income of $1850. Farah works part time and has a net monthly income of $2200. Their monthly expenses, not including housing, are $3000. They have savings of $35 000.
Option 1: Buying a 3-bedroom house
Down payment of $25 000 with weekly mortgage payments of $405
Option 2: Renting a 4-bedroom house
$1595 per month, plus utilities
Is renting or buying the better option for the Feiz family? Justify your answer.

17. Refer to question 16. Suppose Farah has her work hours reduced. Her net monthly income decreases to $1750. Is it still possible for the Feiz family to purchase the home?

18. Caroline is a single parent. Her annual net income is $42 000. She has $25 000 in savings. Caroline spends $125/week on food, $400/month on transportation, $120/month on clothing, $75/2 weeks on entertainment, and $200/month on miscellaneous expenses.
Option 1: Buying a 3-bedroom townhouse
Down payment of $18 000
Bi-weekly mortgage payments of $640
Monthly maintenance fees of $250
Option 2: Renting a 3-bedroom townhouse
$1025 per month, plus utilities
Which housing option would you recommend to Caroline? Justify your answer.

19. List 3 topics you have learned about in this chapter that a person would use in a career related to financial mathematics or in life outside of work.

Practice Test

Multiple Choice: Choose the correct answer for questions 1 and 2. Justify each choice.

1. The Marcella family spends $250 per week on groceries. How much should they budget monthly for their grocery costs?
   A. $1000
   B. $1083.33
   C. $541.67
   D. $500

2. Which is not a tenant's responsibility?
   A. Pay rent on time
   B. Give 60 days written notice before moving out
   C. Pay property tax
   D. Repair any damage he or she causes

Show your work for questions 3 to 6.

3. **Communication** Explain the difference between fixed and variable costs in relation to renting or owning your own home. Include examples.

4. **Knowledge and Understanding** Raj is moving to London to attend college. He finds two housing options near the campus.
   a) What would be the annual rent for each option?
   b) What additional expense will Raj have if he chooses Option 1?
   c) Which option would you recommend to Raj? Justify your answer.

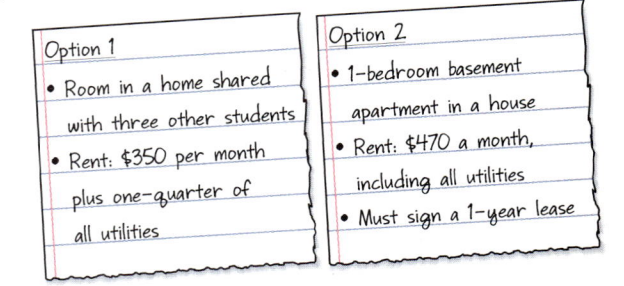

   Option 1
   - Room in a home shared with three other students
   - Rent: $350 per month plus one-quarter of all utilities

   Option 2
   - 1-bedroom basement apartment in a house
   - Rent: $470 a month, including all utilities
   - Must sign a 1-year lease

5. **Thinking** Create a monthly budget that could reflect your circumstances six months after you leave high school. Estimate all income and expenses that you predict you will have.

6. **Application** The Wang family owns a 3-bedroom condominium. Both parents work full time. Their combined take-home pay is $5000 each month. Here are the expenses that the family has on a regular basis:
   a) Express each expense as a monthly expense.
   b) Create a monthly budget for the Wang family. Determine the amount of money left at the end of each month.
   c) Mr. Wang is laid off. Even with Employment Insurance benefits, the Wang's monthly income decreases by $1000. Revise the budget to reflect this change.

   Wang Family Expenses
   - Mortgage: $1400 per month
   - Condo fees: $275 per month
   - Utilities: $350 per month
   - Property tax: $2500 per year
   - Condo insurance: $560 per year
   - Food: $200 per week
   - Day care: $225 per week
   - Transportation: $600 per month
   - Vacations: $1500 per year
   - Entertainment: $150 per week

Chapter Problem: Preparing with Financing

Your plans for next year may include a college or apprenticeship program, or working full time. It's important for you to consider the financial aspects of achieving your goals.

Create a financial plan for your goals, by:
- Describing your goals
- Estimating expenses
- Planning your finances

A spreadsheet may help you summarize your findings.

Goals: A statement of what you want to accomplish; for example, live at home and work for a year while you save for college, or start an apprenticeship and share an apartment, or work and save money for a down payment on a car.

Expenses: What will it cost to achieve your goal? Plan for living expenses as well as the cost of the goal itself. For example, if your goal is attending college, find out about tuition, residence, books, lab and other student fees, as well as living expenses.

Finances: How much can you expect to earn through summer, part-time, or full-time employment? Will you need a loan to cover the extra amount needed? If so, what interest rate might you expect to pay and what will the monthly payments be? If your goal involves buying a car, would leasing be a better plan?

French scientist Louis Pasteur is believed to have said "Luck favours the prepared mind." How can a financial plan help you make important decisions about your future?

You may find it helpful to use the resources on the Ontario School Counsellors' Association Web site.

PROJECTS

C Will Women Eventually Run as Fast as Men?

D A Home of Their Own

Your teacher may give you an expanded version of either project.

What You'll Apply

Use graphical and algebraic modelling to describe a trend and make predictions for Project C, or determine savings from annuities, research costs pertaining to purchasing and running a home, then make and justify a recommendation about purchasing a home for Project D.

And Why

Using technology to represent running times, then predicting results, shows how a mathematical model can be relevant in a sport. Connecting investments with a down payment, researching costs for buying and owning a home, and preparing a budget provide experience for making personal financial decisions.

Will Women Eventually Run as Fast as Men?

The men's and women's 100-m dash are two of the most popular events at the summer Olympics.

The winners of these races are often called "the fastest man/woman on Earth." The winning times in the 100-m dash have decreased over time, but women's times have decreased at a faster rate than men's times. Will women eventually run as fast as men? In groups or as a class, discuss these questions.

1. Why have the winning times decreased over time?
2. Why do you think women's winning times have decreased at a faster rate than men's winning times?
3. How might winning times change over the next 50 years?
4. Do you think women will eventually run as fast as men? Explain.

Year	Men's time (s)	Women's time (s)
1928	10.8	12.2
1932	10.3	11.9
1936	10.3	11.5
1948	10.3	11.9
1952	10.4	11.5
1956	10.5	11.5
1960	10.2	11.0
1964	10.0	11.4
1968	9.9	11.0
1972	10.14	11.07
1976	10.06	11.08
1980	10.25	11.06
1984	9.99	10.97
1988	9.92	10.54
1992	9.96	10.82
1996	9.84	10.94

CAREERS

- Data analyst
- Certified personal trainer
- Exercise physiologist
- Fitness and lifestyle manager

Math Focus

Interpret graphs to describe trends, compare graphs using rates of change, solve problems by modelling relationships graphically and algebraically, solve problems using formulas arising from real-world applications (Chapter 5 Graphical Models, Chapter 6 Algebraic Models)

PROJECT D

A Home of Their Own

Jason and Sheila are planning to buy a home. They have many factors to consider and decisions to make before they make this big purchase.

Jason and Sheila are asking you, their financial advisor, for help. They would like you to prepare a report that analyses their finances and recommends the housing and mortgage alternatives best suited to their budget.

In groups or as a class, discuss these questions.
1. What housing choices are available in your community?
2. What costs are involved in purchasing a house?
3. What costs should be included in a homeowner's monthly budget?
4. Where can you research the information in questions 1 to 3?

HOME for

CAREERS

- Realtor
- Financial advisor
- Mortgage broker
- Insurance agent

Math Focus

Use technology to solve problems involving annuities, including mortgages, gather, interpret, and compare information related to the costs of home ownership, design and justify a budget (Chapter 7 Annuities and Mortgages, Chapter 8 Budgets)

Chapters 1–8 Cumulative Review

CHAPTER 1

1. Which angle in each triangle can be calculated using the cosine ratio? Explain your choice. Use the ratio to determine the angle measure.
   a) Triangle XYZ with XY = 14.3 cm, XZ = 18.7 cm, right angle at Y.
   b) Triangle JKL with JK = 100 m, KL = 85 m, right angle at K.

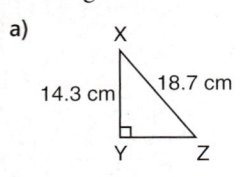

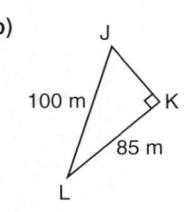

2. A metal cutter uses this scale drawing to cut a triangular sheet of metal. What are the lengths of the two unmarked sides?
   (Right triangle with 15° angle and opposite side 40 cm)

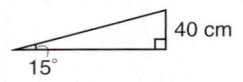

3. The measure of $\angle P$ is between 0° and 180°. Determine all possible values of $\angle P$.
   a) $\sin P = 0.35$
   b) $\sin P = 0.41$
   c) $\cos P = -0.92$
   d) $\tan P = -0.87$

4. Solve each triangle.
   a) Triangle KLS with $\angle K = 15°$, KL = 2.23 cm, $\angle S = 111°$
   b) Triangle FEG with FE = 2.3 mi, FG = 2.3 mi, GE = 3.0 mi

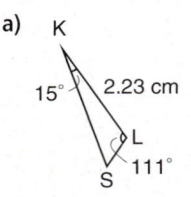

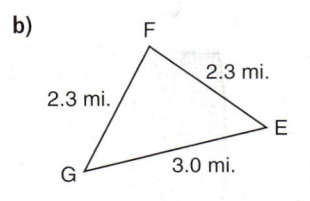

5. A cruise ship leaves dock T and sails south 125 km. Then it travels 235 km on a bearing of 230° to island V. What is the distance from the island to the dock?

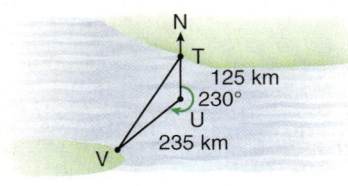

CHAPTER 2

6. Jean-Pierre creates this stencil out of plastic. What is the area of the plastic? The inner design is an equilateral triangle, inside a larger equilateral triangle, inside a square. The side length of each shape is given.
   (8.7 cm, 15.0 cm, 23.0 cm)

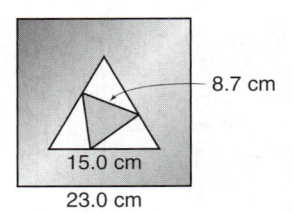

7. Sharon is painting the outside of this toy box. What is the surface area that she will paint?
   (2 ft, 2 ft, 3 ft)

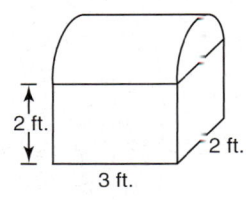

8. Ryan uses 48 interlocking foam tiles to create a rectangular play area along one side of a toy shop wall. Each tile is a square with side length 1 foot. He will tape the edges away from the wall to the floor. What is the minimum length of tape he will need?

9. Geneviève is constructing a rectangular prism with surface area exactly 294 m². It will have the greatest possible volume.
   a) Describe the prism. What will be its dimensions?
   b) What will be its volume?

10. What are the dimensions of the cylinder with volume 282 cubic inches and the least surface area?

530 Chapters 1–8

11. State whether each situation involves one-variable data or two-variable data.
   a) Kaleigh heard that more people are born in August than in any other month. She would like to check this.
   b) A team of scientists in Antarctica concluded that there is a connection between atmospheric carbon dioxide levels and global temperatures.

12. As an incentive for travellers to purchase their tickets early, airlines offer discounts on seats that are booked far in advance of the departure date. This table shows the cost of an airline ticket over time.

Weeks remaining before departure	Price of economy class seat ($)
8	154
7	154
6	193
5	240
4	315
3	489

   a) Create a scatter plot for the data. Include time values up to 11 weeks before departure.
   b) Draw a line of best fit.
   c) Use the line of best fit to predict the cost of a ticket purchased 10 weeks before to departure. Do you think the prediction is reasonable? Why or why not?
   d) Are you satisfied with the way your line of best fit represents the data? Why or why not?

13. A new stop sign is put up in a neighbourhood of about 200 houses. A resident wants to determine what her neighbours think about the sign. Which of the sampling techniques described below is most appropriate? Justify your choice.
   i) She decides to sample 25% of the houses in the neighbourhood. She randomly selects a house to start at and then goes to every fourth house and asks her survey question.
   ii) She stands for an hour in the morning and an hour in the evening at the intersection that has the new stop sign. When people stop, she asks them her survey question.
   iii) She takes a page out of the city phone book and calls every 10th name to ask her survey question.

14. The 2006 UBS *Prices and Earnings* report includes a comparison of net wages in 71 cities. The base wage is the wage in New York. This graph shows data for 10 cities.

Net Wages Around the World

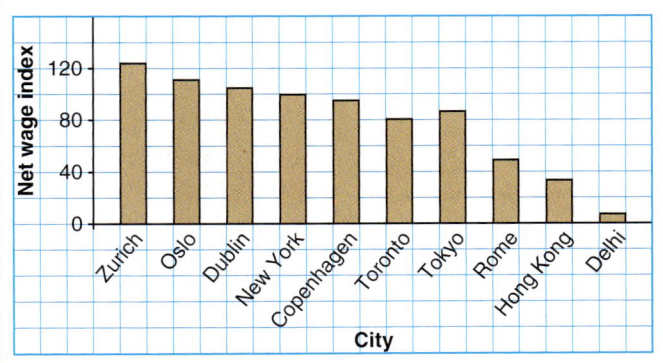

   a) In which cities are net wages greater than those in New York? Justify your answer.
   b) In which cities are net wages less than those in New York? Justify your answer.

Cumulative Review

15. The graph shows the growth of a fruit fly population over 50 days under controlled laboratory conditions.

Fruit Fly Population Growth

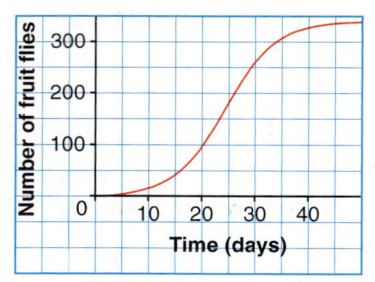

a) Describe the trends in the graph.
b) Estimate the average rate of change in the fruit fly population from day 25 to day 30. Explain the units of the rate of change. What does the rate of change tell you about the fruit fly population?
c) Determine the rate of change in the fruit fly population from day 45 to day 50. Did you have to calculate the rate of change? Explain.

16. The formula $A = P(1 + i)^n$ can be used to model the growth of money when the interest is compounded annually. In the formula, P dollars is the principal invested, i is the annual rate of interest as a decimal, and A dollars is the amount of the investment after n years. Which relation, quadratic, linear, or exponential, results under each condition?
a) $P = 100$ and $i = 0.06$
b) $P = 100$ and $n = 2$
c) $i = 0.08$ and $n = 1$
Justify your answers.

17. The table gives the world population between 1986 and 1991.

Year	1986	1987	1988	1989	1990	1991
Population (billions)	4.936	5.023	5.111	5.201	5.329	5.422

a) Calculate the rate of change in the world population from 1986 to 1991. Interpret the meaning of the rate of change.
b) Which model, linear, quadratic, or exponential, do you think best fits the data? Justify your answer.
c) Determine the equation of the regression model you chose in part a.
d) Use the model to predict the world population in 2007. What factors may affect the reliability of your prediction? Explain.

18. A child's wading pool and a bucket used to fill the pool are shown. How many buckets of water are needed to fill the pool to a height of 15 cm? Justify the formulas used.

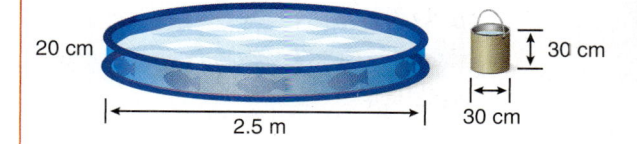

20 cm
2.5 m
30 cm
30 cm

19. The formula $S = 180(n - 2)$ gives the sum, S, of the interior angles of a polygon with n sides.
a) Rearrange the formula to solve for n. Use an arrow diagram to justify your choice of inverse operations.
b) The sum of the interior angles of a polygon is 720°. How many sides does the polygon have?

20. Evaluate without a calculator.
  a) $3^{-6} \times 3^4$
  b) $\left(\frac{2}{3}\right)^4$
  c) $\frac{(2^2)^3}{2^0}$
  d) $\frac{3^{-2} \times 5^3}{3^2 \times 5}$
  e) $144^{\frac{1}{2}}$
  f) $(-64)^{\frac{2}{3}}$

21. Simplify and evaluate for $a = 2$, $b = -4$, and $c = 3$.
  a) $(2a)^b$
  b) $\frac{a^3 b^2 c^{-1}}{a^{-1} b^0 c^{-1}}$

22. Solve for x. Explain your strategy. How do you know your answers are correct?
  a) $x^3 = 27$
  b) $2^x = 8$
  c) $4^x = 25$
  d) $27^{2x+1} = 9$
  e) $x^{\frac{3}{2}} = 64$
  f) $100(1.02)^x = 250$

23. The formula $i = \left(\frac{A}{P}\right)^{\frac{1}{n}} - 1$ gives the annual interest rate, i, required for a principal, P dollars, to grow to an amount, A dollars, in n years when the interest is compounded annually. Determine the interest rate required for an investment of $500 to grow to $629.86 in 3 years.

24. The population, P wolves, in a conservation area can be modelled by the equation $P = 80(1.02)^t$, where t is the number of years since 2000.
  a) Predict the wolf population in 2010.
  b) What is the doubling time for the population?
  Justify your answers.

25. Dominic works after school to save money for college. He makes monthly deposits of $250 into an account that earns 4% per year, compounded monthly. What is the amount of these deposits after 5 years?

26. Jenny wins a lottery. Starting one month from now, she will receive $1000 a month for life.
  a) What is the present value of the lottery payments if Jenny lives another 40 years? Assume that money can be invested at 8% per year, compounded monthly.
  b) Would the present value in part a double if the monthly payments were $2000 instead of $1000? Explain.

27. Suppose you want to have $1 000 000 in a retirement account when you turn 65.
  a) How much would you need to deposit at the end of each week to accumulate this amount if money can be invested at a rate equivalent to 10% per year, compounded weekly.
  b) How much of the $1 000 000 comes from your weekly deposits? How much is interest?

28. a) Determine the regular quarterly payment on a $10 000 loan to be repaid over 5 years at 14% per year, compounded quarterly.
  b) Hillary thinks that if the quarterly payment is doubled, it will take half as long to repay the loan. Is she correct? Justify your answer.

Cumulative Review

29. The Shahs arrange a $200 000 mortgage at 6% per year, compounded semi-annually for 25 years.
   a) Calculate the monthly payment and the total interest paid over the life of the mortgage.
   b) How much interest would the Shahs have saved if they had arranged a 20-year mortgage?
   c) What else could the Shahs do to reduce the interest they will pay over the life of the mortgage? Why will these reduce the interest costs?
   d) Suppose the Shahs had chosen to make accelerated bi-weekly payments instead of monthly payments. How long would it have taken them to repay the mortgage? How much interest would they have saved?

30. Use a spreadsheet to create an amortization table for the Shahs' 25-year mortgage in question 29.
   a) How much interest and principal is paid in the first 6 payments?
   b) How much do the Shahs still owe at the end of 5 years?

31. Faris listed his housing costs. Calculate his total housing cost for one year.

Description	Frequency	Estimated cost ($)
Rent	monthly	825.00
Content insurance	monthly	20.00
Cable/Internet/phone	monthly	124.00
Utilities	bi-monthly	220.00

32. Eiko rents an apartment on a month-to-month lease. She decides to move to a different apartment at the end of the month. Her landlord tells her she must pay rent for the following month or find another tenant to replace her. Is her landlord correct?

33. Jeremy plans to work on a cruise ship for a year to save money for college. He will earn $1500 per month as an ordinary seaman. His accommodations and food are free. He expects to spend about $100 a month on personal expenses, $10 every 2 weeks on laundry, and $150 a week on entertainment.
   a) Design a monthly budget using the data provided. Show your calculations.
   b) How much money does Jeremy expect to save each month?
   c) The tuition cost of the one-year college program Jeremy is interested in is $3300. How much money will he have left over to cover his housing and living expenses while he is in school? Would you recommend he also get a part-time job? Explain.

34. Maral is moving to St. Catharines to start a new job. Her annual net income will be $33 000. Maral has $10 000 in savings. She estimates her monthly expenses, not including housing, will be $1400.
   Option 1: Buying a 1-bedroom condo
   Down payment of $8 000
   Bi-weekly mortgage payments of $415
   Monthly maintenance fees of $220
   Option 2: Renting a 1-bedroom apartment
   $850 per month, plus utilities
   Which housing option would you recommend? Justify your answer.

Glossary

acute angle: an angle between 0° and 90°

acute triangle: a triangle where all angles are less than 90°

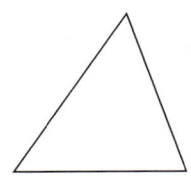

amortization period: the length of time over which a loan is paid off

amortization table: a detailed outline showing how much of each equal payment repays interest and principal and the loan balance after each payment

amount of an annuity: the ending balance after a designated amount of time; it is the principal plus interest; see *future value*

angle of depression: the angle between the horizontal and a sightline to a point below eye level

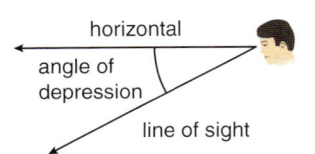

angle of elevation: the angle between the horizontal and a sightline to a point above eye level

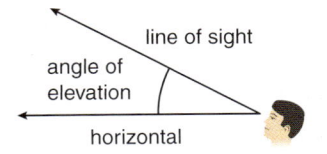

angle of inclination: see *angle of elevation*

annuity: a series of regular, equal payments paid into, or out of, an account

area: the number of square units needed to cover a surface; common units used to measure area include square centimetres and square metres

average: a single number that represents a set of numbers: see *mean, median, mode*

balance: for a budget, the difference when total expenses are subtracted from total income. Adjusting the balance so that expenses meet income is called balancing the budget.

bar graph: a graph that displays data using horizontal or vertical bars whose lengths are proportional to the numbers they represent

Average Rainfall in Toronto

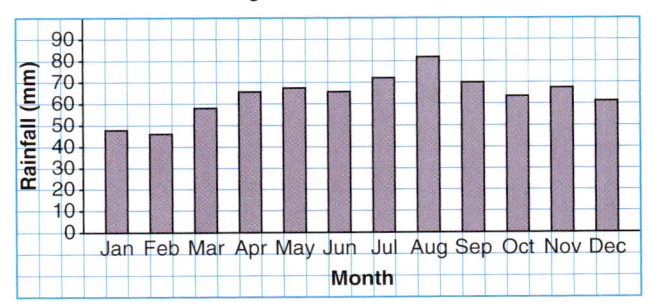

base: the side of a polygon, or the face of a solid or object, from which the height is measured; also, in an expression of the form b^n, b is the base; see *exponent, power*

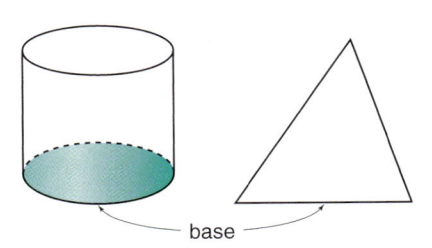

base value: for an index, a set value used for comparisons across the index

bearing: the angle describing an object's position as measured clockwise from North, usually expressed using 3 digits

bias: an emphasis on characteristics that are not typical of the entire population

budget: a written plan to outline how money will be spent

GLOSSARY **535**

capacity: the measure of how much liquid a container can hold. It can be measured in imperial or metric units. Imperial measures of capacity include fluid ounces (fl. oz.), pints (pt.), quarts (qt.), and gallons (gal.)

categorical data: data that are grouped by categories

census: the collection of data about every individual in a population

circle: the set of points in a plane that are a given distance (the radius) from a fixed point (the centre)

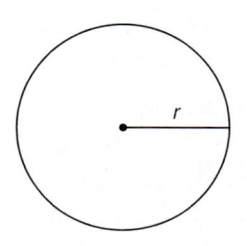

circle graph: a diagram that uses parts of a circle to display data, sometimes called a pie chart

Favourite Colours of People in My Class
Purple 10%
Yellow 5%
Blue 28%
Green 14%
Black 24%
Red 19%

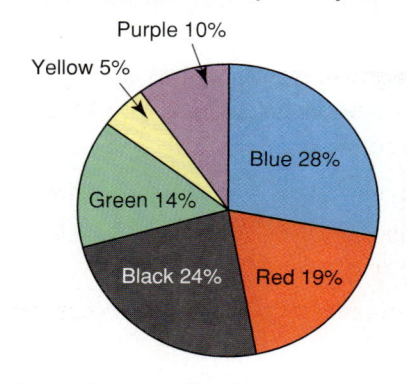

circumference: the distance around a circle; the distance around any region whose boundary is a simple closed curve

cluster sampling: sampling in which the data are organized into representative groups and one group is chosen as a sample

composite figure: a figure made up of two or more other figures

composite object: a 3-dimensional structure or object made up of two or more objects

compound interest: interest earned via a method of calculating interest in which the interest due is added to the principal and thereafter earns interest; calculated using the formula $A = P(1 + i)^n$, where A is the amount of the compound interest, P is the principal, i is the interest rate per compounding period as a decimal, n is the number of compounding periods

concept map: a graphic organizer that illustrates the connections among different terms or concepts

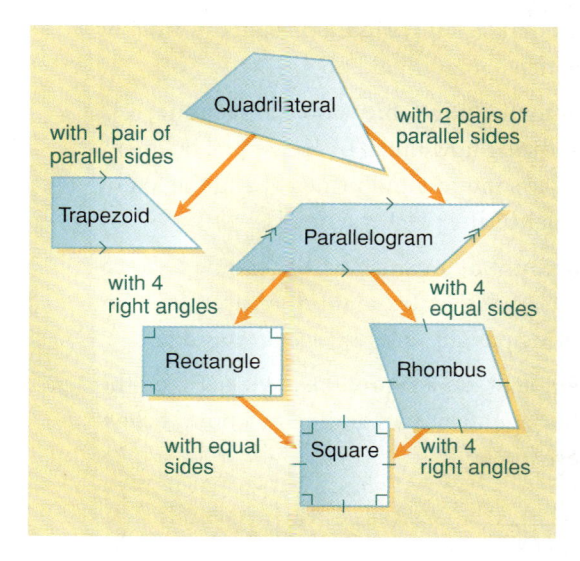

cone: a solid that is formed by a region (base of the cone) and all the line segments joining points in the base to a point not in the base

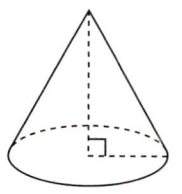

congruent: having the same size and shape, but not necessarily the same orientation

constraint: a condition that limits or restricts options

convenience sampling: sampling in which individuals who are easy to sample are chosen

coordinate axes: the horizontal and vertical number lines on a grid that represents a plane

coordinates: also called Cartesian coordinates; the numbers in an ordered pair that locate a point in the coordinate plane

correlation: the strength of a linear relationship between two variables

correlation coefficient, r: a measure, between -1 and 1, of how closely data can be described by a certain type of function; the closer the value of r to 1 or -1, the more closely the data fits the function

cosine: for an acute $\angle A$ in a right triangle, the ratio of the length of the side adjacent to $\angle A$, to the length of the hypotenuse; written cos A

$$\cos A = \frac{\text{length of side adjacent to } \angle A}{\text{length of hypotenuse}}$$

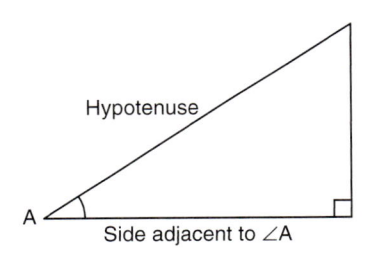

Cosine Law: In any $\triangle ABC$, $c^2 = a^2 + b^2 - 2ab \cos C$

This can also be written as: $\cos C = \frac{a^2 + b^2 - c^2}{2ab}$

cube: a rectangular prism whose length, width, and height are all equal; see *rectangular prism*

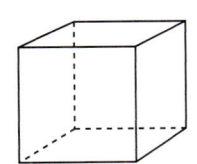

cube number: a power with exponent 3; for example, 8 is a cube number because $2^3 = 8$

cube root: a number which, when raised to the power 3, results in a given number; for example, 3 is the cube root of 27, and -3 is the cube root of -27

curve of best fit: for a given scatter plot, the curve that passes most closely to the majority of points; the curve of best fit can be obtained by exponential or quadratic regression

cylinder: a solid with two parallel, congruent, circular bases

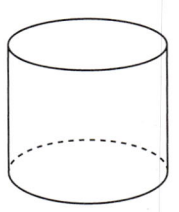

equation: a mathematical statement indicating that two expressions are equal

equilateral triangle: a triangle with three equal sides; each angle is 60°

expenses: items that must be paid from income; for example, food, shelter, transportation

exponent: in an expression of the form b^n, n is the exponent; exponents that are positive integers indicate the number of times a factor is repeated in a product; for example, in 3^4, the exponent 4 indicates that the base 3 is used as a factor 4 times; see *base, power*

exponential equation: an equation in the form $y = ab^x$; a is the initial value; b is the growth or decay factor

exponential regression: the process of identifying an exponential curve of best fit for a given set of data

expression: a meaningful mathematical phrase made up of numbers and/or variables which may include operation symbols

extrapolate: to estimate a value that lies beyond the known values

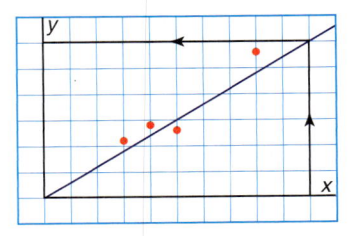

face: a flat surface of a 3-dimensional object

fixed cost: the same amount charged at regular intervals

GLOSSARY **537**

formula: a rule that is expressed as a mathematical equation that relates two or more variables

Frayer model: a graphic organizer with 4 sections which can hold a definition, characteristics or facts, examples, and non-examples of a word or a concept

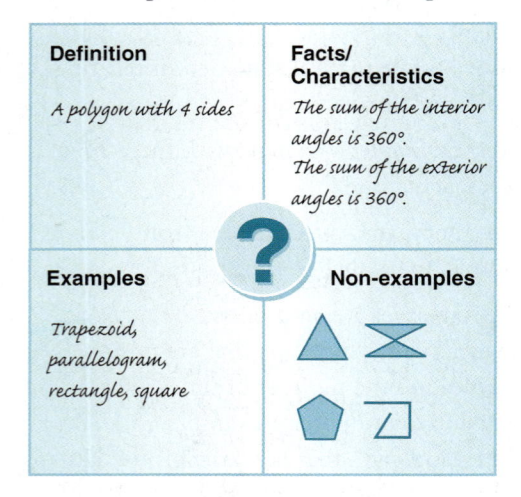

Definition	Facts/Characteristics
A polygon with 4 sides	The sum of the interior angles is 360°. The sum of the exterior angles is 360°.
Examples	**Non-examples**
Trapezoid, parallelogram, rectangle, square	

function: a rule that gives a single output number for each input number

future value: the principal and interest due when an investment matures; also referred to as *amount*
Amount = Principal + Interest, or $A = P + I$

horizontal intercept: the horizontal coordinate of a point where the graph of a relation intersects the horizontal axis

hypotenuse: the side opposite the right angle in a right triangle

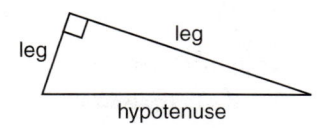

income: the money you earn

income tax: money paid as tax to the federal and provincial governments based on the amount of income earned

index: a comparative set of data used to track trends or establish guidelines for a given situation; each value may be given as a percent of the base value (as in a stock index); or calculated using a formula (as in the body mass index)

inflation: the continuing rise in the general price of all goods and services; it is usually attributed to an increase in the volume of money and credit relative to available goods and services

interest: the fee paid by a borrower for the use of a lender's money

interest rate: the amount earned or paid for the use of money; usually given as a percent of the amount invested or borrowed per year

interpolate: to estimate a value between two known values

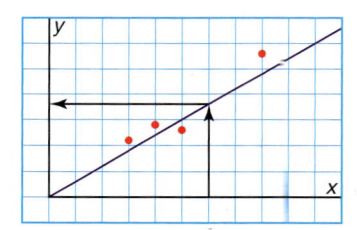

inverse operation: an operation that reverses another operation

inverse ratios: $\sin^{-1}$, $\cos^{-1}$, and $\tan^{-1}$ which are used to determine the measure of an angle when its trigonometric ratio is known

judgement sampling: sampling in which the person doing the sampling uses her or his judgement to create a representative sample

landlord: the owner of a property. He/she receives compensation from the tenant for use of his/her property.

lease: to rent an item from the owner; the lease payments cover the depreciation of the item over the course of the lease plus interest on the outstanding balance of the full purchase price

line of best fit: a line that passes as close as possible to a set of plotted points

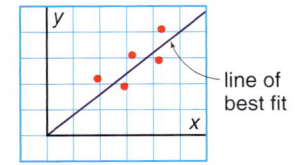

linear correlation: a trend where points may lie in the general direction of a line

linear regression: the process of identifying a line of best fit for a set of data

margin of error: the proportion that we add to and subtract from a result to create a range of possible values between which the result could lie

mass: a measure of the amount of material in an object: common units are grams or kilograms

matrix: a graphic organizer used to list and compare characteristics of different items

mean: one measure of the average of a set of numbers; to find the mean, divide the sum of the data by the number of data

measure of central tendency: the *mean*, *median*, or *mode* of a data set

median: the middle number of a set of data arranged in numerical order; if there are two middle numbers, the median is their average

mode: the most frequently occurring value in a set of data

mortgage: a long-term loan on real estate that gives the person or firm providing the money a claim on the property if the loan is not repaid

numeric data: data that always involve numbers

oblique angle: an angle that is not a multiple of 90°

oblique triangle: a triangle that does not contain a 90° angle

obtuse angle: an angle between 90° and 180°

obtuse triangle: a triangle with one obtuse angle

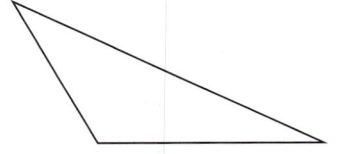

one-variable data: a set of data that describes one attribute per item in a sample

optimization: the process of finding the most efficient use of available materials within given constraints

ordered pair: a pair of numbers, written as (x, y), that represents a point on the coordinate plane; see *coordinates*

ordinary simple annuity: an annuity in which payments are made and interest is compounded with the same frequency; see *annuity*

outlier: an observed value that differs markedly from the pattern established by most data in a set

parabola: the graph of a quadratic relation; see *quadratic regression*

parallel: describing lines lying on the same plane that do not intersect

parallelogram: a quadrilateral with opposite sides parallel

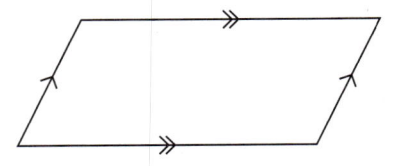

percentile: tells approximately what percent of the data are less than a particular data value

perimeter: the distance around a closed figure

perpendicular: intersecting at right angles (90°)

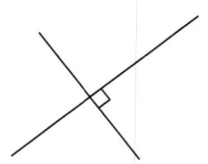

pi (π): the ratio of the circumference of a circle to its diameter; π ≐ 3.1416

poll: a survey or an investigation of a topic to find out people's views

polygon: a closed figure that consists of line segments that only intersect at their endpoints

population: the set of all things or people being considered

power: an expression of the form b^n, where b is the base and n is the exponent; for positive integer exponents, powers are a shortcut for repeated multiplication – for example, $(-4)^3 = (-4) \times (-4) \times (-4)$; see *base, exponent*

present value: the principal that must be invested today to obtain a given amount in the future; compare to *future value*

present value of an annuity: the principal that must be invested today to provide the regular payments of an annuity

primary trigonometric ratios: sine, cosine, and tangent

prime number: a whole number with exactly two factors, itself and 1; for example, 2, 3, 5, 7, 11, 29, 31, 43…

principal: the sum of money invested or borrowed

prism: a solid with two congruent and parallel faces (bases); all other faces are parallelograms

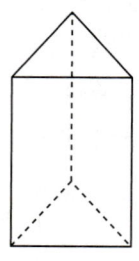

property tax: an amount that property owners pay to their municipal government

pyramid: a solid with one face that is a polygon (the base) and the other faces that are triangles with a common vertex

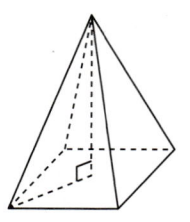

Pythagorean Theorem: for any right triangle, the area of the square on the hypotenuse is equal to the sum of the areas of the squares on the other two sides; $a^2 + b^2 = c^2$

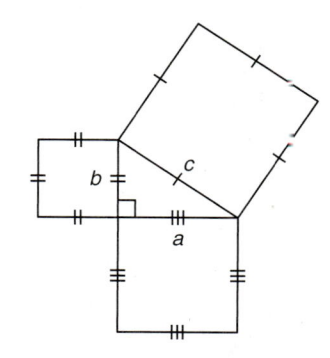

quadrants: one of the four regions into which the coordinate axes divide the plane

	y	
Quadrant II	Quadrant I	
		x
Quadrant III	Quadrant IV	

quadratic regression: a process of identifying the parabola of best fit for a given set of data

quadrilateral: a polygon with four sides

quartile: any of three numbers that separate a sorted data set into four equal parts

radical form: a number written using the root symbol: $\sqrt{}$ or $\sqrt[n]{}$, where $n = 2, 3, …$

rate: a certain quantity or amount of one thing considered in relation to a unit of another thing

GLOSSARY

rate of change: the rate at which something is changing; it is often indicated by the slope of a graph

ratio: a comparison of two or more quantities with the same unit

rectangle: a quadrilateral with four right angles

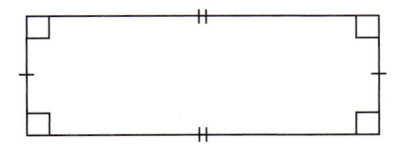

rectangular prism: a prism with rectangular faces; see *prism*

regression: the process of identifying a curve or a line of best fit for a set of data

representative sample: a sample which in certain respects is typical of the population from which it is chosen

rhombus: a parallelogram with four equal sides

right angle: a 90° angle

right triangle: a triangle with one right angle

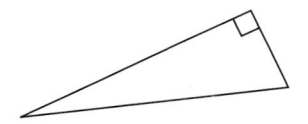

root of a number: the *n*th root of a number is a number which, when *n* copies of the number are multiplied, results in a given number; for example, 3 is a cube root of 27

sampling technique: the process used to select the individuals from a population who will be studied; see *cluster sampling, convenience sampling, judgement sampling, simple random sampling, stratified sampling, systematic sampling, voluntary sampling*

savings: money set aside for future use; in a budget, it is shown as an expense

scatter plot: a graph of data that are a series of points

Height (cm)	154	162	172	178
Mass (kg)	56.3	60.1	72.2	64.3

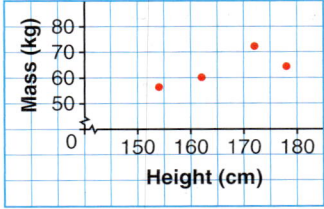

sightline: the line from an observer's eye to a specific object

simple interest: interest earned only on the principal, calculated using the formula $I = Prt$, where I is the simple interest, P is the principal, r is the annual interest rate as a decimal, t is the time in years; compare to *compound interest*

simple random sampling: sampling in which individuals are chosen randomly from the entire population

sine: for an acute $\angle A$ in a right triangle, the ratio of the length of the side opposite $\angle A$, to the length of the hypotenuse; written sin A

$$\sin A = \frac{\text{length of side opposite } \angle A}{\text{length of hypotenuse}}$$

Sine Law: In any $\triangle ABC$,

$$\frac{a}{\sin A} = \frac{b}{\sin B} = \frac{c}{\sin C} \text{ and } \frac{\sin A}{a} = \frac{\sin B}{b} = \frac{\sin C}{c}$$

slope: a measure of the steepness of a line; calculated as slope $= \frac{\text{rise}}{\text{run}}$

solve a triangle: to determine the measures of all unknown sides and angles in a triangle

sphere: the set of points in space that are a given distance (the radius) from a fixed point (the centre)

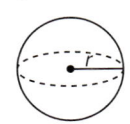

GLOSSARY **541**

square: a rectangle with four equal sides

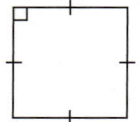

square of a number: the product of a number multiplied by itself; a number to the power of 2

square root: a number which, when multiplied by itself, results in a given number; for example, 5 and −5 are the square roots of 25

stratified sampling: sampling in which data are grouped and a few individuals from each group are selected randomly

supplementary angles: 2 angles whose sum is 180°

surface area: the measure of the area of all the faces of an object

systematic sampling: sampling in which every nth individual is selected

tangent: for an acute $\angle A$ in a right triangle, the ratio of the length of the side opposite $\angle A$, to the length of the side adjacent to $\angle A$; written tan A

$$\tan A = \frac{\text{length of side opposite } \angle A}{\text{length of side adjacent to } \angle A}$$

tenant: the user of a property owned by another person; he/she pays the landlord for use of the property

torus: an object shaped like a doughnut

trends: patterns of change; trends are often used to justify decisions and make predictions

triangle: a polygon with three sides

triangular prism: a prism with triangular bases, see *prism*

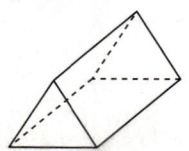

trigonometric ratios: see *cosine*, *sine*, and *tangent*

two-variable data: a set of data that gives measures of two attributes for each item in a sample

utilities: services such as heat, water, and electricity

valid conclusion: a conclusion that is supported by unbiased data that has been interpreted appropriately

variable: a letter or symbol used to represent a quantity that can vary

variable cost: operating costs that change depending on variables; compare to *fixed cost*

Venn diagram: a graphical organizer with loops that group items to show similarities and differences

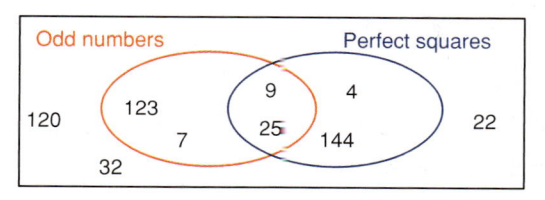

vertical intercept: the vertical coordinate of a point where the graph of a relation intersects the vertical axis

volume: the amount of space occupied by an object; measured in cubic units

voluntary sampling: sampling in which participants volunteer to be included in the sample

x-axis: the horizontal number line on a coordinate grid

x-intercept: the *x*-coordinate of a point where a graph intersects the *x*-axis; see *horizontal intercept*

y-axis: the vertical number line on a coordinate grid

y-intercept: the *y*-coordinate of a point where a graph intersects the *y*-axis; see *vertical intercept*

Answers

Final answers only, for *Review* and *Practice* questions that have a single correct answer.
Explanatory or descriptive answers may vary.

Chapter 1 Trigonometry

Activate Prior Knowledge

The Pythagorean Theorem, page 2

1. a) About 257 km
   b) About 15.39 m
2. QR ≐ 9.5 m
3. About 101 m

Metric and Imperial Unit Conversions, page 3

1. a) 72 mm   b) 9.215 km
   c) 9350 m   d) 8.32 m
   e) 87 900 cm   f) 0.065 m
2. a) 84 in.   b) 84 ft.
   c) 42 240 ft.   d) 321 yd.
   e) 281 in.   f) 4 ft.
3. a) About 502 mm or about 0.50 m
   b) 6 ft. or 2 yd.

1.1 Trigonometric Ratios in Right Triangles, page 8

1. a) i) s or PR
      ii) k or LM
   b) i) p or RS
      ii) m or KL
   c) i) r or PS
      ii) l or KM
2. a) $\sin A = \dfrac{BC}{AB}$
   b) $\cos A = \dfrac{AC}{AB}$
   c) $\cos B = \dfrac{BC}{AB}$
   d) $\tan B = \dfrac{AC}{BC}$
3. a) Sine
   b) Tangent
   c) Cosine
4. a) $e \doteq 1.4$ m
   b) $q \doteq 24$ ft.
   c) $g \doteq 59$ in.
5. a) $\tan A \doteq 1.9230$, $\angle A \doteq 63°$
   b) $\tan A = 1$, $\angle A = 45°$
   c) $\tan A \doteq 0.3801$, $\angle A \doteq 21°$
6. a) $\angle A \doteq 32°$, $\angle B \doteq 58°$, $b \doteq 11.5$ ft.
   b) $\angle R \doteq 64°$, $\angle P \doteq 26°$, $r \doteq 27.9$ cm
   c) $\angle E \doteq 47°$, $\angle D \doteq 43°$, $f \doteq 45$ in. or 3 ft. 9 in.
7. a) About 3 ft.
   b) About 9 ft.
8. About 59°
9. About 2 mi. (about 9471 ft.)
10. About 5.2 m
11. About 87.7 m
12. a) Yes   b) About 326°
13. a) $AB \doteq 4.00$ m
    b) Angle of inclination; $\angle A \doteq 36°$
14. a) About 28°
    b) About 123°
16. About 3.2 km, about 2.0 km

1.2 Investigating the Sine, Cosine, and Tangent of Obtuse Angles, page 19

1. a) Quadrant I
   b) Quadrant II
   c) Quadrant II
2. a) Positive
   b) Positive
   c) Positive
3. a) Negative
   b) Positive
   c) Negative
4. a) Negative
   b) Positive
   c) Negative
5. a) Positive
   b) Negative
   c) Negative
   d) Positive
   e) Negative
   f) Positive
6. a) Acute
   b) Obtuse
   c) Acute or obtuse

1.3 Sine, Cosine, and Tangent of Obtuse Angles, page 23

1. a) $\sin 110° \doteq 0.9397$, $\cos 110° \doteq -0.3420$, $\tan 110° \doteq -2.7475$

    b) $\sin 154° \doteq 0.4384$, $\cos 154° \doteq -0.8988$, $\tan 154° \doteq -0.4877$
    c) $\sin 102° \doteq 0.9781$, $\cos 102° \doteq -0.2079$, $\tan 102° \doteq -4.7046$

2. a) Positive, about 0.5736
   b) Negative, about −0.4877
   c) Negative, about −0.6947

3. a) $\sin A \doteq 0.6$
   b) $\cos A \doteq 0.2425$
   c) $\tan A \doteq 0.5$
   d) $\cos A \doteq -0.6402$
   e) $\sin A \doteq 0.8682$
   f) $\tan A \doteq -0.2857$

4. a) 167°
   b) 101°
   c) 145°
   d) 175°

5. a) 132°
   b) 56°
   c) 103°
   d) 104°

6. a) 25°
   b) 122°
   c) 145°
   d) 123°

7. a) Acute
   b) Obtuse
   c) Acute or obtuse

8. a) Yes
   b) No
   c) Yes
   d) Yes

9.

	Acute	Obtuse
Sine	Positive	Positive
Cosine	Positive	Negative
Tangent	Positive	Negative

10. a) 38° or 142°
    b) 148°
    c) 12°
    d) 162°
    e) 135°
    f) 50° or 130°

11. Equations a and b will result in two different values for angle Y.

12. a) 27° or 153°
    b) 117°
    c) 24°

13. 0.8930

14. −0.3846

15. a) $\sin 90° = 1$, $\cos 0° = 1$

Chapter 1 Mid-Chapter Review, page 26

1. a) $c \doteq 44$ cm, $b \doteq 30$ cm, $\angle B = 43°$
   b) $CD \doteq 30$ yd., $\angle C = 26°$, $\angle D = 64°$

2. a) $\angle Y = 37°$, $z \doteq 4.5$ cm, $y \doteq 2.7$ cm
   b) $\angle X = 38°$, $\angle Y = 52°$, $y \doteq 32$ in., or 2 ft. 8 in.

3. About 41°

4. About 43.6 m

5. a) 6.5 mi.
   b) About 157°

6. a) Positive, positive, positive
   b) Positive, negative, negative
   c) Positive, negative, negative

7. a) Acute
   b) Obtuse
   c) Acute or obtuse
   d) Obtuse

8. a) About 167°
   b) About 169°
   c) About 122°
   d) About 158°

9. a) About 32° or about 148°
   b) About 115°
   c) About 8°
   d) About 3° or about 177°

1.4 The Sine Law, page 31

1. a) $\dfrac{\sin 52°}{x} = \dfrac{\sin 55°}{2.5}$
   b) $\dfrac{\sin 105°}{3} = \dfrac{\sin 23°}{t}$
   c) $\dfrac{\sin 33°}{23} = \dfrac{\sin 19°}{p}$

2. a) About 2.4 cm
   b) About 1.2 ft.
   c) About 14 m

3. a) About 16.6 m
   b) About 505 in.
   c) About 10.6 km

4. a) $\dfrac{\sin 21°}{x} = \dfrac{\sin 21°}{y} = \dfrac{\sin z}{5}$

$\dfrac{\sin 11°}{f} = \dfrac{\sin 120°}{e} = \dfrac{\sin g}{2.4}$

$\dfrac{\sin 69°}{k} = \dfrac{\sin 76°}{m} = \dfrac{\sin L}{2.2}$

c) i) 138°, XZ = ZY ≐ 3 in.
ii) 49°, GE ≐ 0.6 km, GF ≐ 2.8 km
iii) 35°, ML ≐ 3.6 m, KL ≐ 3.7 m

5. a) 7°, YZ ≐ 2.3 m, XZ ≐ 3.3 m
b) 30°, YZ ≐ 23.3 mi., XZ ≐ 12.0 mi.
c) 20°, YZ ≐ 33 cm, XZ ≐ 24 cm

6. a) 41°, YZ ≐ 5 cm, XZ ≐ 1 cm
b) 30°, YZ ≐ 2 ft., XZ = 1 ft. 2 in.
c) 20°, YZ ≐ 157 mm, XZ ≐ 136 mm

7. Use the Sine Law to solve the triangles.

8. a) Yes
b) No
c) Yes

9. a) About 8.4 cm
c) About 20 m

10. a) 13°
b) About 67 yd.

11. a) About 11.2 cm

12. a) $x \doteq 9.1$ cm, $z \doteq 17.7$ cm

13. a) $e \doteq 6.3$ in.

15. ∠Q = 47°, QR ≐ 0.6 m, PQ ≐ 0.2 m

16. ∠B = 45°, BC ≐ 3.6 km, AB ≐ 1.6 km

17. LR ≐ 56 ft., LS ≐ 25 ft.

1.5 The Cosine Law, page 38

1. a) $z^2 = 18^2 + 11^2 - 2 \times 11 \times 18 \times \cos 32°$
b) $v^2 = 9.7^2 + 3.0^2 - 2 \times 9.7 \times 3.0 \times \cos 22°$
c) $o^2 = 36^2 + 134^2 - 2 \times 36 \times 134 \times \cos 115°$

2. a) About 10 ft.   b) About 7.0 m
c) About 153 in., or 12 ft. 9 in.

3. a) About 0.8 m
b) About 5.3 ft.
c) About 1.3 cm

4. a) About 18.8 cm
b) About 18.9 cm
c) About 5.8 in.

5. a) $4.3^2 = 5.0^2 + 3.2^2 - 2 \times 5.0 \times 3.2 \times \cos B$
b) $145^2 = 111^2 + 35^2 - 2 \times 111 \times 35 \times \cos R$
c) $6.23^2 = 4.11^2 + 2.78^2 - 2 \times 4.11 \times 2.78 \times \cos M$

6. a) 58°   b) 164°
c) 128°

7. a) 87°   b) 23°

c) 128°

8. a) 117°   b) 80°
c) 55°

9. a) Cosine Law
b) Sine Law
c) Sine Law

10. a) The bearings are shown as angles measured from North, clockwise.
b) About 9.6 mi.

11. About 3 ft. 4 in.
12. About 31°
13. About 19°, about 32°
14. 120°, about 22°, about 38°
15. a) 116°
17. a) About 40 ft.
b) About 36°
18. a) About 48 m

1.6 Problem Solving with Oblique Triangles, page 47

1. a) Sine Law
b) Sine Law
c) Sine Law

2. a) About 20 mi.
b) About 8 in.
c) About 3.7 km

3. a) Sine Law; $t \doteq 7.9$ mi., $v \doteq 7.0$ mi.
b) Cosine Law; about 3.9 km
c) Cosine Law; about 21 in., or 1 ft. 9 in.

4. a) Cosine Law
b) Both
c) Both

5. a) About 112°
b) About 160°
c) About 34°

6. a) Cosine Law: about 10°
b) Cosine Law: about 72°
c) Cosine Law: about 111°

7. About 54 m
8. Yes
10. a) About 44°
b) About 96°
11. a) About 102°
b) About 39°
12. a) Meaford to Christian Island: about 34 mi.; Christian Island to Collingwood: about 31 mi.; about 85 mi. altogether
b) About 30 mi.

ANSWERS 545

13. About 112°, 40°, 28°
14. a) About 94°; answers may vary by as much as 8°, depending on the method used.
16. About 28 km

Chapter 1 Review, page 54

1. a) $a \doteq 0.8$ m, $b \doteq 1.5$ m
   b) $\angle Y \doteq 34°$ ($\angle Z \doteq 56°$)
2. About 1.2 m
3. About 24°
4. a) $\angle Y \doteq 52°, \angle Z \doteq 38°$
   b) About 1370 ft.
5. About 300°
6. a) Positive
   b) Negative
   c) Positive
7. a) Acute
   b) Obtuse
   c) Acute
   d) Either
8. a) $\cos(180° - \angle A) = 0.94$
   b) $\sin(180° - \angle A) = 0.52$
   c) $\tan(180° - \angle A) = -0.37$
9. a) About 112°
   b) About 124°
10. a) Obtuse
    b) Either
11. a) About 170°
    b) About 21° or about 159°
12. a) About 22°
    b) About 47° or about 133°
13. a) $x \doteq 30$ mm
    b) $y \doteq 32$ mm
14. a) $p \doteq 5$ ft., or 1 yd. 2 ft.,
       $q \doteq 12$ ft., or 4 yd.
15. a) $n \doteq 36$ m, $m \doteq 28$ m, $\angle L = 27°$
    b) $i \doteq 306$ in., or about 25 ft. 6 in.; $j \doteq 262$ in., or about 21 ft. 10 in.; $\angle I = 102°$
16. a) About 40°
    b) About 80°
17. About 2.1 m, about 3.5 m
18. c) $q \doteq 3.0$ m
19. About 10.8 km, $\angle D \doteq 18°, \angle B \doteq 31°$
20. About 87°
21. b) i) $\angle N \doteq 157°$
       ii) $\angle D \doteq 49$
22. About 43°, about 76°
23. About 63°, about 72°, about 45°
24. About 27 m
25. About 5.8 m
26. b) About 1622 m
27. $t \doteq 5.9$ m, $s \doteq 1.7$ m

Chapter 1 Practice Test, page 57

1. B
2. A
3. $\angle A \doteq 28°, \angle B \doteq 13°, AC \doteq 2.9$ m
5. About 114°
6. a) About 326°
   b) About 41 km

Chapter 2 Geometry

Activate Prior Knowledge

Metric and Imperial Units of Length, page 60

1. a) About 5.18 m
   b) About 295 ft. and about 394 ft.
   c) About 75 mi.
   d) About 8.5 in.
2. a) About 11.5 ft. by 13.8 ft.
   b) About 6 in. and 10 in.

Perimeter and Area, page 62

1. a) $P = 21.6$ cm
      $A = 28.8$ cm^2
   b) $P = 30.0$ m
      $A \doteq 38.4$ m^2
   c) $P \doteq 28.3$ in.
      $A \doteq 63.6$ sq. in.
2. $P = 344$ yd.
   $A = 6240$ sq. yd.
3. $C \doteq 47.1$ ft.
   $A \doteq 176.7$ sq. ft.
4. a) About 2.4 m
   b) $P \doteq 47.5$ m; $A \doteq 28.2$ m^2
   c) Perimeter

Metric and Imperial Units of Capacity, page 63

1. a) About 2.8 L
   b) About 17.6 gal.
   c) About 119.3 mL
   d) About 181.8 L
   e) About 14.0 fl. oz.

2. Canada; 21.2 U.S. gallons is about 80.2 L, which is less than the consumption in Canada.

Volumes of Prisms and Cylinders, page 64

1. a) Base is a rectangle;
$A = 51.75$ sq. in.;
$V = 77.625$ cu. in.
b) Base is a triangle;
$A = 6$ cm^2;
$V = 36$ cm^3
c) Base is a circle;
$A \doteq 15.9$ m^2;
$V \doteq 133.6$ m^3
2. The base of a triangular prism is a triangle.

Surface Areas of Prisms and Cylinders, page 65

1. a) 151.5 sq. in.
b) 84 cm^2
c) About 150.6 m^2
2. A rectangular prism has 6 faces.
A triangular prism has 5 faces.

2.1 Area Applications, page 71

1. a) A rectangle and a semicircle
b) A square with a circle cut out
c) A parallelogram and a right triangle
2. a) Add the area of the rectangle to the area of the semicircle.
b) Subtract the area of the circle from the area of the square.
c) Add the area of the parallelogram to the area of the triangle.
3. a) [trapezoid with 40 cm top, 25 cm bottom; parallelogram 40 cm × 30 cm]
b) [triangle with base 8.7 m and height 3.5 m; rectangle 8.7 m × 4.1 m]
c) [large circle with $6\frac{3}{4}$ in. radius and smaller inner circle with $1\frac{1}{2}$ in. radius]
d) [rectangle 36 in. × 15 in. with a semicircle of diameter 15 in. on the right]

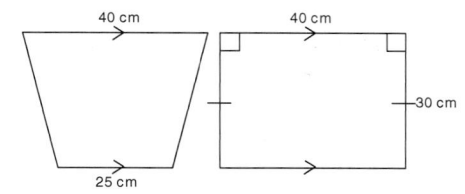

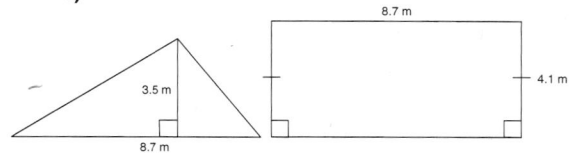

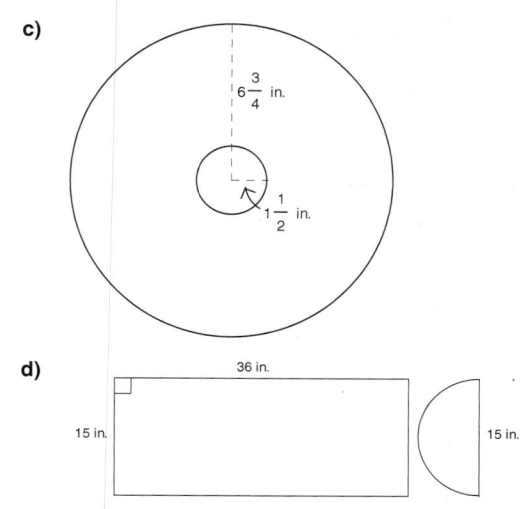

4. a) 2175 cm^2 **b)** About 50.9 m^2
c) About 136.07 sq. in. **d)** About 451.6 sq. in.
5. Both of them are correct.
6. a) A semicircle with a smaller semicircle cut out;
$A \doteq 18.85$ sq. ft.
b) A rectangle on top of a triangle; $A = 510$ sq. in.
c) A semicircle on top of a trapezoid;
$A \doteq 207.2699$ cm^2
7. a) The number of cans equals the square of the layer number.
b) Square **c)** 100 cans
d) About 15 625 cm^2; the cans are stacked without any space in-between.
8. a) [diagrams of stacked circles in triangular layers]

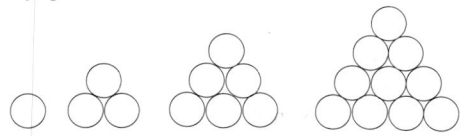

The number of cans in each layer starting at the top is: 1, 3, 6, 10…
b) 55 cans
c) About 6766 cm^2; the cans are stacked without any space in-between.
d) Fewer cans
9. a) $A \doteq 7.955$ m²; $A \doteq 82.4$ sq. ft.
b) 2 cans
10. $A \doteq 4301.87$ m^2
12. a) About 315 sq. ft.
b) About 29 m^2
13. a) Split the composite figure into a rectangle and 2 congruent trapezoids. $A \doteq 1931$ cm^2
14. a) $A \doteq 34.0$ m^2 **b)** About 3.14 m^2

ANSWERS **547**

c) $A \doteq 237.7 \text{ m}^2$

16. a)

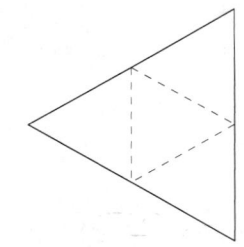

b) $A \doteq 259.3 \text{ cm}^2$

2.2 Working with Composite Objects, page 81

1. a) A large cylinder with a smaller cylinder on top
   b) A large rectangular prism with a smaller rectangular prism on top
2. a) A rectangular prism with a half cylinder on top
   b) A large rectangular prism with a smaller rectangular prism cut out
3. a) A rectangular prism with a triangular prism on top
   b) A rectangular prism with a half cylinder on top and a right triangular prism on the side
4. a) $SA = 1600.5 \text{ cm}^2$
   b) About 3400 cm^2
   c) About 2000 mL
5. a) About 15 567 sq. in.
   b) $V \doteq 10\,800$ cu. in.
   c) About 212.6 L
6. Yes
7. a) The volume of the quarter cylinder is $\frac{1}{4}$ of the volume of a cylinder with the same radius and height.
   b) $V \doteq 1018$ cu. in.
   c) $SA \doteq 611.85$ sq. in.

Face	Shape	Number	Area of each face (square inches)
Front and back	Quarter circle	2	$\frac{1}{4}(\pi \times 12^2) =$ 113.0973
Bottom and side	Rectangle	2	$9 \times 12 = 108$
Curved surface	Unroll to form rectangle	1	$\frac{1}{4}(2\pi \times 12 \times 9)$ = 169.646

d) The front and back are quarter circles, the side and bottom are rectangles, and the curved surface is a quarter of the curved surface of a cylinder.

8. a) The volume of the base is about $11\,946 \text{ cm}^3$; the volume of the top layer is about 5309 cm^3; and the total volume is about $17\,255 \text{ cm}^3$.
   b) About 4298.5 cm^2
   c) About 3767.6 cm^2
9. a) The volume of the base is 784 cubic inches. The volume of the top layer is 400 cubic inches. The total volume is 1184 cubic inches.
   b) 680 sq. in.
   c) 580 sq. in.
10. The rectangular prism cake is a better deal. It has greater volume.
11. a) $V \doteq 198.4 \text{ m}^3$
    b) Yes
12. The sheet metal mailbox has greater volume.
14. a) The dimensions for the shipping crate should be slightly greater than the dimensions of the console.
       The smallest shipping crate is 31 in. by 16 in. by 33 in.
    b) 16 368 cu. in.
    c) 5058.27 cu. in.
15. About 4611.7 cm^3
16. About 396.72 sq. in.
17. About $16\,682 \text{ cm}^2$
18. The wooden mailbox has a smaller surface area. The difference in surface areas is about 3008.3 cm^2.
19. a) Part a: $V \doteq 1512.5$ cu. in. or about $24\,785.4 \text{ cm}^3$
       Part b: $V \doteq 37\,876.4 \text{ cm}^3$
    b) Part b
    c) Part a: $SA \doteq 783$ sq. in. or about 5052 cm^2
       Part b: $SA \doteq 32\,165.4 \text{ cm}^2$
    d) Part b
20. a) 31 104 washers
    b) About 513.22 sq. in.
    c) $SA \doteq 53\,424$ sq. in

Chapter 2 Mid-Chapter Review, page 86

1. a) Equilateral triangle cut out of a square; $A \doteq 14$ sq. in.
   b) Rectangle and right triangle; $A \doteq 723 \text{ cm}^2$
   c) Semicircle cut out of a parallelogram; $A \doteq 6 \text{ m}^2$
2. a) $A \doteq 347$ sq. ft.

b) The cover needs about 32 m² and it will cost about $80.66.
3. $V \doteq 27.14$ cm³; $SA \doteq 54.29$ cm²
4. The simple objects that make up the object are a rectangular and triangular prism.
$V \doteq 45\,000$ cu. in.
$SA \doteq 8186$ sq. in.
5. No, the volume is about 2.2 m³.
6. $V \doteq 265.30$ cm³
$SA \doteq 289.13$ cm²

2.3 Optimizing Areas and Perimeters, page 94

1. The rectangle with the maximum area is a square.
   a) Side length: 10 cm;
      area: 100 cm²
   b) Side length: 27.5 ft.;
      area: 756.25 sq. ft.
   c) Side length: 6.25 m;
      area: 39.0625 m²
   d) Side length: 21.75 in.;
      area: 473.0625 sq. in.
2. The rectangle with the minimum perimeter is a square.
   a) Side length: 5 ft.; perimeter: 20 ft.
   b) Side length: 9 m; perimeter: 36 m
   c) Side length: 12 cm; perimeter: 48 cm
   d) Side length: 13 in.; perimeter: 52 in.
3. The rectangle with the minimum perimeter is a square.
   a) Side length: about 5.5 ft.;
      perimeter: about 21.9 ft.
   b) Side length: about 8.1 m;
      perimeter: about 32.3 m
   c) Side length: about 11.1 cm;
      perimeter: about 44.5 cm
   d) Side length: about 15.8 in.;
      perimeter: about 63.2 in.
4. a) 2 m
   b) 8 m
   c) 4.5 m
5. a) $A = 20$ m²
   b) $A = 32$ m²
   c) $A = 33.75$ m²
6. $A = 10\,000$ sq. ft.
7. a) 200 m
   b) Maximum area: 2500 m²
8. Greatest area: 1600 sq. in.
9. a) Arrange the tiles to form a 10-tile by 10-tile square.
   b) Minimum perimeter: 2000 cm;
      $A = 250\,000$ cm²
10. a) Arrange the tiles to form a 8-tile by 10-tile rectangle.
    b) Minimum perimeter: 1800 cm;
       $A = 200\,000$ cm²
11. Yes
12. a) The minimum perimeter is 22 m.
       The dimensions are 10 m by 6 m or 12 m by 5 m.
    b) The minimum perimeter is about 21.91 m.
       The dimensions are about 5.45 m by 11 m.
13. a) Greatest area: 5000 m²
       The dimensions are 100 m by 50 m.
    b) Greater
14. About 6.831 m²
15. Circular design; about 19.42 m
16. a) Maximum area: 253 sq. ft.
       The dimensions are 22 ft. by 11.5 ft. or 23 ft. by 11 ft.
    b) $d \doteq 28.65$ ft.; $A \doteq 322.29$ sq. ft.
    c) The semicircle design has a greater area; about 69.2 sq. ft.
17. Rectangle
18. Maximum area: 67 500 m²
19. Regular dodecagon

2.4 Optimizing Area and Perimeter Using a Spreadsheet, page 101

1. a) Change the formula in cell B4 to "10 – A4".
      Change title to 20 m.
      In column A, fill down from A5 to A22.
   b) Change the value in A4 to 1 and change the formula in cell A5 to "A4 + 1".
   c) Change the formula in cell B4 to "19 – A4".
      Change title to 38 m.
      In column A, fill down from A5 to A40.
   d) Change the value in A4 to 0.25 and change the formula in cell A5 to "A4 + 0.25".
2. a) 7 m
   b) 0.25-m increments
   c) Maximum area is 3.06 m². Change the value in A4 to 0.25, the formula in cell A5 to "A4 + 0.25", and the formula in cell B4 to "3.50 – A4".
3. Maximum area: 100 m²
4. A square with side lengths 12 ft.
   Change title to 144 sq. ft. and the formula in cell B4 to "144/A4".

ANSWERS **549**

5. a) Minimum perimeter: 28 ft.
   b) 48 is not a perfect square.
6. 20 ft. by 10 ft.
   Change cell D4 to "A4 + 2*B4".
7. a) Maximum area: 9 m^2
   b) Maximum area: 18 m^2
   c) Maximum area: 36 m^2
8. a) Minimum length: 80 m
   b) Minimum length: about 56.57 m
   c) Minimum length: 40 m

2.6 Optimizing Volume and Surface Area, page 110

1. a) The rectangular prism is a cube with edge length 4 cm.
   b) $V = 64$ cm^3
2. a) The rectangular prism is a cube with edge length 9 in.
   b) $SA = 486$ sq. in.
3. a) $SA = 248$ cm^2
   b) $SA = 328$ cm^2
   Prism a has the least surface area.
4. a) 10 in. by 10 in. by 10 in.
   b) 15 cm by 15 cm by 15 cm
   c) 20 in. by 20 in. by 20 in.
5. a) 1 m by 1 m by 1 m
   b) 50 cm by 50 cm by 50 cm
   c) 2 ft. by 2 ft. by 2 ft.
6. a) $V = 1728$ cm^3
   b) The rectangular prism has the shape of a cube with edge length 12 cm.
7. a) 1387.10 cu. in.
   b) The minimum surface area is about 746 sq. in. and the container would be about 11.15 in. by 11.15 in. by 11.15 in.
8. a) A cube with edge length 65 cm
   b) 25 350 cm^2
   c) No
9. a) 160 sq. in.   b) 10 in. by 10 in. by 4 in.
10. 4 cubes by 6 cubes by 6 cubes
12. a) The dimensions of the cylinder should be $r \doteq 4.9$ cm and $h \doteq 9.94$ cm.
    b) $SA \doteq 456.98$ cm^2
13. About 819.3 sq. ft.
15. Cylinder
16. a) $s = b\sqrt{2}$
    b) $V \doteq 184.9$ cm^3
    c) $b \doteq 8.6$ cm, $h \doteq 5.0$ cm
17. b) The side length of the base is 10 cm and $h \doteq 5.8$ cm.
    c) $SA \doteq 259.81$ cm^2
    d) The height of the prism decreases as the side length of the base increases.
18. a) The can with the least surface area has radius 3.8 cm and height about 7.8 cm.
    b) The minimum amount of cardboard that can be used is about 1384.8 cm^2 for a case that has 3 layers of 2 rows by 2 cans each.
19. a) Rectangular prism with all sides measuring 14.7 in.; cylinder with radius 8.3 in. and height 16.55 in.; triangular prism with base length 22 in. and height 13.3 in.
    b) The cylinder holds the most material.

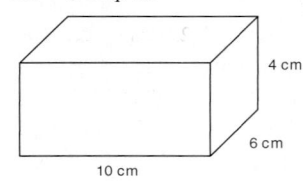

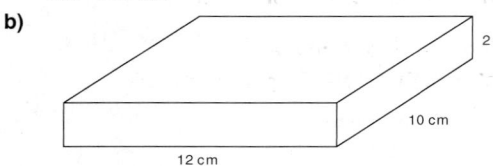

Chapter 2 Review, page 120

1. a) $A = 34$ sq. in.
   b) $A \doteq 402.5$ cm^2
   c) $A = 2.04$ m^2
2. a) $A \doteq 1834.61$ sq. in.
   b) 2 cans
3. a) 22.5 in. by 20.5 in. 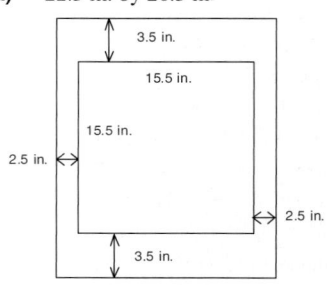
   b) $A = 461.25$ sq. in.
4. a) $V \doteq 388\ 065$ L

    b) $SA \doteq 236.09$ m^2
    c) 4 cans
5. **a)** $SA = 1456$ sq. ft.
    b) $V \doteq 648$ cu. ft.
6. **a)** $V = 2976$ cu. in.
    b) About 6.4 cu. yd.
    c) $SA \doteq 1469$ sq. in.; omit the bottom face.
    d) 2 cans
7. **a)** $V \doteq 3087.9$ cm^3
    b) $SA \doteq 1783$ cm^2
8. **a)** $V = 67.5$ cu. ft.
    b) $SA = 22.5$ sq. ft.
9. The rectangle with the maximum area is a square.
    a) Side length: 10 cm; area: 100 cm^2
    b) Side length: 27.5 ft.; area: 756.25 sq. ft.
    c) Side length: 6.25 m; area: 39.0625 m^2
    d) Side length: 21.75 in.; area: 473.0625 sq. in.
10. The rectangle with the minimum perimeter is a square.
    a) Side length: 5 ft.; perimeter: 20 ft.
    b) Side length: 9 m; perimeter: 36 m
    c) Side length: 12 cm; perimeter: 48 cm
    d) Side length: 13 in.; perimeter: 52 in.
11. $A = 0.5$ km^2 or 500 000 m^2
12. **a)** 5 ft. by 11 ft.
    b) 10 ft. by 11 ft.
13. **a)** Circle
    b) About 170.77 m^2
15. **a)** The prism is a cube with edge lengths 6 m.
    b) $SA = 216$ m^2
16. **a)** The prism is a cube with edge lengths 5 cm.
    b) $V = 125$ cm^3
17. **a)** $SA = 24$ sq. in.
    b) 2 in. by 2 in. by 2 in.
    c) About 391 notes
18. **a)** $r \doteq 6$ cm; $h \doteq 11.9$ cm
    b) $V \doteq 1346.4$ cm^3
19. **a)** Radius is about 4.2 cm.
    The height of the bottle is about 8.12 cm.
    The surface area is about 325 cm^2.
    b) The base is an equilateral triangle with side lengths about 12 cm.
    The height is about 7 cm.
    The surface area is about 384 cm^2.

Chapter 2 Practice Test, page 123

1. B
2. B
3. **a)** $V \doteq 184.3$ cm^3;
    $SA \doteq 226.4$ cm^2
    b) $V \doteq 942$ cu. in.;
    $SA \doteq 696$ sq. in.
4. 12 tiles by 5 tiles or 10 tiles by 6 tiles
5. **a)**

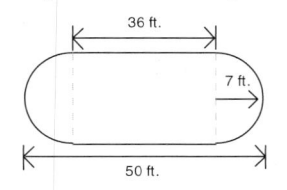

    $A \doteq 657.9$ sq. ft.
    b) 731 plants
    c) About 4.07 kg
6. **a)** About 22.5 cm by 22.5 cm by 22.5 cm;
    $SA \doteq 3040.2$ cm^2
    b) $r \doteq 12.2$ cm; $h \doteq 24.38$ cm;
    $SA \doteq 2804.04$ cm^2
    c) About 236 cm^2

Chapter 3 Two-Variable Data

Activate Prior Knowledge
Interpreting Data Graphs, page 126

1. **a)** 3 runners
    b) 160 to 169 bpm
    c) Minimum heart rate: 120 bpm;
    maximum heart rate: 180 bpm
2. **a)** The percent of students who travel to school by school bus.
    The light blue area represents less than half the graph.
3. **a)** Each point shows the price of a laptop computer relative to its mass.
    b) Four laptop computers have mass less than 4000 g. One costs less than $1250.
    c) Five laptop computers have mass greater than 4000 g. All five cost less than $1250.
4. **a)** The sum is 223.
    b) There were some students who picked more than one leisure activity.
    c) A histogram

ANSWERS   **551**

Working with Slope and Line Graphs, page 128

1. **a)** $\frac{3}{4}$, or 0.75;
   $y = 0.75x + 5$
   b) $-\frac{1}{2}$, or -0.5;
   $y = -0.5x + 6$

3.1 One- and Two-Variable Data, page 133

1. **a)** **i)** One-variable data
   ii) One-variable data
   iii) One-variable data
   iv) Two-variable data
2. **a)** **i)** One-variable data
   ii) One-variable data
   iii) Two-variable data
3. **a)** Purchases made and reward points earned
   b) Brain mass and IQ
   c) Mosquito population and average rainfall
4. **a)** **i)** One-variable data
   ii) Two-variable data
   iii) One-variable data
5. **a)** Histogram
   b) Circle graph
   c) Scatter plot
6. **a)** A histogram would be best for showing the frequency of each sample size.
   b) A scatter plot would be best to display all the data in the table.
   c) Most concern: (50, 2); least concern: (200, 3)
7. **a)** Only one attribute (height) is measured.
   b) Compare mean/median/mode height among the students in English class and Math class.
8. **a)** *Study Time and Test Scores*
   b) *Study Time and Test Scores* provides information about a possible relationship. The variables involved are Study time and Mark on test.
10. **a)** One-variable statistics
    b) One-variable data; histogram
    c) Two-variable data; scatter plot

3.2 Using Scatter Plots to Identify Relationships, page 142

1. **a)** Shoe size of a child relative to her/his age
   b) Child A wears the smallest shoe size.
   Child D wears the biggest shoe size.
   c) Children C and D have the same age.
   d) Children C and F wear the same shoe size.
2. Shoe size is the dependent variable.
3. **a)** Company A
   b) Companies E and F use the largest trucks.
   c) Companies A and B use the smallest trucks.
   d) 3 different truck sizes
   e) Company A for a small load; company E for a very large load
4. Moving cost is the dependent variable; the moving costs increase as truck size increases.
5.

Point	Building Number
A	5
B	6
C	1
D	3
E	2
F	4

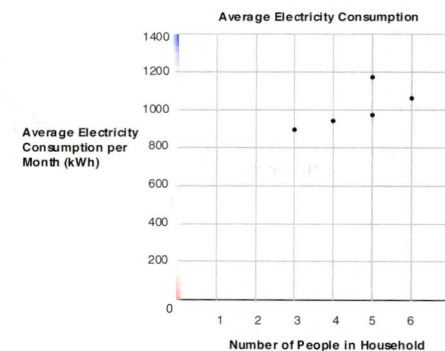

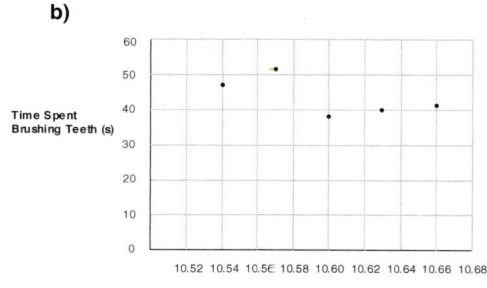

6. **a)** Positive correlation
   b) No correlation
   c) Negative correlation
   d) Positive correlation
   e) No correlation
7. **a)** Average Electricity Consumption

b)

8. a) Positive correlation
   b) No correlation
   c) Negative correlation
   d) Negative correlation
9. a) Number of hours worked
   b) Size of backyard garden
   c) Time to travel to a specified destination
10. a) There are two sets of points plotted. Car A and car B are also labelled on the legend.
    b) 7 laps
       The two cars use the same track each time.
    c) Graph 3
    d) Car A has the greater average speed.
    e) Graph 1
11. a) Car B   b) Graph 2
    c) Graph 3   d) Graph 1
12. a) The hourly cost increases as the surface area increases. The points on the scatter plot should go up to the right. There should be a positive linear correlation.
    b) [Scatter plot: Surface Area and Hourly Cost; Surface Area (sq. ft.) vs Hourly Cost (cents)]
    c) Yes, the scatter plot shows a positive correlation between the two variables.
    d) 4944 ¢ or $49.44
13. a) Speed is the independent variable.
    b) The fuel consumption increases as the speed increases. The points on the scatter plot should go up to the right. There should be a positive correlation.
    c) [Scatter plot: Speed and Fuel Consumption; Speed (km/h) vs Fuel Consumption (L/100 km)]
    d) There is a positive correlation between the two variables.
    e) About 6.7 L/100 km
15. a) Reasonable
    b) Not reasonable
    c) Not reasonable
    d) Reasonable
16. a) Temperature is the independent variable.
    b) [Scatter plot: Temperature and Relative Humidity; Temperature (C°) vs Relative Humidity (%)]
    c) There appears to be a correlation between the two variables. As the temperature increases, the relative humidity decreases.
17. a) About 90%
    b) Yes, it is very likely to rain overnight.

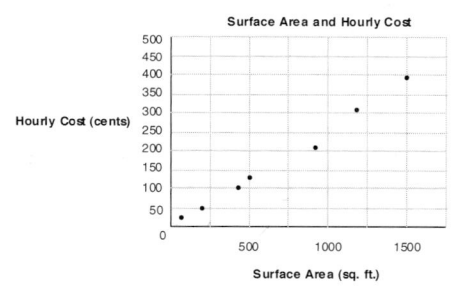

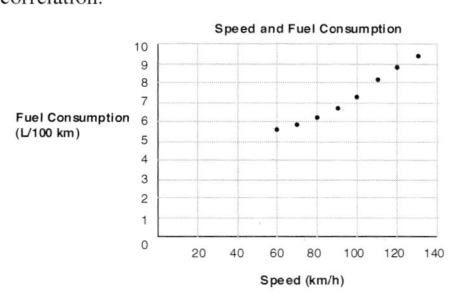

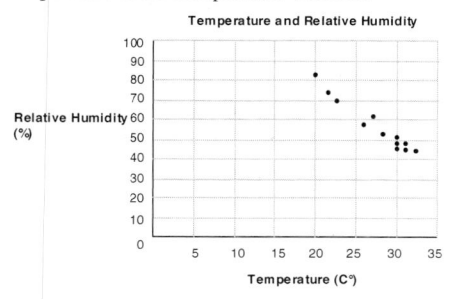

3.3 Line of Best Fit, page 153

1. a) Graph B
   b) Graph D
2. Outlier A
3. a) Interpolation
   b) Extrapolation
   c) Extrapolation
   d) Interpolation
4. a) i) [Scatter plot with line of best fit: Retail Cost of Evergreens; Height (ft.) vs Cost ($)]

ANSWERS 553

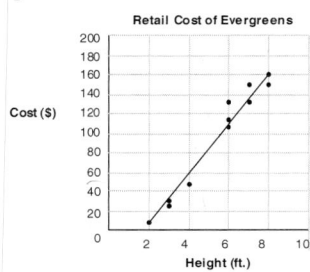

ii) *Eruptions of Cerro Negro Volcano since 1900*

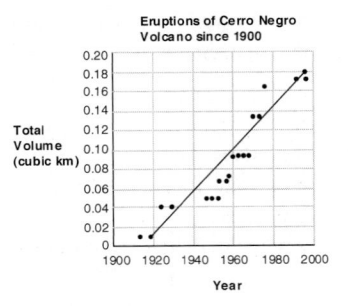

b) i) $y = 26.271x - 47.364$
ii) $y = 0.0019x - 3.615$

5. a) The line of best fit is not in agreement with the trend of the points on the scatter plot.
b) The line of best fit would have a negative slope and lie close to the four points.

6. There is no clear pattern to these data. If a weak correlation exists, it is difficult to determine whether it is positive or negative.

7. a) i) Fairly strong positive correlation
ii) Strong positive correlation
iii) Fairly strong negative correlation
iv) Weak positive correlation followed by a fairly strong positive correlation, then a strong positive correlation
b) i) Positive linear
ii) Non-linear
iii) Negative linear
iv) Non-linear

8. a) i) About 29 games ii) About 11 games

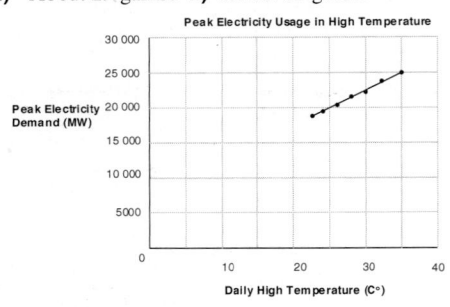

9. a) *Peak Electricity Usage in High Temperature*

b) The equation for the line of best fit is:
$y = 538.32x + 6256.4$
c) The peak electricity demand for a daily high temperature of 28°C is about 21 329 MW.
d) Unlikely; the peak electricity demand for a daily high temperature of 37°C is about 26 174 MW.

e) On days with temperatures below 20°C, the peak electricity demand would likely fall below 17 000 MW, so the power company can take one of the generators offline.

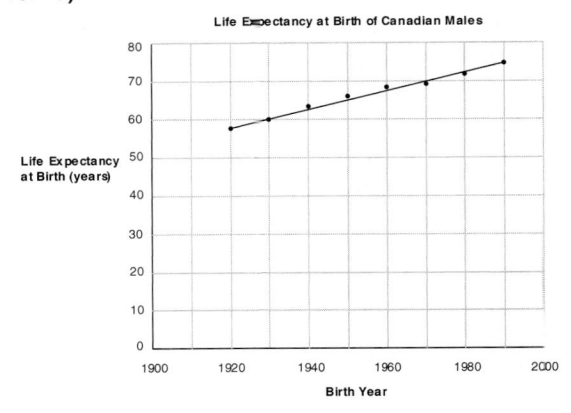

10. a) *Life Expectancy at Birth of Canadian Males*

b) $y = 0.2286x - 380.36$
c) About 71 years
d) About 77 years
e) Males born after 2013 will have a life expectancy of 80 years.

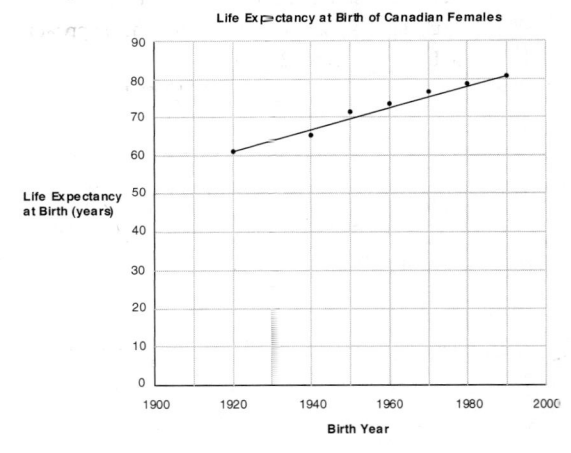

11. b) *Life Expectancy at Birth of Canadian Females*

$y = 0.3071x - 529.21$
The life expectancy of a female born in 1975 is about 77 years.
The life expectancy of a female born in 2000 is about 85 years.
Females born after 1983 will have a life expectancy of 80 years.

13. There seems to be a strong positive relationship between the fuel consumption for city and highway

driving. The relationship does not appear to be linear. A quadratic model for the data would provide a better fit.

Chapter 3 Mid-Chapter Review, page 158

1. **a)** One-variable data
   b) One-variable data
2. **a)** Negative correlation
   b) No correlation
   c) Positive correlation
   d) No correlation
3. Graph A
4. **a)**

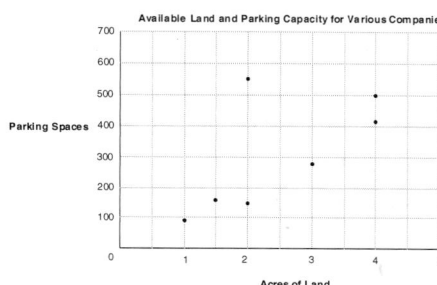

   Available Land and Parking Capacity for Various Companies (scatter plot: Parking Spaces vs. Acres of Land)

   b) The parking capacity increases as the available land increases. The data in the scatter plot show a fairly strong positive correlation. There appears to be an outlier at (2, 550).
5. **a)**

   Tire Pressure and Temperature (scatter plot: Tire Pressure (psi) vs. Outside Temperature (F°))

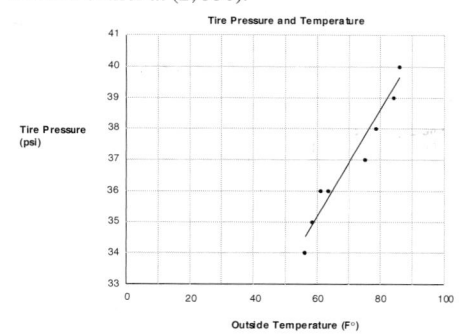

   b) There is a strong positive correlation between the outside temperature and the tire pressure.
   c) **i)** 37 psi
   ii) 32 psi

Chapter 3 Review, page 186

1. **a)** One-variable data
   b) Two-variable data
   c) One-variable data
2. **a)** Air pressure and height above Earth's surface
   b) Rainfall and crop yield
   c) Cooking time and mass of a turkey
3. **a)** Negative correlation
   b) No correlation
   c) Positive correlation
   d) No correlation
   e) Negative correlation
4. **a)** Independent variable: volume of aquarium; dependent variable: cost
   b) Independent variable: rain; dependent variable: time people spend watering their lawns
5.

Cylinder number	Point
1	D
2	B
3	C
4	A
5	F
6	E

6. **a)**

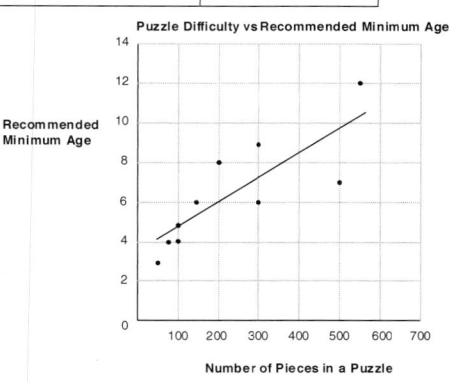

   Puzzle Difficulty vs Recommended Minimum Age (scatter plot: Recommended Minimum Age vs. Number of Pieces in a Puzzle)

7. The number of readings taken testing only one vehicle model is not sufficient to get conclusive data.

ANSWERS 555

8. **a)** Weak negative correlation
   b) Strong positive correlation
   c) Strong negative correlation

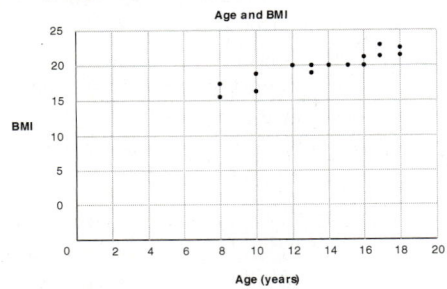

9. **a)** [Scatter plot: Age and BMI]

   A few of the outliers are (12, 20), (10, 18).

   b) $y = 0.6342x + 11.164$

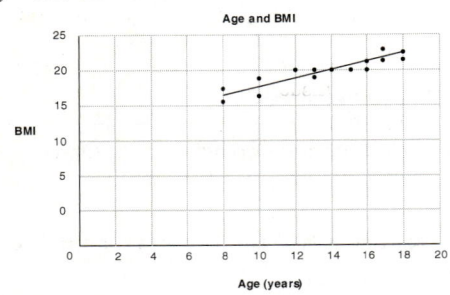

   [Scatter plot: Age and BMI with line of best fit]

   c) Positive linear correlation
   d) 14 people; data are insufficient to draw a conclusion; the sample size is very small.

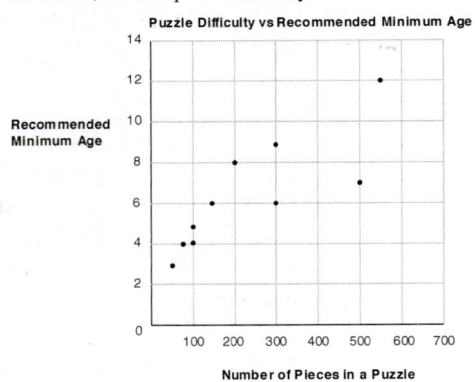

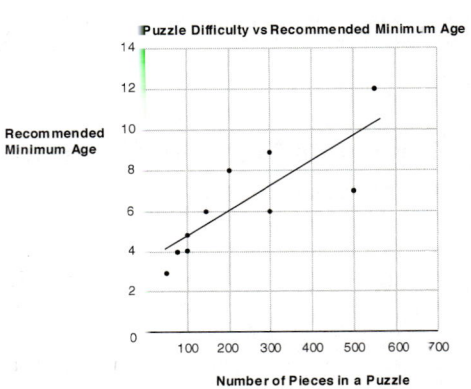

10. **a)** [Scatter plot: Puzzle Difficulty vs Recommended Minimum Age]

    b) [Scatter plot: Puzzle Difficulty vs Recommended Minimum Age with line of best fit]

    c) 516 pieces
    d) 8 years

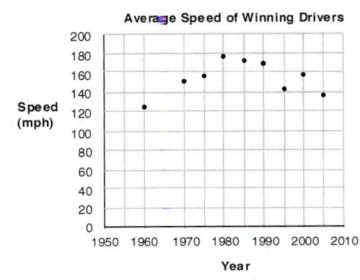

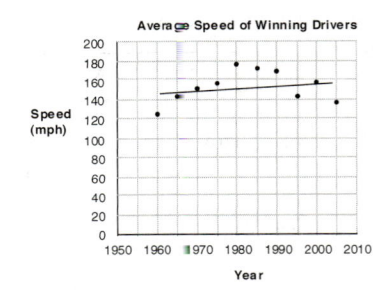

11. **a)** [Scatter plot: Average Speed of Winning Drivers]

    b) [Scatter plot: Average Speed of Winning Drivers with line of best fit]

    The speed of the winning driver in 2010 would be about 159 mph.
    $y = 0.2158x - 275$

    c) The prediction is not very reliable. 2010 is not within the domain, so the answer is acquired through extrapolation.

12. a)

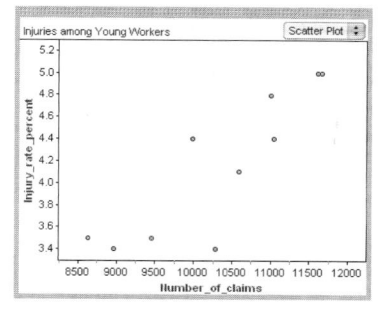

b) About 0.86

c) The coefficient correlation suggests a strong positive correlation between the number of claims and the injury rate.

Chapter 3 Practice Test, page 189

1. C
2. B
3. **a)**

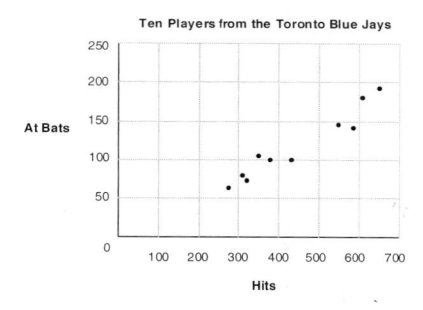

   b) Yes, there is a strong positive linear correlation.

   c)

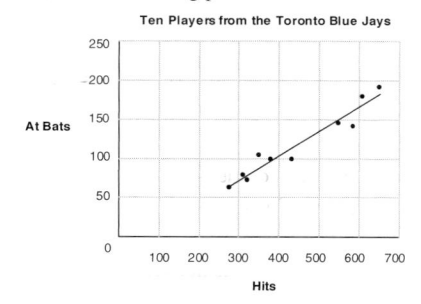

   A player might have about 10 hits after 100 times at bat. $y = 0.3149x - 21.586$

6. **a)** Strong positive correlation
   b) Yes
   c) (98, 15)

Chapter 4 Statistical Literacy

Activate Prior Knowledge

Ratios, page 192

1. **a)** 20:30
   b) **i)** 4:6
   ii) About 67:100
2. **a)** 400 girls and 300 boys
   b) 16 girls to 12 boys
3. The number of people who liked the taste of cheese out of 100 people who tried it. Or: The percent of people who liked the taste of cheese out of the people who tried it.

Measures of Central Tendency, page 193

1. **a)** Mean: about 19.5; median: about 19.4; no mode
   b) 6.5
2. Since there are no outliers, either the mean or the median is appropriate.

Percent Increase and Decrease, page 194

1. About 15.1%
2. About 8.3%

4.1 Interpreting Statistics, page 201

1. **a)** 8, 8, 9, 9, 10, 11, 11, 12, 12, 12, 14, 15
   b) 150, 154, 162, 163, 164, 165, 165, 168, 170, 172, 180
2. **a)** In 2000, 2 in 5 students owing money had repaid their debt 5 years after graduations.
   b) In 2000, 56% of graduates had no debt.
   c) Almost half of the graduates who are still in debt have trouble paying off their loans. Only 1 in 5 graduates who had paid off their loans by 2005 had the same problem.
3. Part ii
7. **a)** Mean list price: $334 466.67
   Median list price: $324 500.00
   Mean sale price: $326 666.67
   Median sale price: $315 000.00
   List price range: $79 100.00
   Sale price range: $75 000.00
   b) Due to the small sample size, the presence of outliers cannot be determined easily. One should use either the median or mean.

ANSWERS **557**

8. a) i) 75th percentile
        ii) 5th percentile
        iii) 95th percentile
    b) Between 50th and 75th percentile
9. a) Mean: 68.65; median: 69; modes: 65, 75
    b) 1st quartile: 61.5; 3rd quartile: 76
    c) 37% of the people that wrote the test received a mark below Vince's. Vince's mark is 65.
10. a) "Do you support the legislation that would ban smoking in cars and other private vehicles where a child or an adolescent under 16 of age is present?"/"Do you smoke?"
    c) Poll results are not always accurate, but 19 times out of 20 (95%), the results are within 2.7% of the true public opinion.
11. a) Are you likely to/certain to/unlikely to/certain not to avoid toys made in China because of concerns about health or safety risks?
    b) 55%
    d) "Majority of Canadians Likely to Avoid Toys Made in China"
12. a) Part i
    b) 15%
13. a) House 1 list price: $284 500.00
        House 1 sale price: $275 000.00
        House 2 list price: $316 000.00
        House 2 sale price: $307 000.00
        House 3 list price: $276 900.00
        House 3 sale price: $272 000.00
    b) Mean list price: $292 500.00
        Median list price: $284 500.00
        List price range: $39 100.00
        Mean sale price: $285 000.00
        Median sale price: $275 000.00
        Sale price range: $35 000.00
14. a) 1st quartile: 70, 2nd quartile: 74.5, 3rd quartile: 77
    b) 85 km/h

4.3 Surveys and Questionnaires, page 214

1. a) 20%
    b) 90%
    c) About 58%
2. a) 35 people
    b) 93 people
    c) 118 people
    d) 1036 people

3. a) About 11%
    b) About 7%
    c) About 4%
4. a) Written form
    b) Written form
    c) Personal interview
    d) Personal interview
    e) Written form
    Personal interviews were selected for parts c and d because people are likely to feel comfortable discussing their colour preference or shampoo.
5. a) Biased
    b) Unbiased
    c) Unbiased
    d) Biased
6. i) Biased
    ii) Unbiased
9. No. The sample only represents those who texted or e-mailed the radio station. The sample could be improved by surveying those directly affected by the proposed change, such as students, teachers, parents, etc.

Chapter 4 Mid-Chapter Review, page 222

1. Quartiles: 66, 72.5, 84.5
2. a) Half of the salaries at the firm are below $85 000, and half are over $85 000.
    b) 95 out of every 100 people preferred the new cereal.
    c) The local athlete was in the top 25% of all the athletes that took the fitness challenge.
    d) On average, each person in Ontario used 11 996 kWh of electricity last year.
    e) 1 quarter of Canadian adolescents are overweight.
4. a) Written form
    b) Neither
    c) Written form
    d) Personal interview
5. a) The senior residents of the town
    b) The assistant will be polling non-senior residents, too; the sample population should be seniors only.
6. a) Invalid due to bias; the reporter talked to dog-owners only.
    b) The study should include opinions from individuals who do not own dogs.

4.5 The Use and Misuse of Statistics, page 229

1. **a)** Part i
   b) Part ii
2. **a)** The data on the bar graph is difficult to read accurately.
   b) The vertical scale makes changes in temperature look more dramatic than they really are.
3. **a)** Part ii
   b) Part i
4. Part ii; an outside agency will have a more objective (unbiased) opinion. It will also have the expertise required in statistical surveys.
5. **a)** Yes   **b)** Yes
   c) No   **d)** No
6. **a)** As the number of students increases, the number of teachers increases. Positive correlation, cause-and-effect relationship
   b) The number of vacation days increases as the person works at a company longer. Positive correlation, cause-and-effect relationship
   c) The population of a town does not affect the amount of precipitation.
   d) Height does not affect a person's performance/marks in mathematics.
7. **a)** Additional information about the person's credentials.
   b) Additional information about the party who conducted the study, so that one can decide about its bias.
8. Parts ii and iii
9. Misleading report; the sample size is too small and the survey question is biased.
10. Yes
11. The error is assuming that a strong positive correlation means necessarily a cause-and-effect relationship. The high r-value means that the variables are strongly connected, but it could be in response to an increase in the population. It is unlikely that hockey is contributing to an increase in crime rate.
12. Yes
13. **a)** Survey only the few residents whose properties border the highway. Include a prefacing statement that will influence their decision.

4.6 Understanding Indices, page 237

1. **a)** The change in the price of fruit from 1990 to 2006
   b) 1997
   c) i) About $84
       ii) About $104
2. **a)** 50%
   b) 25%
   c) 100%
   d) 6%
3. **a)** 26%
   b) 50%
   c) 7%
   d) 80%
4. Denmark and Switzerland, Austria and Iceland, The Bahamas, Finland, and Sweden, Bhutan, Brunei, and Canada
5. **a)** 88.9%
   b) 97.8%
   c) CPI increased by 8.9% between January 1996 and January 2001.
   d) 1.8%
6. **a)** The FPPI for fruit increases with time.
   b) The trend indicates that the same amount of money buys less fruit every year.
7. **a)** Wood Buffalo National Bark (B)
   b) Cochrane District, Ontario (D)
   c) Terra Nova National Park (E)
9. **a)** 1986
   b) i) 130   ii) 140
       iii) 164
   c) i) 30%   ii) 7%
       iii) 17%
   d) 1986 to 1992 had the greatest increase. 1992 to 1998 had the least increase.
10. **a)** The overall change in EPI: about 62%
    The average annual change: about 3.6%
    b) About 187.2%
11. **a)** Both graphs use 1986 as the base year.
    c) The instructional supplies index change between 1986 and 2003: about 105%
    The average annual rate of increase in price: about 6.2%
12. **a)** Zurich, Oslo, Tokyo
    It is more expensive to buy food in these cities than in New York.
    b) It is about 12.1% more expensive to buy food at Oslo than New York. It will cost about 64.9% less to purchase food in Delhi than in New York.

ANSWERS   **559**

13. a)

City	Food Price Index (Toronto = 100)
Zurich	143.1
Oslo	138.7
Dublin	107.2
New York	123.8
Copenhagen	123.1
Toronto	100.0
Tokyo	161.3
Rome	108.7
Hong Kong	107.2
Delhi	43.4

b) Delhi; the cost of food in Delhi is about 43% the cost of food in Toronto.
c) Tokyo has the highest food price index. If Tokyo was the base value, all other cities would be below 100.

Chapter 4 Review, page 251

1. a) The average is the mean of a set of values. As the values of the set increase, the average increases as well.
2. a) 0.167, 0.178, 0.204, 0.208, 0.216, 0.233, 0.236, 0.238, 0.238, 0.240, 0.242, 0.245, 0.251, 0.262, 0.277, 0.289, 0.291, 0.297
   b) 1st quartile: 0.21
      2nd quartile: 0.239
      3rd quartile: 0.262
   c) 0.277
4. a) $64 500; 25% of those surveyed had a salary below this value.
   b) $96 000; 75% of those surveyed had a salary below this value.
5. a) 62 students   b) 124 students
   c) 186 students   d) 372 students
6. a) People who don't have Internet access or a cell phone cannot vote.
   b) Allow viewers to vote by phone.
7. a) Only people that visit the community centre can vote. The sample is biased.
   b) Poll a wider variety of people using a variety of methods.
9. a) False   b) True
   c) True   d) False
10. a) Part i   b) Part i
13. a) Costa Rica
    b) United States, Mexico, Belize, Guatemala, Honduras, El Salvador, Nicaragua, Panama
14. a) Computer prices between 2002 and 2006
    b) 2001
    c) $80.00
       The cost of computers is 20% lower than in 2001.
    d) Computer prices continue to drop throughout the years.
15. a) January 2004
    b) Overall change: 65.1%; average annual decrease: 13.02%

Chapter 4 Practice Test, page 254

1. D
   The 10th percentile indicates that 10% of the data is below this point.
2. B
3. a) 3 movies per capita (mpc)
      Enough people visited the movies in 1996 that each person in Canada could have gone to the movies 3 times that year.
   b)

Year	Movies per capita (mpc)
1996	3.0
1997	3.2
1998	3.6
1999	3.9
2000	3.8

   c) Yes
      There was a lapse in 2000, when the number of movies per capita dropped. More data are required for an accurate conclusion.
   d) Per capita rate takes into account the changes in both variables.
4. b) One must survey a significant portion of the student population. 10% or 100 students represent an appropriately-sized sample. It is important to ensure that the sample is representative. Survey 25 students from each grade.
6. b) Between 2005 and 2006   c) About 14%

Cumulative Review Chapters 1–4, page 262

1. a) About 0.7 m
   b) About 70°
2. a) ∠A = 15°, ∠B = 75°, ∠C = 90°,
      $a \doteq 2$ cm, $b \doteq 8$ cm, $c = 8$ cm

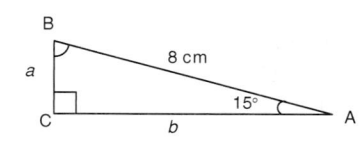

   b) ∠C $\doteq$ 41°, ∠D $\doteq$ 49°, ∠E = 90°,
      $c = 9.2$ yd., $d \doteq 10.6$ yd., $e = 14.0$ yd.

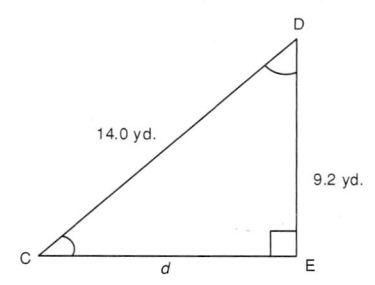

   c) ∠X = 23°, ∠Y = 67°, ∠Z = 90°,
      $x \doteq 9$ m, $y = 21$ m, $z \doteq 23$ m

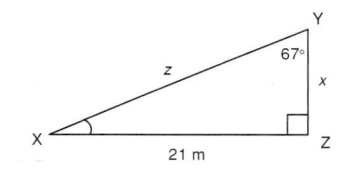

   d) ∠P = 90°, ∠Q = 39°, ∠R = 51°,
      $p \doteq 238$ mm, $q = 150$ mm, $r \doteq 185$ mm

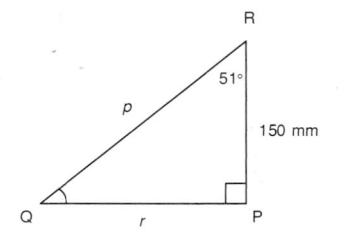

   e) ∠G $\doteq$ 51°, ∠H $\doteq$ 39°, ∠I = 90°,
      $g = 1.5$ m, $h = 1.2$ m, $i \doteq 1.9$ m

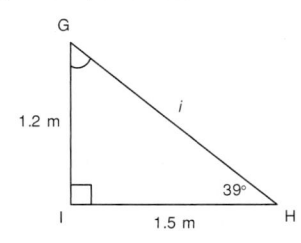

3. a) ∠D $\doteq$ 153°
   b) ∠D $\doteq$ 102°
   c) ∠D $\doteq$ 157°
   d) ∠D $\doteq$ 143°
   e) ∠D $\doteq$ 172°
   f) ∠D $\doteq$ 140°
4. a) Sine Law; ∠R = 135°, $r \doteq 10.0$ km, $p \doteq 5.5$ km
   b) Cosine Law; ∠T = 32°, ∠U $\doteq$ 9°, $t \doteq 11$ in.
5. $z \doteq 1.7$ ft.
6. a) About 76 mi.
   b) 177°
7. a) About 2.3 sq. in.
   b) About 36.9 cm^2; about 7.2 cm^3
8. a) 7 m by 7 m; 49 m^2
   b) 11 in. by 11 in.; 121 sq. in.
   c) 2.5 cm by 2.5 cm; 6.25 cm^2
   d) 23.5 ft. by 23.5 ft.; 552.25 sq. ft.
9. a) Rectangles: 2 in. by 8 in., or 4 in. by 6 in.
      Triangles: 4 in., 8 in., 8 in., or
      6 in., 6 in., 8 in.
   b) A rectangle with dimensions closest to a square:
      4 in. by 6 in.; maximum area: 24 sq. in.
10. a) 4 ft. by 4 ft. by 4 ft.; SA: 96 sq. ft.
    b) 9 m by 9 m by 9 m; SA: 486 m^2
    c) About 6.1 cm by about 6.1 cm by about 6.1 cm;
       SA: about 222 cm^2
    d) About 14.4 in. by about 14.4 in. by about
       14.4 in.; SA: about 1248 sq. in.
11. a) 1 in. by 1 in. by 66 in.,
       2 in. by 2 in. by 32 in.,
       4 in. by 4 in. by 14 in.,
       5 in. by 5 in. by 10 in.,
       8 in. by 8 in. by 3 in.
    b) Maximum volume: 250 cu. in.
12. a) Vertical bar graph
    b) No, Avery is incorrect. The graph is displaying
       one-variable data. Gender represents categories of
       data, not one of the variables being measured.
13. a) Positive correlation
    b) Negative correlation
    c) Negative correlation
    d) Positive correlation
14. a) Price of gasoline
    b) The remaining value of the car
    c) Probability of developing lung cancer
15. Graph B has the best line of best fit. The line's path
    has shifted slightly downwards in response to the
    3 outliers below the main data cluster.
16. a) i) There is a weak/moderate negative correlation.

ANSWERS   561

ii) There is a strong negative correlation; the data points are much closer, indicating a linear correlation.
b) i) One should not try to model a line of best fit for graph in part i. The linear model would not provide accurate predictions.
ii) A linear model would provide accurate predictions.
17. a) First quartile: 1.5
Second quartile: 5
Third quartile: 7.5
b) First quartile: 104
Second quartile: 108
Third quartile: 110
18. a) 3 people
b) 18 people
c) 54 people
d) 1725 people
19. a) Only students in certain classes are surveyed. The sample is not representative of the entire school.
b) Ask every 5th person in the yearbook or out of a list of students provided by the administration.
c) Part b; every grade is represented equally. Part a, Carmelo's method, includes students from his class only.
d) The students may feel intimidated. Carmelo may ask the students to complete a written survey.
20. What does the vertical axis measure?
What does Taste Test Phase 1 and 2 indicate?
What happens between the 2 tests?
What breeds of dogs were tested?
What was the age distribution?
What does the horizontal axis measure?

Chapter 5 Graphical Models

Activate Prior Knowledge

Linear, Quadratic, and Exponential Graphs, page 267

1. a) 60
The distance travelled each hour is 60 km.
b) −200
The value of a computer decreases by $200 each year.
2. a) (150, 22 500)
Maximum revenue is generated when 150 tickets are sold.
b) (2, 22)
A maximum height of 22 m is reached after 2 seconds.

3. The initial value appears as the vertical intercept in the graph. When the constant factor is greater than 1, the graph curves up. When the constant factor is between 0 and 1, the graph goes down and levels off.

5.1 Trends in Graphs, page 273

1. Part c
2. Part a
3. Part a
4. Part b
5.

	a)	i	Fallen dramatically
	b)	iii	Fallen steadily
	c)	ii	Remained constant
	d)	i	Fluctuated

6. a) The immigrant population: increased rapidly from 1901 to 1911; remained fairly constant from 1911 to 1931; decreased steadily from 1931 to 1951; remained fairly constant from 1951 to 1991; increased slightly from 1991 to 2001
b) The number of births: decreased slowly from January to February; increased rapidly from February to March; remained fairly constant from March to September; decreased steadily from September to December
c) The exchange rate fluctuated slightly from January 1970 to 1977; decreased steadily from 1977 to 1985; increased from 1985 to 1992; decreased from 1992 to 2004; increased rapidly from 2004 to present
d) Maximum safe heart rate during exercise decreased steadily with age.
7. a) Power increased very slowly for wind speeds of up to 5 m/s, then slowly, and then very rapidly.
b) 0 kW
No power is generated when there is no wind.
c) 100 kW
d) No
8. a) The power capacity fluctuated with a slow increasing trend.
c) Demand was constant Monday to Tuesday; increased steadily from Tuesday to Friday; decreased rapidly first, then slowly from Friday to Sunday.
d) Demand exceeded capacity on Friday.
9. a) The area increases as the length of the third side increases, reaches a maximum at 100 m, and then decreases rapidly.
b) i) 50 m or 130 m

ii) 100 m

10. a) i) Cost increased steadily as the number of T-shirts produced increased.

ii) As the number of T-shirts produced increased, the profit increased rapidly at first, then slowly, reached a maximum, then decreased slowly, and then rapidly.

11. a) The mortality rate was high for newborns; decreased rapidly for the first two years of life; remained fairly constant until mid-teenage years; increased steadily until age 20; remained constant until age 30; increased steadily to age 50.

b) (9, 10)
The mortality rate for 9-year-olds is about 10 deaths per 100 000.

c) The mortality increased rapidly from age 15 to age 20.

13. a) The fuel economy increased slowly as the speed increased from 40 km/h to 70 km/h, and then decreased steadily as the speed increased beyond 70 km/h.

b) Based on a fuel cost of $1.30; about $16

c) About 5.14 h

14. a) The global population increased very slowly from 1800 to 1880, then increased slowly from 1880 to 1930, and finally increased rapidly after 1930.

5.2 Rate of Change, page 284

1. a) Independent variable: Hours worked
Dependent variable: Earnings

b) Independent variable: Pages printed
Dependent variable: Cost

c) Independent variable: Distance driven
Dependent variable: Fuel used

2. a) $/hour; hourly wage

b) $/page; per-page cost

c) L/km; fuel consumption rate

3. a) $8/hr

b) $0.0225/page

c) 0.6 L/km

4. a) Independent variable: Depth
Dependent variable: Temperature

b) Independent variable: Time
Dependent variable: Distance

c) Independent variable: Year
Dependent variable: Interest earned

5. a) °C/m; the change in temperature per metre of depth

b) m/s; distance travelled per second

c) $/year; amount of interest earned per year

6. a) 10°C/m

b) 12 m/s

c) $0/year

7. a) i) 0:00 to 6:00, 8:00 to 17:00, and 20:00 to 24:00

ii) 17:00 to 20:00

iii) 6:00 to 8:00

b) i) 0°C/h

ii) About −1.33°C/h

iii) 2°C/h

8. a) The attendance increased steadily from 1976 to 1983; remained constant from 1983 to 1989; then increased steadily from 1989 to 1998.

b) About 8000 students/year

c) The rate of change in attendance was 0 students/year.

9. a) Simple interest: $40/year, $40/year, $40/year
Compound interest: $47/year, $68.8/year, $101.4/year

b) The rate of change for simple interest was the same for each interval. The interest increased by equal amounts each interval.
The rate of change for compound interest was not the same for each interval. The interest increased by larger amounts during the later intervals.

c) Simple interest: straight line
Compound interest: exponential increase

10. a) i) 5 m/min, 392.6 m^2/min
The rate of change measures how quickly the radius of the spill is growing.

ii) 5 m/min, 1962.6 m^2/min
The rate of change measures how quickly the area of the spill is growing.

b) Constant increase

c) Exponential increase

11. a) First differences: 15, 29, 41, 60, 76, 75, 73, 57, 40, 34, 14
The first differences show how much more electric energy was generated in each 5-year period compared to the previous 5-year period.

b) From 1970 to 1975

c) The electricity generated increased at a constant rate.

d) The electric energy increased slowly initially, then rapidly, and then slowly again.

12. The sunflower increased steadily in the first 2 weeks, then increased rapidly from week 2 to week 5, and then increased slightly over the next 5 weeks.

13. a) From 30 to 36 months

b) About 75 words/month

c) About 41 words/month

ANSWERS **563**

15. a) *Heights of Two Sunflowers* (graph: Height (cm) vs Weeks, 0–12)

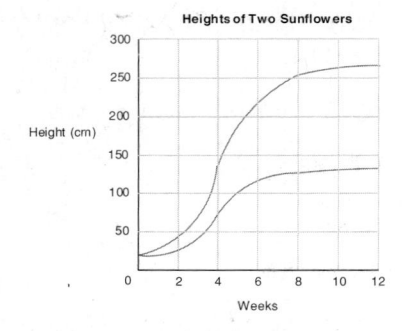

b) *Heights of Two Sunflowers* (graph: Height (cm) vs Weeks, 0–18)

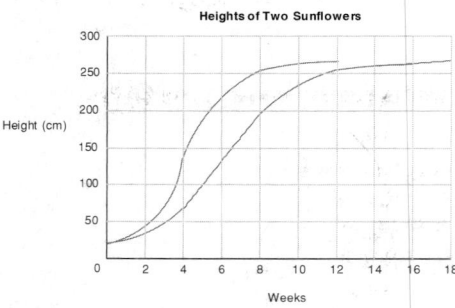

16. *Temperature of the Sunroom over a Day* (graph: Temperature (°C) vs Time (h))

7. a) Temperature decreases at a constant rate as the distance from the inside increases.
   b) $-2°C/cm$
   c) $T = 20 - 2d$

8. a) Both graphs would be straight lines with vertical intercept of 0, but different slopes.
   b) (graph)
   c) Spring 1: 0.1 N/cm; Spring 2: 0.5 N/cm
   d) Low spring constant

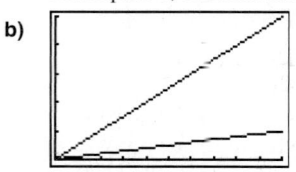

9. a) They start at the same initial position with the same speed of 2 m/s.
   b) One friend stops walking after 10 seconds.

10. a) $500; overhead costs of the prom if no tickets are sold
    b) $15/person; additional cost per person
    c) *Cost and Revenue from the Prom* (graph)
    d) 100 tickets
    e) The graph would be steeper and the break-even point would be 50 tickets.

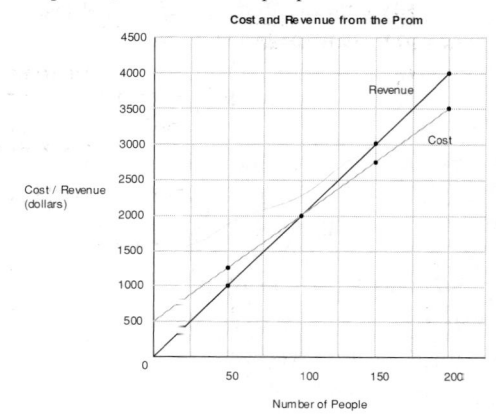

11. a) *Motion of Four Cars* (graph: Distance (m) vs Time (s), showing Car A, Car B, Car C, Car D)

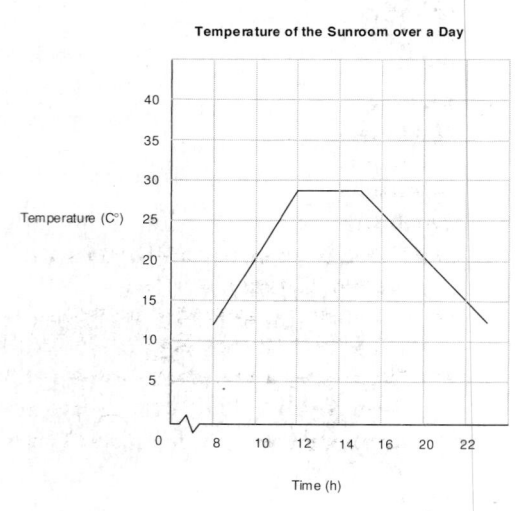

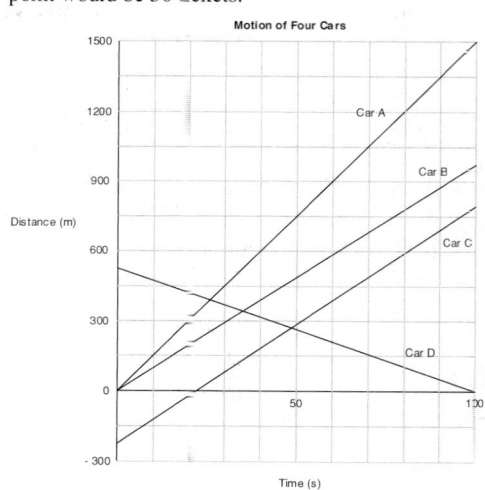

5.3 Linear Models, page 293

1. Part a; equal first differences
2. Parts a and c
   Graph is a straight line.
3. Parts a and c
   Relation is of form $y = mx + b$.
4. a) Part ii b) Part i
   c) Part iii
6. a) 60 km/h
   No, the rate of change is the same between any two points.
   b) The average distance travelled per hour, or speed

564 ANSWERS

b) Car A = 15 m/s; Car B = 10 m/s;
Car C = 10 m/s; Car D = –5 m/s

12. a)

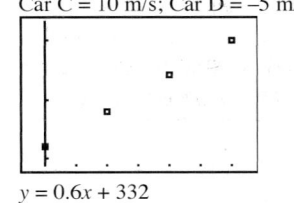

$y = 0.6x + 332$

b) i) 341 m/s **ii)** 362 m/s
iii) 330.2 m/s

13. a) Yes **b)** Yanxia
c) Alea: $y \doteq 15.2 + 0.3x$
Yanxia: $y = 20 + 0.5x$
d) The line of best fit is perfect for Yanxia and a very good fit for Alea.
e) Alea: about 0.3 lbs/month
Yanxia: about 0.5 lbs/month

14. a) $y \doteq 101.4x - 4.6$
b) 401 km **c)** 2429 km

16. a) –0.15°C/min
b) Bedroom Temperature graph showing Higher Initial Temperature and Original Temperature lines.

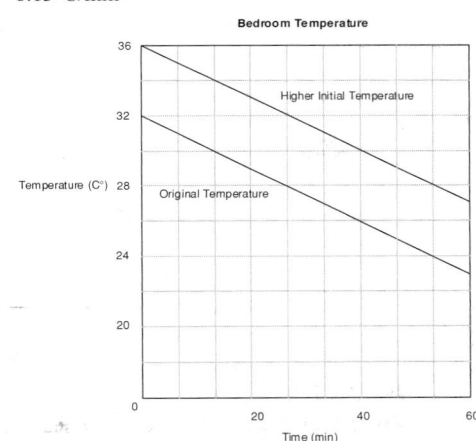

c) Bedroom Temperature graph showing A/C on Low and Original Temperature lines.

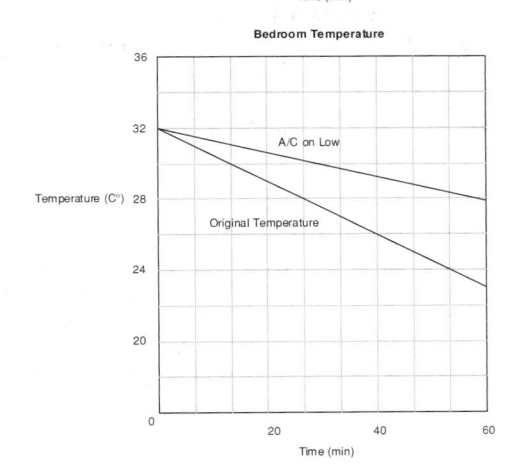

17. Price of Gasoline (Jan. to Dec.) graph.

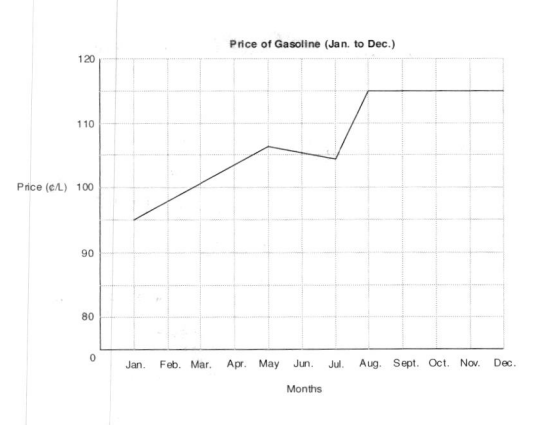

5.4 Quadratic Models, page 303

1. Part b
2. Parts a and c
3. Part b
4. a) vi
b) v
c) ii
d) iii
e) i
f) iv
5. a) True
b) False
c) False
d) True
e) False
6. a) The height increases from 0 s to 0.4 s and decreases from 0.4 s to 0.88 s.
b) The height changes rapidly from 0 s to 0.3 s and 0.6 s to 0.88 s and slowly from 0.3 s to 0.5 s.
7. a) The revenue increased rapidly and then more slowly from 0 to 1750 T-shirts; decreased slowly and then more rapidly from 1750 to 3500 T-shirts.
8. a) r
b) h
c) Volume of a Cylinder graph.

ANSWERS **565**

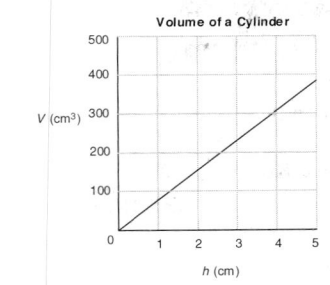

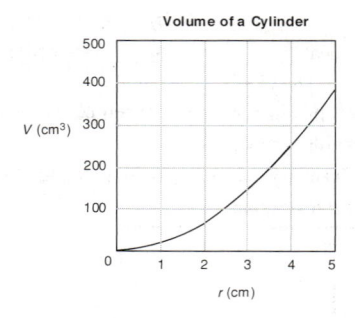

9. **a)** s
b) h

10. a) The consumption of Car A is reasonably constant, with variations between 8 L/100 km and 10 L/100 km. The consumption of Car B is high at low speeds, drops dramatically to about 6 L/100 km at moderate speeds, and begins to rise at greater speeds.
b) Fuel efficiency increases as fuel consumption decreases.
Car A: The fuel consumption decreased rapidly between 24 km/h and 40 km/h, increased rapidly between 100 km/h and 120 km/h, and increased at a constant rate between 100 km/h and 120 km/h.
Car B: The consumption decreased most rapidly between 10 km/h and 20 km/h, increased most rapidly between 80 km/h and 100 km/h, and increased at a constant rate between 100 km/h and 120 km/h.
c) i) Car B
ii) Car A

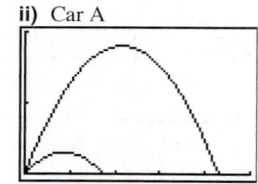

11. a) [graph]
b) Right hand: about 1.8 m; left hand: about 0.29 m
c) Right hand: about 0.86 s; left hand: about 0.35 s
d) Right hand

12. b) $y \doteq -90.5x^2 + 3109x - 6218$
c) About $20 475
d) About $5600

13. b) $y \doteq 1437x^2 - 5\,736\,296x + 5\,723\,797\,471$
c) About 152 695 males; lowest number of males registered in the apprenticeship programs
d) About 254 374 males

14. a) $y \doteq 0.0009x^2 - 3.6x + 3507.6$
b) i) $1.06
ii) $4.04

c) The 1987 estimate; interpolation

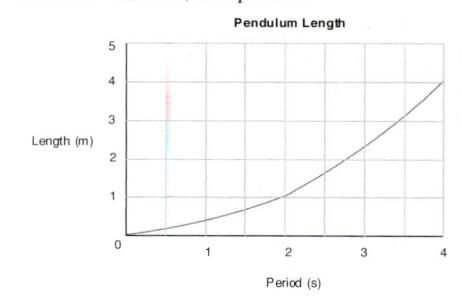

16. a) [graph: Pendulum Length]
b) The relationship between length and time is quadratic.
c) The graph would remain quadratic but open up more slowly.
d) There is a linear relationship between l and g.

Chapter 5 Mid-Chapter Review, page 308

1. a) Quadratic
b) Linear

2. a) Discount increases with order size, first rapidly and then more slowly.
b) Order size should be greater than 150 for a discount of 8%.

3. The hockey player's plus/minus score was constant from games 1 to 4, changed from games 4 to 15 and was zero in games 1-4, 8, 11 and 15.

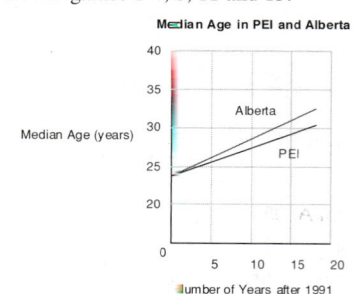

4. a) [graph: Median Age in PEI and Alberta]
The graphs both start at a median age of 24 in 1991. The graph for Alberta rises more quickly.
b) The graph would start at 30 instead of 24.

5. a) r
b) h

6. b) $y \doteq -342.9x^2 + 1\,374\,637x - 1\,377\,338\,500$

5.5 Exponential Models, page 315

1. a) 1.03
b) 1.05
c) 1.12

2. a) Linear
b) Exponential

    c) Exponential
    d) Linear
3. Parts a and b
4. Parts a and c
5. Part c
6. a)

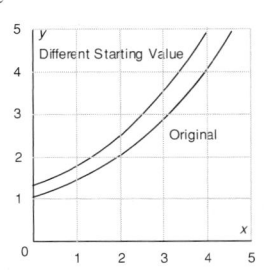

(graph showing "Different Starting Value" curve above "Original" curve)

b)

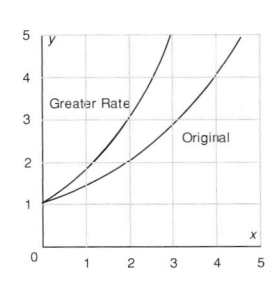

(graph showing "Greater Rate" curve above "Original" curve)

c)

(graph showing "Original" curve above "Lesser Rate" curve)

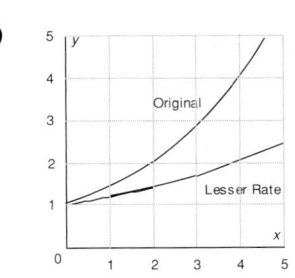

7. a) Plan A: $50
       Plan B: $40
   b) Plan A: linear
       Plan B: exponential
8. a) Both graphs start at 250 μg/mL. The graph for the patient with kidney disease decayed more slowly.

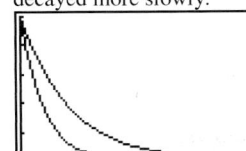

(decay graph)

   b) The new graph would have a y-intercept of 1000 μg/mL, compared with a y-intercept of 250 μg/mL for the original graph. Also, the new graph would decrease at a faster rate than the original graph.

9. a) t **b)** A_0
10. a) Linear   **b)** Exponential
11. a) $y \doteq 18.8(1.47)^x$
   b) a = the number of initial cell phone subscribers
       b = the growth rate
12. a) The charge decreased with time, quickly at first, and then more slowly.

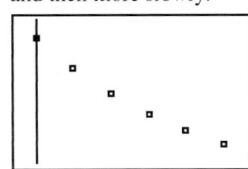

(scatter plot)

   b) $y \doteq 99.95(0.000\ 047\ 5)^x$

(curve fit graph)

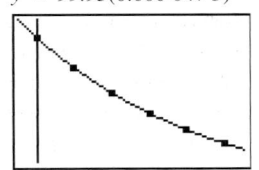

   c) i) About 61 μC   **ii)** About 14 μC
13. a) The volume of the dough increases steadily with time.

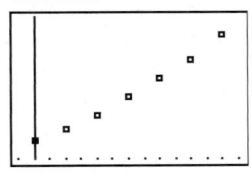

(scatter plot)

   b) $y \doteq 1.5(1.013)^x$

(curve fit graph)

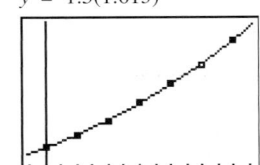

   c) i) About 2.6 L   **ii)** About 4.7 L
15. i) Set n and r constant.
   ii) Set B and r constant; $r = 2$
   iii) Set B and n constant.

5.6 Selecting a Regression Model for Data, page 323

1. a)

r	C	First Differences
0	0	
5	31	$31 - 0 = 31$
10	63	$63 - 31 = 32$
15	94	$94 - 63 = 31$
20	126	$126 - 96 = 32$

The relationship is linear.

ANSWERS   **567**

b)

x	y	First Differences
0	10	
		12.5 − 10 = 2.5
1	12.5	
		15 − 12.5 = 2.5
2	15	
		17.5 − 15 = 2.5
3	17.5	

The relationship is linear.

2. a)

v	E	First Differences	Second Differences
0	0		
		0.5 − 0 = 0.5	
1	0.5		1.5 − 0.5 = 1.0
		2 − 0.5 = 1.5	
2	2		2.5 − 1.5 = 1.0
		4.5 − 2 = 2.5	
3	4.5		3.5 − 2.5 = 1.0
		8 − 4.5 = 3.5	
4	8		4.5 − 3.5 = 1.0
		12.5 − 8 = 4.5	
5	12.5		

The relationship is quadratic.

b)

x	y	First Differences	Second Differences
1	23		
		55 − 23 = 32	
3	55		48 − 32 = 16
		103 − 55 = 48	
5	103		64 − 48 = 16
		167 − 103 = 64	
7	167		

The relationship is quadratic.

3. a)

x	y	Growth Factors
0	0.3	
		$\frac{1.5}{0.3} = 5.0$
1	1.5	
		$\frac{7.5}{1.5} = 5.0$
2	7.5	
		$\frac{37.5}{7.5} = 5.0$
3	37.5	

The relationship is exponential.

b)

x	y	Decay Factors
10	54	
		$\frac{36}{54} \doteq 0.67$
15	36	
		$\frac{24}{36} \doteq 0.67$
20	24	
		$\frac{26}{24} \doteq 1.08$
25	26	

The relationship is not exponential.

4. a) Exponential
   b) Linear
   c) Quadratic
   d) Quadratic
   e) Linear
   f) Exponential
5. Exponential
6. a) Exponential
   b) Quadratic
   c) Linear
   d) Exponential
7. a) Quadratic
   b) Yes
8. a) Exponential
   b) Yes
9. a) Linear
   b) Quadratic
10. a) Linear: $y \doteq 0.063x + 0.53$
    Quadratic: $y \doteq -0.0006x^2 + 0.153x - 2.49$
    Exponential: $y \doteq 1.92(1.013)^x$
    b) Quadratic
12. a) Linear: $y \doteq 0.369x + 1.71$
    Quadratic: $y \doteq 0.001x^2 + 0.019x + 4.04$
    Exponential: $y \doteq 3.37(1.05)^x$
13. a)

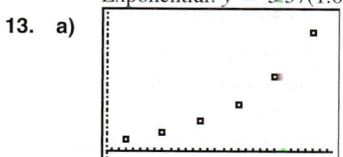

    b) Linear: $y \doteq 5.5x - 31.13$
    Quadratic: $y \doteq 0.25x^2 - 2.95x + 25.2$
    Exponential: $y \doteq 8.76(1.10)^x$
    c) Exponential

14. a) Linear: $y \doteq -5.9x + 20.8$
Quadratic: $y \doteq 2.23x^2 - 14.83x + 25.23$
Exponential: $y \doteq 25.2(0.442)^x$
b) Exponential
c) i) About 0.2°C
ii) 0°C

16. a) [scatter plot]
b) Quadratic: $y \doteq 0.000\,72x^2 + 0.07x + 14.8$
c) The data showed an increase followed by a decrease, which models a quadratic relation.
d) 14.1 h
f) [graph: Hours of Daylight in Dryden, showing Data, Actual, and Quadratic curves vs. Days after May 2]

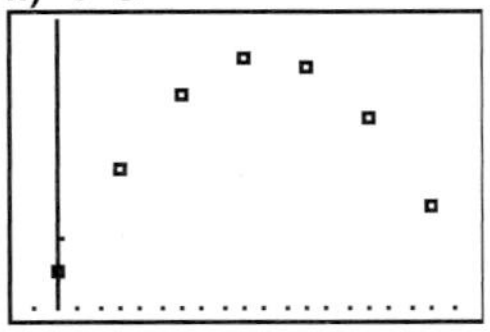

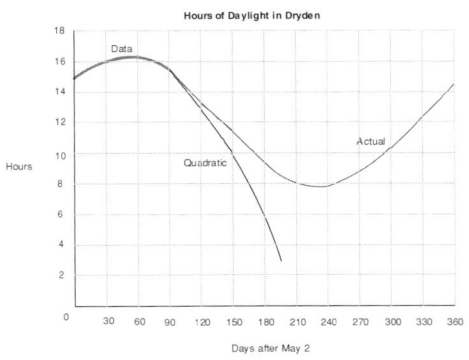

17. a) The price increased steadily with length. 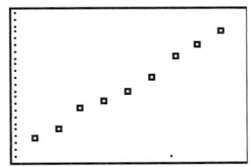
[scatter plot]
b) Linear: $y \doteq 20.75x - 229.47$
Quadratic: $y \doteq 0.76x^2 - 0.66x - 79.77$
Exponential: $y \doteq 0.32(1.445)^x$

Chapter 5 Review, page 332

1. a) The population is constant with time.
b) The population increased with time, rapidly at first, then more slowly.
c) The population increased steadily with time.
d) The population increased with time, slowly at first, then more rapidly.

2. a) Energy consumption was fairly constant from 1990 to 1992; decreased from 1992 and 1996; remained steady through 2000; decreased since, then appeared to have levelled off.
b) 1992
c) About 500 kWh/year

3. a) i) Part a
ii) Part c
iii) Parts b and d
b) People/year

4. a) The rate of change is zero from 0 s to 3 s, then constant and positive from 3 s to 6 s, then constant and negative from 6 s to 8 s.
b) The rate of change is negative at start; positive from November to January; negative from January to May; positive until September

5. a) $1/year
b) The increase in Zoltan's hourly wage with every year of experience with the company
c) No

6. Part a

7. a) Lunch: 15
Appetizers: 10
b) The cost for each additional person attending the event
c) The flat rate cost involved is $100 for both events.

8. a) Consumption increased steadily. 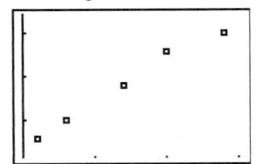
[scatter plot]
b) $y \doteq 0.96x + 22.3$, with $x = 0$ representing 1990
c) About 41.5 g/day

9. Parts a and d

10. b) Linear
d) Quadratic

11. a) Diameter increased slowly at first, and then more quickly. 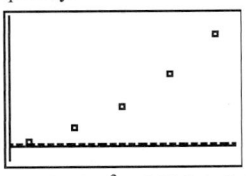
[scatter plot]
b) $y \doteq 3.141x^2 + 0.026x - 0.40$
c) About 46 757 cm^2

12. a) Both isotopes of Uranium start at 100%.
b) U-238

ANSWERS   **569**

13. a) The amount increased at a constant rate.

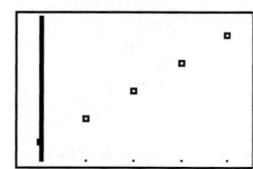

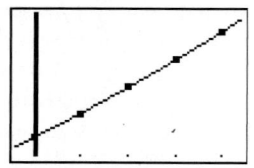

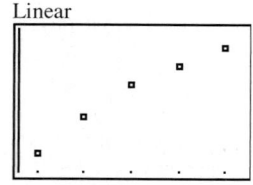

   b) $y \doteq 2500(1.06)^x$

   c) About $4477.11
14. a) Linear
    b) Exponential
    c) Quadratic
    d) Quadratic
    e) Linear
    f) Linear
15. a)

    b) Quadratic
    c) $y \doteq -9.7x^2 + 127.3x + 247.6$
    d) 1995: about –$631 000.00
       2000: about $248 000.00
       2010: about $549 000.00
16. a)

    b) Based on $x = 0$ representing 2000
       Linear: $y \doteq 8.2x + 41.2$
       Quadratic: $y \doteq 0.857x^2 + 4.771x + 42.914$
       Exponential: $y \doteq 42.2(1.153)^x$
    c) Quadratic or exponential
    d) Quadratic:
       2010: About 176 million tonnes
       2020: About 481 million tonnes
       Exponential:
       2010: About 177 million tonnes
       2020: About 736 million tonnes

Chapter 5 Practice Test, page 335

1. C
2. D
3. About 1.72 cm/month
4. i) The savings would grow more slowly.

   ii) The savings would fall more quickly after 6 years. The rate of change would be greater.

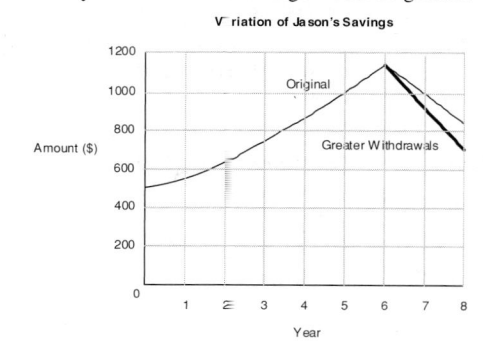

   iii) The initial value on the graph would be greater.

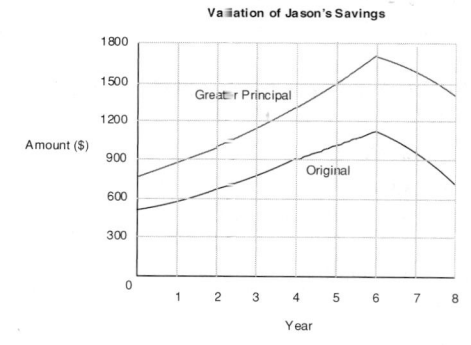

5. a) $y \doteq 2.58x + 15.6$, where x is the number of years since 1990
   b) About 60 000 t
6. a) Plan A: Weight decreased most rapidly in the first 2 months, then decreased more slowly, then increased slowly.
   Plan B: Weight showed a rapid decrease in the first 2 months, then decreased more slowly, then increased slowly.
   Plan C: Weight decreased in the first 2 months,

then increased slowly, then increased more rapidly.

Chapter 6 Algebraic Models

Activate Prior Knowledge

Square Roots, page 338

1. a) 7
   b) −8
   c) About 3.16
   d) −9
   e) About 5.29
   f) −3
   g) 12
   h) About 1.60
2. a) $81 = 9^2$, $82 \doteq 9.06^2$
      9.06 is not an integer.
   b) 1, 1; 4, 2; 9, 3; 16, 4; 25, 5; 36, 6; 49, 7; 64, 8; 81, 9; 100, 10; 121, 11; 144, 12
3. 0.9 s

Solving Linear Equations, page 339

1. a) $x = 7$
   b) $x = 18$
   c) $x = 3$
   d) $x = -7$
2. a) $x = 2$
   b) $x = 4$
   c) $x = -2$
3. 2 km

Evaluating Powers with Integer Exponents, page 340

1. a) 8
   b) 64
   c) 25
   d) $\frac{1}{9}$
   e) 1
   f) $\frac{1}{8}$
   g) $-\frac{1}{7}$
   h) $\frac{25}{9}$
2. a) 0.70
   b) 0.24
   c) 243.33
   d) 583.18
   e) 0.09
   f) 1.73

3. a) $3^2 = 3 \times 3 = 9$; $2^3 = 2 \times 2 \times 2 = 8$
   b) $4^3 = 4 \times 4 \times 4 = 64$;
      $(-4)^3 = (-4) \times (-4) \times (-4) = -64$
   c) $5^2 = 5 \times 5 = 25$; $5^{-2} = \frac{1}{5 \times 5} = \frac{1}{25}$

6.1 Using Formulas to Solve Problems, page 346

1. a) 40 m^2
   b) 48 cm^2
   c) 39.9 m^2
   d) 60.48 cm^2
2. a) 20 g/cm^3
   b) About 6.43 g/cm^3
   c) 3 kg/L
   d) About 1.85 kg/L
3. a) 349.5 m/s
   b) 322.5 m/s
   c) 337.5 m/s
   d) 316.5 m/s
4. a) 25°C
   b) 100°C
   c) 10°C
   d) −20°C
5. a) About 222.22 kPa
   b) About 2666.67 kPa
   c) About 611.11 kPa
   d) About 370.37 kPa
6. About 99 cm^3
7. a) 15 games
   b) No
8. a) $1700
   b) $2700
9. a) About 76.18 m^2
   b) $209.94
10. a) About 0.77 m^2
    b) About 15 mg
11. 230 full cones
12. 6300 t
13. a) About 33.51 cm^3
    b) No
    c) Price proportional to volume: $5.00
14. a) 450 000 L
    b) 2025 g
    c) About $10.00
15. About 0.36 m^3
17. a) $\frac{1}{3}$
       The dose increases as age increases.
    b) Yes
18. a) 2
    b) 2

ANSWERS 571

c) 2
$V + F - E = 2$

6.2 Rearranging Formulas, page 354

1. a) $L = A - E$ b) $E = A - L$
2. a) $R = P + C$ b) $C = R - P$
3. a) $b = \dfrac{A}{h}$ b) $h = \dfrac{A}{b}$
4. a) $M = DV$ b) $V = \dfrac{M}{D}$
5. a) $s = 32$ b) $a = 63$
6. a) 35 h   b) 10.5 h
   c) 42 h
7. a) $15.50/h
8. a) 1. Multiply by 0.6. 2. Add 331.5.
      T → S
      2. Divide by 0.6. 1. Subtract 331.5.
   b) $T = \dfrac{1}{0.6}(S - 331.5)$
   c) i) 20°C
      ii) 7.5°C
      iii) −5°C
9. 9.5 in.
10. Up to 10 pots
11. a) 66 words/min   b) 330 words
12. a) About 3.14 h, or 3 h 9 min
13. a) $t = 5$ cm   b) $l = 4$ m
14. About 12.4 cm
15. About 12 m/s
16. a) $r = \sqrt{\dfrac{V}{\pi h}}$ b) About 2 m
    c) About 30 cm
18. a) About 93 km/h   b) About 10 m
19. a) $\cos A = \dfrac{b^2 + c^2 - a^2}{2bc}$
    b) About 78°

6.3 Laws of Exponents, page 362

1. a) 2^7
   b) $\dfrac{1}{3^3}$
   c) $(1.05)^1$
   d) c^9
   e) $\dfrac{1}{2^8}$
   f) a^3
2. a) 4^3
   b) $\dfrac{1}{5^4}$
   c) $(1.02)^3$
   d) d^4
   e) $(-3)^{11}$
   f) h^{10}
3. a) 5^6 b) $\dfrac{1}{3^{10}}$
   c) $\dfrac{1}{(-2)^{12}}$ d) m^{20}
   e) r^{20} f) a^9
4. a) 10 000   b) 1
   c) $\dfrac{1}{9}$ d) $\dfrac{1}{8}$
   e) $\dfrac{16}{81}$ f) 25
5. a) 19 683   b) $\dfrac{1}{16}$
   c) $\dfrac{1}{16}$ d) −16
   e) 4   f) $\dfrac{8}{125}$
   g) About 3.73   h) −1
6. a) d^3 b) $\dfrac{1}{x^{10}}$
   c) c^{14} d) z^{18}
   e) $\dfrac{1}{n^2}$ f) $\dfrac{1}{w^2}$
   g) s^{12} h) $\dfrac{1}{t^{26}}$
7. a) 3200   b) 400
   c) 50
8. a) 4500 transistors   b) 1125 transistors
   c) 1 207 959 552 000 transistors
9. a) $\dfrac{1}{16}$ b) $\dfrac{1}{125}$
   c) $\dfrac{1}{8}$ d) 9
10. a) 8   b) $x^3 = 8$
11. c) $(a \times b)^n = a^n b^n$
    d) i) $16f^4$ ii) $a^{12}b^4$
       iii) $s^{-15}v^{20}$ iv) $\dfrac{1}{25h^2}$
12. a) 288   b) $-\dfrac{500}{3}$
    c) −120   d) $-\dfrac{177\ 147}{10}$
13. a) −15 360
14. a) $V = 28\pi x^3$
    b) $V \doteq 10\ 996$ cm^3; $V \doteq 152\ 003$ cm^3
16. b) About 81%
17. a) $P = \dfrac{A}{(1+i)^n}$ b) $P = A(1+i)^{-n}$
    c) $747.26

6.4 Patterns in Exponents, page 369

2. a) 3   b) 7
   c) 8   d) 3
   e) −2   f) 10

572   ANSWERS

3. a) i) 5 ii) 25
        iii) 125 iv) 625
        v) 3125
    b) The answers in part a are consecutive powers of 5.
    c) $25^{\frac{6}{2}} = 15\,625$; $25^{\frac{7}{2}} = 78\,125$; $25^{\frac{8}{2}} = 390\,625$
4. b) $100^{\frac{3}{2}} = (100^{\frac{1}{2}})^3$; $100^{\frac{5}{2}} = (100^{\frac{1}{2}})^5$; $100^{\frac{7}{2}} = (100^{\frac{1}{2}})^7$
    c) 1000, 100 000, 10 000 000
5. a) $\sqrt[5]{32} = 2$
    b) $\sqrt[4]{81} = 3$
    c) $\sqrt{16^3} = 64$
    d) $\sqrt{9^5} = 243$
    e) $\sqrt{100^3} = 1000$
    f) $\sqrt[4]{16^3} = 8$
    g) $\sqrt[3]{8^4} = 16$
    h) $\sqrt[3]{27^3} = 27$
6. b) About 1.4
    c) $2^{\frac{3}{2}}$ is between 2 and 4.
        $2^{\frac{5}{2}}$ is between 4 and 8.
        $2^{\frac{7}{2}}$ is between 8 and 16.
    d) 2.8; 5.7; 11.3
7. a) About 93%
    b) About 81%
    c) About 91%

Chapter 6 Mid-Chapter Review, page 371

1. a) 6 m^2
    b) 1598 cm^2
3. a) 66 m^2
    b) $47.97
4. $93 374.73
5. a) 1. Multiply by 3. 2. Add 5.

        x → y

        2. Divide by 3. 1. Subtract 5.
    b) $x = \dfrac{y-5}{3}$
6. a) $d = 225$ m
    b) $a = 8$ m/s^2
    c) $t = 64$ s
7. 5.25%
8. a) 9
    b) -9
    c) $-\dfrac{1}{9}$
    d) $\dfrac{1}{9}$
9. a) p^2
    b) $\dfrac{1}{p^5}$
    c) $\dfrac{1}{p^{10}}$
10. a) $\dfrac{59\,049}{16}$
    b) 291 600
    c) 2916
11. a) $4^{\frac{5}{2}}, 4^3, 4^{\frac{7}{2}}$
    b) 1, 2, 4, 8, 16, 32, 64, 128

12. a) 4
    b) 4
    c) 125

6.5 Rational Exponents, page 376

1. a) 6
    b) 9
    c) 12
    d) 0.5
2. a) 2
    b) 4
    c) 0.3
    d) -5
3. a) $64^{\frac{1}{2}}$
    b) $1.21^{\frac{1}{2}}$
    c) $216^{\frac{1}{3}}$
    d) $(-343)^{\frac{1}{3}}$
4. a) 8
    b) 1.1
    c) 6
    d) -7
5. a) 2
    b) 0.4
    c) -3
    d) 2
6. a) $\sqrt[5]{243} = 3$
    b) $\sqrt{9^3} = 27$
    c) $\sqrt[3]{8^5} = 32$
    d) $\sqrt[4]{81^3} = 27$
    e) $\sqrt[4]{0.0625} = 0.5$
    f) $\sqrt[5]{(-32)^3} = -8$
    g) $\sqrt{0.01^3} = 0.001$
    h) $\sqrt[3]{(-27)^4} = 81$
7. a) 64
    b) 64
8. a) $D = 0.099\sqrt[10]{M^9}$
    b) i) About 2.43 L/day
        ii) About 27.54 L/day
        iii) About 0.003 963 L/day
9. Maria's
10. a) 49
    b) 3
    c) 4
    d) 4
    e) $\dfrac{3}{4}$
    f) 125
11. About 10.41%
12. a) 1.5 ha
    b) 13.5 ha
13. About 9.85 cm
14. a) About 0.704 kg
    b) About 0.006 169 kg
    c) About 3.43 kg
16. a) $r = \sqrt{\dfrac{2P}{s^3}} = \left(\dfrac{2P}{s^3}\right)^{\frac{1}{2}}$
    b) $s = \sqrt[3]{\dfrac{2P}{r^2}} = \left(\dfrac{2P}{r^2}\right)^{\frac{1}{3}}$
17. a) About 4.4 m/s
    b) 20 cm

6.6 Exponential Equations, page 384

1. a) $x = 3$
    b) $x = 2$
    c) $x = 7$
    d) $x = \dfrac{3}{2}$
2. a) $x = 15$
    b) $x = 2$
    c) $x = 2$
    d) $x = -12$

ANSWERS **573**

3. **a)** $x = 5$ **b)** $x = 1$
   c) $x = 2$ **d)** $x = \frac{9}{2}$
4. **a)** 6^2 **b)** 2^4
   c) 5^3 **d)** 10^3
5. **a)** $3^x = 3^2$; $x = 2$ **b)** $3^x = 3^{-2}$; $x = -2$
   c) $3^{2x} = 3^4$; $x = 2$ **d)** $3^{x+5} = 3^3$; $x = -2$
6. **a)** $x = 4$ **b)** No
7. **a)** $x = 3$ **b)** $x = \frac{3}{2}$
   c) $x = 2$ **d)** $x = 3$
   e) $x = \frac{3}{2}$ **f)** $x = \frac{-1}{2}$
   g) $x = -3$ **h)** $x = -2$
9. **a)** **i)** 30 is between 2^4 and 2^5; 4 and 5
   ii) 100 is between 5^2 and 5^3; 2 and 3
   iii) 75 is between 3^3 and 3^4; 3 and 4
   iv) $\frac{1}{5}$ is between 2^{-2} and 2^{-3}; -2 and -3
   b) **i)** $x = 4.91$ **ii)** $x = 2.86$
   iii) $x = 3.93$ **iv)** $x = -2.32$
10. **a)** $x = 2$ **b)** $x = -5$
    c) $x = -2$ **d)** $x = 1$
    e) $x = \frac{-9}{5}$ **f)** $x = -7$
12. $x = 1.8$
13. **a)** $x = 2.40$ **b)** $x = 2.37$
    c) $x = 1.70$ **d)** $x = 5.17$
    e) $x = 2.60$ **f)** $x = 11.90$
14. **a)** Approximate **b)** About 1.68
15. **a)** $6400 = 100(2)^t$ **b)** $t = 6$
16. **a)** **i)** $x = 0$ **ii)** $x = 1.2$
    c) $x = 1.2$

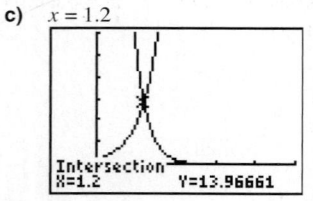

18. **a)** 2010
19. **a)** 1 min, 9°C

6.7 Applications of Exponential Equations, page 391

1. **a)** $\frac{1}{2} = (0.84)^t$ **b)** $\frac{3}{4} = (0.84)^t$
   c) $\frac{3}{8} = (0.84)^t$ **d)** $\frac{7}{8} = (0.84)^t$
2. **a)** 4 years **b)** 2 years
   c) 6 years **d)** 1 year
3. **a)** $2.5 = 2.4(1.017)^t$ **b)** $2.7 = 2.4(1.017)^t$
   c) $3 = 2.4(1.017)^t$
4. **a)** About 1990 **b)** About 1994
   c) About 2001

574 ANSWERS

5. **a)** Part ii **b)** Part iii
   c) Part i
6. **a)** **i)** $y = 1$ **ii)** $y = 8$
   b) **i)** $x = 3$ **ii)** $x = 1$
7. **a)** 100 bacteria **b)** 2 h
8. **a)** $600 = 500(1.08)^n$ **b)** $1000 = 500(1.08)^n$
   c) $1500 = 500(1.08)$
9. **a)** $50 = 100(0.98)^n$
   b) 34 washings
10. **a)** $2000 = 100(1.4)^t$
    b) About 9 years

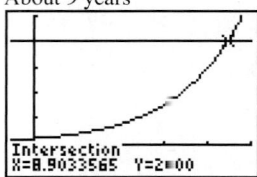

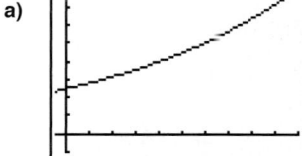

11. **a)**
    b) **i)** About 3 years
    ii) About 12 years
    iii) About 15 years
12. **a)** Decreased
    b) 2010
    c) 2021
13. **a)** 8.7 min
    c) About 17 min
    d) Decrease
14. **a)** About 42.04 kPa
    b) About 8 km
    c) Part b
15. **a)** $A = 500(0.5)^{\frac{t}{12.3}}$
    b) About 82 years
16. About 1300 years old
17. About 3 h
19. **a)** About 56.24 years
    b) $P = 29.6(2)^{\frac{t}{56.24}}$
    c) 2020

Chapter 6 Review, page 400

1. $3300
2. $334.32
3. $1.40
4. **a)** 17 Wh
   b) About 233.33 Wh

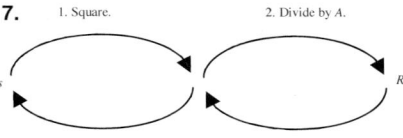

    c) 13.7 h
5. a) $d = \dfrac{C}{\pi}$
    b) About 2.55 m
6. About 6.94 Ω
7. 1. Square. 2. Divide by A.
    2. Take the square root. 1. Multiply by A.
    a) $s = \sqrt{RA}$
    b) 9 ft.
8. a) $P = \dfrac{A}{1+rt}$
    b) $1702.13
9. a) −25
    b) $\dfrac{1}{25}$
    c) 25
10. a) $10^5 = 100\,000$
    b) $7^8 = 5\,764\,801$
    c) $1.12^6 \doteq 1.97$
    d) $2^0 = 1$
11. a) kl
    b) a^2
    c) $\dfrac{1}{x^2 y^4}$
    d) $\left(\dfrac{s}{n}\right)^7$
12. a) $\dfrac{1}{125}$
    b) −243
    c) $\dfrac{9}{25}$
    d) 729 000
13. a) 4
    b) −3
    c) 2
    d) 32
    e) 125
    f) 0.729
14. a) About 2.94 mg
    b) About 2.47 mg
    c) About 0.88 mg
15. a) 10
    b) 4
    c) 81
    d) 4
16. About 4.5%
17. a) $B = 0.4089(\sqrt[4]{M})^3$
    b) i) About 1.26 m³/day
        ii) About 1.95 m³/day

    c) About 11.8 kg
    d) No
18. a) $2^2, 2^4, 2^5$
    b) $5^{-1}, 5^2, 5^3$
    c) $3^2, 3^4, 3^5$
19. a) 3   b) 4
    c) $\dfrac{1}{2}$ d) 4
    e) 9   f) 8
21. a) 2.1
    b) 1.6
    c) −2.9
    d) 15.9
    e) 7.4
23. a) $1 = 300(0.8)^t$
    b) About 25.56 h
24. About 17 200 years old
25. a) $7549.74 = 36\,000(0.8)^n$
    b) About 7 years old

Chapter 6 Practice Test, page 403

1. B
2. D
3. a) i) $\dfrac{-1}{8}$
       ii) $\dfrac{5}{3}$
       iii) 27
    b) $\dfrac{25}{8}$
5. a) 32 000 represents the population in 1990. 1.09 represents an increase in the population by 9% each year.
    b) 2004
6. a) About 0.63 g
    b) About 99 kg
    c) Yes

Chapter 7 Annuities and Mortgages

Activate Prior Knowledge

Simple and Compound Interest, page 406

1. $4584.38
2. $11 596.93
3. a) Yes   b) No
    c) No

Present Value, page 407

1. $5365.95
2. $2387.21

3. If the final amount is given and you need to find the amount invested then solve for the present value. If the initial value is given and you need to find the final value or earnings then solve for the amount.

7.1 The Amount of an Annuity, page 415

1. a)

Year	Starting balance	Interest earned	Deposit	Ending balance
1	$0.00	$0.00	$1000.00	$1000.00
2	$1000.00	$80.00	$1000.00	$2080.00
3	$2080.00	$166.40	$1000.00	$3246.40
4	$3246.40	$259.71	$1000.00	$4506.11

b)

Month	Starting balance	Interest earned	Deposit	Ending balance
1	$0.00	$0.00	$100.00	$100.00
2	$100.00	$0.50	$100.00	$200.50
3	$200.50	$1.00	$100.00	$301.50
4	$301.50	$1.51	$100.00	$403.01

2. a) $i = 0.03; n = 7$ b) $i = 0.045; n = 24$
   c) $i = 0.006; n = 32$ d) $i = 0.015; n = 60$
3. a) $630.50 b) $8922.80
   c) $9388.46
4. a) N: 7; I%: 3; P/Y: 1; C/Y: 1
   b) N: 24; I%: 9; P/Y: 2; C/Y: 2
   c) N: 32; I%: 2.4; P/Y: 4; C/Y: 4
   d) N: 60; I%: 18; P/Y: 12; C/Y: 12
5. a) $300.00   b) Monthly
   c) 6   d) 8%
   e) Monthly   f) $1830.27
6. a) $41 449.34 b) $14 767.57
   c) $69 362.25 d) $32 302.36
7. a) $11 449.34 b) $4367.57
   c) $17 162.25 d) $5302.36
9. No, Shen Wei will have saved only $9636.38.
10. a) $37 086.78 b) $15 486.78
    c) $3712.06
11. a) $3818.16
    b) i) Yes   ii) No
       iii) No
    c) Part iii
       Part i gives an amount of $7636.32, part ii gives $4053.56, and part iii gives $8122.26.
12. a) Both Jackson and Abina invested $14 400, but Jackson's annuity has an amount of $40 180.60 and Abina's annuity has an amount of $23 265.50.
    b) Jackson's: $25 780.60
       Abina's: $8865.50
    c) Financial planners recommend saving for retirement from an early age because more interest is earned that way, therefore less investment is required to save a given amount.
13. a) Annuity 1; deposit: $6000; amount: $6629.90
       Annuity 2; deposit: $6000; amount: $6605.70
    b) The deposits are made more frequently and interest is calculated more often.
14. a) Part iii
    b) i) Deposit: $10 000; amount: $10 724.93
       ii) Deposit: $5000; amount: $5750.74
       iii) Deposit: $10 000; amount: $11 731.39
       iv) Deposit: $10 000; amount: $10 825.40
16.

Annuity Growth for 30 Years

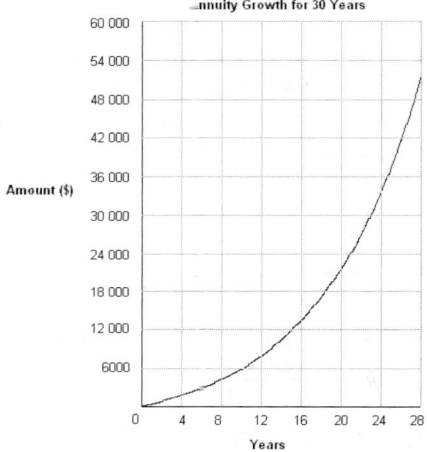

    b) Yes
17. a) Kishore will have a greater amount because he started investing sooner so his money had more time to collect interest.
    b) Kishore: investment: $20 000;
              amount: $95 760.96
       Giselle: investment: $60 000;
              amount: $38 988.05
    c) Yes;
       Kishore has saved far more money than Giselle, even though he invested less.

7.2 The Present Value of an Annuity, page 423

1. a) 851.13
   b) 25 669.07
   c) 26 681.04
   d) 471.93
2. a) $544.65   b) $5206.37

    c) $7931.51
3. a) $350   b) Monthly
    c) 32 withdrawals   d) $10 678.37
4. a) $6373.02   b) $23 589.62
    c) $5936.30   d) $13 081.15
6. a) $468.87
    b) $4345.77
    c) $105 628.79
    d) $805.12
7. a) $281.13
    b) $2854.23
    c) $106 871.21
    d) $94.88
8. $11 495.78
9. $205 218.51
10. Isabel should choose the plan that receives monthly payments because she will earn more money over the ten years.
11. a) $13 419.22   b) Part ii
12. a) $2499.92   b) $206.80
13. No; Becky's annuity had more compounding periods, therefore it earned more interest and had a lower present value than Angela's annuity.
15. $35 714.48
16. a) $1356.45
    b) $720.34
    c) $49.70

7.3 The Regular Payment of an Annuity, page 430

1. a) $R = \dfrac{Ai}{[(1+i)^n - 1]}$
    b) $R = \dfrac{PVi}{[1 - (1+i)^{-n}]}$
2. a) 456.51
    b) 180.49
3. a) $3500
    b) 21%
    c) $131.86
    d) 36
    e) 3 years
4. a) $299.48
    b) $1481.93
    c) $244.63
    d) $185.05
6. a) Present value
    b) Amount
    c) Amount
    d) Present value
7. a) $1535.13   b) $92 107.80
    c) $17 107.80

8. a) $419.36
    b) i) Yes   ii) No
    iii) No
9. a) $2928.73
    b) $5712.70
10. a) 36-month loan: $579.96; 48-month loan: $455.66
    b) $993.12
11. a) David's: $638.79; Ulani's: $669.07
    b) $1090.08
13. Lincoln should make monthly deposits into the account with 7.8% interest compounded monthly.
14. a) $325.91
    b) $1035.91
    c) $18 040.90
16. $1841.77

Chapter 7 Mid-Chapter Review, page 442

1. $5237.09
2. a) i) $39 087.32
    ii) $136 842.87
    iii) $992 763.04
    b) i) $14 087.32
    ii) $86 842.87
    iii) $892 763.04
3. $476 931.87
5. $197.12
6. a) 7 years: $409.88; 10 years: $324.86
    b) $4553.28
7. a) $2076.26
    b) i) $415.25
    ii) $15.55
8. a)

Payment number	Payment	Interest paid	Principle paid	Outstanding balance
0				$8000
1	$177.96	$80.00	$97.96	$7902.04
2	$177.96	$79.02	$98.94	$7803.11
3	$177.96	$78.03	$99.02	$7703.18
⋮	⋮	⋮	⋮	⋮
58	$177.96	$5.23	$172.72	$350.64
59	$177.96	$3.51	$174.45	$176.19
60	$177.96	$1.76	$176.19	$0.00

    b) $3780.38

7.6 What Is a Mortgage?, page 445

1. a) $38 750   b) $116 250

ANSWERS   **577**

2. **b)** A greater down payment will decrease the amount of interest paid on the mortgage.
3. $7362.50
4. **a)** $4650
5. **a)** Kyle and Tea have borrowed $139 500. They will pay it back over a 25-year period. Every 5 years, they will have to renew their mortgage based on current interest rates.
   c) Weekly, bi-weekly, semi-monthly, monthly

7.7 Amortizing a Mortgage, page 450

1. **a)** $195 000
   b) 5.8%
   c) $1224.54
   d) 300 monthly payments
   e) Nadia makes monthly payments and the interest is compounded semi-annually.
2. **a)** i) $225 000   ii) 240 payments
       iii) $347 596.80   iv) $122 596.80
   b) i) $80 000   ii) 180 payments
       iii) $132 553.80   iv) $52 553.80
3. **a)** $479.73   **b)** $872.41
   c) $1517.34   **d)** $1764.44
4. **a)** $21 351.40   **b)** $111 723.00
   c) $174 161.60   **d)** $346 198.40
5. **a)** $125 000   **b)** $799.76
   c) $617.33   **d)** $183.33
   e) $124 450.00
   f) Total interest is about 3 times as large as the total principal paid in the first 3 payments.
6. **a)** Principal borrowed: $120 000; monthly payment: $1104.62
   b) 15 years
   c) Interest: $738.54; principal: $366.08
   d) $78 830.44
   The amount of interest paid is about $\frac{2}{3}$ of the principal originally borrowed.
   e) $118 894.99
7. **a)** The amount of interest paid is much greater than the amount of principal paid.
   b) The amount of interest paid is much less than the amount of principal paid.
   c) Interest: about 67%; principal: about 33%
   Interest:
   $(738.54 + 736.29 + 734.02) \div (1104.62 \times 3)$
   $\doteq 0.6665$
   Principal:
   $(366.08 + 368.33 + 370.60) \div (1104.62 \times 3)$
   $\doteq 0.3334$

8. **a)** Down payment: $26 250; principal: $148 750
   b) $1134.77
   c) $204 258.60
   The total amount paid is about $55 500 more than the principal originally borrowed.
10. **a)** N = 300
    I% = 6.00
    PV = 185 000.00
    PMT = −1183.64
    FV = 0.00
    P/Y = 12
    C/Y = 2
    b) Interest: $906.91; principal: $276.73
    c) $182 802.34   **d)** About 1.8%
11. **a)** No
    b) 15-year amortization period: $1802.05
    30-year amortization period: $1234.59
12. **b)**

Payment number	Monthly payment	Interest paid	Principal paid	Outstanding balance
0				$200 000.00
1	$1197.60	$873.73	$323.87	$199 676.13
2	$1197.60	$872.32	$325.28	$199 350.85
3	$1197.60	$870.90	$326.70	$199 024.15
4	$1197.60	$869.47	$328.13	$198 696.02
5	$1197.60	$868.04	$329.56	$198 366.46
6	$1197.60	$866.60	$331.00	$198 035.46

13. **c)** $2343.00

7.8 Using Technology to Generate an Amortization Table, page 459

1. **a)** $800.91   **b)** $1068.68
   c) $1974.69
2. **a)** $843.33   **b)** 0.412%
   c)

Payment number	Monthly payment	Interest paid	Principal paid	Outstanding balance
0				$145 000.00
1	$843.33	$597.97	$245.36	$144 754.64
2	$843.33	$596.96	$246.37	$144 508.27
3	$843.33	$595.94	$247.39	$144 260.88
:	:	:	:	:
298	$843.33	$10.34	$832.99	$1674.70
299	$843.33	$6.91	$836.42	$838.28
300	$841.74	$3.46	$838.28	($0.00)

   d) $107 997.41
4. **a)** $1339.65

578 ANSWERS

b)

Payment number	Monthly payment	Interest paid	Principal paid	Outstanding balance
0				$200 000.00
1	$1339.65	$1068.95	$270.70	$199 729.30
2	$1339.65	$1067.50	$272.15	$199 457.15
3	$1339.65	$1066.05	$273.60	$199 183.55
4	$1339.65	$1064.58	$275.07	$198 908.48
5	$1339.65	$1063.11	$276.54	$198 631.94
6	$1339.65	$1061.64	$278.01	$198 353.93

   c) $1646.07
   d) $6391.83

5. a) $1419.24
   b)

Payment number	Monthly payment	Interest paid	Principal paid	Outstanding balance
0				$245 000.00
1	$1419.24	$1160.14	$259.10	$244 740.90
2	$1419.24	$1158.91	$260.33	$244 480.57
3	$1419.24	$1157.68	$261.56	$244 219.01
4	$1419.24	$1156.44	$262.80	$243 956.21
5	$1419.24	$1155.19	$264.05	$243 692.16
6	$1419.24	$1153.94	$265.30	$243 426.86

   c) $6942.30

6. a) $1059.08
   b)

Payment number	Monthly payment	Interest paid	Principal paid	Outstanding balance
0				$168 000.00
1	$1059.08	$624.17	$434.91	$167 565.09
2	$1059.08	$622.56	$436.52	$167 128.57
3	$1059.08	$620.94	$438.14	$166 690.43
⋮	⋮	⋮	⋮	⋮
22	$1059.08	$588.95	$470.13	$158 049.39
23	$1059.08	$587.20	$471.88	$157 577.51
24	$1059.08	$585.45	$473.63	$157 103.88

   c) i) Interest: $14 521.80; principal: $10 896.12
      ii) $157 103.88
      iii) 6.49%
      At this rate, it would take 30 years and 10 months to repay the mortgage. In reality it does not take this long because over time the percentage of the monthly payment contributing to paying off the principal increases.

7.9 Reducing the Interest Costs of a Mortgage, page 466

1. a) $1116.13
   b) $558.07
   c) $279.03

2. a) The regular payment decreases and the total interest paid increases.
   b) The regular payment decreases and the total interest paid decreases.
   c) The regular payment decreases and the total interest paid decreases.

3. Monthly payment: $1145.25
   Total interest: $164 575.00

4. a) Monthly payment: $1274.82
      Interest saved: $37 618.20
   b) Monthly payment: $1118.79
      Interest saved: $7938.00
   c) Weekly payment: $263.79
      Interest saved: $648.00
   d) Accelerated bi-weekly payment: $572.63
      Interest saved: $30 460.92

5. Part a

Chapter 7 Review, page 468

1. a) $23 970.46
   b) $13 177.77
   c) $62 246.69

2. a) $6120.46
   b) $2137.77
   c) $17 246.69

3. b) Yvonne: $8452.49; Teresa: $8387.68

4. a) Carlos: $465 297.31; Renata: $235 065.28
   b) No

5. a) $19 195.82
   b) $51 821.34

6. a) $5304.18
  b) $33 978.66

7. a) $1470.49
   b) $257.51

8. a) Annuity A: $706.56
      Annuity B: $1413.13
      Annuity C: $1263.36
   b) Part i

9. $2158.58

10. a) First annuity payments are $2478.12 quarterly for 6 years at 7.5% per year compounded quarterly and amount to $74 250. Second annuity payments are $4969.91 bi-yearly for 6 years at 7.8% per year compounded semi-annually and amount to $74 250.

b) Tarak should choose the first annuity because he invests less money and ends up with the same future value.

11. a) $90.34
    b) $211.80
12. a) $57.33
    b) $208.98
    c) 6 years and 3 months
13. a)

Payment number	Monthly payment	Interest paid	Principal paid	Outstanding balance
0				$8000
1	$138.32	$50.00	$88.32	$7911.68
2	$138.32	$49.45	$88.87	$7822.81
3	$138.32	$48.89	$89.43	$7733.38
:	:	:	:	:
70	$138.32	$2.56	$135.76	$274.07
71	$138.32	$1.71	$136.61	$137.46
72	$138.32	$0.86	$137.46	$0.00

b) $1959.10
This is about 25% of the amount originally borrowed.

14. RESP
 – saving for education
 – tax deductible
 – save using annuities
 RRSP
 – saving for retirement

15. A closed mortgage with a 30-year amortization period means that it will take 30 years to repay the mortgage. Bi-weekly payments mean that payment will be made every other week. The interest rate is 6% and the compounding period is 6 months. The interest rate will be reevaluated every 5 years.

16. a) Monthly payment: $536.87
       Total interest: $64 061.00
    b) Monthly payment: $820.83
       Total interest: $51 999.20
    c) Monthly payment: $2198.90
       Total interest: $188 802.00
    d) Monthly payment: $1686.08
       Total interest: $307 988.80

17. a) $150 000.00
    b) $959.71
    c) $737.53
    d) $149 340.00
    e) $4428.43
18. $1077.84; $296.61; $779.94; $178 231.91
19. a) $1521.73

b)

Payment number	Monthly payment	Interest paid	Principal paid	Outstanding balance
0				$190 000.00
1	$1521.73	$822.30	$699.43	$189 300.57
2	$1521.73	$819.27	$702.46	$188 598.11
3	$1521.73	$816.23	$705.50	$187 892.61
:	:	:	:	:
10	$1521.73	$794.58	$727.15	$182 867.88
11	$1521.73	$791.43	$730.30	$182 137.58
12	$1521.73	$788.27	$733.46	$181 404.12

c) Total amount: $273 911.40
   Interest paid: $83 911.40

20. a) 24 payments
    b) 26 payments
    c) 52 payments
    d) 26 payments
21. a) $342 743.60
    b) Option IV

Chapter 7 Practice Test, page 471

1. B
2. B
3. $2419.44
4. a) $80 000.00
   b) Principal: $45.22; interest: $589.05
   c) $634.27; $588.72; $45.55
   d) $148 337.20
5. Option A

Chapter 8 Budgets

Activate Prior Knowledge

Fixed and Variable Costs, page 475

1. $2788.78
2. $9152.18
3. a) Fixed expenses: obedience training course (if one time only), annual vet visits and shots, buying the dog and dog licence
   Variable expenses: food, vet visits for illness
   b) Buying the dog and dog licence
4. Fixed expenses: bicycle, bicycle shoes, helmet, running shoes, sunglasses, wet suit, swim goggles, membership in triathlon club, fitness training, and race entrance fees
   Variable expenses: bike maintenance and parts, biking and running clothes, and bathing suits

580 ANSWERS

8.2 The Costs of Owning or Renting a Home, page 484

1. a) $15 000 b) $25 364.56
   c) $656.28 d) $395.40
2. a) Fixed cost b) Variable cost
   c) Fixed cost d) Variable cost
3. a) $738 b) $1159
   c) $370
   d) $285
4. Ashley is responsible for replacing the window.
5. No;
   Paul would have to give his tenants a 90-day written notice after they have rented the place for 12 months.
6. Fixed expenses: mortgage payments, home insurance, property taxes, phone and TV
   Variable expenses: water, sewer, electricity, and natural gas
7. $15 802.54
8. $20 271.28
9. a) About $58.01 b) About $66.77
   c) The plan provides stability in the monthly expenses.
10. a) About $391.54/month; about $395.96/month if they include water heater
    b) Yes, the rent they plan to charge is reasonable, maybe on the low side.
11. a) Option 1: $7620
       Option 2: $7800
    b) Electricity
12. a) Option 1: $8988
       Option 2: $7548
    b) Option 1: electricity
       Option 2: electricity (includes heat) and transportation

Chapter 8 Mid-Chapter Review, page 492

1. a) Electricity, gas, water, TV/cable, phone, internet
   b) Property tax, home insurance, mortgage, maintenance
2. a) Rent, cable, phone, internet, utilities
3. a) $18 032.56 b) $359.40
   c) $9180 d) $855.20
   e) $275.80
4. No
   Karim's landlord would have to give him a 90-day written notice after he has rented the place for 12 months.
5. $14 790
6. a) $4500 b) $1320
   c) Yes;
      Melissa and Farideh would still save $525.

8.4 Designing Monthly Budgets, page 501

1. a) $1615
   b) Tuition, residence, and meals
   c) No
2. a) Expense b) Expense
   c) Income   d) Expense
   e) Expense f) Income
3. a) About $2708.33
   b) Ingrid can save about $308 monthly.
4. a) About $606.67/month
   b) About $4666.67/month
   c) About $137.50/month
   d) About $266.67/month
5. a)

Income	Monthly amount ($)
Student loan ($5000 ÷ 12)	417.00
Part-time job ($260 × 52 ÷ 12)	1127.00
Total income	1544.00
Expenses	
Rent and utilities	550.00
Transportation	65.00
Food ($85 × 52 ÷ 12)	368.00
Cell phone service	25.00
Books and supplies ($1200 ÷ 12)	100.00
Miscellaneous ($250 × 26 ÷ 12)	542.00
Total expenses	1650.00
Balance (Total income − total expenses)	−106.00

   b) No

6.

Income	Monthly amount ($)	
Income	5417.00	Fixed
Investments	75.00	Fixed
Expenses		
Mortgage payment and property tax	1463.00	Fixed
Home insurance	54.00	Fixed
Car insurance	88.00	Fixed
Utility costs	230.00	Variable
Phone/cable	75.00	Fixed
Vehicle lease	410.00	Fixed
Gasoline costs	175.00	Variable
Retirement savings plan	488.00	Fixed
Grocery costs	693.00	Variable
Clothing costs	250.00	Variable
Entertainment costs	520.00	Variable
Charitable donations	83.00	Fixed
Miscellaneous costs	325.00	Variable

ANSWERS **581**

7. a)

Income	Monthly amount ($)		
	Fixed	Variable	Total
Salary	5417.00		5417.00
Investments	75.00		75.00
Total income	5492.00		**5492.00**
Expenses			
Housing			
Mortgage payment/ property tax	1463.00		1463.00
House insurance	54.00		54.00
Utilities		230.00	230.00
Phone/cable	75.00		75.00
Subtotal	1592.00	230.00	**1822.00**
Transportation			
Vehicle lease	410.00		410.00
Gas costs		175.00	175.00
Insurance	88.00		88.00
Subtotal	498.00	175.00	**673.00**
Food			
Groceries		693.00	693.00
Subtotal		693.00	**693.00**
Other			
Clothing		250.00	250.00
Entertainment		520.00	520.00
Charitable donations	83.00		83.00
Miscellaneous expenses		325.00	325.00
Subtotal	83.00	1095.00	**1178.00**
Savings	488.00		488.00
Total Expenses	2661.00	2193.00	**4854.00**
Balance (Total income – total expenses)			638.00

b) Yes

8. No

9. a)

Income	Monthly amount ($)
Part-time job ($425 × 52 ÷ 12)	1842.00
Total income	1842.00
Expenses	
Rent and utilities	650.00
Transportation (bus pass)	80.00
Groceries ($95 × 52 ÷ 12)	412.00
Education (community college course)	100.00
Miscellaneous ($150 × 26 ÷ 12)	325.00
Total expenses	1567.00
Balance (Total income – total expenses)	275.00

b) Yes **c)** Yes

10. a) Each month, Nihal will have $299 left over, even with his savings for education built into the budget.

Income	Monthly amount ($)
Part-time job ($350 × 52 ÷ 12)	1517.00
Total income	1517.00
Expenses	
Rent	450.00
Utilities	120.00
Transportation (bus pass)	60.00
Groceries	320.00
Clothing	30.00
Entertainment	40.00
Education (tuition and equipment) 900 ÷ 8	113.00
Phone/Internet	60.00
Contact lenses	25.00
Total expenses	1218.00
Balance (Total income – total expenses)	299.00

b) He will need to use $1105 of his savings.

Income	Monthly amount ($)
Employment Insurance ($2310 ÷ 3)	770.00
Total income	770.00
Expenses	
Rent	450.00
Utilities	120.00
Transportation (bus pass)	60.00
Groceries	320.00
Clothing	30.00
Entertainment	40.00
Phone/Internet	60.00
Contact lenses	25.00
Total expenses	1105.00
Balance (Total income – total expenses)	−335.00

c) Yes

8.5 Creating a Budget Using a Spreadsheet, page 508

1. B7 calculates the sum of all of the monthly income items. B20 calculates the sum of all of the monthly expense items.

B22 calculate the monthly balance by subtracting total expenses from total income.

3.

Rebecca's Monthly Budget	
Description of income	**Monthly amount ($)**
Employment income	2085.00
Total	2085.00
Description of expenses	**Monthly amount ($)**
Rent and utilities	480.00
Food expenses	325.00
Entertainment	173.00
Transportation	65.00
Miscellaneous	433.00
Savings for education	175.00
Total	1652.00
Money left at end of month	433.00

   a) $175
   b) $1299

4. a)

Rebecca's Monthly Budget	
Description of income	**Monthly amount ($)**
Employment insurance (EI)	1147.00
Total	1147.00
Description of expenses	**Monthly amount ($)**
Rent and utilities	480.00
Food expenses	325.00
Entertainment	173.00
Transportation	65.00
Miscellaneous items	433.00
Total	1477.00
Money left at end of month	−330.00

5. a) $176

Oliver's Monthly Budget	
Description of Income	**Monthly amount ($)**
Employment income	596.00
Total	596.00
Description of Expenses	**Monthly amount ($)**
Room and board	130.00
Cell phone expenses	30.00
Entertainment	108.00
Miscellaneous items	152.00
Total	420.00
Money left at end of month	176.00

   b) No

8.6 Making Decisions about Buying or Renting, page 513

1. $31 850
2. $3475

Chapter 8 Review, page 520

1. a) One-bedroom condo downtown
   b) Two-bedroom apartment/townhouse/house, city/suburbs
   c) Three-bedroom home and double-car garage, suburbs
2. a) Mortgage payment, home insurance, utilities, property taxes, and regular repairs and upgrades
3. a) $10 500
   b) $16 222.44
   c) $537.10
   d) $455.40
4. a) $780
   b) $1040
   c) $350
   d) $380
5. $15 468
6. Yes
10. a) The greatest expense is the mortgage: $1200
    The least expense is the insurance: $100
    b) $800
11. a) Income; $130
    b) Expense; about $333.33
    c) Expense; $97.50
    d) Income; about $67.92
12. a)

Income	Monthly amount ($)
Take-home pay	2400.00
Total income	2400.00
Expenses	
Rent and utilities	975.00
Phone plan	50.00
Car payment	325.00
Gas ($50 × 52 ÷ 12)	217.00
Food ($120 × 52 ÷ 12)	520.00
Car and contents insurance ($30 × 52 ÷ 12)	130.00
Entertainment ($170 × 26 ÷ 12)	368.00
Miscellaneous	200.00
Total expenses	2785.00
Balance (Total income − total expenses)	−385.00

   b) No

ANSWERS   **583**

13. a)

Phillip and Teresa's Monthly Budget	
Description of income	Monthly amount ($)
Employment income - Phillip	2752.00
Employment income - Teresa	3250.00
Total	6002.00
Description of expenses	**Monthly amount ($)**
Mortgage payment	1380.00
Food and restaurants	943.00
Utilities and services	550.00
Transportation	700.00
Property tax	250.00
Vacation expenses	160.00
Entertainment	390.00
Miscellaneous	500.00
Insurance and RSPs	400.00
Total	5273.00
Money left at end of month	729.00

   b) Yes

14. a)

Phillip and Teresa's Monthly Budget	
Description of income	Monthly amount ($)
Employment income – Phillip	2752.00
Employment income – Teresa	3250.00
Total	6002.00
Description of Expenses	**Monthly amount ($)**
Mortgage payment	1380.00
Food and restaurants	943.00
Utilities and services	550.00
Transportation	700.00
Property tax	250.00
Vacation expenses	160.00
Entertainment	390.00
Miscellaneous	500.00
Insurance and RSPs	400.00
Total	5273.00
Money left at end of month	729.00

Chapter 8 Practice Test, page 523

1. B
2. C
4. a) Option 1: $4200
       Option 2: $5640
   b) Utilities

6. a)

Expenses	Monthly amount ($)
Mortgage	1400.00
Condo fees	275.00
Utilities	350.00
Property tax	208.00
Condo Insurance	47.00
Food	867.00
Day care	975.00
Transportation	50.00
Vacations	125.00
Entertainment	650.00

   b) $53

	Monthly amount ($)
Income	
Income	5000.00
Total income	5000.00
Expenses	
Mortgage	1400.00
Condo fees	275.00
Utilities	350.00
Property tax	208.00
Condo insurance	47.00
Food	867.00
Day care	975.00
Transportation	50.00
Vacations	125.00
Entertainment	650.00
Total expenses	4947.00
Balance (Total income – total expenses)	53.00

   c)

Expenses	Monthly amount ($)
Income	
Income	4000.00
Total income	4000.00
Expenses	
Mortgage	1400.00
Condo fees	275.00
Utilities	350.00
Property tax	208.00
Condo Insurance	47.00
Food	867.00
Daycare	975.00
Transportation	50.00
Vacations	125.00
Entertainment	650.00
Total expenses	4947.00
Balance (Total income – total expenses)	−947.00

Cumulative Review Chapters 1–8, page 530

1. a) $\angle X \doteq 40°$
   b) $\angle L \doteq 41°$
2. a) About 155 cm, about 149 cm

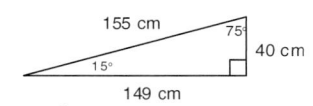

   (triangle: 155 cm hypotenuse, 40 cm vertical, 149 cm base, 75° top, 15° base)

3. a) About 20° or about 160°
   b) About 24° or about 156°
   c) About 157°
   d) About 139°
4. a) $l \doteq 2.57$ m, $k \doteq 0.71$ m; $\angle S = 54°$
   b) $\angle F \doteq 82°$, $\angle G \doteq 49°$, $\angle E = 49°$
5. About 330 km
6. About 464.3 cm^2
7. $SA \doteq 39$ sq. ft.
8. About 20 ft.
9. a) Cube with edge length 7 m
   b) 343 m^3
10. The cylinder has radius about 3.55 in. and height about 7.1 in.
11. a) One-variable data
    b) Two-variable data
12. a) *Price of Airfare versus Weeks before Departure* (scatter plot)

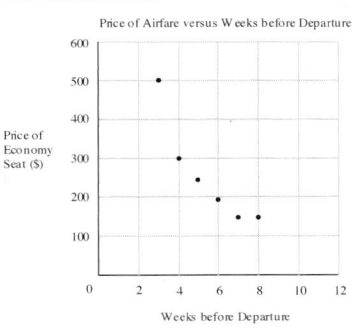

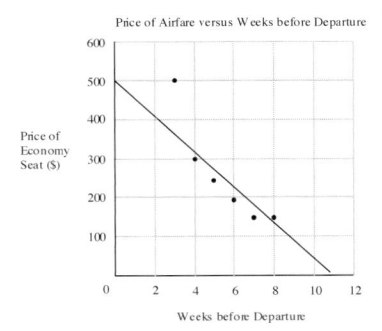

    b) *Price of Airfare versus Weeks before Departure* (scatter plot with line of best fit)

    c) The cost is about $26.
       No, the price is not reasonable.
       The cost would never reach zero after 10 weeks – it is not realistic.
    d) No, the line of best fit does not represent the data.
       The data curve, but the line does not.
       The correlation is non-linear, except for the middle range values (from weeks 4 to 7).
13. Part i
14. a) Zurich, Oslo, Dublin
    b) Copenhagen, Toronto, Tokyo, Rome, Hong Kong, New Delhi
15. a) The population increases slowly at first, then rapidly, then slowly again before leveling off.
    b) The population increases by about 16 fruit flies per day.
    c) The rate of change in the fruit fly population is about 0 flies per day.
       No calculation was required.
       The graph is almost horizontal.
16. a) $A = 100(1.06)^n$ Exponential
    b) $A = 100(1 + i)^2$ Quadratic
    c) $A = P(1.08)^1$ Linear
17. a) About 0.0972 billion/year
       The world's population is increasing by about 97 million people each year.
    b) The scatter plot looks nearly linear.
       A linear model should fit the data well.
    c) $y = 0.098x + 4.925$
    d) The population in 2007 (year 21) is about 6.988 billion.
18. About 35 buckets
19. a) $n = \dfrac{S}{180} + 2$

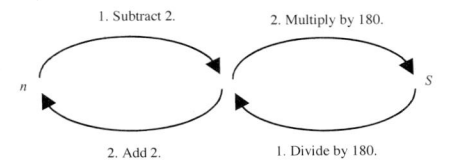

    (diagram: n → 1. Subtract 2. → 2. Multiply by 180. → S; reverse: 2. Add 2. ← 1. Divide by 180.)

    b) 6 sides
20. a) $\dfrac{1}{9}$
    b) $\dfrac{16}{81}$
    c) 64
    d) $\dfrac{25}{81}$
    e) 12
    f) 16

21. a) $\dfrac{1}{256}$
    b) 256
22. a) $x = 3$
    b) $x = 3$
    c) $x \doteq 2.322$
    d) $x = -\dfrac{1}{6}$
    e) $x = 16$
    f) $x \doteq 46.27$
    To check if the solutions are correct, substitute the value of x into the left side of the equation. If the answer is correct, the value of the left should equate the right.
23. About 8%
24. a) In 10 years, the wolf population will be 98 wolves.
    b) The doubling time for the wolf population is about 35 years.
25. The amount of these deposits after 5 years is $16 574.74.
26. a) PV = $143 820.39
    b) Yes, the value would double.
       PV ($2000 monthly payments) = $287 640.78
28. a) Regular quarterly payment: $703.61
    b) No, it will take about 2 years to repay the loan.
29. a) Monthly payment: $1279.61
       Total interest paid: $183 885.31
    b) Interest saved: $42 035.67
    c) It would take 21 years to repay the mortgage, and they would have saved $34 035.35.
30.

Principal	$200 000.00
Annual interest rate	6.00%
Equivalent monthly rate	0.49%
Amortization period in years	25
Number of payments	300
Monthly payments	$1279.61

Payment number	Payment	Interest paid	Principle paid	Outstanding balance
0				$200 000.00
1	$1279.61	$987.72	$291.89	$199 708.11
2	$1279.61	$986.28	$293.33	$199 414.78
3	$1279.61	$984.83	$294.78	$199 120.00
4	$1279.61	$983.38	$296.23	$198 823.77
5	$1279.61	$981.92	$297.69	$198 526.08
6	$1279.61	$980.45	$299.16	$198 226.92

a) Interest: $5904.58
   Principal: $1773.08
b) $198 526.08
31.

Description	Total costs ($)
Rent	9900.00
Content insurance	240.00
Cable/Internet/phone	1488.00
Utilities	1320.00

Total housing costs for one year: $12 948
32. Yes
33. a) Personal expenses (monthly): $100.00
       Laundry (monthly): $22.00
       Entertainment (monthly): $650.00
       Total expenses: $772.00
    b) Jeremy will save $728 each month.
    c) He will have saved $8736 after 1 year.
       He will have $5436 to cover his housing and living expenses ($453/month).
       He will likely have to get a part-time job.
       It is unlikely he will be able to live on $450/month.
34. Annual net income: $2750/month
    Expenses: $1400
    Housing expenses: $1350
    Option 1 cost: $1139 monthly ($231 remaining)
    Option 2 is recommended.
    Maral's annual income is not enough for her to comfortably afford Option 1.

Technology Index

TI-83 or TI-84

accessing the Stat List Editor, 163, 165
accessing the Y= Editor, 160, 161, 322, 388
calculating
    a cube root, 106
    a power with a rational exponent, 368
changing graph dimensions, 92, 109, 162, 293, 322, 383
changing table settings, 92, 109, 389, 390
creating an exponential curve of best fit, 314, 321, 322
creating a line of best fit, 161, 292, 321, 322
creating a parabola of best fit, 302, 321, 322
creating a scatter plot, 88, 160, 292
    deleting or adding a point, 163
    hiding a plot, 160
creating a table of values
    using equations, 92, 109, 389, 390
    using lists, 88, 89, 160
evaluating a function
    algebraically, 162, 293
    using the TRACE feature, 162, 293
performing regressions
    performing an exponential regression, 314, 321
    performing a linear regression, 161, 165, 292, 321
    performing a quadratic regression, 302, 321
    storing the regression equation, 161, 292, 302, 321
    viewing the correlation coefficient, 161, 165
setting the number of decimal places to display, 416

using equations
    entering, 92, 109, 160, 383, 388
    graphing, 92, 109, 383
    hiding, 160, 322
using lists
    clearing all lists, 88, 89
    clearing a list, 160
    entering formulas, 88, 89
    entering values, 88, 89
using the INTERSECT feature
    to determine a point of intersection of two graphs, 383
using the MAXIMUM feature
    to determine the coordinates of the vertex, 302
using the MINIMUM feature
    to determine the coordinates of the vertex, 92, 109, 322
using the TRACE feature
    to determine the coordinates of a plotted point, 106
    to evaluate a function, 162
using the TVM Solver
    for mortgages, 451, 452
    for mortgages with accelerated payments, 469
    interpreting the variables, 416
    solving for FV (amount or future value), 417, 418
    solving for N (number of payments), 469
    solving for PMT (payment), 432, 433, 452
    solving for PV (present value), 425, 426
    with the ΣInt command (total interest), 417, 426, 452

TI-89 CAS

simplifying expressions involving powers, 359, 364

Scientific Calculator

calculating
   an angle measure using an inverse ratio, 6
   a cube root, 106, 354
   a negative power, 340
   a negative power of a sum, 411, 424, 425
   a power of a sum, 410, 415
   a square, 2, 36, 69
   a square root, 2, 338, 353
dividing by a sum, 432
multiplying or dividing by π, 69, 353, 354
multiplying or dividing by the sine, cosine, or tangent of a given angle, 5, 7, 28, 36

Microsoft Excel

calculating the sum of a range of cells, 442, 462
converting dimensions, 99
copying formulas, 98, 100, 116, 117, 438, 441, 461
drawing a trend line on a scatter plot, 168
   displaying the equation, 168
entering formulas, 97, 100, 116, 437, 440, 461, 506
inserting a chart, 98, 167
   changing the scale of an axis, 167
   entering titles, 167
   hiding legend, 167
   selecting data to graph, 98

moving among pages, 115
using $ in a formula, 438
using the PMT function, 440, 461
viewing formulas, 461

Fathom

accessing the Graph menu, 174
creating a collection of data
   renaming a collection, 173
creating a graph, 173
   deleting a point, 176
   drawing a line of best fit, 174
creating a statistical model for data, 174
   performing linear regression, 174
   simplifying information, 174
creating a table, 173
   changing an attribute name, 173
   resizing a table, 173
importing data from E-STAT, 178

The Geometer's Sketchpad

deselecting an object, 14
moving among pages, 103
using the selection arrow tool, 14

The Internet

searching for information, 51, 247, 248, 397, 517
   useful search words, 51, 52, 128, 195, 247, 397, 443, 447, 476, 478
using E-STAT, 169, 170, 176–178, 241–244, 488–490
using online financial calculators, 440, 445, 450, 451, 460, 467, 468

Index

A
acre, 95
acute angle, 13–18, 21, 22, 27, 35
acute triangle, 28, 35, 36
algebraic expressions,
    simplifying, 360, 362
amortization period, 448, 449
    adjusting, 461, 462
amortization table, 449
    creating with *Microsoft Excel*,
    456–458
angle of depression, 11, 30
angle of elevation (*also* angle of
    inclination), 7
angles,
    acute, 13–18, 21, 22, 27, 35
    measuring in triangles, 6, 7,
    45, 53
    obtuse, 13–18, 20–22, 27, 35
    right, 2, 4–7
annual interest rate,
    407, 412, 462
annuity,
    amount of, 410–414
    calculating amounts of,
    409–414, 467
    investigating with *Microsoft
    Excel*, 433–438
    present value of, 419–422
    regular payment of, 426–429
arboriculture, 4
area, 61
    finding optimals with
    graphing calculator, 87
    finding optimals with
    manipulatives, 87
    finding optimals with
    spreadsheets, 97–99
    of a circle, 353
    of a composite figure, 67–69,
    119
    of a rectangle, 79
    of a triangle, 79, 343, 344
average rate of change, 279–281,
    331

B
balance, 496, 506
bar graph, 126, 130

base of a power, 360, 374,
    381, 399
base value, 234
bearing, 6
bias, 211–213, 225, 227
budget, 493–500, 519
    adjusting, 507
    balancing, 496, 497
    creating with *Microsoft Excel*,
    505–508

C
capacity, 63
cause-and-effect relationships,
    141, 142, 228
centimetre (cm), 3
centimetre cubed (cm³; *also*
    cubic centimetre), 64, 118
centimetre squared (cm²; *also*
square centimetre), 81
circle,
    area of, 61, 353
    circumference (*also*
    perimeter) of, 61, 62
    enclosing areas of, 93
circle graph, 127
circumference, 62
clinometer, 4
composite figure, 67–69, 119
composite object, 76–81, 119
compound interest, 406, 407
computer algebra system (CAS),
    359
cone,
    volume of, 377
conjecture, 176
Consumer Price Index, 234, 235
correlation, 228
    negative, 140, 185
    positive, 140, 185
correlation coefficient, 161,
    174, 228
cosine (cos), 5, 14–18, 35–37, 53
    signs of, 20–22
Cosine Law, 35–37, 53
    solving oblique triangles
    with, 42–46
costs, 474, 488–491
cube root, 354

cubic feet, 64, 77
cylinder,
    surface area of, 65, 77
    volume of, 64, 77, 345

D
data,
    applying trends in, 328–330
    collecting by experimentation,
    180–184
    E-STAT, 169–171, 176–178
    graphing with TI-83/84
    graphing calculator, 159–164
    linear, 151, 152
    mean, 193
    median, 193
    mode, 193
    non-linear, 151
    one- and two-variable,
    130–133
    range, 193
    reliability of, 151
    setting a regression model to,
    319–322
    spread of, 151
    statistics, 196–200
decay factor, 320
dependent variable, 139, 140,
    174, 220, 279, 283, 290

E
equivalent ratio, 192
E-STAT,
    analysing data from,
    169–171, 176–178
    living costs data, 488–491
    price indices, 241–245
expenses, 494–500, 505–508, 519
    predicted *vs.* actual, 496
exponent rules, 360, 399
    for rational exponents, 373
exponential decay, 266, 267, 390
exponential equations (*see also*
    formulas), 380–383
    applications of, 387–390
exponential growth, 266, 311,
    390
exponential regression,
    fitting to data, 314

INDEX **589**

exponential relations, 266, 310–314, 331, 320
  comparing pairs of, 313
exponents (*see* integer exponents)
extrapolation, 150, 162, 185

F
Fathom, 172–179
financial mathematics,
  researching courses and occupations, 516–518
financial savings, 476
first differences, 282, 283, 289, 290, 299, 300, 331, 320
fixed cost, 474, 497, 519
fluid ounce, 63
foot, 3, 60
formulas (*see also* exponential equations), 342–345
  Euler's, 349
  for amount, 427, 428
  for amount of an ordinary simple annuity, 411
  for body surface area, 347
  for compound interest, 406
  for density, 346
  for present value, 427, 428
  for pressure, 346
  for simple interest, 406
  for speed of animals, 372
  for speed of sound in air, 346
  for surface area of a tank, 347
  for temperature conversion, 346, 350, 352
  for volume of wood in a log, 349, 358
  Heron's, 343
  involving powers, 353, 354
  rearranging, 350–354, 399
  relations in, 300, 301
  Young's, 343

G
gallon, 63, 118
Geometer's Sketchpad,
  exploring trigonometric ratios with, 13–16
  investigating optimal measures with, 103, 104
graphs, 126–128
  assessing data accuracy with, 225, 226

bar, 126, 130
circle, 127
extrapolation, 150, 162
histogram, 126
interpolation, 150, 151
line, 28
line of best fit, 148–152
relations and trends in, 266, 269–272
scatter plots, 127, 130, 138–142, 148–152
slope, 266
solving exponential equations with, 383
growth/decay factors, 311, 312, 331, 387–390

H
histogram, 126
hypotenuse, 2, 5, 14, 17, 53

I
imperial units of capacity, 63
imperial units of length, 60
inch, 3, 60
income, 494–500, 505–508, 519
independent variable, 139, 140, 174, 220, 279, 283, 290
indices (*also* indexes), 233–236, 250
  and E-STAT, 241–245
indirect measurement, 41
inflation, 235
integer exponents, 340, 358–362, 399
  patterns in, 366–368
interest (investment), 406
  cost of mortgage, 461–465
interpolation, 150, 151, 185
inverse operations, 350, 366, 367
  of squaring, 338, 399
inverse trigonometric ratios, 6, 22, 53
isosceles triangles, 70

K
kilometre (km), 3, 60

L
landlord, 481, 482
length,
  measuring in triangles, 5, 7, 44, 53

units of, 3, 60
linear data, 151, 152
linear equation (*see also* formulas),
  solving, 339
linear growth, 311
linear regression, 291–293
linear relation, 266, 288–293, 331
  rate of change, 281
  vs. exponential relations, 311
line graph, 128
line of best fit (*also* least-square line regression line; trend line), 148–152, 161, 170, 174 185
litre (L), 63

M
margin of error, 200
mathematical modelling,
  researching courses and occupations, 396–398
mathematical models,
  exponential, 310–314, 331
  linear, 288–292, 331
  quadratic, 298–302, 331
mean, 193
median, 134, 193, 197, 198, 489
metre (m), 3, 60
metre squared (m^2; *also* square metre), 81
metric units of capacity, 63
metric units of length, 60
Microsoft Excel,
  analysing scatter plots with, 166–171
  creating amortization tables with, 456–458
  creating budgets with, 505–508
  investigating annuities with, 433–438
  investigating optimal areas and optimal perimeters with, 97–100
  investigating optimal surface areas with, 115–118
mile, 3, 60
millilitre (mL), 63
millimetre (mm), 3, 60
mode, 193
mortgage, 443–444, 467, 481, 482
  amortizing, 446–450, 456–458

590 INDEX

N

negative correlation, 140, 185, 220
negative integer exponent, 340, 399, 360, 361
negative slope (*see also* negative correction), 128
net, 75
net income, 493
non-linear relation, 290, 299–302
    average rate of change, 281
non-random sampling, 211

O

oblique triangle, 28
    solving with Cosine and Sine Laws, 42–46
obtuse angle, 13–18, 20–22, 27, 35
obtuse triangle, 28, 35, 36
one-variable data, 131–133
optimizing dimensions, 87–93, 103–110, 119
    investigating with *Geometer's Sketchpad*, 103, 104
    investigating with *Microsoft Excel*, 97–100, 115–118
    investigating with TI-83/84 graphing calculator, 87, 88, 91, 92, 109, 110
    with constraints, 89, 90, 91, 107, 108, 119
ordinary simple annuity, 410, 411, 467
    present value of, 420
outlier, 149, 150, 163, 185

P

parabola, 266, 299
parabola of best fit, 301, 302
parallelogram,
    area of, 61, 68, 69
    perimeter of, 61
percent increase/decrease, 194
percentiles, 197, 198
perfect squares, 338
perimeter, 61
    finding optimals with graphing calculators, 88
    finding optimals with manipulatives, 87
    finding optimals with spreadsheets, 100
pint, 63
polls, 200
positive correlation, 140, 185, 228
positive integer exponent, 340, 360, 361, 399
positive slope (*see also* positive correlation), 128
power, 353, 354, 380–383
    dividing, 359, 399
    multiplying, 358, 359, 399
    of a power, 358, 359, 374, 399
    with rational exponents, 374
present value, 407, 427, 428, 467
    of an annuity, 419–422
price indices (*also* price indexes), 233–236, 250
    on E-STAT, 242–245
primary trigonometric ratios, 5, 13–18, 53
    signs of, 20–22
principal, 406, 407, 412
prisms,
    surface area of, 65
    volume of, 64
property (home),
    costs of renting *vs.* purchasing, 477–483, 511–513, 519
Pythagorean Theorem, 2, 80

Q

quadrants, 13–17, 21, 22
quadratic regression, 301, 302
quadratic relations, 266, 298–302, 321, 322, 331
quart, 63
quartiles, 197, 198

R

random sampling, 211
range, 193
rate of change, 278–283
    in a linear model, 289, 331
ratio, 192
rational exponents, 366–368, 372–375, 399
reciprocal, 340, 360
rectangle,
    area of, 61, 68–70, 79
    enclosing areas of, 93
    finding optimal dimensions of, 90
    perimeter of, 61
rectangular prism,
    optimizing dimensions of, 105–108
    volume of, 79, 345
Registered Education Savings Plan (RESP), 439–441
Registered Retirement Savings Plan (RRSP), 439–441
regression equation,
    for exponential model, 311, 314, 331
    for linear model, 289, 292, 293, 331
    for quadratic model, 299, 302, 331
rent, 477–483
representative sample, 211, 213, 227
Residential Tenancies Act, 482
right triangle, 2, 35, 70
    trigonometric ratios, 4–7
root (*see* cube root; square root)

S

sample size, 211, 213, 227
savings, 495–500
scatter plot, 127, 130
    analysing using *Fathom*, 172–179
    analysing using *Microsoft Excel*, 166–171
    graphing with TI-83/84 graphing calculator, 159–164
    identifying relationships with, 138–142
    line of best fit, 148–152, 161, 170, 174
    outlier, 149, 150, 163
second differences, 299, 300, 320, 331
semi-annual interest rate, 407
semicircle,
    area of, 68, 69
simple interest, 406
sine (sin), 5, 14–18, 27–30, 53
    signs of, 20–22
Sine Law, 27–30, 34, 53
    solving oblique triangles with, 42–46

vs. Cosine Law, 36
slope of a line graph, 128, 185, 266
sphere,
    volume of, 353
square feet, 95
square root, 338, 353
statistics, 196–200
    in the media, 206, 207
    researching courses and occupations, 246–248
    uses and misuses of, 223–228
supplementary angles, 20–22, 53
surface area,
    of a composite object, 80, 81
    of cylinders and prisms, 65, 77
    optimizing for a cylinder, 105, 106
    optimizing for a rectangular prism, 105–108
surveys, 208–213, 227
systematic trial,
    solving exponential equations with, 382

T

tangent (tan), 5–7, 14–18, 53, 70
    signs of, 20–22
tenant, 481, 482
TI-83/84 graphing calculator,
    analysing scatter plots with, 159–164
    calculating annuities with, 412–414
    investigating optimal areas and optimal perimeters with, 87, 88, 91, 92
    investigating optimal surface area with, 109, 110
    solving exponential equations with, 383, 388–390
time management, 66
Time Value of Money (TVM) Solver,
    calculating annuities with, 412–414, 421, 429, 467
    calculating mortgages with, 448
torus, 346
trapezoid,
    area of, 61
    perimeter of, 61
trends in graphs, 269–272
triangle measure,
    angle-angle-side (AAS), 28, 43, 46, 53
    angle-side-angle (ASA), 29, 43, 44, 53
    side-angle-side (SAS), 36, 43, 46, 53
    side-side-side (SSS), 37, 43, 45, 53
triangles,
    acute, 28, 35, 36
    area of, 61, 70, 79, 343, 344
    Cosine Law in, 35–37
    measuring angles and lengths in, 4–7
    methods to solving, 43
    oblique, 42–46
    obtuse, 28, 35, 36
    perimeter of, 61
    right angle, 2, 35
    Sine Law in, 27–30
    solving, 9
triangular prism,
    volume of, 79
trigonometric ratios,
    exploring with *Geometer's Sketchpad*, 13–16
    exploring with pencil and paper, 17, 18
    in right triangles, 4–7
    of supplementary angles, 20–22, 53
trigonometry,
    determining unknown lengths with, 70
    researching applications of, 50–52
two-variable data, 131–133
cause-and-effect relationships, 141, 142
collecting by experimentation, 180–184
collection by survey, 218–221
in scatter plots, 139–142

U

unit conversions of capacity, 63
unit conversions of length, 3, 60
utilities, 481, 495

V

valid conclusion, 225
variable, 131
    dependent, 139, 140
    independent, 139, 140
    isolating in formulas, 351, 352
variable cost, 474, 497, 519
volume,
    of a composite object, 77–79
    of a cone, 377
    of cylinders and prisms, 64, 77, 345
    of a rectangular prism, 79, 345
    of a sphere, 353, 354
    of a torus, 346
    of a triangular prism, 79
    optimizing for a cylinder, 105, 106
    optimizing for a rectangular prism, 105–108

Y

yard, 3, 60

Z

zero exponent, 340, 360, 361, 399

Acknowledgments

The publisher would like to thank the following people and institutions for permission to use their © materials. Every reasonable effort has been made to find copyright holders of the material in this text. The publisher would be pleased to know of any errors or omissions.

Photography

Cover: Holmes/A.G.E. Foto Stock/First Light; p. 1 Flip Nicklin/Getty Images; p. 4 ©Martin Jenkinson/Alamy; p. 7 ©Bill Brooks/Alamy; p. 10 ©Richard Baker/Alamy; p. 11 ©Jeremy Woodhouse/Masterfile; p. 13 ©Stefan Sollfors/Alamy; p. 14 Gary Blakeley/Shutterstock; p. 16 KB Studio/Shutterstock; p. 20 ©Transtock/Corbis; p. 27 Ken Straiton/First Light; p. 34 (left) ©2008 Jupiterimages Corporation; p. 34 (middle left) ©Dex Image/Alamy; p. 34 (middle right) © Simon/Belcher/Alamy; p. 34 (right) Laurence Gough/Shutterstock; p. 35 ©Masterfile; p. 40 ©Skyscan/Corbis; p. 42 Timothy R. Nichols/Shutterstock; p. 43 All Canada Photos/Alamy; p. 49 ©Natalie Tepper/Arcaid/Corbis; p. 50 ©Martin Harvey/Corbis; p. 51 NBAE/Getty Images; p. 52 ©Corbis Premium RF/Alamy; p. 58 ©Jim Vecchi/CORBIS; p. 59 ©Baloncici/Shutterstock; p. 67 James Wilson/Shutterstock; p. 71 ©2008 Jupiterimages Corporation; p. 73 Pixland/Jupiterimages; p. 75 Transtock/Jupiterimages; p. 76 Time & Life Pictures/Getty Images; p. 77 ©Erica Shires/zefa/Corbis; p. 78 Buhantsov Alexey/Shutterstock; p. 87 amana productions inc./Getty Images; p. 93 Adam Borkowski/Shutterstock; p. 95 vera bogaerts/Shutterstock; p. 97 robcocquyt/Shutterstock; p. 99 Alt-6/First Light; p. 101 Adam Przezak/Shutterstock; p. 102 Cheryl Casey/Shutterstock; p. 103 ©Amanda Brown/Star Ledger/Corbis; p. 105 Alexey Stiop/Shutterstock; p. 112 Joe Gough/Shutterstock; p. 115 ©Food Features/Alamy; p. 118 ©2008 Jupiterimages Corporation; p. 122 Mircea BEZERGHEANU/Shutterstock; p. 124 altrendo images/Getty Images; p. 125 ©Doug Houghton/Alamy; p. 130 DARRYL DYCK/The Canadian Press; p. 137 Michelle Donahue Hillison/Shutterstock; p. 138 MAXIMILIAN STOCK LTD/Jupiterimages; p. 140 Galyna Andrushko/Shutterstock; p. 145 ©Bruce Benedict/Transtock/Corbis; p. 146 A.G.E. Foto Stock/First Light; p. 148 Wayne Hanna/The Canadian Press; p. 155 Karen Andersen/The Canadian Press; p. 157 ©imac/Alamy; p. 159 Ken Straiton/First Light; p. 164 Alan Marsh/First Light; p. 166 Jerry Kobalenko/First Light; p. 171 Elke Dennis/Shutterstock; p. 172 ©David LeBon/Transtock/Corbis; p. 175 ©Klaus Hackenberg/zefa/Corbis; p. 176 Peter Cade/Getty Images; p. 179 ©Andrew Turner/Alamy; p. 180 ©Natural Visions/Alamy; p. 181 BelleMedia/Shutterstock; p. 182 Vladimirs Koskins/Shutterstock; p. 183 Michele Rasotto/Shutterstock; p. 184 Danny E Hooks/Shutterstock; p. 191 Francisco Turnes/Shutterstock; p. 196 Amore/Shutterstock; p. 201 ©Alex Hinds/MaXx Images; p. 202 © 2008 Jupiterimages Corporation; p. 205 ©Robert W. Ginn/MaXx Images; p. 206 Feng Yu/Shutterstock; p. 208 Nigel Carse/Shutterstock; p. 211 Darcy Cheek/The Canadian Press; p. 218 ©2008 Jupiterimages Corporation; p. 223 ©2008 Jupiterimages Corporation; p. 227 ©2008 Jupiterimages Corporation; p. 233 6377724229/Shutterstock; p. 241 Feverpitched/Shutterstock; p. 241 ©Charles Gullung/zefa/Corbis; p. 242 sculpies/Shutterstock; p. 245 ©Atlantide S.N.C./MaXx Images; p. 246 National Geographic/Getty Images; p. 247 ©Lloyd Sutton/Masterfile; p. 256 Patrick Hermans/Shutterstock; p. 257 Workbook Stock/Jupiterimages; p. 257 (inset) ©Josh Gosfield/Corbis; p. 258 (inset) redcover.com/Getty Images; pp. 258–259 ©Jean-Yves Bruel/Masterfile; p. 259 (inset, top) ©Gary W. Carter/CORBIS; p. 259 (inset, bottom) ©Birgid Allig/zefa/Corbis; p. 260 ©Banana Stock/Jupiterimages; p. 261 Phanie/First Light; p. 261 (inset) ©Martin Shields/Alamy; Hermans/Shutterstock; p. 265 ©Masterfile; p. 278 ©Creatas/MaXx Images; p. 280 Andrew Burns/Shutterstock; p. 285 ©Hu Zhao/Alamy; p. 286 Ariel Bravy/Shutterstock; p. 288 ©Antonio López Román/MaXx Images; p. 296 ©Dan Pressman; p. 297 ©Jeff Greenberg/MaXx Images; p. 298 ©Jon Hrusa/epa/Corbis; p. 304 amlet/Shutterstock; p. 305 Andreas Stirnberg/Getty Images; p. 306 ©Zoran Milich/Masterfile; p. 310 ©Masterfile; p. 313 Dr Kari Lounatmaa/Photo Researchers, Inc; p. 317 ©Dan Pressman; p. 319 ©2008 Jupiterimages Corporation; p. 326 M. Moita/Shutterstock; p. 327 Thomas Kitchin/First Light; p. 328 Ant Clausen/Shutterstock; p. 330 ©2008 Jupiterimages Corporation; p. 336 ©2008 Jupiterimages Corporation; p. 337 Michael Dwyer/Alamy; p. 342 Bennett Egypt Arch/Alamy; p. 348 Dewayne Flowers/Shutterstock; p. 350 design pics/First Light; p. 356 Roger Bamber/Alamy; p. 358 David R. Frazier Photolibrary, Inc./Alamy; p. 361 Jose Gil/Shutterstock; p. 363 Ruslan Kokarev/Shutterstock; p. 366 Supri Suharjoto/Shutterstock; p. 369 Frank Whitney/Brand X Pictures/Jupiterimages; p. 370 Marie C. Fields/Shutterstock; p. 372 Stock Connection Distribution/Alamy; p. 375 Robyn Mackenzie/Shutterstock; p. 377 Tischenko Irina/Shutterstock; p. 380 PHOTOTAKE Inc./Alamy; p. 386 ©2008 Jupiterimages Corporation; p. 387 J.D. Dallet/MaXx Images; p. 393 Tom Oliveira/Shutterstock; p. 395 FRED CHARTRAND/The Canadian Press; p. 396 ©2008 Jupiterimages Corporation; p. 397 Tonis Valing/Shutterstock; p. 398 ©2008 Jupiterimages Corporation;

p. 404 Reimar Gaertner/Alamy; p. 405 ©Image State/Alamy; p. 409 ©2008 Jupiterimages Corporation; p. 412 Colin & Linda McKie/Shutterstock; p. 416 ZOOM(189) Friends-Campus Life/Getty Images; p. 419 Toronto Star/First Light; p. 426 aceshot1/Shutterstock; p. 432 design pics/First Light; p. 433 ©Photodisc/Alamy; p. 435 ©ImageSource; p. 439 Ariel Skelley/CORBIS; p. 440 Pierre Desrosiers/First Light; p. 443 image100/First Light; p. 444 ©2008 Jupiterimages Corporation; p. 446 radius images/First Light; p. 450 Kirk Peart Professional Imaging/Shutterstock; p. 456 ILYA GENKIN/Shutterstock; p. 459 ©2008 Jupiterimages Corporation; p. 460 MWProductions/Shutterstock; p. 461 Seth Joel/Getty Images; p. 462 ©2008 Jupiterimages Corporation; p. 472 John A. Rizzo/A.G.E. Foto Stock/First Light; p. 473 ImageMore/Jupiterimages; p. 477 © Gordon M. Grant/Alamy; p. 479 ©2008 Jupiterimages Corporation; p. 484 ©Lee Sullivan/Alamy; p. 485 Michael Shake/Shutterstock; p. 486 ©Brownstock Inc./Alamy; p. 488 Andre Blais/Shutterstock; p. 491 ©paolo siccardi/MaXx Images; p. 493 ©2008 Jupiterimages Corporation; p. 496 ©Daniel Mirer/Corbis; p. 499 ©Tim McGuire/Corbis; p. 502 Michael-John Wolfe/Shutterstock; p. 505 Kazakov/Shutterstock; p. 506 David Lee/Shutterstock; p. 507 ©Don Johnston/Alamy; p. 509 ©Southern Stock Corp/Corbis; p. 511 ©MARKOS DOLOPIKOS/Alamy; p. 512 Devis Da Fre'/Shutterstock; p. 515 Gina Sanders/Shutterstock; p. 516 ©2008 Jupiterimages Corporation; p. 517 ©Jon Feingersh/CORBIS; p. 521 ©Keith Leighton/Alamy; p. 524 Joy Brown/Shutterstock; p. 525 Patrik Giardino/Getty Images; p. 525 (inset) ColorBlind Images/Getty Images; p. 526 ©Bettmann/CORBIS; p. 527 ©Peter Griffith/Masterfile; p. 527 (inset) ©Peter Griffith/Masterfile; p. 528 (inset, top) Ryan McVay/Getty Images; p. 528 (bottom) First Light; pp. 528–529 ©Steven Puetzer/Masterfile

Illustrations
Neil Stewart/NSV Productions

p. 238 *Fire Weather for 15 June 2007*. illustration courtesy of the Canadian Forest Service, reproduced with permission

p. 242 Screen Capture: "E-STAT: Table Contents", from the Statistics Canada Web site

p. 243 Screen Capture: "Table: New housing price indexes-Monthly", from the Statistics Canada Web site, CANSIM in E-STAT, Table 327-0005, Series 21148172, v21148181, and v21148238

p. 244 Screen Capture: "Graph: New housing price indexes; Total (house and land)", from the Statistics Canada Web site, CANSIM in E-STAT, Table 7-0005, Series v21148172, v21148181, and v2114838

p. 275 *Summary for Week ending August 26, 07*. illustration courtesy of Independent Electric System Operator (IESO)

p. 372 from The University of Sheffield, Sorby Geology Group, *Dinosaur Speed Calculator*. Modified and reprinted with permission

Fathom™ *Dynamic Statistics* technology references and screen shots reprinted with permission from Key Curriculum Press, 1150 65th Street, Emeryville, CA 94608

The Geometer's Sketchpad® technology references and screen shots reprinted with permission from Key Curriculum Press, 1150 65th Street, Emeryville, CA 94608

Microsoft® product technology references and screen shots reprinted with permission from Microsoft Corporation

Statistics Canada information is used with the permission of Statistics Canada. Users are forbidden to copy the data and redisseminate them in an original or modified form, for commercial purposes, without permission from Statistics Canada. Information on the availability of the wide range of data from Statistics Canada can be obtained from Statistics Canada's Regional Offices or the Statistics Canada Web site.

TI-30XIIS, TI-83 Plus, TI-84, and TI-89 images and references courtesy of Texas Instruments

Course Study Guide

Units of Measure Used in Canada

Metric units

unit	measure of	symbol
ampere	electric current	A
becquerel	radioactivity	Bq
coulomb	electric charge	C
degree Celsius	temperature	°C
hectare	area	ha
hour	time	h
joule	energy	J
kilogram	mass	kg
litre	volume or capacity	L
metre	length	m
minute	time	min
newton	force	N
ohm	electric resistance	Ω
pascal	pressure	Pa
second	time	s
tonne	mass	t
volt	electric potential	V
watt	power	W
watt hour	electrical energy	Wh

Other units

unit	measure of	abbreviation or symbol
acre	area	–
degree Fahrenheit	temperature	°F
foot	length	ft. or ′
fluid ounce	volume or capacity	fl. oz.
gallon	volume or capacity	gal.
inch	length	in. or ″
mile	length	mi.
parts per million by volume	gas concentration	ppmv
pint	volume or capacity	pt.
pounds per square inch	pressure	psi
quart	volume or capacity	qt.
yard	length	yd.

Conversion Factors

Metric prefixes

name	symbol	multiply by…
micro-	μ	0.000 001
milli-	m	0.001
centi-	c	0.01
kilo-	k	1000
mega-	M	1 000 000
exa-	E	10^{18}

Units of length

Metric conversions
1 cm = 10 mm
1 m = 100 cm
1 km = 1000 m
1 mm = 0.1 cm
1 cm = 0.01 m
1 m = 0.001 km

Imperial conversions
1 foot = 12 inches
1 yard = 3 feet
1 mile = 1760 yards
1 yard = 36 inches
1 mile = 5280 feet

Imperial to metric
1 inch = 2.54 cm
1 foot = 30.48 cm
1 foot = 0.3048 m
1 mile ≐ 1.6093 km

Metric to imperial
1 cm ≐ 0.3937 inch
1 m ≐ 39.37 inches
1 m ≐ 3.2808 feet
1 km ≐ 0.6214 mile

Units of volume or capacity

Metric conversions
1 mL = 1 cm^3
1 L = 1000 cm^3
1 m^3 = 1000 L

Imperial conversions
1 gallon ≐ 277.42 cubic inches
1 cubic foot ≐ 6.2288 gallons

Imperial to metric
1 fluid ounce ≐ 28.4131 mL
1 pint ≐ 0.5683 L
1 quart ≐ 1.1365 L
1 gallon ≐ 4.5461 L
1 gallon ≐ 4546.1 cm^3
1 U.S. gallon ≐ 3.785 L

Metric to imperial
1 mL ≐ 0.0352 fluid ounce
1 L ≐ 1.7598 pints
1 L ≐ 0.8799 quart
1 L ≐ 0.2200 gallon

Other units

Area
1 ha = 10 000 m^2
1 acre = 43 560 square feet

Mass
1 mg = 1000 μg
1 kg = 1000 g
1 t = 1000 kg
1 g = 0.001 kg

Time
1 h = 60 min

Course Study Guide continued

Sum of the Angles in a Triangle
In any △ABC:

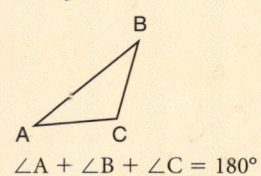

$\angle A + \angle B + \angle C = 180°$

Pythagorean Theorem
In right △ABC with hypotenuse c:
$c^2 = a^2 + b^2$

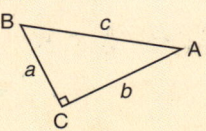

Primary Trigonometric Ratios
When ∠A is an acute angle in a right triangle:

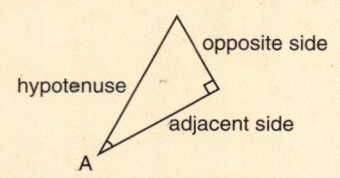

$\sin A = \dfrac{\text{length of side opposite } \angle A}{\text{length of hypotenuse}}$

$\cos A = \dfrac{\text{length of side adjacent to } \angle A}{\text{length of hypotenuse}}$

$\tan A = \dfrac{\text{length of side opposite } \angle A}{\text{length of side adjacent to } \angle A}$

Trigonometric Ratios of Supplementary Angles
For an acute angle, A, and its supplementary obtuse angle, (180° − A):
$\sin A = \sin(180° - A)$
$\cos A = -\cos(180° - A)$
$\tan A = -\tan(180° - A)$

Sine Law
In any △ABC:

$\dfrac{a}{\sin A} = \dfrac{b}{\sin B} = \dfrac{c}{\sin C}$

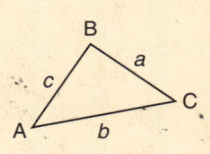

Cosine Law
In any △ABC:
$c^2 = a^2 + b^2 - 2ab \cos C$

Area and Perimeter (Circumference)

Square: $A = s^2$
$P = 4s$

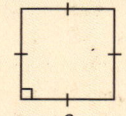

Rectangle: $A = \ell w$
$P = 2\ell - 2w$

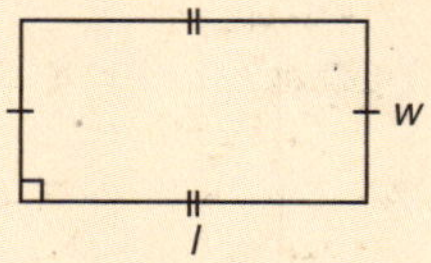

Triangle: $A = \dfrac{1}{2}bh$
$P = a + b + c$

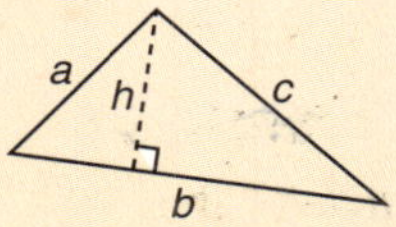

Parallelogram: $A = bh$
$P = 2b + 2c$

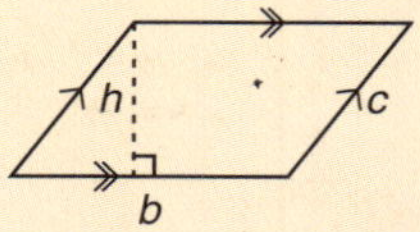

Trapezoid: $A = \dfrac{1}{2}(a + b)h$
$P = a + b + c + d$

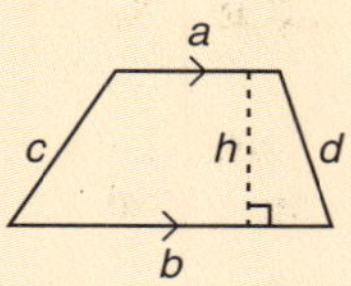

Circle: $A = \pi r^2$
$C = 2\pi r$

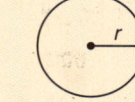

Volume and Surface Area

Cube: $V = s^3$
$SA = 6s^2$

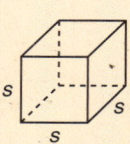

Sphere: $V = \dfrac{4}{3}\pi r^3$

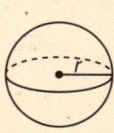

Prism or cylinder: $V = \text{base area} \times \text{height}$

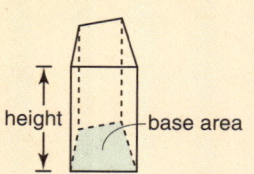

Pyramid or cone: $V = \dfrac{1}{3} \times \text{base area} \times \text{height}$

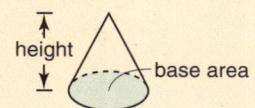